MARKETING OF AGRICULTURAL PRODUCTS

NINTH EDITION

Richard L. Kohls
Joseph N. Uhl
Purdue University

Prentice Hall
Upper Saddle River, NJ 07458

Library of Congress Cataloging-in-Publication Data
Kohls, Richard L. (Richard Louis)
Marketing of agricultural products / Richard L. Kohls, Joseph N. Uhl.—9th ed.
p. cm.
Includes bibliographical references and index.
ISBN 0-13-010584-8
1. Produce trade. 2. Farm produce—Marketing. 3. Produce trade—United States. 4. Farm produce—United States—Marketing. I. Uhl, Joseph N. II. Title.

HD9000.5 .K57 2002
381'.41-dc21 2001033197

Editor-in-Chief: Stephen Helba
Executive Editor: Debbie Yarnell
Associate Editor: Kimberly Yehle
Production Liaison: Eileen O'Sullivan
Production Editor: Pine Tree Composition
Director of Manufacturing & Production: Bruce Johnson
Manufacturing Buyer: Cathleen Petersen
Marketing Manager: Jimmy Stephens
Production Manager: Mary Carnis
Formatting/Page Makeup: Pine Tree Composition, Inc.
Design Director: Cheryl Asherman
Cover Artist: Blair Brown
Senior Cover Designer: Miguel Ortiz
Printer/Binder: VonHoffman

Prentice-Hall International (UK) Limited, *London*
Prentice-Hall of Australia Pty. Limited, *Sydney*
Prentice-Hall Canada Inc., *Toronto*
Prentice-Hall Hispanoamericana, S.A., *Mexico*
Prentice-Hall of India Private Limited, *New Delhi*
Prentice-Hall of Japan, Inc., *Tokyo*
Pearson Education Asia Pte. Ltd.
Editora Prentice-Hall do Brasil, Ltda., *Rio de Janeiro*

10 9 8 7 6 5 4 3 2 1
ISBN: 0-13-010584-8

CONTENTS

PREFACE

This ninth edition represents an updating of *Marketing of Agricultural Products,* first published in 1955. As with previous editions, the goal has been to keep the text up to date with the real world in which food marketing students and managers find themselves. More than ever before, the appropriate organization and functioning of the world's food marketing system are of concern not only to those producers and marketing firms directly involved but to consumers and the general public as well. Increasingly, foreign trade in agricultural and food products ties domestic agricultural markets into the worldwide scene.

No book can be all things to all people. This book, like the previous editions, is written for those who are beginning their study of the food marketing system. It is designed for students who have had little or no previous contact with marketing or economics. The book presents the starting points for learning and discussion and leaves to the teacher the task of adjusting the levels of knowledge and achievement to the particular class involved. To aid in this, bibliographic references to key government and commercial references are supplied. It is hoped that these will encourage further study and call attention to the immense volume of marketing literature now available.

Textbooks are not novels! Learning requires work and discipline. However, it has been our pleasure over the years to hear students' and instructors' favorable comments on the readability of this book. Some of our colleagues have commented that it is too elementary and easy. This, however, we have accepted as a compliment rather than a criticism. Textbooks should be written for students, not our professional colleagues.

The approach of this edition continues that of earlier editions. It attempts to look at the food marketing system from several angles: the functional approach, the institutional approach, the micro-firm and macro-system approaches, and the commodity approach. Each of these approaches provides unique and complementary perspectives of the food industry. The book also blends the descriptive, analytical, and normative approaches to understanding the food marketing system.

New material has been added to update important market developments and events, particularly in the areas of vertical market coordination, risk management, farm policy, international trade and globalization, biotechnology, e-commerce, and value-added marketing strategies. Mini-cases have also been added to illustrate the controversial nature of food marketing and food marketing public policy concerns. There are also many new references to valuable public and private food marketing world wide web sites.

The objectives of the book continue to be to assist students and managers in understanding the structure and workings of the food marketing system, to examine how this system affects farmers, consumers, and middlemen, and to illustrate how this dynamic market system has responded to technological, social, economic, and political forces over time. The book reflects the adage, *Nothing is changing, except everything.*

While the focus of the book remains on the economics of the food system, there are liberal references to the social, political, and historical aspects of food marketing. The authors are aware of the use of this text throughout the world. The book emphasizes the U.S. food marketing system, but there are extensive references to marketing and market development in other countries.

In any book that has an extended history such as this one, many people have made valuable comments and suggestions for improvement. Indeed, in a survey book like this, it is difficult to cite all the original sources of material. We have borrowed shamelessly from our book users and competitors in writing the text. The authors would be happy to receive suggestions and critical comments from readers (uhl@agecon.purdue.edu). The author's class at Purdue that uses this text can be visited at the web site: http://www.agecon.purdue.edu/academic/agec220/index.html

We wish to acknowledge, in particular, the following reviewers who provided valuable insight and helpful feedback in the preparation of this revision: Dragan Miljkovic, Southwest Missouri State University; Wendy Umberger, University of Nebraska-Lincoln; Anthon H. Turley, Ricks College; Robert Herrmann, Penn State University; Kevin J. Bacon, Central Missouri State University; John E. Cottingham, University of Wisconsin-Platteville; Douglas J. Miller, Iowa State University; and Raymond T. Folwell, Washington State University.

We are grateful for the administrative support of the Department of Agricultural Economics at Purdue and to our families.

Richard L. Kohls
Joseph N. Uhl

ABOUT THE AUTHORS

Richard Kohls, a pioneer in agricultural marketing, is Emeritus Professor of Agricultural Economics at Purdue University. Born in 1921 in Kentland, Indiana, he received his degrees in Agricultural Economics from the University of Missouri and Purdue. He wrote the first edition of this pathbreaking book in 1955, when interest in agricultural marketing was beginning to increase. After teaching agricultural marketing for several years and conducting research and extension programs in the area, he served as Dean of Agriculture at Purdue from 1968 to 1980. Professor Kohls was awarded the American Agricultural Economics Association Outstanding Teacher Award in 1966.

Joe Uhl has been a professor of food marketing at Purdue University since 1966. He was born in Lima, Ohio in 1939. He teaches agricultural and food marketing classes, including the class that uses this text. He also counsels students and does research in food marketing. He served on the staff of the National Commission of Food Marketing in 1966, and he has lectured widely in Eastern Europe. He began collaborating with R. L. Kohls on this book in 1980. Professor Uhl has won both student counseling and teaching awards, the most recent for Distinguished Undergraduate Teaching from the American Agricultural Economics Association in 1989.

PART I

THE FRAMEWORK OF THE MARKETING PROBLEM

CHAPTER 1

INTRODUCTION TO FOOD MARKETING

The task of marketing is to convert society's needs and wants into profitable opportunities.

OBJECTIVES

After reading this chapter, you will be able to:

1. Define marketing and explain why we have a food marketing system.
2. Describe the organization of the food industry, the major players, and the nature of the food marketing process.
3. Explain how the food marketing system adds value to farm products by creating time, form, place, and possession utility.
4. Appreciate the key factors affecting the organization and performance of the food marketing system and better understand the historical development of the food marketing system.

Why should you read this book and learn more about the food marketing system? Most of us already know quite a bit about the food industry, either as food producers, workers, managers, or consumers. What is so difficult about producing food? Generally the food industry works so well and silently that we often take it for granted. How can we be sure that enough food will be produced, that stores will be stocked with adequate food supplies, and that we will have the food we want when we want it, and in the forms we like? For much of mankind's history, just getting enough food produced was the major problem. Transportation, grading, refrigeration, packaging, labeling, advertising, processing, and other modern marketing techniques were the least of the food problems of people even 200 years ago.

Today, the food production and marketing process is much different. The way that we produce and market food today is not the same as it was for your parents and grandparents. And, the future food production and marketing system will be different from the one we have today. There are a number of reasons then for you to learn about the food marketing system and how it works today. First, you may be preparing for a career in food marketing, and your success will depend on your knowledge of the field. Second, you may plan to be a food producer who will need to understand the changing nature of the marketing system which will influence your sales, prices, and income. Third, you will most certainly be a food consumer whose food supply and prices will depend upon the food production and marketing system. Finally, you will be a citizen with responsibilities to shape and regulate the food industry in ways that serve the public interest.

Still not convinced of the value of studying food marketing? Fine. Skeptical students make the best learners.

WHAT IS MARKETING?

Food marketing has many faces. It is a farmer and his wife discussing over morning coffee when and where to sell their crops; a rancher monitoring livestock prices on a computer; a group of farmers touring the Far East to better understand the needs of foreign markets; brokers on the Board of Trade furiously signaling their buy and sell orders; a tramp steamer hauling U.S. grain to Singapore; a fast-food crew at rush hour; a grocer deciding how to merchandise and price products; a shopper pushing a cart through supermarket aisles picking and choosing from among literally thousands of different food products; a student buying a late-night snack from a vending machine. These aspects and others are part of the fascinating food marketing scene that we will examine in this book.

An overview of the food marketing system is shown in Figure 1-1. It is composed of alternative product flows (called marketing channels), a variety of firms (middlemen), and numerous business activities (referred to as marketing functions). Within this food marketing system a myriad of decisions are made that influence the quality, variety, and cost of the nation's food supply as well as the prices and profits of these products.

This book is about this food marketing system. It is called a *system* because it consists of interrelated component parts that contribute toward overall firm, industry, and social goals. The food marketing system encompasses two major types of activities. One is concerned with the physical handling, storage, processing, and transfer of raw and finished goods as they move from producers to consumers. The other is concerned with the exchange and price-setting processes in the market system. This latter economic aspect of the food marketing system is less tangible than that of the physical distribution aspect, but no less important. Both the physical distribution and economic roles of markets will be examined and evaluated in this book.

MARKETING IS COMPLEX AND COSTLY

The food and fiber sector is one of the largest in the United States and has a marketing system larger than any other industry. This system, which brings the vast array of food products together and places them at the disposal of over 281 million Ameri-

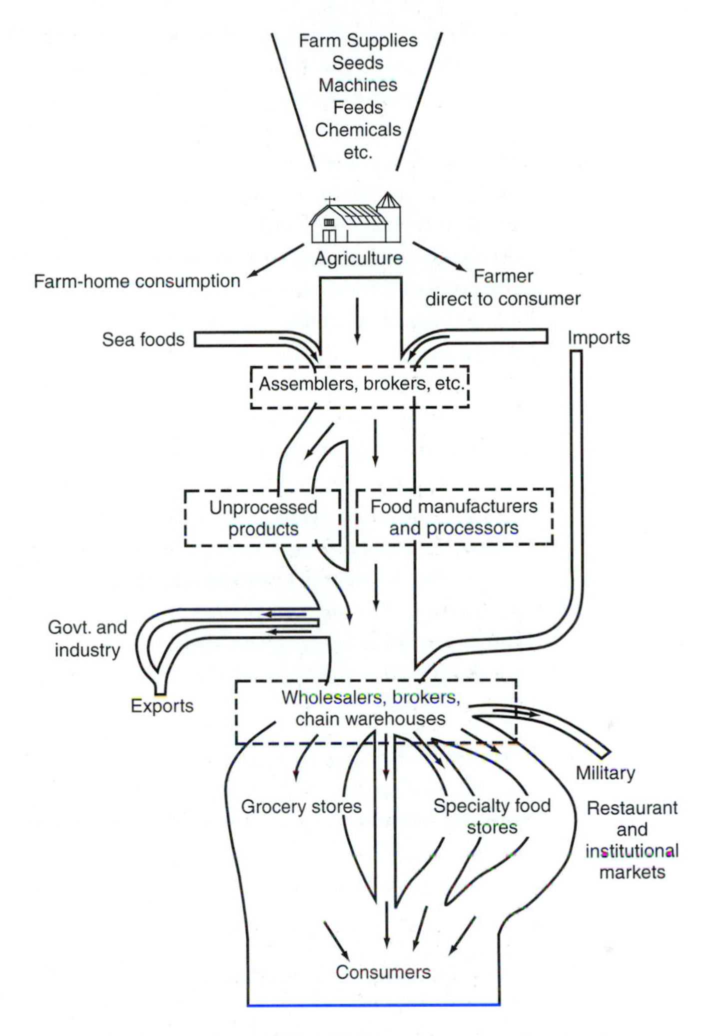

FIGURE 1–1 Flow of food from sources to destinations. (U.S. Department of Agriculture.)

cans and countless overseas customers, is complex, expensive, and a significant component of the national economy.

The initial production of raw materials for these food products takes place on some 2 million U.S. farms with an average size of about 470 acres. The marketing task begins here, on these diverse farms. Some farms produce small amounts of a great many commodities. Other specialized operations produce large amounts of only a single commodity. Farm products are not only perishable, but they vary in quality. Production is highly seasonal and geographically concentrated in areas that are often located some distance from consumers. Farm commodities must be collected, sorted, and swiftly moved to market, or stored for later use. These production and commodity characteristics give rise to the basic marketing activities, such as storage, transportation, processing, and the like. Few products make more stringent demands on their marketing systems.

A large number of diverse firms market the nation's agricultural products. According to the U.S. Census of Business, there were more than 730,000 food marketing establishments in 1998, owned by perhaps one-half as many firms. This included 10,000 raw farm product assemblers, 21,000 food manufacturing plants, 41,000 food wholesaling facilities, 96,000 grocery stores, and 485,000 retail eating and drinking places. Each of these operations makes a valuable contribution to markting the nation's food supply.

The food marketing process is expensive. According to the U.S. Department of Agriculture, U.S. consumers spent $691 billion for food in 1999. This amounted to 10.4 percent of their disposable income. Seventy-nine percent of consumer food dollars paid for the off-farm marketing activities, leaving the farmer 21 percent. Since 1977, food marketing firms' labor costs have exceeded the total value of commodities sold by farmers. Today, marketing costs have as great an influence on retail food prices as do the farm prices of foods, thus warranting greater attention to the food marketing system on the part of consumers, farmers, and legislators.

The food marketing system makes a substantial contribution to the national economy. The contributions of various agricultural and food marketing sectors to the national economy are shown in Table 1-1. This system accounted for 25 million jobs and 15.6 percent of the gross national product in 1998. The contributions of

Table 1-1 Contributions of the Food and Fiber Producing and Marketing System to the U.S. Economy, 1998

	Employment		*Contribution to Gross National Product*	
Sectors	*Million*	*Percent of Total*	*$Billion*	*Percent of Total*
Farming	1.8	7%	74.3	5%
Food processing, manufacturing	1.4	6	166.9	12
Related manufacturing	2.2	9	215.3	16
Wholesaling, retailing	8.4	34	397.0	29
Eating places	6.8	27	188.1	14
Supporting activities	4.2	17	325.6	24
Total food and fiber system	24.8	100%	1,367.2	100%
Food and fiber system as a percent of the U.S. economy:				
1947		41.0		40.0
1967		22.9		24.0
1975		21.4		20.4
1985		18.5		17.5
1993		17.2		14.2
1999		18.0		15.6

SOURCE: U.S. Department of Agriculture, Economic Research Service.

the off-farm sectors to employment and output are about nine times greater than the farming contributions. While the relative importance of the food and fiber sector in the economy has declined over time, aggregate employment and output are generally more stable in this industry than in other sectors.

MARKETING DEFINED

Marketing means different things to different people. To the consumer, marketing may refer to the weekly food shopping trip to the supermarket—the most visible tip of the food marketing iceberg. The farmer deals primarily with local farm product buyers and may associate marketing with the loading of hogs into the pickup for the trip to market, or with the calling of local elevators to determine which is offering the highest price for grain. In contrast, food middlemen such as retailers, wholesalers, and processors may view marketing as a process for gaining competitive advantage over market rivals, improving sales and profits, and satisfying consumers. Each group has only a partial concept of the total food marketing process.

For our purposes, we will define food marketing as *the performance of all business activities involved in the flow of food products and services from the point of initial agricultural production until they are in the hands of consumers.* Several key points of this definition should be noted.

First, the definition does not limit marketing to all the nonfarm activities in the food industry or to everything that happens to food after it leaves the farm gate. Because no product should ever be produced unless it has a market, marketing begins with production decisions on the farm. The farm gate has become an ambiguous dividing line between farming and food marketing. More and more, farmers are performing many of the traditional food marketing functions on the farm. Through cooperatives and other arrangements, farmers are also extending their operations to off-farm marketing activities. At the same time, food marketing firms are crossing over the farm gate to engage in food production. Hence, care must be taken in drawing a hard and fast line between farming and food marketing at the farm gate.

This definition of marketing also suggests a mutual interdependence between farmers and food marketing middlemen. The food production process does not stop at the farm gate. The food marketing activities complement the agricultural production process. Although it is true that there would be no food without farmers, it is also true that consumers rely on the food marketing system to complete the food production process begun on the farm. The relationship between farmers and food marketing firms is at the same time competitive and complementary.

Food marketing is neither a mechanical nor an automatic operation. The essence of marketing is management decision making. All business activities involve interpersonal relations and decisions: What is the correct buying and selling price? Should you sell now or store for later sale? How much should be spent on advertising and new product development? Correct answers to these questions influence the profit and loss statement. The quality of food marketing management decisions influences to a great extent the cost and efficiency of the food marketing system.

There was once a debate about whether the farm supply industries—such as the feed, fertilizer, farm machinery, and seed industries—should be included as part of the food marketing system. Today these farm supply or input markets must be considered a vital part of the food industry. Many of the new food production technologies originate here. How well a farmer purchases in these input-procurement markets may influence his profits just as much as how well he makes product marketing decisions.

The food industry is shaped by and serves three key players: food producers, food marketing firms, and food consumers. Government, media, food and agricultural organizations, and allied industries such as transportation, communication, energy, and other sectors also have an interest in the food industry. Conflicts can arise among these food industry players.

Consumers are interested in securing the highest food value at the lowest possible price. Farmers want the highest possible returns from the sale of their products. Food marketing middlemen seek to earn the greatest profit possible. One of the primary tasks of the food marketing system is to reconcile these sometimes conflicting demands. It is not always an easy or fully appreciated task.

We will define a *market* as an arena for organizing and facilitating business activities and for answering the basic economic questions: what to produce, how much to produce, how to produce, and how to distribute production. A market may be defined by (1) a location (for example, the St. Louis market); (2) a product (for example, the grain market); (3) a time (for example, the May soybean market); or (4) an institutional level (for example, the retail food market). The choice of market definition depends on the problem to be analyzed. Sometimes it is desirable to study the farm price of corn in a local market; at other times it is necessary to analyze the world price of corn. Markets join together the various components of the food industry: the farm supply sector, the farm sector, the food marketing system, and national economies. The input supply and farm markets are often referred to as *agribusiness.*

In this book, we shall examine in detail the many functions that markets play in the food economy. First, they facilitate the exchange of products and money between buyers and sellers. In doing so, they make both parties to a transaction better off. Markets also create value by encouraging competing firms to improve their prices, services, products, and values for consumers. Through competition, markets can also contribute to efficiency as well as economize on the efforts of buyers and sellers. Markets also place values on economic activities, creating both rewards for correct decisions and punishments for inefficient decisions. Perhaps most importantly and least appreciated by many, markets assist in the efficient allocation of resources in the food industry, which in turn can improve the living standards of a society.

MARKETING AS A VALUE-ADDED PROCESS

We will define *production* as the creation of utility—the process of making useful goods and services. The utilities created in the productive processes are further classified into *form* utility, *place* utility, *time* utility, and *possession* utility.

Food and farm products are not always the same. The miller and baker who make bread from wheat and flour are creating form utility, as is the breakfast cereal firm that produces corn flakes. Dairies change raw milk into cheese and butter. Canners and freezers alter the form of raw farm products. The packer who slaughters hogs and cuts them into pork carcasses also adds form utility. All of these change the form of raw materials and create something useful that consumers are often willing to pay for.

Food is often produced some distance from consumers. The railroad or trucker adds place utility by moving the hogs from Iowa to the packing plant, and then, after processing, moving the cuts of pork to wholesalers, retailers, and finally to consumers. The food processor, wholesaler, and retailer create place utility by transferring foods from production to consumption areas. Food exporters and importers specialize in place utility.

Food production is often seasonal while consumers eat products year round. Time utility is created when the timing and availability of the product is altered by marketing activities. Meat packers may freeze some of the pork products for later use. The pork is more useful by being held from periods of relative plenty to periods of relative scarcity. Grain elevator and warehouse operators—and even supermarket operators, through their inventory holdings—add time utility to products.

Possession utility is created by marketing activities that assist the consumer in aquiring and taking title to desired products. For example, supermarkets make it relatively easy for shoppers to purchase a wide assortment of foods in one stop. Vending machines increase the convenience of purchasing certain foods. Advertising that assists consumers in shopping for food and selecting various items for purchase also create possession utility.

Most people accept the activities of the farmer and the manufacturer as being productive. They create visible changes in products. However, middleman activities should also be considered productive. Those engaged in the marketing process, too, are producers in the sense that they add usefulness or utility. To argue which group is more important is rather senseless. Both groups—the producers of raw products and the marketing agencies—are necessary to the creation of the final products for consumption. Both create something useful for which society will pay a price. Both groups are productive in the real sense of the word. Thus, the twin food problems are often described as, first, producing enough food and, second, distributing this food to those who need it most and are willing and able to pay for it.

The value-adding productive processes in the food industry are illustrated in Figure 1-2. The food industry is divided into three components: (1) the input sector, which provides machinery, fertilizer, seeds, and other farm supplies; (2) the farm sector; and (3) the product market sector. In 1997, farmers purchased $164 billion worth of farm supplies. They added $59 billion of value to these purchased supplies through their productive activities and sold farm products valued at $223 billion. Food marketing firms then added another $444 billion of value to farm commodities, bringing the total retail value to $667 billion. Thus, food "production" can be viewed as a sequential and value-adding process, having its origin in purchased and farm-supplied resources and ending with the meal on the table.

FOOD MARKETING CAREER OPPORTUNITIES

The food industry is one of the nation's largest, and numerous jobs and career opportunities are available in this sector. While it is not the fastest-growing or highest-paying sector in the economy, employment is generally stable, salaries are competitive, and there is excellent opportunity for career advancement.

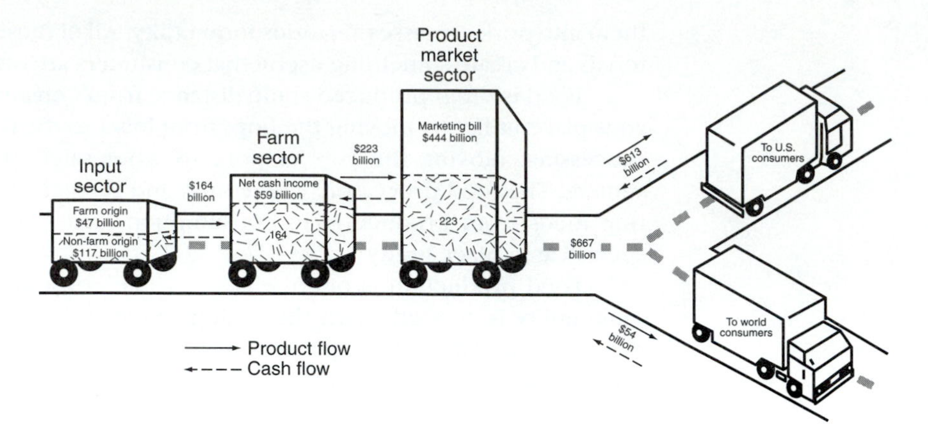

FIGURE 1–2 How products and cash flow through the food and fiber system, 1997. (U.S. Department of Agriculture, Economic Research Service data.)

Table 1-1 shows that many of the jobs in the food industry are located in the wholesale and retail sectors. A 1999 study projected some 57,800 new employment opportunities per year will be available to college graduates with agricultural, forestry, natural resources, and veterinary medicine expertise over the 2000 to 2005 period. Food marketing, merchandising, sales, and management positions will account for 41 percent of the projected new jobs. While the number of farm and ranch operators is expected to continue declining, good career opportunities will be available for students interested in farm management and farm and business services. These students will also require marketing, as well as communication and leadership skills.

Some food marketing positions include market analyst, merchandiser, advertising or market communciations manager, market economist, foodstore manager, food service manager, food or commodity broker, sales management, purchasing agent, public relations, new product development manager, logistics or transportation manager, export-import manager, product or brand manager, and marketing researcher.

THE MARKETING PROCESS

If we take our definition of marketing as the business activities involved in the flow of goods and services from the point of initial production until they reach the ultimate consumer, two essential characteristics of the process become evident. First, the marketing process is one of movements; it is a *series of actions and events* that take place in some *sequence.* Second, some form of *coordination* of this series of events and activities is necessary if goods and services are to move in some orderly fashion from the hands of producers into the hands of consumers.

Figure 1-3 brings more clearly into focus some of the more pertinent aspects of this marketing process. It shows the agricultural marketing system starting with

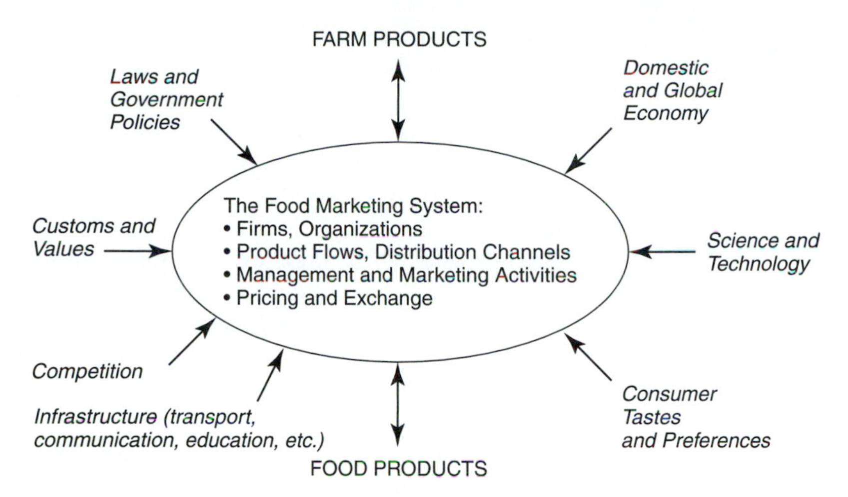

FIGURE 1–3 The changing food market system.

the farmer. The nature and way in which this production is initially offered to the marketing system has a major influence on the organization and operation of the system itself. At the same time, the dynamics of the marketing process may have a direct influence on agricultural production. A good example of this two-way flow of effects can be seen in the dairy industry. The extreme perishability and bulkiness of milk once required a costly assembly system of trucks picking up the milk in cans at the farm each day. However, the invention of large cooling tanks to provide storage for milk on the farms, combined with the development of bulk tank trucks, provided another possible assembly method. Because this method required large equipment investments, the small dairy farmer was at a disadvantage. Here, then, was a marketing technology that encouraged the reorganization of dairy farms into larger, more specialized units.

At the other end of the sequence of marketing activities is the consumer. Here again the path of influence is a two-way one. Certainly consumers' preferences and behavior dictate to a major extent the activities of the marketing process. Similarly, marketing firms expend a great deal of effort in trying to influence and change consumers' behavior and wants to the marketers' advantage.

Between these two forces—the agricultural producer and the consumer—is the marketing system. This complex system is composed of business firms engaged in physical, technological, and economic activities and run by managers who make the necessary decisions and direct people. Another part of the food marketing system is made up of firms and organizations whose activities contribute to the pricing of food products and to establishing the various arrangements, contacts, and procedures that will ensure an orderly and purposeful flow of goods and services. Here we see the complex operations of a large terminal market, a trade association, or a private or government market news agency.

Today's food system is the product of many forces operating over many years. The complex influences shaping the organization and behavior of the food sector are shown in Figure 1-3.

The U.S. food industry is not an economic island. It is part of a larger U.S. economy and is therefore affected by macroeconomic trends in growth, employment, inflation, and interest rates. In turn, the U.S. economy is highly integrated into the global economy and impacted by world trade patterns, foreign exchange rates, political and world economic events. Today's marketing student and manager must understand the complex economic interrlationships of these economic systems.

Science and technology are also major influences on the U.S. food system. Food production is increasingly becoming industrialized, and new technologies in food processing, packaging, and marketing have given rise to new products, companies, and industries. Technology has become a powerful engine of change in the food industry, as illustrated by recent biotechnology developments.

Food consumers and their wants and needs are also reshaping the nature of the food system. Today's consumer demands increasingly diverse, healthy, conveniently prepared, and economical food products. As food production and marketing have become more customer-driven, the relationships between food producers, processors, and marketing firms have changed. Today's successful food marketing manager must understand and anticipate the needs and wants of consumers. Market research is a key to this success.

We sometimes neglect or take for granted the role that *social capital* or infrastructure plays in the development of an efficient food marketing system. These are the resources created by society for the benefit of all. A complex food industry cannot exist without well-developed transporation, communication, and educational systems. Government, in both its regulatory and facilitating functions, is part of this social capital. A marketing system cannot function well without laws, courts, and policies to promote the public interest.

Competition is a major influence on the organization and behavior of the food industry. Through their attempts to improve profits, satisfy consumers, and gain a competitive edge, food producers and marketing firms continually search for new and different ways to market their products. Some of these innovations succeed and others fail, but the competitive process is never quiet for long and is a frequent source of market change.

The behavior of the marketing system is also limited by the rules, customs, and values of a society. In the United States, firms cannot employ children regardless of how little they might cost. We do not advertise liquor and tobacco on television, though it is acceptable to advertise beer and wine. In our country the kickback, or collusion in business dealings, is not acceptable, although in certain other countries such practices are carried on without public censure. Though such rules may be changed by society, at any given time the marketing system must work out its activities within the currently existing framework.

ALTERNATIVE PERSPECTIVES OF FOOD MARKETING

Today's student of marketing must understand two different views of the food marketing world. One perspective gives the "big picture" of food marketing, sometimes called *macromarketing*. It looks at how the food system is organized, how well it performs its economic and social tasks, and how the food system is changing over time. This is often the view taken by economists, industry analysts, and government officials. Often, economic theories and concepts are used to answer the question, How well is the marketing system contributing to the public interest and serving all the actors in the process?

The *micromarketing*, or business management, view of marketing, in contrast, is that taken by an individual decision maker in the food industry. This could be a food producer, a business manager, or a food consumer who is making choices and decisions about how, when, where, and what to buy or sell. Using the tools and principles of marketing management, firms develop marketing strategies that are designed to satisfy customers at a profit. This is the world of competition, new product development, market research, demand stimulation, creative pricing and distribution strategies, customer service, and other marketing management tactics.

These two marketing worlds are closely related. To be successful, marketing managers must be aware of how the changing macromarketing environment is affecting their decisions and creating new market opportunities. They must also be concerned with the changing public perception of the ultimate performance of the food system. At the same time, managers must constantly find new and better ways to attract and satisfy customers, outmaneuver their competitors and make a profit in doing so. Through the complex market economy, each individual firm or consumer decision contributes to the larger market picture and the larger economy shapes each decision. Students who hope to understand—and possibly find employment in—the food system will be well advised to study both of these marketing perspectives.

MINI CASE 1: FOOD FOR THOUGHT AT THE COFFEE SHOP

Farmer Fred, Consumer Connie, and Middleman Mike were seated next to each other one morning at the coffee shop. The farmer overheard Connie complain to her friend, "Isn't it terrible how high food prices are?" "Yes," said her friend, "Farmers and food middlemen are charging us too much." Choking on his breakfast, Farmer Fred leaned over and said, "Excuse me, ladies, but you shouldn't criticize farmers with your mouths full. I don't set the price of my products, prices this year are not covering my costs of production, and we farmers get only about 22 percent of the money that you spend on food. The middlemen take the other 78 percent of your food dollar." Spilling his coffee, Middleman Mike jumped into the conversation to defend himself, "Food prices are not our fault. We just give consumers what they want, including ready-to-eat, safe, wholesome, packaged, processed, and high-quality products year-round. This costs money, but our food system is so efficient that Americans are well fed and only pay 11 percent of their income for food, the lowest level in history in the world."

Who is right here: Fred, Connie, or Mike? Why do they have such different views of food prices and marketing?

GROWTH AND ROLE OF MARKETING

Marketing has developed in importance and complexity as economic development and specialization have increased our productive capacity and separated food producers from consumers. The pioneers of our country did not have to concern themselves with marketing problems. Each family grew its own food and fiber and built its own shelter. Producers and consumers, if not actually the same individuals, lived near to each other.

Very early in the development of any community, however, people realized that some were better adapted to certain kinds of activities than others. Thus, they specialized in their work. This specialization increased the output of goods, but it also broke down the self-sufficiency of the family unit. As people specialized in different activities, they began to produce more than was needed for home consumption. Markets then developed to facilitate the exchange of this *marketable surplus* between rural and urban areas. Here, then, was the beginning of the marketing task and the middlemen who specialized in its performance.

The growth of urban areas fostered further specialization. With the disappearance of the necessity for people to produce all of their own basic needs, they were able to leave the land and congregate in larger groups where work could be carried on more efficiently. The people remaining on the farms could produce food and fiber more efficiently.

Throughout most of our early history, one of the pressing agricultural marketing problems was that of providing adequate transportation facilities at a reasonable cost to move the increasing output from the farms to the consumers in our growing cities. Here we find an early interest of government in helping the marketing system function adequately. First turnpikes, then canals, and then railroads were subsidized by government. This public concern has continued even to the present day as government has developed the interstate highways and subsidized the continued upgrading of our air terminals.

A complex and costly marketing system is not necessary in situations where the volume of production and trade are limited. On the other hand, assembly-line mass production is not feasible until the marketing system opens the doors to the broad mass market. Many people have viewed with alarm the impersonal relationships of our huge factories, the growing proportion of our population living in cities, and the increasing numbers who are engaged in the marketing trades and services. Such developments, however, are usually the marks of the more productive countries with increasing standards of living.

This interrelationship between the increasing productivity of the agricultural production process and the development of an efficient marketing system is not always well understood. Since the end of World War II there has been great interest in improving the economic status of the developing countries of the world. At first, it was widely thought that if these countries would only apply the known improved production technologies, such as fertilizers, better seeds, and cultivating machinery, advancing output and progress would be forthcoming. In order for farmers to adopt such changes, however, they must be able to reap some benefits from the increased output—they must be able to sell their products profitably to someone else. Without modern marketing systems, including communications, transportation, storage facilities, and financing arrangements, this is not possible. Once again it is demonstrated that agricultural production and food marketing must develop hand-in-hand. They are partners in a progressive system.

FOOD MARKETING IN THE MARKET ECONOMY

Marketing, as defined above, is a universal characteristic of all economies and societies. Every country must have a food marketing system that moves and transforms products from producers to consumers. However, there are differences in how market activity is organized and conducted in market economies versus centrally planned economies. It has often been observed that output, efficiency, and standard of living are higher in market economies, and transitions to market economies frequently produce dramatic increases in the economic performance of nations. As a result, many former command economies are attempting to make a transition to the free market economy. What are the differences in the two alternative ways of organizing market activities?

There are a number of necessary conditions for a market economy to exist. First, market economies generally recognize private property and provide income and profit incentives for individuals and companies to improve their economic status. Second, market economies tend to rely on a decentralized price system to assist individuals in making decisions and allocating resources to their best uses. Prices are influenced by supply and demand factors and determine how risks and rewards are shared in the market economy. Third, market economies emphasize competition as a way of encouraging efficiency and improved performance. Fourth, market economies permit a relatively high level of freedom of choice for producers and consumers, including the freedom to fail. Fifth, and most importantly for our study of food marketing, market economies place a high value on satisfying consumers through a wide variety of marketing activities and strategies that are often criticized in non-market economies. Finally, all of these conditions for a market economy imply a limited role of government in the economy. In place of government output and marketing decisions, the market economy generally relies on Adam Smith's "invisible hand" to channel individual, profit-seeking behavior to serve the public interest.

This is not to say that the free market economy is automatic, free of government influence, or even entirely free. The market economy has its faults and limitations. It can lead to excesses, resource waste, and pollution. It provides for freedom of opportunity but not equal results for all. It does not eliminate unemployment, inflation, or economic recession. It doesn't always lead to what everyone would consider fair prices and incomes. When the market economy fails to meet the public's expectations, government intervention often occurs. Maintaining a balance between government involvement and private freedoms is a continuing challenge for market economies. The market economy is sometimes said to be a terrible way to organize an economy—but preferable to any of the alternatives.

The American food marketing system must be studied in the context of a market economy. It is difficult to understand one without knowledge of the other. As you read this book, consider the perennial economic problems of what to produce, how much to produce, how to produce, and how to distribute production. Further, consider how the market economy answers these questions in a way that attempts to minimize costs, maximize output, satisfy consumers, and provide freedoms for producers, consumers, and profit-seeking firms. Then, consider the validity of the belief of many that even an imperfect market economy is preferable to a planned, centrally controlled economy.

HISTORICAL BENCHMARKS IN FOOD MARKETING

The history of the growth and development of agricultural marketing is a rich and exciting one. Only a broad, sweeping picture of it can be presented here. But the study of marketing history serves a very useful purpose. Knowing the marketing problems confronted by our grandparents and great-grandparents will help prevent us from believing that our current problems are the only difficult ones that have ever begged solutions. The study of past developments also helps emphasize that changes are continually underway. New developments from many different sources have always challenged the existing marketing organization to adapt or die.

Until the end of the first half of the nineteenth century, the pattern was one of small industry with transportation and communication and other marketing problems of a largely local nature. Tobacco, as one of our first surplus crops, presented one of the earliest marketing problems. In seeking solutions, colonial governments and growers tried price controls, production regulations, and struggled with grading problems. Colonial flour milling also had problems of rate fixing, monopoly, and product adulteration.

Immediately following the Civil War, a whole new set of problems developed as the foundations were laid for the modern nationwide commercial marketing system. Agricultural production grew tremendously with the opening of new western lands. Land in farms jumped from about 294 million acres in 1850 to 536 million acres in 1880. The number of farms increased from about 1.5 million to 4 million during the same period. Outlets for this potential production bonanza had to be found in both our own growing cities and in foreign countries.

About this time many of the large food industries were established. In 1865, the Union Stockyard at Chicago was formed and soon was the largest livestock market in the world. The Chicago Board of Trade was founded in 1848 and quickly developed into the nation's leading grain market. By 1870, ice was being used to preserve meat, and by 1880, the use of the refrigerated railroad car made the large national packers possible. By the late 1880s, Minneapolis was the milling center of the country, and the pattern of organization for the modern giant milling and grain corporations was established. With the introduction of tin cans and other canning equipment in 1880, large-scale canning of fruits and vegetables got underway.

While these changes were occurring, transportation also expanded tremendously. During the decade of the 1870s, the mileage of railroads almost doubled. The two coasts were joined and transcontinental shipments became possible. Telegraph communications rapidly expanded. Extensive highway development, however, was to await the automobile.

This was a turbulent period. The extraordinary expansion of farm production suddenly made available vast amounts of agricultural raw materials. The marketing system was put under great pressures to move this productive capacity into the hands of consumers. The rapid growth of food processing firms and their search for sales led to cries of monopoly and unethical practices. To this situation was added a severe agricultural price depression during much of the latter part of the century.

Farmers organized to protest these situations. Railroads were bitterly attacked for charging exorbitant and unfair rates. Bitter attacks were made on the alleged evils of the middlemen. Demands were made for all kinds of corrective and regulative measures. The immediate result of this agitation was federal regulatory action. The Act to Regulate Commerce that, among other things, authorized the Interstate Commerce Commission and the surveillance of interstate freight rates, was passed in 1887. In 1890, the passage of the Sherman Act laid the basis for our antimonopoly policies and made private business activities a matter of public concern.

With an improvement in general economic conditions, the tension eased. And in the ensuing years until World War I, the marketing system was allowed time to grow up to its job. Agricultural production continued to expand, but there were no upheavals of the extent of the post-Civil War period. This period was one of changing emphasis in marketing. Previously, much of the effort of those working in agricultural marketing had been directed toward developing foreign markets for our exportable surplus. Now the emphasis shifted toward developing the domestic outlets of our growing country. Recognition of this changing emphasis came in 1908, when the name of the Division of Foreign Markets in the U.S. Department of Agriculture was changed to the Division of Production and Distribution. In 1914, Congress officially set up the Office of Markets to collect and disseminate information concerning the marketing of farm products.

After World War I, the attitude toward agricultural marketing took a still different perspective. It appeared that our production capacities had outrun our consumption capabilities. Throughout the 1920s, many schemes were proposed to permit us to dump our excess production abroad. Domestically, however, improvement in the marketing machinery was to be the answer. Governmental blessings were given the cooperative movement as an approach to more effective and orderly marketing. Such regulatory laws as the Packers and Stockyards Act and the Commodity Exchange Act were passed to prevent abusive practices by marketing agencies. Increased marketing efficiency was the goal.

The depression decade of the 1930s further increased the difficulties of moving agricultural products to market at satisfactory prices. Attention was now turned away from marketing to production. Developments in the marketing area could not solve the problem. We simply produced too much. Public attention was now directed toward perfecting various schemes that would reduce the amount of production.

With the demands of World War II, however, efforts were made to produce all that our resources would permit. After the wartime demands had receded, the marketing machinery again came in for increasing public attention and criticism. There was widespread belief that improvements in production had outdistanced improvements in marketing. Ways had been found to make two blades of grass grow where one did before, but no effective way had been found to market the extra blade. The feeling was that all that could be produced could be consumed at satisfactory prices if the marketing system was functioning well.

In 1946, Congress officially gave recognition to the renewed emphasis on agricultural marketing when it passed the Research and Marketing Act. Under this Act, additional funds for expanding research into all phases of marketing were authorized. The post–World War II years were full of far-reaching changes in almost every phase of marketing. The perfected quick-freezing of foods found rapid consumer acceptance. Rising living standards led to demands for more services and more processing of food. More and more food preparation and processing were done by marketing firms, and less and less in the kitchen. The retail food store experienced revolutionary changes as the mass, self-service merchandising techniques of the supermarket were perfected. With improved transportation and communication, most of the marketing activity became oriented to a national rather than a regional or local viewpoint. The great central market institutions established almost a hundred years earlier gradually were replaced by a more decentralized activity. Increasing labor costs focused attention on ways to mechanize and standardize the operation of the marketing machinery. Great faith was placed in increased market efficiency as a major solution for the problems of agriculture.

The 1970s and 1980s brought a new set of food marketing developments and issues. The struggle to keep food supplies and demand in balance at prices acceptable to farmers and consumers led to new farm policy instruments such as the farmer-owned reserve, the payment-in-kind program, and target price supports and deficiency payments. As U.S. agricultural exports spurted in the early 1970s, U.S. farmers became more export-minded and U.S. food prices were more closely linked with the world economy. The subsequent decline in agricultural exports, farm prices, and land values in the 1980s presented new challenges for farmers. The introduction of agricultural options in the mid-1980s widened farmers' marketing alternatives. Foreign investment, mergers, diversification, and new product developments drastically altered the food marketing system in the 1980s.

The 1990s were marked by changes in food consumption preferences, technological developments, increased concerns with diet and health, growing regulations on the uses of chemicals and additives in the food supply, and increased concerns with the impacts of the food industry on the environment and resource conservation. The growth and market power of large, multinational food companies and their impacts on food producers and consumers was also a matter of concern. In addition, the American food system became more closely linked with the global food economy as foreign investment into and outside the American food system grew.

Changes in agricultural production and food marketing continue at a rapid rate. New marketing developments and challenges are presented by biotechnology, trade and globalism, organic and natural food products, functional foods with health benefits, specialty and niche farm products, home meal replacements, dashboard dining, minority-owned farms, and farmer-direct marketing, to name a few. Beyond a doubt, the American food marketing system is a dynamic operation that is furnishing an ever-increasing variety of new and improved products at a high level of economic efficiency. However, the domestic problem of how to obtain satisfactory prices for the steadily increasing potential of agricultural output still remains. Attention is focused, on the one hand, on how best to organize and control agricultural production, and, on the other, on how to expand the market potential for these products. In addition, there is continuing concern about the growing potential power of giant firms in the food industry.

SUMMARY

Food marketing encompasses all the business activities involved in the flow of food products and services from producers to consumers. The food marketing system is a complex and expensive network of channels, middlemen, and marketing activities that facilitates the production, distribution, and exchange of the nation's food supply. This system has increased in size and importance with the growth of specialized agriculture, urbanization, and other trends. Marketing is a productive process that adds form, place, time, and possession utility to farm commodities. The value added in the food marketing process complements the productive processes in farming. Marketing forms a vital bridge between food producers and consumers and is influenced by social resources, technology, and the laws and norms of society. The food marketing system has evolved in response to the changing needs of farmers, food consumers, and society. The market economy with its emphasis on profit incentives and efficiency has been extraordinarily successful in increasing food production and feeding a growing world population.

KEY TERMS AND CONCEPTS

agribusiness
food marketing system
infrastructure
market
marketable surplus
marketing
marketing utility
middlemen
specialization
value-added

DISCUSSION QUESTIONS

1. How would you explain food marketing to someone who is unfamiliar with the subject?

2. Some people associate food marketing with everything that happens outside the farm gate. Do you agree or disagree?

3. What does it mean to say that farmers and food marketing firms are mutually interdependent?

4. How do improved transportation and communication influence the food marketing process?

5. What do you think the food production and marketing system will be like in the year 2010? What are you assuming about farming and consumers?

6. Can you think of any industry that has no specialized middlemen? Why is this so rare?

7. Why has the food marketing system become more complex and expensive with time?

8. Which of the four marketing utilities do farmers provide? Consumers? How are these provided?

9. Suppose you are hired as a consultant to a developing country that wishes to expand its food industry. How would you divide the available development resources between production agriculture and the food marketing system?

10. What are the pros and cons of a market economy? What are the alternatives to a market economy?

SELECTED REFERENCES

Cochrane, W. W. (1979). *The Development of American Agriculture, A Historical Analysis.* Minneapolis: University of Minnesota.

Current Issues In Economics of Food Markets (October 2000). Economic Research Service, U.S. Department of Agriculture. http://www.ers.usda.gov/epubs/ pdf/aib747/

Dimitri, Carolyn (1999). Order Out of Chaos? The Evolution of Marketing Institutions. *Choices.* Ames, Iowa: American Agricultural Economics Association.

Economic Research Service, U.S. Department of Agriculture website (January 2001). http://www.ers.usda.gov/

Food Market Indicators Briefing Room. (June, 2000). Economic Research Service, U.S. Department of Agriculture. http://www.ers.usda.gov./briefing/FoodMark/

Goecker, A. D., C. M. Whatley, and J. L. Gilmore (1999). *Employment Opportunities for College Graduates in the Food and Agricultural Sciences, 2000-2005.* Published by U.S. Department of Agriculture and Purdue University. Available in pdf form at http://faes.tamu.edu/supplydemand.

Lipton, K. L., W. Edmondson, and A. Manchester (1998). *The Food and Fiber System: Contributing to the U.S. and World Economies.* Agriculture Information Bulletin No. 742. Washington D.C.: U.S. Department of Agriculture.

Manchester, A. C. (September 1992). *Rearranging the Economic Landscape: The Food Marketing Revolution, 1950-1991,* Agricultural Economic Report No. 660. Washington, D.C.: U.S. Department of Agriculture.

Meulenberg, M. (Ed.) (1993). *Food and Agribusiness Marketing in Europe*. Binghamton, NY: International Business Press.

Padberg, D. I., C. Ritson, and L. M. Albisu (1997). *Agro-Food Marketing.* New York: CAB International.

Rhodes, V. J., and J. L. Dauve (1998). *The Agricultural Marketing System,* 5th edition. Arizona: Holcomb Hathaway Publishers.

U.S. Department of Agriculture. (November 1998). *Agriculture Fact Book 1998.* Washington, D.C.: U.S. Department of Agriculture. http://www.usda.gov./ news/pubs/fbook98/contents.htm

CHAPTER 2

ANALYZING AGRICULTURAL AND FOOD MARKETS

You can eliminate the middlemen, but not their marketing functions.

OBJECTIVES

After reading this chapter, you will be able to:

1. Distinguish between the functional, institutional, behavioral science, and marketing management approaches to understanding and analyzing the food marketing system.
2. Explain why we have specialized food marketing firms and why it is so difficult to "eliminate the middleman."
3. Identify the major elements of a firm's marketing strategy: product differentiation, market segmentation, target marketing, and the marketing mix.
4. Discuss the sovereign role of consumers in determining the success of all food marketing decisions.
5. Question and evaluate the performance of the food system, using such criteria as efficiency, variety, consumer satisfaction, farmer returns, pollution, food safety, and other factors.

The difference between an orderly closet in which you can find what you want with a minimum of effort and a disorderly one in which nothing can be easily found can be traced to an adequate system of hooks, hangers, and shelves. The study of complex marketing systems and problems can be frustrating and confusing without a system that organizes our observations, thoughts, and judgments. We now turn to building this organizational framework.

APPROACHES TO THE STUDY OF FOOD MARKETING

There are several ways to approach the study of food marketing and each orientation provides a unique perspective on the nature and workings of the food marketing process. Some of the approaches are purely descriptive, whereas others are more analytical and attempt to evaluate the food marketing system and suggest recommendations for improving it.

The Functional Approach

One method of classifying the activities that occur in the marketing processes is to break down the processes into *functions.* A marketing function may be defined as a major specialized activity performed in accomplishing the marketing process.

Any listing of functions must be recognized as arbitrary. Authors list from as few as eight to as many as three or four times that number. Each composer of a list, of course, believes that list to be best. Others disagree and propose lists of their own. We are looking for hooks and shelves on which to arrange our ideas. The exact terminology of the list is not of great importance as long as the scope of the individual functions is understood. We shall follow a fairly widely accepted classification of functions, as shown in Figure 2-1.

The exchange functions are those activities involved in the transfer of title to goods. They represent the point at which the study of price determination enters the study of marketing. These functions are never performed in our economy without a judgment of the value, usually expressed at least partially as a price, being placed on the goods. Both the buying and selling functions have as their primary objective the negotiation of favorable terms of exchange.

The *buying* function is largely one of seeking out the sources of supply, assembling the products, and performing the activities associated with purchase. This function can involve either the assembling of the raw products from the production areas or the assembling of finished products into the hands of other middlemen in

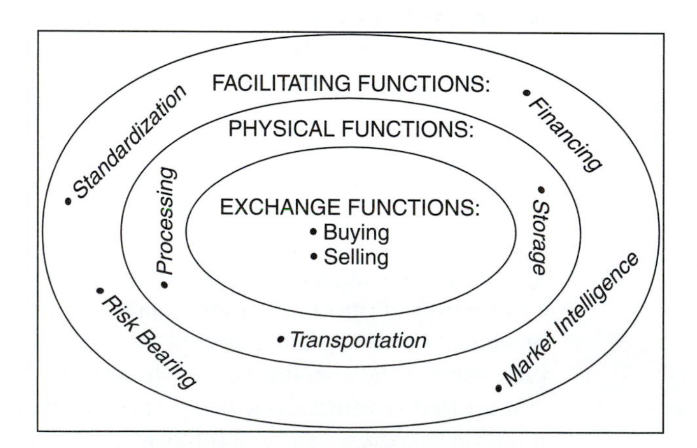

FIGURE 2–1 The marketing functions.

order to meet the demands of the ultimate consumer. These procurement activities can be just as important to the profits and success of a farm or food marketing firm as sales activities.

The *selling* function must be broadly interpreted. It is more than merely passively accepting the price offered. In this function can be grouped all the various activities that sometimes are called merchandising. Most of the physical arrangements of display of goods are grouped here. Advertising and other promotional devices to influence or create demands are also part of the selling function. The decision as to the proper unit of sale, the proper packages, the best marketing channel, and the proper time and place to approach potential buyers are all decisions that can be included in the selling function.

The physical functions are those activities that involve handling, movement, and physical change of the actual commodity itself. They help solve the problems of when, what, and where in marketing.

The *storage* function is primarily concerned with making goods available at the desired time. It could be a farmer storing grain. It may be the activities of elevators in holding large quantities of raw materials until they are needed for further processing. It may be the holding of supplies of finished goods as the inventories of processors, wholesalers, and retailers.

The *transportation* function is primarily concerned with making goods available at the proper place. Adequate performance of this function requires weighing alternative routes and types of transportation as they might affect transportation costs. It also includes the activities involved in preparation for shipment, such as crating and loading.

The *processing* function is sometimes not included in a list of marketing functions because it is essentially a form-changing activity. However, in the broad view of agricultural marketing this activity cannot be omitted. The processing function would include all those essentially manufacturing activities that change the basic form of the product, such as converting live animals into meat, fresh peas into canned or frozen peas, or wheat into flour and finally into bread.

The *facilitating* functions are those that make possible the smooth performance of the exchange and physical functions. These activities are not directly involved in either the exchange of title or the physical handling of products. However, without them the modern marketing system would not be possible. They might aptly be called the grease that makes the wheels of the marketing machine go round.

The *standardization* function is the establishment and maintenance of uniform measurements. These may be measurements of both quality and quantity. This function simplifies buying and selling, because it makes the sale by sample and description possible. It, therefore, is one of the activities that makes possible mass selling, which is so important to a modern economy. Effective standardization is basic to an efficient pricing process. A consumer-directed system assumes that the consumers will make their wants known to producers through price differentials. These differentials must then be passed back through the marketing channel so that marketing agencies and producers can know what is wanted. Only if a commodity is traded in well-defined units of quality and quantity can a price quotation do this job effectively. Standardization also simplifies the concentration process, because it permits the grouping of similar lots of commodities early in movement from the producing points. After their establishment, the use of standards must be policed. Such activities as quality control in processing plants and inspections to maintain the standards in the marketing channel can be considered part of this function.

The *financing* function involves the use of money to carry on the various aspects of marketing. To the extent that there is a delay between the time of the first sale of raw products and the sale of finished goods to the ultimate consumer, capital is tied up in the operation. Anywhere that storage or delay takes place, someone must finance the holding of goods. The period may be for one year or more, as in the operations of the canning industries, or a relatively short time, as in the marketing of perishables. Financing may take the easily recognizable form of credit from various lending agencies or the more subtle form of tying up the owner's capital resources. In either instance, it is a necessary activity in modern marketing.

The *risk bearing* function is the accepting of the possibility of loss in the marketing of a product. Economic risks can be classified into two broad classifications—physical risks and market risks. The physical risks are those that occur from destruction or deterioration of the product itself by fire, accident, wind, earthquakes, or other means. Market risks are those that occur because of changes in value of a product as it is marketed. An unfavorable movement in prices might result in high inventory losses. A change in consumer taste might reduce the desirability and price of the product. A new marketing strategy of competitors might result in a loss of customers for other firms. All these risks in varying degrees must be considered in the marketing of a product. Risk bearing may take a more conventional form, such as the use of insurance companies in the case of physical risks or the utilization of futures exchanges in the case of price risks. Or, the entrepreneur may bear the risk without the aid of any of these specialized agencies. The function of risk bearing is often confused with the function of finance. Their differences can be kept clear, however, if it is remembered that the need for financing arises because of the time lag between the purchase and sale of products, whereas the need for risk bearing arises because of the possibility of loss during the holding period.

The *market intelligence* function is the job of collecting, interpreting, and disseminating the large variety of data necessary to the operation of the marketing processes. Efficient marketing cannot operate in an information vacuum. An effective pricing mechanism is dependent on well-informed buyers and sellers. Successful decisions on how much to pay for commodities or what kind of pricing policy to use in their sale require that a large amount of market knowledge be assembled for study. Profitable storage programs, an efficient transportation service, and an adequate standardization program all depend to a considerable extent on good information. As with other functions, this function may be performed by those who specialize in its performance. On the other hand, everyone in the marketing structure who buys and sells products evaluates available market data and therefore performs this function to some degree.

Uses of the Functional Approach

Analyzing the functions of various middlemen is particularly helpful in evaluating marketing costs. Retailing is usually much more costly than wholesaling. The functional approach, however, points to the greater complexity of retailing. Cost comparisons are meaningful only when they are related to the job done. Retailer A may operate at lower costs than retailer B, but does retailer A perform the same functions as B? Perhaps A is a cash-and-carry merchant whereas B extends credit and delivers. As such, A probably performs considerably less of the functions of financing, risk bearing, and transportation than B.

The functional approach is also useful in understanding the difference in marketing costs of various commodities. For example, a perishable product is often more costly to market than one that is less perishable. Much of this difference may be because of the greater difficulty in the performance of the transportation, storage, and risk bearing functions.

The breaking down of a complex marketing task into its component functions greatly aids in efforts to improve the performance of the marketing machinery. Again in reference to our retailer, perhaps retailer B is losing money even though other retailers having similar operations are not. A function-by-function study of B's business might show that the cost of its credit function is unduly high because of unpaid accounts. Or a careful analysis of A's selling function may show that the firm has not kept up with new methods in merchandising products and thus is losing out to competitors.

There are three important characteristics of these marketing functions. First, the functions affect not only the cost of marketing food but the value of food products to consumers. Processing, transportation, and storage provide form, space, and time utility for consumers. The exchange and facilitating market functions grease the wheels of the marketing machinery and perhaps provide services at costs lower than farmers and consumers can perform them. In evaluating marketing functions, consideration must be given to both the costs and benefits of the functions. The value added by a marketing function may be greater or less than the cost of performing that function.

Second, although it is frequently possible to "eliminate the middleman," it is not possible to eliminate marketing functions. Usually, eliminating the middleman involves the *transfer* of marketing functions—and costs—to someone else. For example, farmers may assume the storage, selling, and transportation functions, eliminating brokers and commission men. A neighborhood group of consumers can eliminate the food retailer by purchasing in large lots from wholesale food outlets, but in doing so they will assume some retailing functions—storage, standardization, and perhaps transportation. And the group often will settle for fewer services—such as check cashing or price marking. The cost of performing a marketing function, then, can be reduced, but the function cannot be eliminated from the marketing process.

The third characteristic of marketing functions is that they can be performed by anyone anywhere in the food system. Conceivably, all the functions could be performed by a single firm that had complete control of food, from farm to fork. On the other hand, there are specialized firms and industries—such as railroads, grain brokers, and speculators—who perform only one marketing function. Grain may be shipped direct from farm to storage in the city, or it may be stored on-farm and shipped to market later in the season. There are some traditional combinations, placements, and timing of food marketing functions. Food processors usually combine the storage, processing, and transportation functions; and many farmers view on-farm storage as an integral part of farming. But in general, a variety of firm combinations and timing of food marketing functions is observed. The functions may be indispensible, but they are quite flexible because they can be performed in various places within the food industry.

Two key issues concerning marketing functions are whether the necessary number of functions is being performed and whether these functions are being performed in the most efficient manner. Because the functions add value as well as cost to food products, simply minimizing the functions is not an acceptable goal. The rule is that additional functions and services should be performed until the costs exceed the values of the functions—a difficult point to determine in practice. It is also important to determine when the shifting of a marketing function from one firm to another, or the combining of additional functions into one firm, is warranted. Some functions are most efficiently performed by specialized firms; others seem appropriately combined in multifunction firms.

Because of their traditional emphasis on the food production process, it is particularly important for farmers to consider whether they should perform more or fewer of the food marketing functions. There is no guarantee that profits will flow to those who assume additional marketing functions, even when the functions are known to be profitable. Some firms can perform the functions more efficiently than others. Food retailers would probably make poor farmers, just as farmers may lack the skills and interest to manage a modern supermarket or a complex railroad network. Then, too, the assumption of additional functions spreads each manager thinner, and some economies of specialization are lost. In the food industry several marketing functions may be combined within a single firm. There may be good reasons for this, especially when marketing functions are so complementary that profits can be made by performing them jointly.

The Institutional Approach

Another method of analysis is to study the various agencies and business structures that perform the marketing processes. Where the functional approach attempts to answer the "what" in the question of "who does what," the institutional approach to marketing problems focuses attention on the "who." Marketing institutions are the wide variety of business organizations that have developed in the food marketing system. The institutional approach considers the nature and character of the various middlemen and related agencies and also the arrangement and organization of the marketing activities.

Marketing Middlemen

Middlemen are those individuals or business concerns who specialize in performing the various marketing functions involved in the purchase and sale of goods as they are moved from producers to consumers. Food marketing middlemen can be classified as follows:

- Merchant middlemen
  1. Retailers
  2. Wholesalers
- Agent middlemen
  1. Brokers
  2. Commission men

- Speculative middlemen
- Processors and manufacturers
- Facilitative organizations

Merchant middlemen take title to, and therefore own, the products they handle. They buy and sell for their own gain. *Retailers* purchase and merchandise food products for final consumers. Their task is to provide a wide variety of products at a single location, making it convenient for consumers to assemble a desired market-basket of goods. Food retailers may include supermarkets; restaurants; convenience food stores; specialty meat and fruit and vegetable stores, dairies, or bakeries. They are the most numerous of the marketing institutions.

Food wholesalers sell to retailers, other wholesalers, and industrial users, but do not sell in significant amounts to final consumers. Wholesalers make up a highly heterogeneous group of firms with varying sizes and characteristics. One numerous group of wholesalers is the local buyers or country assemblers who buy goods in the producing area directly from farmers and ship the products to the larger cities where they are sold to other wholesalers and processors. In this group are such agencies as grain elevators and local livestock buyers. Another group of wholesalers is located in the large urban centers. These may be "full-line" wholesalers who handle many different products or those who specialize in handling a limited number of products. They may be cash-and-carry wholesalers or service wholesalers who will extend credit and offer delivery and other services. Such terms as *jobbers* and *car-lot receivers* are often used synonymously with *wholesalers.*

Agent middlemen, as the name implies, act only as representatives of their clients. They do not take title to, and therefore do not own, the products they handle. Whereas merchant wholesalers and retailers receive their incomes from a margin between the buying and selling prices, agent middlemen receive their incomes in the form of fees and commissions. Agent middlemen provide services to buyers and sellers. In many instances, the principal service of the agent middleman is market knowledge and the know-how necessary to bring the buyer and seller together. Their services are often retained by buyers or sellers of goods who feel that they do not have the knowledge or power to bargain effectively for themselves.

Agent middlemen can be divided into two major groups, commission men and brokers. *Commission men* are usually granted broad powers by those who consign goods to them. They normally take over the physical handling of the product, arrange for the terms of sale, collect, deduct their fee, and remit the balance to the seller. *Brokers,* on the other hand, usually do not have physical control or ownership of the product. They follow the directions of buyers and sellers and have less influence in price negotiations than commission men. Livestock commission firms and grain brokers are good examples of these two classifications of food industry agent middlemen.

Speculative middlemen are those who buy and sell products with the major purpose of profiting from price movements. All merchant middlemen, of course, speculate in the sense that they must face uncertain conditions. Usually, however, wholesalers and retailers attempt to secure their incomes through handling and merchandising their products and holding risks to a minimum. Speculative middlemen, on the other hand, seek out and specialize in taking these risks and usually do a minimum of handling and merchandising. Several names are given to these middlemen, such as *traders, scalpers,* and *spreaders.* They often attempt to earn their profits from short-run fluctuations in prices. For example, the grain scalper may buy and sell grain futures several times within a trading day, hoping to profit from small price moves.

Food processors and *manufacturers* are among the best known food marketing firms. They specialize in adding time, form, place, and possession utility to raw farm products. For example, they convert wheat into flour and bread. They combine food ingredients to make complete frozen dinners. They process livestock into the meat products that consumers desire. Food processors perform a number of marketing functions. They are important buyers of farmers' products, and most also wholesale their finished products to food retailers. Their packaging, branding, and promotional activities may be viewed as a form of market information as well as important influences on consumers' choices.

Facilitative organizations aid the various middlemen in performing their tasks. Such organizations do not, as a general rule, directly participate in the marketing processes either as merchants, agents, processors, or speculators. They may furnish the physical facilities for the handling of food products or for the bringing of buyers and sellers together. They establish the "rules of the game" that must be followed by the trading middlemen, such as hours of trading and terms of sale. They may also aid in grading and in arranging and transmitting payment. They receive their incomes from fees and assessments from those who use their facilities. Examples of this group are the stockyard companies, grain exchanges, and fruit auctions.

Trade associations are another such group. The primary purpose of a large majority of these organizations is to gather, evaluate, and disseminate information of value to a particular group or trade. They may conduct research. They also may act as unofficial policemen in preventing practices that might be considered unfair or unethical. Though not active in the buying and selling of goods, these organizations often have far-reaching influence on the marketing process.

Uses of the Institutional Approach

The institutional approach can help us understand why there are specialized middlemen in the food industry. It is possible to imagine a food system without middlemen. Farmers can, and at times do, perform such middleman activities as storage, transportation, selling, and even processing. By the same token, consumers can assume food middleman functions, such as processing, transporation, and storage. Farmers' markets eliminate the food middleman by transferring marketing functions to farmers and consumers. Why, then, are there so many food middlemen if there are no practical reasons why farmers and consumers could not replace them? The answer is that these specialized firms often can perform the food marketing functions more efficiently than either farmers or consumers.

There are three reasons for the presence of specialized middlemen in the food system. First, the rise of middlemen specializing in such activities as storage, transportation, processing, and retailing is an example of division of labor and specialization. Middlemen free farmers to specialize in agricultural production and free consumers for other activities. Normally, there are gains from such specialization, and division of labor is a common characteristic of industrialized societies. Second, these gains from specialization mean that many of the food marketing functions are marked by economies of scale. That is, the average cost of performing the marketing functions falls as the volume of products handled rises. For example, the freight rate for a full carlot of grain is normally lower per bushel than for a less-than-carlot shipment. Typically, the average cost of processing and packing food declines with increasing plant volume. These scale economies encourage the growth of specialized food middlemen. Finally, middlemen can reduce market search and transactions costs. Markets are not costless. There are expenses associated with finding buyers and sellers and negotiating exchanges between them. By specializing in these functions, food middlemen relieve farmers and consumers of the considerable costs they would otherwise incur for search and transactions activity.

Without food middlemen, farmers and consumers would forfeit these economic gains from specialization and scale economies. This is illustrated in Figure 2-2. In Figure 2-2A, each farmer sells directly to each consumer; all of the marketing functions are performed on small lots of product by either farmers or consumers; farmers and consumers bear the full costs of finding and consummating desirable exchanges. This might represent a farmer's market situation. In Figure 2-2B, specialized food middlemen perform the food marketing functions, allowing farmers and consumers to specialize in other activities and sparing them search and transaction expenses.

Which is the better system? The general rule is that specialized food middlemen will perform the food marketing functions when their efficiency in doing so is greater than that of farmers and consumers. The relevant comparison is the food middlemen's costs versus farmer or consumer costs for performing the same marketing functions, not middlemen's costs versus no costs. Judging by the number and

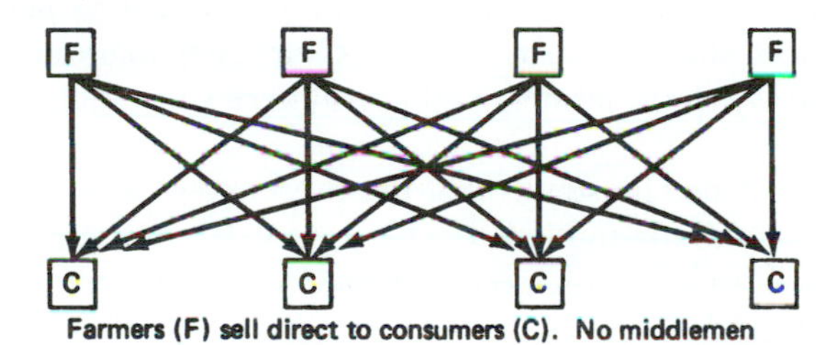

FIGURE 2–2A Farmers (F) sell direct to consumers (C). No middlemen.

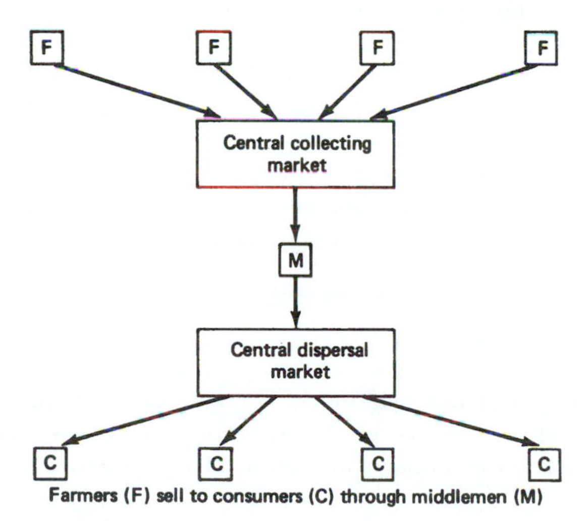

FIGURE 2–2B Farmers (F) sell to consumers (C) through middlemen (M).

MINI CASE 2: WHY DO WE NEED FOOD MIDDLEMEN?

Bill was angry and fed up. He had just delivered his hogs to the local buyer and learned that the price he received didn't even cover his costs of production. His wife, Kathy, sympathized with him saying, "I was just in the supermarket and the price of pork has gone up." "There is something wrong here," they agreed. The next day they met with a number of other unhappy pork producers to discuss the problem. "It's the middlemen," one of them said. "They buy low, sell high and pocket our profit." "What can we do about it?" said Bill. "We need to eliminate those greedy middlemen by operating our own marketing facilities and selling direct to consumers," one of them said. "That way we would get 100 percent of the consumer's dollar." But Bill cautioned, "That would cost a lot of money wouldn't it? Are you sure we want to get into the food processing, wholesaling, and retailing businesses? That is not our business." "It's the food business isn't it, and who knows more about food than farmers?" said Kathy. "I don't know" said Bill. "It sounds risky and complicated."

How would you advise Kathy, Bill and the other producers on their plan for getting into the food marketing business? What are the pros and cons of this proposal to elimate the middlemen?

success of middlemen, this condition appears to be frequently met in the food industry. This is not to say that all middlemen are efficient or necessary, or that farmers and consumers should never perform their own marketing functions. There are many situations where farmers and consumers can efficiently perform the middleman functions.

Figure 2-2 also illustrates that the presence of a few, specialized food middlemen tends to make the food system hourglass-shaped. There are many more farmers and consumers than middlemen, and their numbers will normally be larger in size than those of the farmers and consumers with whom they deal. This can result in imbalances of market power between producer, consumers, and middlemen.

Despite their apparent value, public distrust of middlemen has a long history and continues to this day. This is not unique to America. The *middleman bias* is even more pronounced in other countries. Throughout the world, middlemen are often scorned, sometimes outlawed, and usually merely tolerated. But because marketing is an activity practiced in some form in every country, those who take the time to understand it better usually come away with, if not a fondness for middlemen, at least a greater appreciation of their role in the economy.

The Behavioral Systems Approach

Both the functional and institutional approaches are useful in analyzing the existing marketing activities. However, the marketing process is continually changing in its organization and functional combinations. Understanding and predicting change is a major task.

Either a particular marketing firm or an organization of firms, such as the marketing channel, can be viewed as a system of behavior. Each is composed of people who are making decisions in an attempt to solve particular problems.

In either the firm or the organization of firms, four major types of problems, with their associated behavioral systems, can be identified. The first and perhaps most obvious is the *input-output system.* Each marketing firm or organization of firms is attempting to produce an output of something. This is true whether it is a meat processor, a commission man, or a marketing channel consisting of many firms. Each is using as an input, resources that are costly and scarce. Each hopes to find a satisfactory solution as to how to combine these input resources to secure a profitable output. Here we find the motives to develop and adopt new technology, new products, and different organizations that may be cost reducing or output enhancing. It is in this system context that the discipline of economics and the various physical and engineering sciences make their major contributions to understanding marketing.

Another major concern is the *power system.* All firms and groups of firms have status and a vested interest in the present role they are playing. They may have developed a reputation for quality, being market leaders, having a community conscience, or being the fastest growing. No decisions will be made that might deteriorate their particular niche of power, and means will actively be sought to enhance it. It is in this framework that we can understand the urge of many firms to grow and expand and to be innovators or followers. Economic theories of monopoly and imperfect competition behavior, as well as the political scientists' concern with power behavior, give insights into this system of behavior.

Once a firm or organization develops beyond the simple unit of a unified manager-laborer, the problem of getting appropriate information to the managers and of transmitting their decisions into actions by other workers becomes of increasing importance. Each firm or organization of firms can therefore be viewed as a *communications system.* How to establish effective channels of information and direction is a major problem of large firms and complex organizations. Desirable actions may be frustrated by not receiving the right information or by the misinterpretation of the messages of action. It is in this area that the concern of psychology, sociology, and business management over the proper ways to organize and direct subordinate workers and units becomes of particular relevance.

Finally, if change is the essential characteristic of marketing, then one of the major problems of marketing firms and organizations is how to adapt to these changes. The *behavioral system for adapting to internal and external change* is then a major component of the firm or organization. As a rule, firms desire to survive and are prepared to pay some price to do so. How to operate in a manner that will assure that important changes are identified and adaptive solutions to these changes are effectively evolved is a major area of behavior.

Input-output, power, communications, and adaptive behavior are all components of the operation of the marketing system at any one time. Keeping this in mind may help explain actions that, when viewed only in the context of one of the systems, seem irrational or unintelligent. A firm may forgo the ultimate in an input-output solution because its communications system has broken down or because of considerations of its power situation. For example, a firm may choose to acquire another related firm and integrate its activities in order to enhance its power in the marketplace. Or in another instance, it may acquire another firm because it thinks it foresees changes coming, and the purchase of the know-how and management of the acquired firm may present the most feasible method of adapting to the new conditions.

The behavioral systems approach is useful to an understanding of a major change that is currently underway in the food system. Many observers have noted that the food production and marketing systems are more and more resembling the nonfood, factory sectors of the economy. This is often termed the *industrialization* of the food sector. Changes in consumer demands for foods and new agricultural and food science technology, it is suggested, are forcing closer linkages between formerly independent food producers, processors, and consumers. Farmers must now produce a product that is tailored to meet the specific needs of food market niches. In place of open market arrangements that rely on prices to coordinate the flows of products. Contracts and other forms of market communication are coordinating producer and processor input/output decisions. The process is increasing the market power of farm product buyers and encouraging larger farms. In these ways, food firms continually attempt to adjust their behavior to market forces at the same time that they attempt to influence these forces.

FOOD MARKETING MANAGEMENT

The functional, institutional, and behavioral systems approaches are very useful in understanding the macro picture of food marketing. They help us understand the major forces shaping the organization of the food marketing system and how it is changing over time. However, you may get the feeling that these approaches appear theoretical and not too helpful to someone who is preparing for a career with a food marketing firm. It's not true that they are just theory, but there is another approach to food marketing that is highly useful to a working manager in the food industry.

Marketing management is the process by which firms and organizations develop and implement programs that assist them in satisfying consumers and making a profit. This micro approach to food marketing looks at how firms compete with each other and respond to changing market opportunities. A marketing manager for a food company must analyze the market, plan for the future, develop successful marketing strategies, and focus the entire firm's efforts toward meeting market needs and desires.

The key concepts and theories in food marketing management will be developed throughout this book. Only a few will be discussed here. One important principle of marketing is that firms should "find wants and fill them." The *marketing concept* holds that all the efforts of the firm should be directed toward satisfying customers at a profit. Market-oriented firms begin their production planning with the question, What do consumers want?, not What can we make? A second principle of marketing is that not all consumers want the same product. Firms can distinguish their products in many ways—for example, by varying the quality, the price, the packaging, the brand, or by offering services. This *product differentation* is very common in food marketing. A third principle is that consumers differ in their tastes and preferences. As a result, markets can be divided into groups of consumers, called *market segments,* who share similar needs and wants. Each firm should concentrate its marketing efforts on its *target* market segments, the customers it wishes to best satisfy.

Food firms develop and employ different competitive marketing strategies in appealing to and satisfying these customer market segments. The firm's *marketing mix,* often termed its 4P strategy, consists of its combination of price, product, promotion, and place, or distribution, strategies. Each firm will develop a unique marketing mix to take advantage of its resources, attract its target market customers, and outcompete its rivals. The firm's choices of target markets and marketing mix is called its *marketing strategy.*

The marketing management approach to food marketing is useful in understanding why some firms are more successful and more profitable than others. They apply these marketing principles and satisfy customers better than their competitors. As a result, some firms grow and prosper while others stagnate or fail. This approach also provides insight into why the market economy produces so much product variety and different marketing alternatives. Some firms may select a low-price strategy while another emphasizes high-price-high-quality. Some sell many different versions of processed foods while others sell only one specialized product. Some companies brand their products and others don't. One firm may promote on TV while another chooses only cents-off coupons or point-of-purchase sales materials. A firm may choose to market its food product through supermarkets, discount stores, restaurants, vending machines, mail catalog, or on the internet. This is all part of the search for a competitive marketing strategy.

Though widely practiced in all market economies, marketing management is not without its critics. It is variously charged with causing higher prices, deceiving consumers, excessive variety and duplication, fostering materialism, monopolization, resource exploitation, and other unethical practices. We will examine these charges against marketing management in the food industry throughout this book. We will also weigh them against the positive contributions that marketing management strategies can make to satisfying consumers.

ANALYZING MARKET PERFORMANCE IN THE FOOD INDUSTRY

It isn't possible to study the food marketing system for long without making some judgments about it. Is it competitive? Are prices fairly determined? How well does it serve farmers and consumers? Could it be improved? These are market performance questions. *Market performance* is defined as how well the food marketing system performs what society and the market participants expect of it. Questions of food marketing performance have become increasingly important in recent years.

Evaluating marketing performance raises the question, What do we expect of the food marketing system? Here we find that there are multiple and often conflicting goals for the marketing system. Compromises and tradeoffs are necessary to satisfy consumers', farmers', and society's goals for this system. What is more, the quality of market performance ranges on a scale from "good" to "bad," and there are more gray areas than easy judgments. There also appear to be rising standards for evaluating market performance in our society.

One way to begin the study of food marketing performance is to list some common concerns expressed about the industry. Consumers frequently complain about high and fluctuating food prices, expensive marketing practices such as overpackaging and supermarket games, out-of-stock specials, slack filling of food containers, and deceptive labels and advertising. Farmers voice other complaints: declining numbers of farm product buyers, reduced competition for supplies, large buyers with control over prices, the failure of retail and farm prices to move together, excessive marketing costs and prices, price speculation, and below-cost prices. Society in general might be more concerned with such issues as the food marketing sector's contributions to employment, investment, and economic growth; price stability; the standard of living and quality of life; resource use and conservation; and the overall health and prosperity of the rural economy.

The evaluation of market performance requires specific measures. Trends in retail food prices and consumer food costs are frequently used for this purpose, as are the level and stability of farm prices and income. The share of consumers' income spent for food is also frequently taken as a measure of the food industry's contribution to the standard of living and quality of life. The farm-retail price spreads and the farmers' share of the consumers' food dollar are popular measures of market performance.

Each of these has some value and limitations in the measurement of food marketing performance, but no single one tells the whole story. Care must be exercised in their use and interpretation. Market performance is a complex notion. It is dangerous to oversimplify it to a single market characteristic. The "bigness is bad" attitude is an example of this. Is bigness bad in and of itself, or is it the consequences of bigness that we object to? If the latter, is there a necessary identity between bigness and the undesirable consequences? There is room for debate here. Profit is another controversial market performance measure. We may take a different view of profits if we see them as rewards for superior performance rather than as the unearned booty of monopolistic firms.

The numerous conflicts between these performance criteria mean that compromises must be made in public policies that are designed to improve food marketing performance. A balance must be struck between the demands and dissatisfactions of each group concerned. It is unlikely that all dissatisfaction with the system will ever be eliminated, but this is not a reason to ignore critics or to cease searching for improved market performance. The balance of these criteria is frequently disturbed by a new technology, a new marketing procedure, a change in markets, or a change in political power. Thus, the analysis of food marketing performance is an ever-changing and dynamic area.

Attempts to measure and influence market performance have given rise to another approach to market analysis—the theory of industrial organization. This theory suggests that the performance of an industry is determined by two factors: (1) the *structure* of the industry (numbers and sizes of firms, degree of product differentiation, and conditions of entry); and (2) market *conduct* (firms' price, product, and promotional strategies). This relationship was widely used in the food marketing studies of the 1960s and 1970s conducted by the Federal Trade Commission (FTC), the U.S. Department of Agriculture, and the National Commission on Food Marketing. This approach predicts, for example, that monopoly power, prices, and profits will be greater for industries with high firm dominance, strong product differentiation, and high barriers to entry than for unconcentrated industries with unbranded products and low entry barriers. The market structure-conduct-performance relationship underlies many of the government market policies and regulations in the food industry.

FOOD MARKETING EFFICIENCY

Efficiency in the food industry is the most frequently used measure of market performance. Improved efficiency is a common goal of farmers, food marketing firms, consumers, and society. Most of the changes proposed in food marketing are justified on the grounds of improved efficiency. It is a commonplace notion that higher efficiency means better performance, whereas declining efficiency denotes poor performance.

Efficiency is measured as a ratio of output to input. Automotive efficiency, for example, is expressed as miles (output) per gallon of fuel (input). Student efficiency might be measured by grade level attained per hour of study. Efficiency ratios can be expressed in physical terms or in dollar terms. When dollars are used, the efficiency concept becomes a ratio of benefits to costs.

Food marketing can be viewed as an input-output system. Marketing inputs include the resources (labor, packaging, machinery, energy, and so forth) necessary to perform the marketing functions. Marketing outputs include time, form, place, and the possession utilities that provide satisfaction to consumers. Thus, resources are the costs and utilities are the benefits of the marketing efficiency ratio. Efficient marketing is the maximization of this input-output ratio.

The input cost of marketing is simply the sum of all the prices of resources used in the marketing process. However, it is more problematical to measure the value of the marketing utilities to consumers. This makes it difficult to determine when the marketing efficiency ratio is rising and when it is falling. Probably the best measure of the satisfaction-output of the market process is the price that consumers will pay in the marketplace for foods with different levels of the marketing utilities. If consumers are willing to pay three cents more per orange for orange juice than for fresh oranges, we may infer that the processing of fresh oranges into juice adds three cents of form utility to fresh oranges.

The marketing efficiency ratio can be increased in two ways. Any marketing change that reduces the costs of performing the functions without altering the marketing utilities would clearly be an improvement in marketing efficiency. Alternatively, enhancing the utility-output of the marketing process without increasing marketing costs would also increase efficiency.

Improved *operational efficiency* refers to the situation where the costs of marketing are reduced without necessarily affecting the output side of the efficiency ratio. An example would be a new labor-saving machine that reduces the cost of processing oranges into juice. Other examples of operational efficiency gains would be a new, less expensive method of handling and storing grain; a newly designed retail dairy case that reduces refrigeration energy costs; a lighter shipping carton for lettuce; or a new wrapping film that reduces the spoilage of prepackaged meat in the grocery store.

Operational efficiency is frequently measured by labor productivity or output per labor-hour. Table 2-1 compares the trends in labor productivity of farming, food marketing, and the total private economy over the 1970-1997 period. As a result of rapid technology change, labor productivity in agriculture has generally increased more rapidly than in other sectors of the economy. Productivity has grown erratically for most food processing sectors over the year. Operational efficiency in food wholesaling and retailing has lagged behind that of food processing and farming because of the greater difficulties in mechanizing and automating food wholesaling and retailing.

In practice, both the numerator and the denominator of the marketing efficiency ratio often change at the same time. Many changes in the costs of marketing influence consumers' satisfaction, and efforts to increase the marketing utilities normally affect marketing costs. Thus, a new marketing practice that reduces costs but that also reduces consumer satisfaction may increase *or* decrease the efficiency ratio. By the same token, higher marketing costs might *increase* marketing efficiency if they result in a more than proportionate rise in the marketing utilities. A true evaluation of any marketing change requires consideration of its effect on both the numerator and the denominator of the efficiency ratio.

The compromise that must be achieved between operational efficiency and consumer satisfaction explains the difficulty encountered in improving food marketing efficiency. For example, it is not difficult to lower food marketing costs by reducing the variety of food products and brands, eliminating packaging, or reducing the number of foodstore checkout lanes. The problem is that at some point these actions result in a greater loss of consumer satisfaction than is compensated for by the decline in marketing costs and consumer prices. A key task of the food marketing system, then, is to achieve the appropriate balance between marketing costs and consumer satisfaction.

TABLE 2-1 Labor Productivity in Agriculture and Food Marketing, 1970-1997

	Average Annual Percentage Change in Output Per Employee-Hour		
Sector	***1970–80***	***1980–91***	***1991–97***
Nonfarm economy	1.4%	1.7	1.8
Farming	4.1	3.2	3.8
Food manufacturing:			
Meat packing	3.3	1.3	–0.2
Poultry	3.0	3.4	2.2
Fluid milk	5.5	3.6	0.6
Fruits and vegetables	2.4	1.6	2.1
Grain mill products	2.8	3.7	2.2
Bakery products	0.7	0.9	1.6
Sugar	1.5	1.2	2.3
Retail foodstores	–0.7	–1.2	–1.1
Eating places	–0.1	–0.1	0.3

SOURCE: U.S. Bureau of Labor Statistics.

Pricing efficiency is a second form of marketing efficiency. It is concerned with the ability of the market system to efficiently allocate resources and coordinate the entire food production and marketing process in accordance with consumer directives. Pricing efficiency is less than perfect when prices fail to (1) fully represent consumer preferences; (2) direct resources from lower to higher-valued uses; or (3) coordinate the buying and selling activities of farmers, marketing firms, and consumers. The goal of pricing efficiency is efficient resource allocation.

Competition plays a key role in fostering pricing efficiency. Marketing firms compete for the consumer's favor by lowering marketing costs and increasing operational efficiency wherever possible. At the same time, there is competitive pressure for firms to add more utility to foods in order to gain an increased market share by catering to consumer preferences.

Frequently there are conflicts between these varieties of efficiency. For example, a new technological development may improve a firm's operational efficiency and permit it to grow very large. However, this growth may reduce the number of firms and competition in the industry, thereby lowering pricing efficiency. Or, a requirement of mandatory product grading to improve pricing efficiency might increase industry costs and thus lower operational efficiency. Similarly, the variety of products and brands developed to heighten consumer satisfaction may impose higher costs on the industry. Efficiency conflicts such as these contribute to the difficulties surrounding attempts to "improve" the food marketing system.

Because of the profit opportunities involved, there is a constant market struggle in the food industry between more efficient and less efficient marketing strategies and techniques. The knowledge that more efficient marketing practices tend to replace less efficient practices is useful in predicting and understanding changes in the food system.

CONSUMERS AND FOOD MARKETING

The consumer is the overall ruler and coordinator of marketing activity in the market economy. The goal of the food system is to satisfy consumers. Food marketing firms and food producers serve as means to this ultimate goal. Failure to recognize the primacy of consumer preferences in the economic system has resulted in the downfall of many firms, and even entire industries.

The notion that all business and marketing activity is directed toward the satisfaction of consumers is called the *doctrine of consumer sovereignty.* The statement, "The consumer is king," illustrates this doctrine. Consumers exercise their sovereignty over the food industry by their dollar voting, rewarding firms and activities that please them, and withholding approval from others.

Are food consumers really sovereign? The doctrine of consumer sovereignty is, simultaneously, a description and an economic prescription for the food industry. Consumer preferences and dollar votes are a powerful influence on food producers and marketing firms. Some areas of the food system are more sensitive and more responsive than others to changing consumer desires. But no firm or industry can completely ignore these desires for long without peril.

Food processors might prefer square tomatoes because they are easier to harvest, pack, and handle, but if consumers prefer round tomatoes, we will have round tomatoes. Nutritionists may prefer unrefined flour and sugar because of their nutritional value, but these will fail in the marketplace if consumers want "white" flour and sugar. Consumers may prefer white eggs to brown eggs despite the industry's insistence that they are nutritionally equivalent. Nevertheless, white eggs will be sold and brown eggs will be discounted—except, of course, in New England, where brown eggs are preferred to white! Consumers may be irrational, ignorant, fickle, and capricious in their views and decisions, but, like kings, they are obeyed by successful firms.

This is not to say that all firms follow passively every consumer's whim. Through advertising, packaging, product design, merchandising, and other marketing strategies many firms attempt to educate, influence, and persuade consumers in their buying decisions. Costly advertising and sales promotion programs seek to persuade consumers to prefer one store, one product, or one brand over another. To the extent that these efforts are successful in altering preferences and changing consumer decisions, the consumer loses a degree of sovereignty. As a result, inefficient firms with low-quality products and high prices might gain some freedom from the disciplinary nature of the consumer's dollar. However, competition will eventually work against the firm that is not providing true value for consumers by "finding wants and filling them."

Consumer sovereignty does not guarantee that food consumers will always behave in their own or in the public interest. Indeed, producing and marketing what consumers want may lead to a decline in the nutritional level of our diet, or to consumption patterns that are associated with obesity, heart disease, and other diet-related problems. The rationale for placing the consumer at the center of food industry decisions is that there is no other acceptable judge of these decisions. The phrase *de gustibus non est disputandum* (what is pleasing is not disputable) applies here. Consumers are the only ones with insight into their own preferences and values. It is simply unacceptable for others—nutritionists, food scientists, farmers, or anyone else—to decide what is pleasing to consumers. Of course, we can hope that consumer food choices are rational and in the public interest. And we might even attempt to educate the consumer to make wise buying decisions. But, when all is said and done, the final success of food production and marketing decisions hinges on consumer choices, no matter how irrational, ill-founded, or whimsical they are.

The doctrine of consumer sovereignty is dependent on three basic conditions. First, the consumer must be provided with real alternatives from which to choose. If all product choices are alike, or only trivially different, the consumer's freedom to choose is an illusion. Second, the consumer must have reliable and accurate information in order to accurately match available product choices with preferences. Deceptive or misleading advertising, packaging, or labeling subvert consumers' sovereignty. Last, prices of all foods must fully reflect all private and social costs of producing and marketing products. If consumers prefer foods that have an expensive environmental impact cost, or that impose high medical or productivity costs on the economy, their choices will not necessarily coordinate food production and marketing decisions in a satisfactory way. Generally, when these conditions limit the doctrine of consumer sovereignty, society intervenes through the instrument of government. However, there is substantial disagreement on the severity of these problems in the food industry and on the desirability of solving them through government intervention.

Food marketing firms contribute to consumer sovereignty when they practice the *marketing concept.* This philosophy holds that the most important function of an industry or firm is to satisfy consumers at a profit, and that this goal directs all other company activities, including production, finance, packaging, distribution, and other decisions.

The marketing concept is the third successive business philosophy of the Industrial Revolution. The first and earliest was the production-engineer orientation—What can we make? This was followed by the sales orientation—How can we sell what we make? In contrast, the marketing orientation is illustrated by the questions, What do consumers want, and how can we satisfy these wants profitably? In the food industry, this orientation involves a shift in emphasis from that of marketing what the farmer produces to that of finding out what the consumer wants, and then producing to fill that need.

This shift in business philosophy has been slow and subtle. One major result of the shift has been to give marketing—those activities of companies that are in closest contact with consumers—greater prominence within the firm. The business graveyard is filled with those who tried to sell a product that "couldn't miss," who failed to sense a change in consumer preferences, who produced a product so expensive or so cheap that it lacked a profitable market, or who first produced a product and then went looking for a market. Not all farmers and food marketing firms subscribe to the marketing concept. Some pork producers persisted in marketing fat-type hogs even though consumer preferences had shifted to lean-type pork, and the market for lard and animal fats had declined. For years, the dairy industry refused to produce margarine, despite consumers' increasing substitution of margarine for butter. And many food industry groups have spent millions of dollars unsuccessfully advertising foods with declining consumer demand trends.

FOOD MARKETING TRENDS AND ISSUES

Change and adjustment are the hallmarks of the dynamic food industry. Some trends represent adjustments of food industry firms to changing market conditions and new technologies. Other trends can be traced to the attempts of firms to gain competitive and economic advantage in the marketplace. Still other trends result from public policies designed to foster better market performance in the food industry.

Whatever their source, these market trends affect farmers, food consumers, and the public interest. Typically, change creates both costs and benefits. There is increasing concern with the costs and benefits of the changing food marketing machinery. The major trends and issues are discussed in greater detail throughout this book. For this brief review, they can be grouped into four areas: (1) organization and competitive issues; (2) coordination and control issues; (3) farmer marketing problems; and (4) consumer and public interest issues.

Organization and Competitive Issues

In many parts of the food marketing system there are very few and quite large firms. This "giantism" and centralization of the food flow concern farmers and consumers who are, by comparison, relatively small and unorganized. Americans have a distrust of bigness and concentrated economic power, even though they recognize that some efficiencies may result from large organizations.

Integration and diversification are two related trends in market organization. Multinational and conglomerate food corporations add a new dimension to competition in food markets. There is also a trend toward mergers of similar food companies, and a combining of food marketing levels and functions within a single company. When food marketing firms assume control of some agricultural decisions, the farmer may be trading freedom and independence for improved prices and incomes.

The flow patterns of food within the marketing channels are also changing. Decentralized marketing has replaced the traditional central food markets that in the past played a key role in pricing, standardizing, and product exchange. As a result, farmers are selling more of their products directly to larger buyers who, in turn, are attempting to gain greater control of the flow, quantity, and quality of farm products. This departure from traditional marketing patterns has vastly changed the way farmers market their products today.

How have these trends affected the nature of competition in food markets? Is competition among the few better or worse than competition among the many? Are decentralized, contract markets as competitive as their predecessors? Is there sufficient competitive discipline in the new food marketing system to assure efficient and equitable results for all market parties? These are important questions to ask about the organizational trends in food marketing.

Coordination and Control Issues

There is concern that these organizational trends may have shifted the locus of control and balance of power in the food industry. This raises the issues of freedom, access to markets, and economic power. The questions, Who will control agriculture? and Who runs the food industry? are continuing public policy concerns. The closure and domination of food markets by marketing firms is especially threatening to farmers.

There has been a decline in the coordinating role of prices in the food industry. This involves a separation of pricing from product exchange, a shift of the pricing process from the light of public markets to the secrecy of private negotiations, and a separation of prices from costs in large, multiproduct firms. The capability of prices to perform their traditional market roles under these conditions is in question, and the economic impact of alternative coordinating devices is much debated.

The concern here is with the efficiency and equity of food markets. Without the traditional market price signals, can we be sure that efficient practices are winning over less efficient ones? Can we be confident that the consumer's food dollar is being fairly apportioned among farmers and food marketing firms?

Farmer Marketing Problems

There is a wide and growing gulf between farmers and the food marketing system. The relative numbers and sizes of these two sectors make them unlikely bedfellows, despite their mutual interdependence. There is reason to wonder whether the independent farmer can survive and prosper in this new food industry, regardless of the efficiency of agricultural production.

There is a growing concern regarding the farmer's place in the changing food industry. One school of thought believes that farmers must become as large, powerful, and market-oriented as the major food corporations. Others argue that food marketing firms should be broken up, or be made to behave like small-unit firms, in order to establish a parity of bargaining power between farmers and marketing firms. A third alternative is to foster farmer group action through cooperatives, bargaining associations, and other devices, in an attempt to preserve the present structure of agriculture and provide farmers with countervailing market power.

Food industry trends create other farmer marketing problems. In some cases, farmers have fewer market opportunities than they did formerly. Farmers need more and better market information in order to make production and marketing decisions. They also need periodic educational programs to learn about new market alternatives and choices. Farmers are also increasingly called upon to provide input into the public policy process that influences prices, markets, and regulations.

Consumer Issues

The consumer movement has affected the food industry in many ways. Information programs, such as unit pricing, truth-in-packaging, nutritional labeling, open-code dating, and the like, illustrate the activist role consumers are assuming in the marketplace. Similar examples can be cited in the areas of food safety, biotechnology nutrition, grades and standards, and wholesomeness. Consumers are also becoming involved in farm policies, such as import-export policy, price and income supports, and rural development. "Let the buyer beware" is clearly not an appropriate philosophy for the food industry today. But what is the appropriate role for consumers in the food industry decision-making processes?

Consumer sovereignty and efficiency-utility conflicts are at the heart of consumers' concerns with the modern food industry. The food marketing sector is increasingly merchandise-oriented, as can be seen in its emphasis on new products, packaging, advertising, and promotion. Its allies are food scientists and engineers, packaging experts, and advertising agencies. Consumers are searching for the appropriate environment for providing the necessary freedoms and incentives to develop new products and market programs, and for the regulatory climate necessary to ensure that these efforts are in the public interest. The consumer increasingly will be involved in the food industry, not only as a final buyer but as a regulator and shaper of market decisions. In the future, public food policy issues will be hammered out on a broader anvil of public opinion than in the past.

SUMMARY

There are many useful approaches to the study of food marketing. Each provides a unique perspective of the food marketing system and process. The functional approach emphasizes the "what" of marketing; the institutional approach emphasizes the "who"; and the behavioral systems approach analyzes the input-output processes, power relationships, communications, and techniques for adapting to changes in the marketing system. The marketing management approach teaches us why and how firms develop competitive marketing strategies. Three important characteristics of marketing functions are that they add value as well as costs to farm products; they are difficult to eliminate; and the arrangement of functions within the marketing system is somewhat flexible. A multidimensional approach to evaluating food industry performance is necessary. Frequent compromises are necessary between the market performance criteria. Market efficiency is measured as a ratio of marketing utilities to marketing costs. Market performance can be measured by changes in operational and pricing efficiency. The doctrine of consumer sovereignty and the market concept recognize the primary role of consumer preferences in coordinating farm and food marketing decisions. Several organizational and competitive trends raise concerns about the performance of food markets.

KEY TERMS AND CONCEPTS

agent middleman
behavioral system
commission man
consumer sovereignty
division of labor
economies of scale
input-output system
market conduct
market intelligence
market performance
market segments
market structure
marketing concept
marketing function
marketing institution
marketing management
marketing mix
marketing strategy
middleman bias
operational efficiency
pricing efficiency
product differentiation
risk bearing
search and transactions costs
specialization
speculative middleman
standardization
target market

DISCUSSION QUESTIONS

1. Can you think of any other food marketing activities not included in the list of marketing functions provided in this chapter?

2. Comment on the meaning of the statement, "You cannot eliminate the middleman." Under what circumstances can middlemen be eliminated? Give an example.

3. How do farmers and consumers decide which of the marketing functions to perform for themselves?

4. Why are there so many different kinds of food marketing institutions?

5. Give familiar examples of changes in operational efficiency in the food industry.

6. What are some reasons for the middleman bias in our society?

7. What criteria should be used in evaluating the performance of the food system?

8. How sovereign are food consumers? Is their sovereignty increasing or decreasing?

9. Are consumers really sovereign when some products are addictive or injurious to health and when firms spend millions of dollars to convince shoppers that one product is better than another?

SELECTED REFERENCES

Manchester, A.C. (1992). *Rearranging the Economic Landscape: The Food Marketing Revolution, 1950-91,* Agricultural Economic Report No. 660. Washington, D.C.: U.S. Department of Agriculture.

CHAPTER 3

AGRICULTURAL PRODUCTION AND MARKETING

Market first, then produce the product.

OBJECTIVES

After reading this chapter, you will be able to:

1. Describe the characteristics of farming and agricultural products which make the marketing of food different from other products.
2. Explain why farm and food marketing decisions are so interdependent today.
3. Identify the farm marketing alternatives and the key marketing decisions that farmers make.
4. List some important farm marketing problems and give recommendations for solving these problems.

You may be puzzled by the quotation above. How can you market something before you have produced it? It is, of course, impossible to do so literally. But it is desirable to think about the marketing of the product, in your mind, before you produce it. What do your customers want? Where and how will you sell it? What values and services should you add to the product? What price do you need to cover your costs of production? What marketing functions should you perform? This is what it means to *market before you produce.* It is very important in today's business environment, and many farmers neglect to do it.

Perhaps there has been no greater change in the food industry over the past 100 years than the growing interdependence between farm management decisions and food marketing decisions.

In the past, farmer's production decisions were often treated as independent of food marketing decisions. It was assumed the market would absorb what farmers and Mother Nature produced. This is no longer the case. Today, farmers' production decisions are shaped and closely controlled by marketing firms' and consumers' decisions. Producers must now produce what the market wants and values rather than what they have always produced or are good at producing. More than ever, farmers must subscribe to the marketing concept and honor the doctrine of consumer sovereignty. For this reason, today's food producer must understand more about the marketing system and food consumers' desires. This chapter deals with the interrelationship of farming and food marketing. The next chapter explores how the changing food consumer is affecting farmers' decisions.

THE AGRICULTURAL PLANT

Many people think of farms and land area as synonymous. The United States is a big country, and it is commonly believed that there is a relatively unlimited amount of farmland upon which to produce food. This is not true. Nearly two-fifths of the country's land area receives less than 20 inches of rainfall annually and therefore has very limited agricultural potential. Figure 3-1 shows the uses of the country's land. About 46 percent of the nation's land is used to produce food and fiber, with 20 percent in crops and 26 percent in grassland, pasture, and range.

Technology, not land, is the reason for U.S. agriculture's productive success. As shown in Figure 3-2, U.S. agriculture's productivity increased dramatically during the 1920-2000 period. The output per unit of input of U.S. farms increased 2-3 percent per year during the 1980-2000 period as a result of new farming technologies, better educated farmers, and the substitution of machinery, chemicals, and other inputs for labor on the farm. In contrast to other sectors of the U.S. economy, agricultural productivity growth did not slow down in the 1970-1990 period. As a result,

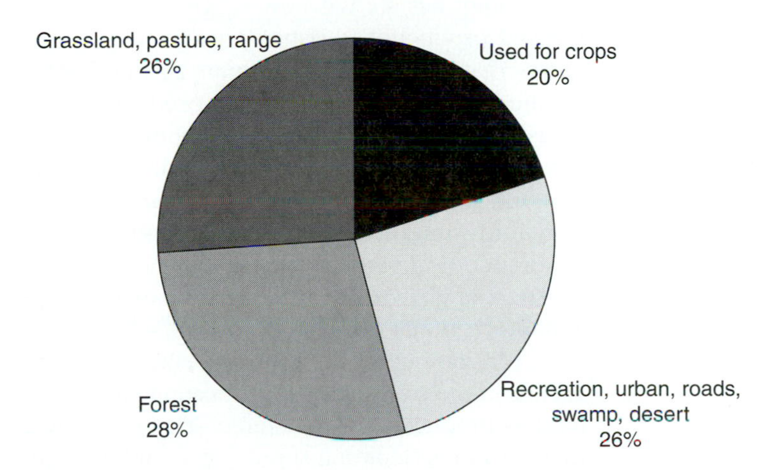

FIGURE 3–1 Major uses of U.S. land, 1997. (U.S. Department of Agriculture.)

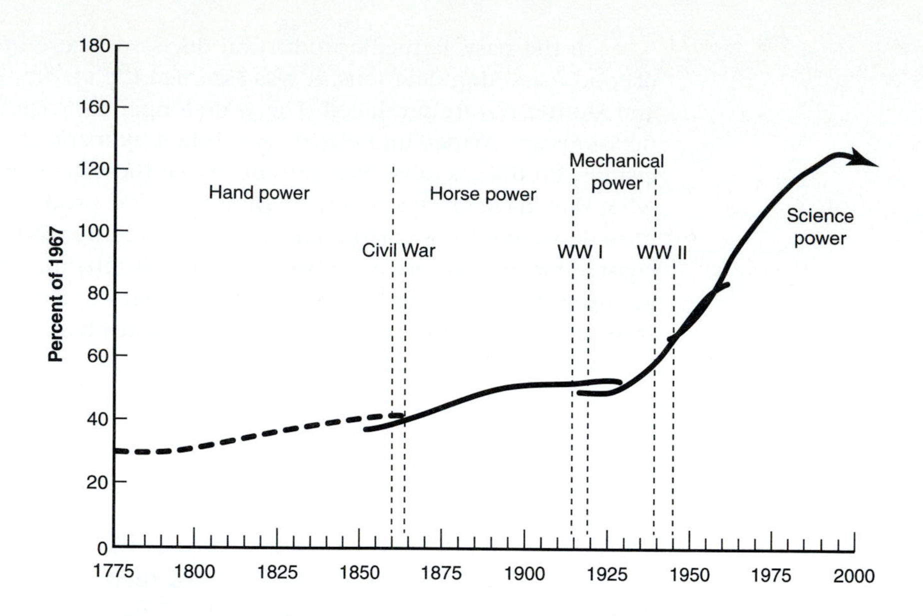

FIGURE 3–2 Trends in agricultural productivity, 1775–2000. (U.S. Department of Agriculture.)

one farmer could feed more than 100 people in 1990, compared to feeding only 5 consumers in 1900 and 26 consumers in 1960. In 1997, 2 percent of all farms produced 50 percent of the value of all agricultural products, compared to 17 percent of farms producing one-half of farm product sales in 1900. While earlier in our history new agricultural technology tended to be labor-saving, many of the newer improvements in technology are yield-enhancing. The food marketing system, of course, has grown along with this increased output of farm products.

Great improvements have been made in seed variety, livestock nutrition, fertilization, and a multitude of other areas, making agriculture more productive. Truly, technology has shown how to grow two plants where one grew before. This increased production capability has been a mixed blessing for the U.S. farmer. It requires that more and more inputs be purchased from off-farm suppliers. It also has resulted in a near-chronic excess production, often depressing farm incomes and increasing the role of government income support activities. There is also greater pressure on the marketing machinery to move this added production into consumption. Not only has the volume of output from farms increased, but also an increasing portion of this output has moved into commercial and global marketing channels. With fewer people living on farms, less of the farm output is consumed at the place where it is grown and more must move to supply the increasing city populations and international markets.

U.S. agriculture is changing in other ways that affect the food marketing system. Trends in the size, ownership, location, and specialization of farms are changing how food is produced and marketed. There is also a trend toward closer ties between farmers and agricultural supply firms and marketing firms. At the same time, changes in the food marketing system are influencing farmers and the nature of farming. Table 3-1 illustrates some general trends in U.S. agriculture during the 1950–1997 period.

TABLE 3-1 Selected Trends in U.S. Agriculture, 1950-1997

	1950	*1960*	*1970*	*1980*	*1990*	*1997*
Million farms	5.6	4.0	3.0	2.4	2.1	2.0
Percent of population on farms	15	9	5	3	2	2
Average farm size:						
Acres	213	297	373	427	461	471
Net Farm Income ($1987)	11,200	10,400	13,600	13,000	18,100	16,000
Assets ($ thousand)	17.2	42.6	89.5	403.1	395.4	497.1
Farm efficiency:						
Output per unit of input (1950 = 100)	100	131	150	174	233	255
Persons fed per farm worker	16	26	98	76	96	100+
Farm prices (1950 = 100)	100	93	106	238	264	298
Gross farm income ($ billion)	33.1	38.9	58.8	149.3	196.4	238.2
Percent of farm income from nonfarm sources	NA	42	55	64	85	88
Percent of farm income from government payments	2	3	11	2	5	3
Per capita farm income as a percent of nonfarm income	55	50	70	81	130	105

SOURCE: U.S. Department of Agriculture.

THE STRUCTURE OF U.S. AGRICULTURE

The structure of U.S. agriculture refers to the number, size, ownership, and specialization of farms. Here we must use averages to generalize about farming, but we should avoid stereotypes and recognize the diversity of U.S. agriculture. There are very large and very small farms; there are family farms and there are giant corporate farms owned by nonagricultural firms; there are wealthy farmers and poor ones; there are full-time and part-time farmers; there are farmers who receive all of their income from farming and there are farmers who receive a good share of their income from nonfarm employment. This diversity of farming greatly complicates the food marketing process.

The dominant characteristic of American agriculture is the prevalence of large numbers of relatively small, owner-operated farms. These small production units buy and sell in a world of large organizations. Because of their size, numbers, and competitive position, farmers have limited influence in the marketplace and generally make fewer marketing decisions than others in the system.

There were an estimated 2 million U.S. farms in 1997 compared to the record high of 6.8 million in 1935. At the current rate of declining farm numbers, it is estimated that there may be only 1.5 million farms by the year 2002. The average farm size is also increasing, as shown in Table 3-1. These trends, however, do not show the great variation in farm sizes or the rising importance of larger farms. There is a tendency for a growing share of our agricultural output to come from a declining number of very large farms. In 1999, farms with annual sales below $50,000 accounted for 74 percent of all farms but only 7 percent of total farm sales (Figure 3-3).

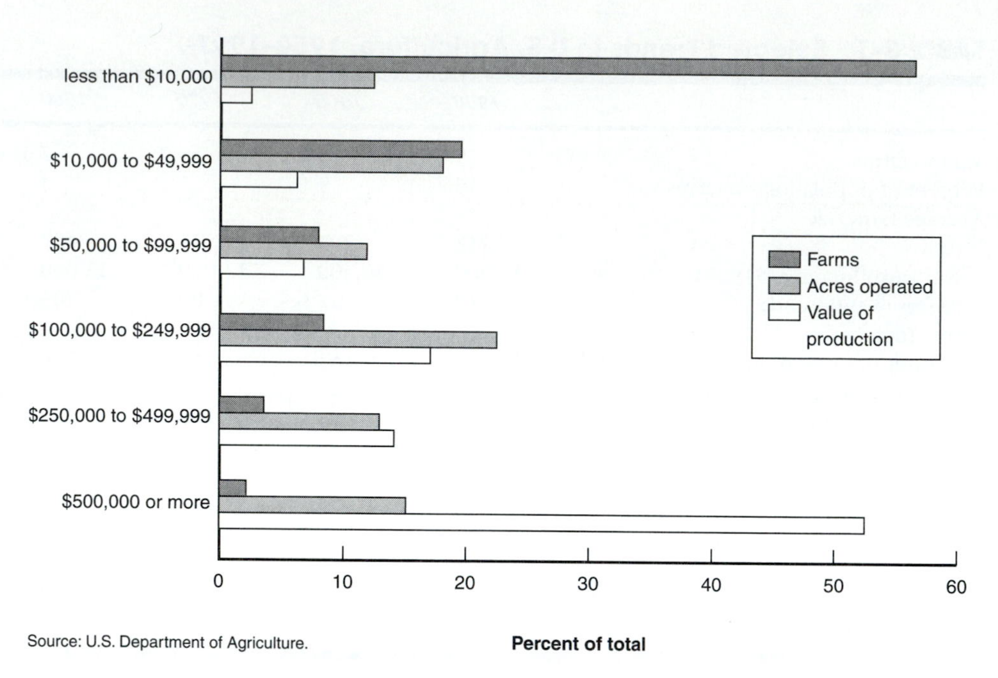

FIGURE 3–3 Numbers and sizes of U.S. farms, 1999. (U.S. Census of Agriculture.)

In contrast, the 4 percent of farms with sales exceeding $500,000 were responsible for 57 percent of all farm sales. Thus, there are really two different farm sectors today: very large commercial farms, which are few in number but account for most of the total farm output, and a large number of very small farms, which represent a small portion of total farm supplies or sales. Family farms are found in both groups. The importance of very large farms varies by commodities, with large farms more common for poultry, beef, vegetables, fruits, nuts, and cotton.

Ownership and operation are not always the same in U.S. agriculture. Of the three forms of business organization—individual proprietorships, partnerships, and corporations—the individual proprietorship is by far the most common in U.S. agriculture (Table 3-2). These farms are typically smaller than partnership and corporation farms. Nevertheless, they dominate farming in most parts of the country.

Many people express concern about the rise of corporate agriculture and the future of the family farm. The family farm is generally defined as a production unit in which family members provide most of the labor, management, and finances. As such, family farms may be organized under any of the three types of business organization indicated in Table 3-2. Many farm corporations are family-held.

Publicly held farm corporations represented less than 1 percent of U.S. farms and acreage in 1997 but, because of their larger size, accounted for 6 percent of total farm sales. These corporate farms are concentrated in California, Florida, and Texas and are frequently engaged in cattle feeding; fruit, vegetable, and sugar cane production; forestry; and the growing of nursery and greenhouse plants. Many nonfarm corporations that entered agriculture in the early 1970s retreated from farming in the early 1980s. Less than 1 percent of U.S. land is owned by foreign individuals or firms.

TABLE 3-2 Farm Ownership Patterns, 1997

	Percentage Distribution				
Farm Ownership	*Number of Farms*	*Acreage*	*Sales*	*Average Farm Acreage*	*Average Annual Sales ($000)*
Individual Proprietorship	86	63	52	356	62.4
Partnership	9	16	18	881	209.7
Corporations:					
Family Held	4	13	23	1,571	603.0
Nonfamily	0.3	1	6	1,507	1,395.0
All Farms	100%	100	100	487	103.0

SOURCE: U.S. Census of Agriculture, 1997.

Farmers are increasingly specialized today. Specialization involves restricting the scope of economic activity and concentrating on doing a few tasks well. Farmers may specialize by commodity, by personnel, or by the number of processes performed. A farmer who discontinues livestock production to concentrate on crop production is specializing by commodity. A farmer who elects to feed hogs but discontinues his sow-farrowing (birthing) operation is specializing in a single stage of the production process. A farm where one person is the crop manager and the other is the livestock manager is considered to be specialized by personnel.

Farm enterprise specialization differs by commodities. Fruit, vegetable, poultry, and dairy farms tend to be highly specialized. However, many farms still produce more than one commodity. The grain-livestock operation illustrates diversified farming, even though the grain may be fed to livestock on the farm. Some farmers have attempted to diversify their operations as a form of protection against the weak prices of traditional commodities. Farmers also differ in process specialization. Many confine their activities to farming, but others increasingly are becoming involved in off-farm marketing activities. It should also be pointed out that farmers receive more than two-thirds of their income from nonfarm wages and salaries or from off-farm businesses.

These facts about the agricultural production plant add up to some very important marketing considerations. First, much of our production is made available to the food marketing system in relatively small lots from a large number of specialized farms. A food marketing firm must often assemble supplies from a great many individual sellers scattered over a relatively large area. The marketing system must serve a wide variety of farmers and their differing needs. The relatively small volumes marketed by a single farmer may also encourage group marketing efforts of farmers, such as marketing cooperatives.

A second observation is that the farmer is by nature, and probably by necessity, a person primarily interested in production, often with less time and interest in marketing. Because farmers are the manager-laborers of a highly diverse production unit, the complexities of modern agricultural production absorb a great deal of their time and energies. As marketers, they either sell very small amounts at a time or very few times a year. This often means that farmers are less informed and skilled in marketing than in food production.

A third observation, and a very important one, is that changes are taking place. Larger, more specialized farm units are developing, and in some areas and products they are developing very rapidly. Accordingly, the interest of the farmer in marketing is increasing. In addition, the assembly and buying methods of marketing firms are changing. For example, a processor may find it feasible to buy more supplies directly from large, specialized producers instead of utilizing the services of country buyers who specialize in assembly of products from many small producers.

A final observation is that the food marketing system must serve two very different agricultural groups: a few, large farm owners who sell the majority of farm products, and a large number of small farmers who have much less to sell. Many feel that the marketing system serves the larger farms better than the smaller farms.

CHARACTERISTICS OF THE PRODUCT

A Raw Material

The output of agriculture is largely a raw material that will be used for further processing. This processing may be limited, as in converting livestock into meat, or it may be highly complex, as in converting wheat into breakfast cereal. Regardless of the complexity, however, the product sold by the farmer soon loses its identity as a farm product and becomes food. The marketing system completes the production process begun on the farm and provides much of the final food value that consumers desire.

This means that there is a considerable distance between the products that farmers produce and the final products that consumers eat. It is both a geographic distance and also an economic distance. As a result, food producers sometimes are led to believe that their task is only to produce raw commodities, and the food marketing system has the sole responsibility for producing and selling the final foods that satisfy consumers. Nothing could be further from the truth. Successful farmers will and must become more market-oriented and learn to produce value added products that meet food consumers', processors', and distributors' needs.

There are increasing opportunities for farmers to add more value to their products. This can be done by altering the production process to differentiate the product, as when farmers engage in low-input, organic, "natural," conservation, sustainable, or animal-friendly agriculture. Producers can also sometimes receive price premiums for specialty products, such as high-oil or white corn or breeds of animals in high demand. Food producers may also move forward in the food marketing channel to add more value to farm products and project the identity of the producer to buyers. Farmer-direct sales to consumers at roadside or farmers' markets are examples. But here the rule holds, producers can eliminate the middlemen only by assuming their functions and costs.

Bulky and Perishable Products

Compared to most other products, agricultural products are both bulkier and more perishable. Bulk affects the marketing functions concerned with physical handling. Products that occupy a lot of space in relation to their value are expensive to transport and store. A truckload of breakfast cereal would be considerably more valuable than a truckload of wheat. Sometimes it is possible to reduce the bulkiness and weight of food products before marketing them. The frozen concentrated juice industry developed in response to a need to ship higher-valued, less bulky products to markets instead of raw farm products.

Perishability also influences the marketing of farm and food products. All biological products ultimately deteriorate. Some agricultural products, like fresh strawberries or fresh peaches, must move into consumption very quickly or they completely lose their value. Products such as hogs or cattle continue to grow and change if "storage" in the form of withholding them from market is attempted. Grains, on the other hand, can be stored for a considerable length of time without much deterioration. Even the most storable agricultural products, however, are usually more perishable than industrial products.

These product characteristics have their effect on the facilities necessary to market farm products. Bulkiness requires large storage capacities. Perishable products require speedy handling and perhaps special refrigeration. Quality control often becomes a real and costly problem. From the farmer's viewpoint, withholding farm products from the market is extremely difficult. When products are ready, they must move to market.

Quality Variation

The quality of agricultural commodities varies from year to year and from season to season. During some years the growing conditions are such that the crop is generally of high quality. In other years, unfavorable conditions may prevail and the crop is of much lower quality. Farmers and food marketing firms rely on standardization and grading to sort out and price farm products of differing qualities.

Variations in quality may also change marketing patterns. For example, during a year in which corn does not mature properly, large amounts of "soft" corn are harvested. The corn will spoil if it is not used before the following spring. Farmers may then buy additional feeder animals in order to utilize this corn. The marketing pattern of these feeders, however, will be different from the usual pattern because the feeding period is adjusted to the condition of the corn.

Significant changes are occurring here also. Increasingly, the quality of farm products may be controlled by following certain production practices. Some control over quality is possible through the control of the breeds and kinds of parent stock used. Crop spraying and other production practices can be used to affect quality. As such developments are perfected, the marketing system may have to contend with less variation in the quality of agricultural products. This possibility of quality control through improved technology and production practices is another factor that is encouraging closer relationships between food marketing and production units.

The development and rapid adoption of genetically modified crops by America's farmers in the 1990s created some marketing problems. Genetically modified organisms (GMOs) have the potential to lower costs of production, increase yields, and change the characteristics of farm products to better meet the needs of processors and consumers. Yet despite the U.S. Food and Drug Administration's findings that the foods are safe for human consumption, biotech foods have not been accepted by all consumers. Some governments have denied imports of GMOs, required segregation of these products from others, or mandated labeling of foods containing GMOs. This is seen as an unwarranted trade barrier by some. Certain American companies have also declined to use GMOs and advertise their products as GMO-free. As a result, separate marketing systems are developing for non-GMO and GMO food products. This has increased the costs of marketing these products and created a two-tiered pricing system with premium prices for non-biotech products.

Despite this quality variation, farm products in general are said to be homogeneous. This means that overall, buyers have little reason to prefer one farmer's product over another. Consequently, each farmer receives about the same price for the same quality of product. Individual farmers can do little to raise the general price level of their commodities, although some farmers with superior marketing skills do receive higher prices than others for like-quality products.

CHARACTERISTICS OF PRODUCTION

Total Output

The long-term trend in U.S. food production is upward, as shown in Figure 3-4. In the developed world, food output has generally grown more rapidly than the population, resulting in increased per capita food supplies and a higher standard of living. Unfortunately, in other portions of the world, population growth often rises faster than food production. Matching world food supplies with needs to prevent hunger and improve living standards is a constant challenge.

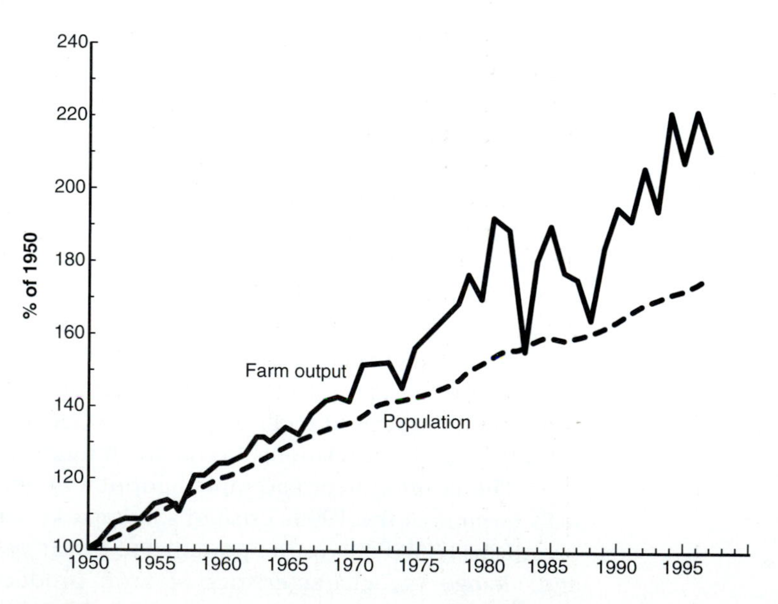

FIGURE 3–4 Farm output and U.S. population, 1950–1997. (U.S. Department of Agriculture, Economic Research Service.)

Annual Variability in Production

Aggregate farm output varies from year to year (Figure 3–4). These variations are caused by farmer responses to prices, government programs, and uncontrollable factors such as weather and disease. Such changes in farm output influence the food marketing process and the use of the food marketing system's capacity.

The annual output variability of individual farm commodities is even greater than that of total farm output. This variability is of critical importance to food marketing firms that, by and large, specialize in the handling of only a few agricultural commodities. A sharp reduction in the wheat crop will concern millers and bakers but is of little interest to meat packers. Year-to-year changes in farm supplies have a significant impact on the purchase prices, need for storage facilities, and plant utilization rates of food marketing firms. As Figure 3–4 shows, annual fluctuations in agricultural output have increased in recent years, largely because of weather variations. The desire to reduce the risks and uncertainties of fluctuating farm supplies is one of the forces creating closer contractual ties between marketing agencies and farmers.

Seasonable Variability in Production

In addition to the annual production variability, much of agricultural production is highly seasonal. Livestock receipts may vary substantially throughout the year. Egg and milk production is larger in the spring and early summer than in fall and early winter. The bulk of the nation's turkey crop moves to market during the last half of the year. And of course the harvest of such crops as wheat, cotton, soybeans, fruits, and vegetables is crowded into a relatively short period.

Some farm products are produced more or less continuously year-round—for example, eggs and milk—while others such as grains are harvested only once each year. Their marketing differs accordingly. If the single-harvest product is storable, storage facilities must be furnished to hold the product until it is consumed. This means that during part of the year, storage will be used at near capacity; at other times it will be almost empty. If the product cannot be stored it must either be processed or consumed immediately. This may result in processing plants running at capacity for some periods and well below capacity, or shut down, for other periods. If the product must move directly into consumption, transportation and refrigeration facilities must be available immediately. All these situations affect the costs of the marketing process.

Some progress has been made toward reducing the seasonal variability of production for some commodities. The seasonal output variation of such products as milk, eggs, and broilers, for example, is considerably less than years ago. Larger, more specialized farms often plan for evenly distributed production to fully utilize labor and facilities. New developments in management, breeding, and nutrition are making more uniform levels of production possible. Moreover, the widespread use of more rapid transportation and refrigeration has tended to reduce the seasonality of available supplies. Bananas today, for example, are plentiful all year, whereas a few years ago they were quite seasonable.

The United States has a wide range of climatic conditions. By employing improved transportation methods, different geographic areas can be used to extend the season of availability for many commodities. For example, there is a relatively uniform supply of such vegetables as lettuce, fresh tomatoes, and green onions now available as area after area can be tapped in its appropriate season. Such operations, of course, mean that the marketing system must be flexible, geared to procurement and movement from a widespread and changing area.

Geographic Concentration of Production

Although a variety of farm products is produced in all states, there is increasing geographic specialization of farm production. Figure 3-5 shows the major types of farming throughout the United States. Each region tends to specialize in the production of commodities for which its resource base is best suited: fruits and vegetables in California, Texas, and Florida; wheat in the Great Plains; poultry and eggs in the

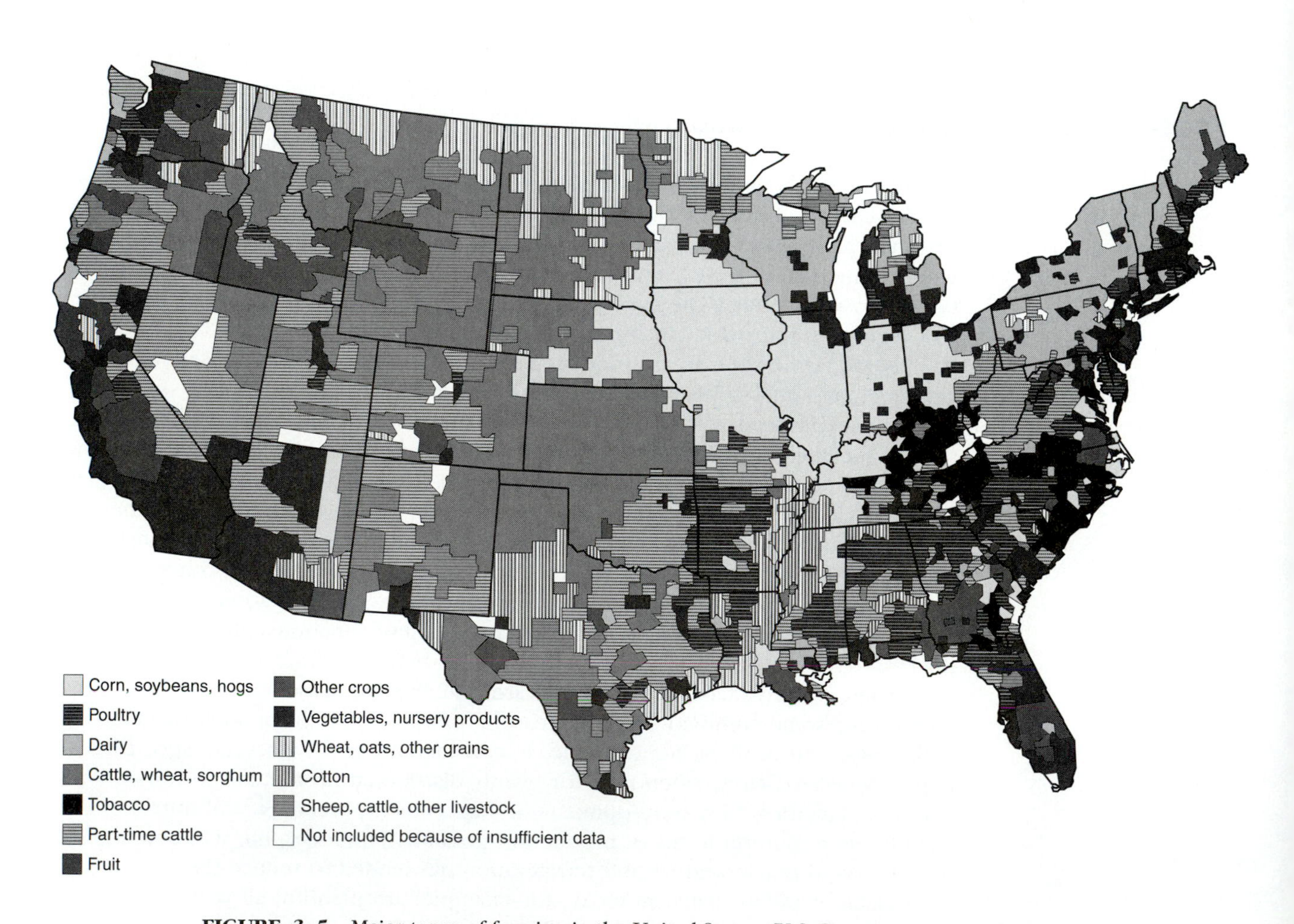

FIGURE 3–5 Major types of farming in the United States. (U.S. Department of Agriculture.)

Southeast; and grains and livestock in the Cornbelt. This specialization has lengthened the food marketing channel and increased the importance of the transportation marketing function. Geographic specialization changes from time to time, as illustrated by the shift of beef production from the Eastern cornbelt to the Western cornbelt, and the marketing system, of course, must adjust to these changing geographic production patterns.

Varying Costs of Production

Farmers' costs of production are affected by climate, technology, farm size, and individual managerial skills. Consequently, the cost of producing a farm commodity varies widely by regions and among farmers. For example, the U.S. Department of Agriculture estimated that the cash costs of producing a bushel of corn in 1998, excluding land costs and management returns, was $1.07 in the Cornbelt, $1.19 in the plains states, and $2.24 in the southeastern states.

Are larger farmers more efficient, lower cost operators than small farmers? Many studies have found that the average cost of farm production falls as small farms grow larger, but there is a point at which average costs do *not* fall further as farm size increases. Further, these studies suggest that most production efficiencies can be attained at relatively modest farm sizes, except for such commodities as poultry, cattle feeding, and lettuce production. There is also some indication that economies of size in marketing are more important than production economies of size for large farms. These farmer marketing economies are associated with cost advantages in buying and selling large quantities of farm supplies and commodities.

This variability in farm costs of production is worth noting for two reasons. First, since all farmers tend to receive about the same price for a commodity, even though their costs may differ, there are wide variations in farm profits and returns to farmers' management skills. Second, at any point in time and at any price there will be some farmers who are breaking even, some who are making money, and some who are losing money. Average profits can be deceptive. The fact that the price of cattle is above or below the average cost for the industry does not mean that all cattle operations are making, or losing, money.

The Farm Supply Industry

Farmers are both buyers and sellers. They participate in procurement as well as product markets. The farm supply industry provides such agricultural inputs as chemicals, seeds, machinery, feeds, capital, labor, credit, and land. In 1997, farmers spent $186 billion for farm supplies (Figure 3-6). Feed, livestock, and seed, of course, are purchased from other farmers, but more than one-half of farm cash receipts are now spent on farm supplies purchased from the nonfarm sector. With increased farm size and enterprise specialization, farmers are purchasing more of their supplies from off-farm sources. This means that increasingly farmers' costs and profits are being determined by forces outside the farm economy.

The growth and importance of the farm input sector affects farmers in several ways. Farm income can be increased by skillful buying of supplies, just as by skillful

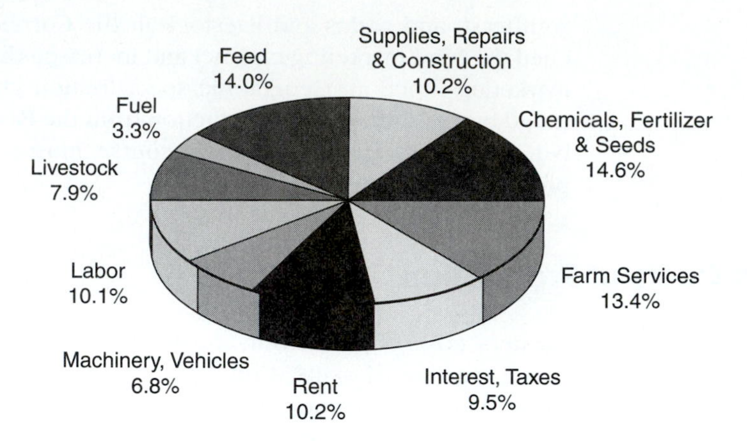

FIGURE 3–6 Farm production expenditures, 1997. (U.S. Department of Agriculture, Economic Research Service.)

sale of farm commodities. The farm input markets have also been responsible for much of the dramatic gain in agricultural efficiency in recent years, especially the chemical and machinery markets. The farm capital or money markets, in turn, have financed mechanization, modernization, and expansion of farming. Overall, the farmers' increasing dependence on off-farm inputs has reduced the self-sufficiency of farmers and tied their economic welfare more directly to the nonfarm economy where many of their inputs originate. The prices that farmers paid for purchased inputs increased faster than the prices they received during the 1970–2000 period, contributing to the familiar cost-price squeeze in agriculture.

After increasing rapidly in the 1960s and 1970s, the market for farm inputs stabilized in the 1980s and 1990s and is no longer growing. This mature industry is now the scene of some major marketing battles. For example, the independent farm supply retailer is threatened by the growth of large farmers' volume purchases direct from manufacturers. Marketing to large farmers is shifting from the retail dealership to the telephone and computer. Many feed and animal health product companies are distributing directly to contract producers. Biotechnology promises to substitute for traditional agricultural chemicals. Mass merchandisers such as K-Mart and Wal-Mart have expanded into farm supplies. Many farmers are also now purchasing farm supplies from websites on the Internet. Marketing strategy has become more important in this industry.

THE FARM MARKETING PROBLEMS

The farm problem is usually associated with unstable and relatively low farm prices and incomes. A related set of farmers' problems can be termed the farm marketing problems, which have several dimensions.

First, farmers do not have control over the output of their production activities to the same degree as nonfarm firms. Agricultural output comes from many small units operated independently. The production is to a great extent dependent on weather and biological patterns of production. Farmers may wish to change their outputs and attempt to do so by planting more or fewer acres or by breeding more or fewer pigs. However, the final output is beyond the farmer's control, as weather, disease, and other relatively uncontrollable factors will affect yields per acre and the productivity of animals. It is not possible to quickly shut off or turn on agricultural production. This means that marketing agencies—and also consumers—in the short run must adjust to farm supplies rather than farmers adjusting to agencies.

A tomato canner must estimate how much product can be sold at a price estimated nearly a year in advance. From this estimate the processor can then contract suitable acreage with producers. If estimations of the market conditions and yields turn out correctly, the canner will pack and sell as planned. But market estimates could be wrong. The weather could be unfavorable. Overestimates of yield may find the canner without enough tomatoes to meet market needs.

Aside from such short-run adjustment problems, it takes long periods to change the production of some commodities. Fruit trees are planted years in advance of their coming into production. The market situation may change during this period. The expansion of milk production is a slow process. Even significantly decreasing farm production is slow and difficult. Once an investment is made in buildings, equipment, and other fixed assets, changes are very difficult and expensive to make.

This inability to adjust quickly to changing conditions creates a high-risk element in agriculture. The market for which a long-time production plan is made may change by the time the product is finally marketed. Changes in consumer tastes may find agricultural resources being devoted to the production of something that is no longer so greatly desired. High prices resulting from shortages of production may reduce the consumer market for that product when it finally arrives in quantity. This relative unpredictability and uncontrollability of output creates many farm marketing problems. Farmers are adopting new risk management production and marketing strategies to cope with their increasing volatile markets, such as diversification, futures market hedging, and contracting.

A related component of the farm marketing problem is the difficulty farmers face in improving their prices through independent or group activities. Farmers are, for the most part, price-takers—they cannot, individually, influence the price of their products through their output decisions. In order to raise prices through the control of supplies or advertising programs, farmers must act as a group. However, the large numbers of farmers and their differing economic circumstances frequently frustrate any attempts to organize and market jointly.

The *free-rider problem* often plagues farmers when they do attempt to organize to influence farm prices. Free riders hamper any group effort that requires each member to sacrifice for the overall welfare of the group when the group benefits go to everyone regardless of their participation. For instance, farmers may try to raise their prices through voluntary supply-control programs, advertising efforts, or bargaining associations. If successful, the resulting price rise benefits all farmers, whether or not they have contributed to the program. Thus, it is sometimes difficult to achieve the group participation and support necessary for success.

Many of these conflicts exist in agriculture where what is in the best interests of a single farmer may differ from the interests of farmers as a group. For example, if prices and profits for a commodity are high, individuals farmers will have an incentive to expand output. A few can do this without consequence. However, if all producers attempt to expand, market prices and profits will fall. In market analysis care must be taken in generalizing from the individual to the group.

The *cost-price squeeze* is another component of the farm marketing problem. The competitive conditions of agriculture tend to keep farm prices close to the costs of production. Falling farm prices would not be so critical if they were accompanied by falling farm costs, or if the farmer could adjust input costs as prices fell. However, the increased dependence of farmers on off-farm produced supplies leaves farmers little leeway in adjusting to falling farm prices. Rising farm prices, on the other hand, attract farmers to more profitable enterprises and tend to bid up the costs of production, especially land—another way that the farmer is caught in the cost-price squeeze.

To many, the *superior bargaining power of the buyers* of farm products as compared with that of farmers is the most serious farm marketing problem. Food marketing firms are usually larger and, because of their national and international activities, normally have better market information than the farmers from whom they buy. In addition, through contracts and other arrangements, food marketing firms are thought to gain some control over farm decisions and farm markets. "Who will control agriculture" has been a major controversy in food marketing.

Changing food market pricing efficiency is still another element of the farm marketing problem. Perhaps at one time farmers did not need to be concerned with food marketing because competitive conditions assured all farmers a fair—or at least equal—price. However, with today's direct negotiations, integration, and contractual arrangements, there is no longer any assurance of a high level of pricing efficiency in food markets. As a result, farmers must be more skilled in their marketing decisions.

Finally, there is growing concern about the *increasing gulf* between the farm sector and the food marketing sector. Farmers retain a commodity orientation, whereas food marketing firms stress a merchandising orientation. Furthermore, some farmers have chosen not to participate in many of the food marketing activities that appear to have the greatest growth and profit potential. By not participating in the value-adding advertising, processing, and merchandising marketing functions, some farmers have accepted the status of raw material producer for the food industry. Food marketing firms have not shown the same restraint in crossing over the farm gate.

All of this suggests that agriculture is changing and today's food producers operate in a different market environment than their predecessors. Some of the major changes in U.S. agriculture which are altering farmers' marketing decisions are shown in Table 3-3.

FARM MARKETING ALTERNATIVES

Today's farmers have many marketing alternatives, and their choices of these affect their prices and incomes. Figure 3-7 summarizes many of the marketing alternatives open to producers: delivery or sales alternatives, storage or time alternatives; product form alternatives; transportation or place alternatives; group marketing alternatives; and pricing alternatives. Each of these will be discussed in later chapters of this text. At this point it is sufficient to state that farmers have many alternative ways of adding time, form, place, and possession utilities to food products.

TABLE 3-3 Changes in Farm and Agribusiness Firms

	Old Concept	*New Concept*
1.	Produce and sell homogeneous commodities	Produce specific attributes and differentiated products
2.	Market staple products	Market specialty, niche products
3.	Sell in open, impersonal markets	Personal, negotiated, closed markets
4.	High farmer independence	High farmer-agribusiness interdependence
5.	Relatively stable prices	High price risk
6.	Agriculture as an art	Science-based agriculture
7.	Emphasis on tradition	Emphasis on innovation
8.	Open information	Closed information systems
9.	Public research and development	Private, proprietary research and development
10.	Resource exploitation	Resource conservation, environmental protection
11.	Emphasis on agronomic, technical skills	Emphasis on interpersonal, managerial skills
12.	Family farming	Industrialized farming
13.	U.S. as primary world food producer	Global food competition
14.	Farm incomes lower than nonfarm incomes	Farm incomes comparable to nonfarm incomes

SOURCE: Boehlje, M. (1995). *The New Agriculture, Choices,* 4th Quarter, pp. 34–35. Reprinted with permission.

DELIVERY (SALES) ALTERNATIVES
1. Processing Plant (5)
2. Commission House, Broker (26)
3. Auction House (23)
4. Terminal Market (12, 26)
5. Farmers' Market, Roadside (3, 4)
6. International Market (7)

STORAGE ALTERNATIVES (19)
1. On-Farm Storage
2. Commercial Facility
3. Government Program (21)

PRODUCT FORM ALTERNATIVES
1. Raw Farm Product
2. Processed Product (5)
3. Branded Product (14)
4. Specialty, Niche Product

GROUP MARKETING
1. Individual Seller
2. Cooperative (13)
3. Marketing Agreement (15)
4. Bargaining Assoc. (15)

TRANSPORT MODES (18)
1. Truck
2. Rail
3. Water
4. Air

PRICING ALTERNATIVES (20)
1. Cash Sale
2. Forward Contract
3. Hedge
4. Options
5. Delayed Pricing
6. Contracting (12)

(The numbers in parentheses indicate chapters in this book where this alternative is discussed.)

FIGURE 3–7 Marketing alternatives for farmers.

MINI CASE 3: HELPING THE JONES FARM ADJUST TO THE NEW FOOD MARKETING ENVIRONMENT

Farmer Matt Jones and his wife, Peggy, operate a grain farm in Iowa. The farm has been in the family for three generations, and the two Jones children plan to return to the farm and operate it when they complete their B.S. degrees from Purdue University next year. The Jones's own 1,000 acres and rent another 1,000 acres. Normally about half the acres are planted to corn and the other half to soybeans. These crops are sold to local elevators and through the grain marketing cooperative which the Jones family is a member of.

Matt and Peggy are considered progressive, profitable farmers. They have a well-planned farm, own and operate the latest equipment, typically are among the highest-yield farmers in their area, and are usually among the first to adopt new technologies among the local farmers.

In preparation for the two children, Bill and Kim, to return to the farm, Matt and Peggy have been reviewing their entire operation to see what their strengths and weaknesses are and how they could improve their profits in the future.

The Jones's have concluded that their farm production is efficient, progressive, and reasonably profitable. However, they believe that more attention must be given in the future to the marketing side of their business. They recently heard a professor say that future farmers must be good business managers and marketers, as well as agronomists. That is why they enrolled Bill in Agricultural Economics and Kim in Business Management, where they would receive a good education in production agriculture but also be trained in modern farm and business management.

They have now asked Bill and Kim to put their education to work advising them of new farm product marketing techniques.

Help Bill and Kim outline the major marketing problems that farmers are now facing and will need to solve in the future. Then, suggest ways that the Jones farm might change its farm product marketing programs in the future to improve their prices and profits.

SUMMARY

The nature of agriculture significantly influences the organization of the food marketing system and the complexity of the food marketing process. In general, fewer, larger, and more specialized farms are producing the nation's food supply. The family farm continues to dominate U.S. agriculture, although there are also large, corporate farms owned and operated by farmers, food marketing firms, and nonfood firms. The key farm product and output characteristics that influence the food marketing process are bulkiness, perishability, quality differences, output variations, and the geographic specialization of individual commodities. Farm input markets are increasing in importance as farmers purchase a growing share of their supplies from off-farm sources. The farm marketing problem has several dimensions, including the difficulty of adjusting farm output to rapidly changing market needs, the price-taking status of farmers, the farm cost-price squeeze, the imbalance of bargaining power between farmers and marketing firms, and declining pricing efficiency in agricultural markets.

KEY TERMS AND CONCEPTS

agricultural structure
corporate farming
cost of production
cost-price squeeze
family farm
farm inputs
farm marketing problems
farm productivity, efficiency
free-rider problem
genetically modified organism
price-taker
product homogeneity
quality control
seasonality
specialization
supply control

DISCUSSION QUESTIONS

1. How is the economic position of the farmer in the food industry like the position of any other raw material producer (for example, the coal miner, the lumberjack)? How does it differ?

2. Suppose that scientists discover a way to produce the world's food supply from crude oil. How would this affect agriculture as we know it?

3. How would you expect the increasing geographic concentration of farm production to affect the cost of producing food?

4. Is it always true that "All farm products are alike, and buyers have no preference for one farmer's product over another"?

5. Why don't farmers perform more marketing functions? What are some consequences of not doing so?

6. A farmer says, "Farmers should get a fair price for their products and a fair income for their efforts." How would you define fair price and income?

7. In what sense have technology and the agricultural productivity record been a problem for farmers?

8. Why are farmers purchasing a growing share of their production supplies from off-farm sources?

9. Give one solution to each of the farm marketing problems mentioned in this chapter.

10. How have farm input suppliers (machinery, feed, seed, and chemical firms) adjusted to the changing farm structure?

SELECTED REFERENCES

Agricultural Outlook. U.S. Department of Agriculture. Economic Research Service. Monthly. Paper or http://usda.mannlib.cornell.edu/reports/erssor/economics/ao-bb/

Agricultural Statistics. (1999). National Agricultural Statistics Service. U.S. Department of Agriculture. Annual, paper or http://www.usda.gov/nass/pubs/agstats.hem

Ahearn, Mary et al. (January 1998). *Agricultural Productivity Growth in the United States.* Agricultural Information Bulletin No. 740. Washington, D.C.: U.S. Department of Agriculture.

Biotechnology. Homepage (June 2000). U.S. Department of Agriculture. http://www.aphis.usda.gov:80/usda_biotech.html

Census of Agriculture, 1997. National Agricultural Statistics Service. U.S. Department of Agriculture. Paper copy every 5 years, ending in 2 or 7 or http://www.nass.usda.gov/census/

Direct Marketing Today: Challenges and Opportunities (February, 2000). Washington, D.C.: Agricultural Marketing Service.

Economic Research Service. Homepage (June 2000). U.S. Department of Agriculture. htttp://www.ers.usda.gov/

Farmer Direct Marketing, Homepage (June 2000). Agricultural Marketing Service. U.S. Department of Agriculture. http://ams.usda.gov/directmarketing/

Farm Structure Briefing Room, Economic Research Service, U.S. Department of Agriculture (January 2001). www.ers.usda.gov/briefing/FarmStructure/Questions/Struct.htm

Greene, C. (April 2000). U.S. Organic Agriculture Gaining Ground. *Agricultural Outlook.* Washington, D.C.: U.S. Department of Agriculture.

Harwood, J. et al. (March 1999). Farmers Sharpen Tools to Confront Business Risks. *Agricultural Outlook.* Washington, D.C.: U.S. Department of Agriculture. http://www.ers.usda.gov/

Hrubovcak, J. et al. (June 1999). *Green Technologies for a More Sustainable Agriculture.* Agriculture Information Bulletin No. 752. Washington, D.C.: U.S. Department of Agriculture.

Krause, K. R. (April 1989). *Farmer Buying/Selling Strategies and Growth of Crop Farms,* Technical Bulletin No. 1756. Washington, DC: U.S. Department of Agriculture.

National Agricultural Statistics Service. Homepage (June 2000). U.S. Department of Agriculture. http://www.usda.gov/nass/

a. Agricultural Graphics, http://www.usda.gov/nass/aggraphs/graphics.htm
b. State Information, http://www.usda.gov/nass/sso-rpts.htm
c. Statistical Highlights: Farm Economics. Annual, http://www.usda.gov/nass/pubs/stathigh/1998/summ1.htm

Small Farm Program. Homepage (June 2000). Coorperative State Research, Education, and Extension Service, U.S. Department of Agriculture. http://www.reeusda.gov/agsys/smallfarm/

U.S. Department of Agriculture. (1992). *New Crops, New Uses, New Markets.* Yearbook of Agriculture. Washington, DC: U.S. Department of Agriculture.

U.S. Department of Agriculture. (1996). Structure of U.S. Agriculture: The U.S. Farm Sector. *Agriculture Fact Book 1998,* annual. Washington, D.C. http//www.usda.gov/news/pubs/fbook98/content.htm

U.S. Department of Agriculture. Homepage (June 2000). http://www.usda.gov/

PART II

FOOD MARKETS AND INSTITUTIONS

CHAPTER 4

FOOD CONSUMPTION AND MARKETING

Consumers are right, even when they are wrong.

OBJECTIVES

After reading this chapter, you will be able to:

1. Better understand why consumer's decisions are so important to farmers and food marketing firms and why it is so difficult to predict consumers' food preferences.
2. Identify the major trends in food spending and consumption.
3. Explain some of the major economic and social reasons for consumers' changing food consumption patterns.
4. Indicate how rising income influences food consumption.

Consumer tastes and preferences shape the nation's food and fiber system in many ways. The doctrine of consumer sovereignty states that the ultimate task of the food marketing system is to deliver the food utilities that consumers desire. This involves much more than just matching the total food supply with the total food demand. It is the process of matching the right form of a product at the right place and time to a particular buyer. The facts that "we all have to eat" and "the stomach can only hold so much" do not tell us very much about food consumption. The interesting questions are *what* do people eat, *why, how much, when, where,* and *how often* and *how* are consumers' food choices changing. These food consumption patterns are influenced by physiological needs, social conditions, and economic factors.

UNDERSTANDING FOOD PREFERENCES

Consumers choose their diet to satisfy their needs and wants. Because these choices condition all of the food industry's production and marketing decisions, it is important to understand the nature of consumer food preferences. Humans are hunter-gatherers whose food preferences are not much different from those of early man, even though the hunting and gathering may now take place at the supermarket. Despite some obvious changes in our food consumption patterns, our diet still consists largely of meat, potatoes, and bread, washed down with milk, coffee, and soft drinks.

It is sometimes said that all animals feed, but humans alone eat. Humans are omnivorous and can thrive on a wide variety of different foods. They are also choosy and no society defines all the potentially edible material in its environment as food. Moreover, humans are social creatures, and their food preferences and eating patterns are culturally bound and socially influenced.

It is not easy to explain the diverse food preferences of various societies. There seems to be no physiological reason why some societies cultivate certain crops for food and others shun these crops; why some people eat insects (an excellent protein source) and others do not; why some eat animal flesh and some do not. Food idiosyncrasies are not limited to cross-cultural comparisons, either. There are substantial dietary variations within most societies, including America.

Each society develops common patterns of dealing with food, which we refer to as *foodways.* These govern how food is acquired, prepared, and eaten. Foodways are complex behavioral patterns that, from the standpoint of food marketing, have four important characteristics. First, no two societies have identical foodways. Second, standardized foodways result in somewhat similar and stable food preferences and eating patterns within a society. Third, foodways defining "how to eat" add social significance to the diet and are taught to each succeeding generation. And fourth, foodways adapt to socioeconomic changes such as urbanization, education, income, technology, and changing lifestyles.

American foodways are the result of five influences: (1) the functional, physiological values of foods (their nutritional contributions to health and survival); (2) the sociopsychological values of foods (status, religion, aesthetics, and lifestyle); (3) the economic values of foods; (4) the availability of foods; and (5) consumers' knowledge and information about foods.

No one of these influences fully explains American food preferences. Consumers are concerned with nutrition, but it is doubtful that our diets maximize the nutritional value of each food dollar. It has been estimated that Americans put billions of dollars worth of food into garbage cans or garbage disposals each year, much of it edible. We are health-conscious, but overeating is as great a problem for many as insufficient food. There are some regional, ethnic, and religious food preferences, but these appear to be breaking down in our mobile society. Food consumers are price conscious, but they do not always buy the least expensive foods. These examples caution against simple analyses of food preferences.

One thing seems clear. In high-income, affluent, urbanized societies food consumers purchase much more than physical farm products. Modern consumers don't want only farm products, they want food, and increasingly they don't want just food, they want a meal. The modern food consumer purchases a whole *bundle of attributes,* which includes, along with farm products, time, form, space, and possession utilities. Even the store where the food is purchased and the setting in which it is served contribute to consumers' satisfaction and must be considered part of the *product bundle.* Both the functional and the sociopsychological values of food are pleasing to consumers. Knowledge of this product bundle of satisfactions is necessary for understanding and predicting food preferences. The task of food marketing firms is to discover the product bundle of attributes that will appeal, profitably, to consumers. This search results in large expenditures for product innovation and design, packaging, merchandising, and advertising.

FOOD CONSUMPTION AND EXPENDITURE PATTERNS

Table 4-1 shows the trends in U.S. food consumption, prices, and expenditures from 1930 to 1999. Total spending and per capita food expenditures rose significantly over this period. However, because food spending rose less rapidly than consumers' income, the share of consumers' income spent for food declined from 24 percent to 10 percent. By way of comparison, it has been estimated that food expenditures represented 61 percent of consumers' income in 1869 and expenditures for all farm-origin products (food, clothing, and tobacco) accounted for 70 percent of income in that year.

Consumers' food expenditures are rising for three reasons: increased population and quantities of food eaten, rising food prices, and consumer preferences for more expensive foods and marketing services. In recent years rising retail food prices have accounted for most of the increase in consumers' food bills. The second most important reason for this rise has been the result of consumers substituting more expensive foods into their diet and eating away from home more frequently. Only a small portion of the rise has been due to increased quantities of food eaten.

TABLE 4-1 Trends in Food Consumption, 1930-1999

	Consumer Food Expenditures		Food Spending			
Year	$ Billion	$ Per Capita	Total Food Spending As a Percent of Disposable Income	Spending for Food Eaten Away from Home As a Percent of Total Food Spending	Retail Food Prices (1967 = 100)	Per Capita Calories Consumed Per Day
1930	$18	$146	24%	17%	46	3,440
1940	16	127	21	19	35	3,340
1950	43	313	20	24	75	3,260
1960	64	415	17	26	88	3,130
1970	102	605	14	33	115	3,300
1980	266	1,357	13	39	255	3,400
1990	489	2,267	11	44	389	3,600
1999	691	2,891	10	48	482	3,654

SOURCE: U.S. Departments of Labor and Agriculture, Economic Research Service.

The food industry is mature with steady but only modest growth over time. Per capita food expenditures, after inflation, grew less than one half of one percent per year from 1970 to 1994 and have held steady since then. This slow growth rate influences the food system in many ways. It limits profit prospects for farmers and food marketing firms, intensifies the competition among commodities and firms for the consumer's dollar, encourages food firms to diversify into more rapidly growing products, and provides an incentive for food firms to attempt to increase their sales overseas.

TABLE 4-2 Per Capita Consumption of Selected Foods, 1972-1996

Item (in pounds unless otherwise designated)	*Per Capita Consumption (average)*		
	1972–1976	*1982–1986*	*1992–1996*
Generally Increasing Trend:			
Pork	43	47	49
Chicken	28	35	49
Turkey	7	9	14
Fish and shellfish	13	14	15
1-2% fat milk (gallons)	6	9	11
Cheese	14	22	27
Yogurt (½ pint)	3	6	9
Salad and cooking oils	18	23	27
Fresh fruit	98	112	126
Fresh vegetables	147	153	175
Processed tomatoes	62	63	75
Frozen vegetables	52	60	78
Flour and cereal products	137	153	192
Corn sweeteners	25	57	80
Corn products	10	16	23
Carbonated soft drinks (gallons)	28	36	51
Vegetable fats	42	49	57
Frozen potato products	36	43	56
No Clear Trend:			
Ice Cream	24	25	23
Butter	5	5	5
Margarine	11	11	10
Fruit juices	106	123	121
Lamb, mutton	2	1	1
Fresh potatoes	52	48	50
Animal fats	12	12	11
Generally Decreasing Trend:			
Beef	82	74	64
Eggs (number)	284	259	236
Whole milk (gallons)	21	14	9
Sugar	96	67	65
Coffee (gallons)	33	27	23

SOURCE: U.S. Department of Agriculture, Economic Research Service.

As Table 4-1 indicates, the caloric intake of each consumer has been rising since its low point in 1957. Within this total, there have been significant changes in the consumption of individual foods as consumers have substituted food products within the diet. For example, diets shifted away from animal products and toward crop products in the 1980s. Other dietary changes are illustrated in Table 4-2. Per capita consumption is increasing for some foods, decreasing for others, and stable for still others. Some direct substitutions can be seen in the table—for example, the increase in lowfat milk use and the decline in whole milk consumption and the opposite trends in fresh and processed potato consumption.

Some interesting and important turning points in the American diet can be identified. In the nineteenth century and until 1950, pork was the most popular meat, but since 1950 Americans have consumed more beef than pork. Consumption of fresh vegetables overtook fresh fruits in the mid-1950s. Poultry consumption first exceeded egg consumption in 1964, and poultry pushed past pork consumption in 1982. The rising soft drink consumption trend crossed the falling milk trend in 1967. These trends indicate that no food has a guaranteed market. Each must earn its place in the mind and stomach of consumers.

These changing consumption patterns are of enormous importance to food producers and marketing firms. Considerable money is spent monitoring these trends and attempting to influence them through new product development and promotional efforts. It is widely debated whether and how marketing programs can alter consumer food preferences, but farm commodity groups and food marketing firms certainly believe they must try to do so in today's marketplace.

THE DEMOGRAPHICS OF FOOD CONSUMPTION

Demography is the study of populations—how many people there are, where they live, and how they live. These trends influence food marketing by affecting the number of mouths to feed, what people eat, where food is sold, and how people buy their food. A number of trends influencing food consumption and the food industry are illustrated in Table 4-3.

Population Trends

The food market expands in proportion to the rate of population growth. The U.S. population grew from 152 million to 268 million people between 1950 and 1997. However, the rate of population growth has declined in recent years after climbing from record low levels in the 1930s: 1.8 percent growth per year in 1950–1955; 1.5 percent in 1960–1965; and below 1 percent in the 1970–2000 period. The United States reached the zero population growth rate in the early 1970s. However, total population is growing at the 2.1 percent replacement rate as a result of increased immigration and fertility.

Immigration has increased American ethnic diversity and the American food industry. Nineteenth-century immigration waves from Europe have been replaced by new immigrants from Latin America and Asia. Hispanics are the fastest growing ethnic population. Ethnic populations not only introduce new foods and food consumption patterns into America, they also create new market opportunities for tradi-

TABLE 4-3 Selected Trends Influencing Food Consumption and Marketing, 1950-1997

Trend	*1950*	*1970*	*1990*	*1997*
U.S. population (millions)	152	204	250	268
Population per square mile	43	57	70	76
Number of households (millions)	40	63	93	101
One- and two-person households (percent of total)	39	46	57	60
Life expectancy (male)	66	67	72	73
Number of marriages (millions per year)	1.5	2.2	2.4	2.3
Number of births (millions)	3.9	3.7	4.2	3.9
Average family size	3.5	3.6	3.2	3.2
Median age of first marriage (female)	20.3	20.8	23.8	NA
Female-headed families (millions)	3.7	5.6	10.9	12.8
Median age	30.2	27.6	32.8	34.9
Percent of population: under age 18	34	34	28	26
over 64 years of age	8	10	11	13
Percent of husband-wife families with wife in labor force	24	41	58	62
Percent living in metropolitan areas	56	69	78	80
Median family income	$3,319	9,876	35,353	42,300
Percent living below poverty level	NA	13	14	14
Percent completing 4 or more years of college	5.0	8.2	18.4	22.4

SOURCE: U.S. Statistical Abstract, Annual, U.S. Government Printing Office.

tional foods. In some cases they have also fostered new forms of food retailing, such as the *bodegas* (small neighborhood foodstores) in large cities.

The mobility of the population also influences the food marketing machinery. About 20 percent of Americans move each year, but only 3 percent move to another state, thus potentially exposing themselves to new food consumption patterns. As a result of this migration, the population has grown most rapidly in the Sunbelt—the south, southwest, and west coast states. Some predominately agricultural states have experienced declining populations. This alters the food distribution network. For example, an increasing amount of Western food production now finds a market nearer to home than previously.

There are differences in food consumption patterns between one region of the country and another that cannot be explained on the basis of income or other factors. Lamb consumption is largely limited to the East and West coasts, with very little being consumed in the interior. Boston housewives prefer brown eggs, whereas New York housewives prefer white ones. Consumers in the South eat relatively more pork and less beef than consumers in other regions of the country. People in San Francisco are relatively large lamb and poultry consumers. Buffalo and Minneapolis consumers are heavy white potato eaters, whereas Birmingham people consume more sweet potatoes.

Urbanization is another important demographic trend. In 1997, 80 percent of the population lived in metropolitan areas compared with 56 percent in 1950. The country-to-central-city migration of earlier years was replaced by a migration out of central cities to the suburbs in the 1950s and 1960s. The 1970s saw a back-to-the-country movement as rural areas and small towns grew more rapidly than cities and suburbs. The population of nonmetropolitan U.S. counties has been increasing more rapidly than that of metropolitan counties. These migrations have shaped the modern food distribution network. The supermarket, for example, is a suburban product. The regional movement of population has required new investments in food marketing facilities and provided opportunities for some food marketing firms to gain a competitive advantage over others. The geographic distribution of the population influences the efficiency of food marketing because it is usually more economical to service highly concentrated populations than scattered consumers.

The age and education of the population also influence food consumption. With the aging of the post-World War II baby boomers and reduced family sizes, the percentage of the population under 18 years of age is declining. In contrast, the elderly population is growing rapidly. The median age was 35 in 1997 and is expected to rise to 40 by year 2030. Youthful and elderly consumers eat differently from others. Recent declines in milk consumption and increases in red meat, cheese, and potatoes have been attributed to the changing age distribution of the population. Food marketing firms have developed special foods, packages, and promotional messages for different target market age groups.

Household Food Consumption

The household is the basic unit of food consumption. Household members typically pool their income, buy as a unit, and share somewhat similar food preferences. The number of households rose from 40 million to 101 million between 1950 and 1997. However, not all households are families. There has been a rapid growth in the proportion of households with only one or two people, including young married and retired couples. More than half of all households have only one or two people in them. These trends have resulted in increased demand for smaller food packages and more prepared foods.

The average family size fell from 3.6 in 1970 to 3.2 in 1997. Studies show that there are some economies of scale in consumption so that food spending per capita falls as the number of children increases. Smaller families typically spend more per capita for food, eat away from home more often, and purchase more food marketing services than larger families.

Other important household trends are the increasing number of female-headed households, the prevalence of working wives in wife-husband households, an increasing number of households headed by unmarried men, and the growing number of individuals with more than one job. Only about 6 percent of all households today resemble the traditional household of a married couple with two or more children and only one earner in the workplace. These trends have contributed to the popularity of convenience foods and the growing away-from-home food market.

Household appliances also influence food consumption. Practically all families have some form of mechanical refrigeration, and many have a separate food freezer, food processor, and microwave oven. Air conditioning has also changed food patterns. Undoubtedly, the rising number of multi-car families, the exposure to television and other mass communications, and the popularity of outdoor grills and microwave ovens have also affected food purchases and preferences. It is rumored that the automobile industry is considering offering a microwave oven in the glove box. The food industry is developing dashboard foods to be eaten in the automobile.

The importance of better understanding the needs, wants, and motivations of food consumers have led food marketing firms to look beyond economics and demographics in analyzing consumer behavior. Social psychology has been helpful in examining food fads and fashions, food and diet attitudes, food symbolism, and the concept of consumer lifestyles. In lifestyle analysis, consumers sharing similar attitudes and behavior are grouped into target market classes. For example, groups of meat consumers have been identified as either creative cooks, people with active lifestyles, or health-oriented individuals. By developing unique products and promotional messages for each of these groups, meat marketing firms hope to increase their sales and profits.

INCOME AND FOOD CONSUMPTION

Historically, population and income growth have been the two major sources of growth in food consumption. The number of people determines the total need for food, whereas their income determines their ability to pay for it. Population growth has slowed, but income growth is expected to continue to increase the demand for food. Consumer incomes in the United States have grown as a result of technological change, increased worker productivity, and general economic growth. Half of U.S. families had an income greater than $42,000 in 1997 compared with the median income of $9,867 in 1970. The rise is much less—but still evident—when this is adjusted for inflation.

Effective food demand consists of both needs and the ability and willingness to satisfy those needs with income. A need does not register in the marketplace unless it is backed up with purchasing power. Many low-income countries and people have a great need for food but lack the income to make this an effective demand in the marketplace. High-income countries, such as the United States, are generally characterized by a strong effective demand for food.

The responsiveness of food consumption to an increase in income is called *income elasticity.* The more responsive consumption is to income changes, the greater the income elasticity. If the quantity of a food rises along with income, the product is termed a *normal good.* If quantity falls as income rises, the food is termed an *inferior good.* Livestock products generally are normal goods and have a higher income elasticity than crop products. There are few inferior foods; perhaps grits and lard are in this category. In general, the income elasticity of food is lower than for other products consumers buy. Moreover, the income elasticity of raw farm products is lower than the income elasticity of the utilities added in the marketing process. This means that the demand for food marketing services grows more rapidly than the demand for farm products. As one of the first claims on the consumer's budget and one of the last items to be adjusted in difficult times, food expenditures also tend to be more stable than expenditures for more discretionary items.

There are, of course, wide variations in consumers' incomes, and these influence food consumption. The poor do not eat or shop like the rich. In recent years there has been great emphasis on improving the diets of the poor. According to Table 4-3, about 14 percent of the population was classified as poor in 1997. Studies suggest that the higher the income, the "better" the diet. However, other factors also influence diet adequacy, and high income itself does not assure a good diet.

Rising consumer incomes affect food consumption patterns in several ways. First, consumers do not buy many more pounds of food as incomes rise. Instead, they "upgrade" their diet by substituting more expensive foods, often meats, for staple items. Although consumption of crop products is now rising faster than meat consumption, increased income was a major factor in the growth demand trend for meat and livestock products in the past in America and it remains a characteristic feature of most other societies.

The second result of rising consumer incomes is to reduce the consumers' share of income spent for food. Food expenditures do not normally increase as rapidly as income, so the ratio of food spending to income falls with rising income. This is Engel's law, named after a German statistician who first observed this tendency in Europe in 1857. Sometimes this falling food share of income is cited to imply that "food is a bargain," and it is true that a declining food share of income represents a growing share of income available for nonnecessities with which to increase the consumer's standard of living. However, as shown in Table 4-1, the food share of income can fall while food prices are rising. Figure 4-1 illustrates the share of total expenditures going for food in different countries. The U.S. food share of spending is an average, of course. There are American families that spend more than 50 percent of their income for food and, probably, some who spend less than 1 percent.

A third consequence of rising income is a broadening of the product bundle of attributes most valued by consumers. At low-income levels, price and perhaps nutrition are paramount concerns. As incomes rise, consumers add quality, variety, convenience, and service to the desired product-attribute bundle. More recently, consumers have become concerned about the impacts of food production and marketing on the environment and the welfare of farm animals. Thus, in high-income societies and populations, the product bundle of attributes becomes more complex.

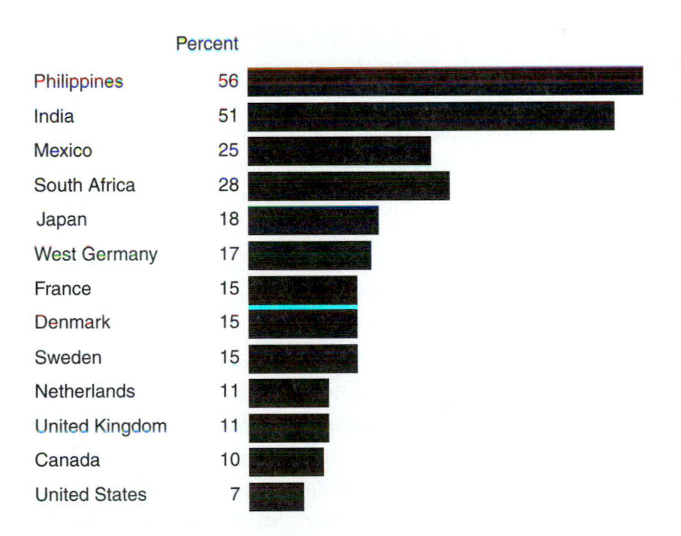

FIGURE 4–1 Food spending as a percent of total expenditures, 1994. (Organization for Economic Cooperation and Development; United Nations.)

These affluent attributes are often added in the food marketing process, and they become equally important, if not more important, than the farm-originating product attributes. Hence, the scope of food marketing activities is much greater in high-income than in low-income societies. The rapid rate of new product development, the growth in marketing services, and the rising sophistication with which the American food supply is marketed are evidence of the broadening of the consumer's preferred food product-attribute bundle.

Perhaps most importantly, rising consumer incomes increase the value of consumers' time and expand the demand for convenience foods with built-in maid service. A *convenience food* is a product that reduces the amount of time, effort, or additional ingredients required of the consumers in preparing food. When they buy convenience foods, consumers are transferring marketing functions from the kitchen to the food marketing system. They are buying time with money. Rising education levels, a desire for more leisure time, working spouses, busy lifestyles, urbanization, and a number of other socioeconomic trends favor this transfer of marketing functions outside the home.

Americans' use of highly processed convenience foods has been criticized as unnecessarily expensive and detrimental to the quality of the diet. Some of these foods are referred to as "junk foods." However, there is evidence that some convenience foods are less expensive than their home-prepared counterparts and that highly processed foods are not necessarily more expensive sources of nutrients than unprocessed foods. Convenience foods should not be condemned without considering the rising costs of home food preparation, particularly time and energy costs. It is possible that the food marketing system can perform some food preparation tasks more efficiently than the household today. Of course, many people receive satisfaction from "home cooking" and others avoid highly processed foods for a variety of reasons. Providing that they have freedom of choice from among real alternatives and are aware of the consequences of their decisions, each individual consumer is probably the best judge of whether to purchase convenience or unprocessed foods.

How does the growth in convenience foods affect farmers? Because the highly processed products, such as potato chips and snacks, probably do not reduce the amount of fresh potatoes eaten, there is no reason to believe that farmers will sell fewer potatoes as a result of increased use of "convenience" forms of potatoes. On the contrary, frozen potato products have expanded the total market for fresh potatoes. Nor is there any particular reason why farmers would receive a lower (or higher) price for potatoes that are to be processed rather than sold fresh.

The convenience food trend should not be exaggerated. About 40 percent of all farm commodities are marketed in fresh or so-called low-convenience forms, such as fresh meats, produce, milk, eggs, and the like. More than 90 percent of grain products are processed whereas less than 40 percent of milk, fruits, and vegetables are processed. The marketing task varies considerably for these, but it is not necessarily less expensive to market unprocessed products. With increasing health and diet concerns, consumption of many fresh, unprocessed foods has grown rapidly in recent years. The "salad revolution" is an example of this.

Rising incomes and educational levels intensify consumers' concern with the quality of food markets and the social impact of the food industry. Affluent consumers have been in the vanguard in supporting such consumer causes as unit pricing, nutritional labeling, and open-code dating. These informational programs al-

MINI CASE 4: THE AMERICAN CONSUMER'S DIETARY DILEMMA

Barbara Jones, mother of four young children, wife of Bob, and a full-time manager of her own business, was frustrated. She loved to cook and she knew she should be feeding her family more home-prepared, nutritious foods. But more often they went to a restaurant, ordered take-out foods or popped a frozen dinner into the microwave. Their busy schedules meant that many evenings each member of the family ate at different times. Often, they skipped some meals. The kids seemed to snack all day long, and she noticed that they were drinking much more cola than milk or fruit juices. She also noticed that the family preferred high-salt and sweet foods, and they were becoming overweight. She had just read a news report that 55 percent of Americans are overweight, and they are consuming several hundred more calories per day than they did in 1957. "There has to be a better way," said Barbara. "Next week we are going to eat as a family, three square meals a day, and only home-cooked, nutritious, and healthy foods."

Do you think the Jones family did it? Why don't people eat more like they want to and know they should? What role does the food marketing system play in this problem?

legedly improve consumer food choices and contribute to more competitive food markets. And increasingly, many consumers are concerned with food-related social problems: rural and urban poverty, farm labor, environmental issues, and so on. In their consumption decisions, affluent food consumers face difficult political and economic choices insofar as these also relate to energy use, environmental impact, and welfare programs. The recent food debates in the United States and elsewhere over the safety and desirability of biotechnology and genetically modified organisms (GMOs) reflect consumers' concerns for a wide range of issues beyond the supply and prices of foods. Both science and value judgments are involved in these debates, and the outcomes will affect farmers and food marketing firms alike.

DIET, HEALTH, AND FOOD MARKETING

In recent years, many Americans have become more concerned with their food choices. There is a growing awareness of the connection between diet and health. Prompted by government consumer education programs, health professionals, and the media, some consumers have altered their diets toward eating healthier foods and many others wish to do so. This has changed the demand for some farm products, notably grains, fruits and vegetables, eggs, dairy products, fats and oils, sugar, and meat types. It has also affected the marketing programs of food firms.

Functional foods are foods which are believed to provide medical or health benefits, including disease prevention or treatment, beyond the general health benefits of any foods. The trend toward healthier eating is resulting from the aging population and a growing American lifestyle which places greater responsibility on the individual for personal health. It is not yet clear whether this is a fad or a long-term change in behavior.

Functional foods, sometimes called farmaceuticals, have presented the food industry with new marketing opportunities. Rather than make difficult dietary changes, consumers in some cases have asked the food industry to alter their products and provide healthier choices. In response, food industry manufacturers and distributors have added many new products to their lines. These include food products with improved nutritional contents (e.g., vitamin fortified), altered food values (e.g., low-fat, diet, low-cholesterol, sugar-free, leaner meats), and changed ingredients (e.g., poultry-based hot dogs, soymilk). The production of organic foods, for which no chemicals or additives are used either in farming or manufacturing, is another response of the food industry to the environmental and healthy eating trends. Once confined to specialty health stores and small sections of the grocery store, functional foods are now widely available in most grocery stores. Although often more expensive than traditional foods and to some not as tasty, many consumers are choosing these foods as part of their desired product bundle of attributes.

A new food category like this can present marketing problems. Consumers must be educated about the product and its benefits. New quality standards may be needed. Consumer deception and fraud are possibilities. All food health claims used in advertising and labeling foods must be approved by the U.S. Food and Drug Administration. Only a few disease prevention claims have been approved. The U.S. Department of Agriculture's nutrition education program, including the dietary guidelines represented in the food pyramid, have helped educate consumers. Functional foods present the food industry with a dilemma. Aren't all foods "healthy" and "organic"? If some foods are labeled as "healthy" or "organic," what will consumers think of foods not so labeled? The U.S. Department of Agriculture issued its organic food standards in 1999 in an attempt to help consumers distinguish organic foods from other foods.

Despite Americans' interest in diet and health and some recent improvements, there is still a gap between public health dietary recommendations and food consumers' practices. Dietary patterns are difficult to change. People may say one thing and do another. Not all consumers are choosing healthier foods. Food choices have moved only slowly toward the dietary recommendations put forward in the food pyramid. Food marketing firms can reinforce these healthy eating trends in their product development and advertising programs. Farmers can also benefit from these trends by altering their production plans and practices to provide these new value-added products and capitalize on dietary trends. The marketing concept and the doctrine of consumer sovereignty hold here: Successful firms are those who produce what consumers want.

THE AWAY-FROM-HOME FOOD MARKET

Increased consumer incomes, working spouses, and changed lifestyles have contributed to a rapid growth in the away-from-home food market, sometimes referred to as the institutional or foodservice market. Food eaten away from home accounted for 48 percent of all food expenditures in 1999.

The foodservice sector includes restaurants; school, factory, and hospital cafeterias; hotel and motel eating places; government foodservice operations; and military feeding establishments. As shown in Table 4-4, fast-food establishments increased their market share from 14 percent in 1967 to 36 percent in 1999, while the market share of table-service restaurants declined.

The trend toward eating out must be kept in perspective. Many people are required to eat out because of the distance they travel to work. Further, the difference between eating at home and away from home is not so clear in this day of grocery store delicatessens and restaurant drive-through windows. It is also instructive to differentiate between "eating out" and "dining out." The fast-food restaurants sell low-service, economy-oriented meals. Sit-down restaurants are selling service and atmosphere as much as food. For many families and individuals the added expense of eating away from home can be budgeted as much to leisure and recreation as to the food category of the budget.

The away-from-home food market is quite different from the home food market. Prices tend to be more stable and there is a higher ratio of marketing-services-to-food in the foodservice industry than in the home food industry. For many restaurants, 45 to 65 percent of the price charged to consumers represents nonfood costs of service and profits, whereas about 80 percent of supermarket prices represent food costs. Because of the higher ratio of marketing services to food, farmers receive a smaller share of the away-from-home food dollar than they receive for food purchased at grocery stores.

The eating-out trend has markedly affected sales of hamburgers, chicken, pasta, potatoes, and ketchup. In some cases, major foodservice firms, like McDonald's, have significantly affected the food marketing system by reorganizing the distribution channels and altering the quality of products provided by farmers and marketing firms. The importance of the away-from-home food market to farmers is illustrated by the pork industry's extensive efforts to develop and promote a fast-food pork sandwich. Foodservice managers are quite price conscious and have somewhat different preferences for the product bundle of attributes than homemakers. These managers are interested in standardization, portion control, and labor-saving foods. Precut butter, prepattied hamburger, bulk frozen french fries, and indi-

TABLE 4-4 Foodservice Sales, by Type of Outlet, 1929-1999

	Percent of Sales of Meals and Snacks Eaten Away From Home							
Year	*Restaurants, Caterers*	*Fast Food Places*	*Hotels, Motels*	*Schools and Colleges*	*Stores, Bars, Vending Machines*	*Recreational Places*	*Military Outlets, Others*	*Total*
1929	51%	9	10	5	19	1	5	100%
1939	47	7	11	7	21	2	5	100
1948	48	8	9	10	18	1	6	100
1967	46	14	6	14	12	2	6	100
1977	39	28	5	11	9	3	5	100
1987	40	33	6	8	8	2	3	100
1999	36	36	6	7	8	4	3	100

SOURCE: U.S. Department of Agriculture, Economic Research Service.

vidual servings of jelly and syrup are examples of the types of services demanded by these food buyers.

PUBLIC FOOD PROGRAMS

Despite its productive and relatively low-cost food supply, there is hunger and malnourishment in America. The U.S. Department of Agriculture estimated that 90 percent of American households were food-secure in 1998, leaving 10 million households with inadequate access to enough food to fully meet their needs at all times. This is primarily a result of limited incomes. The food industry works best for those with high effective demands. A number of public and private programs feed the needy. These are important in the study of food marketing because they increase the demand and spending for farm and food products.

The U.S. Department of Agriculture administers several government food assistance programs. These include the food stamp program; child nutrition programs (school breakfast and lunch, the special milk program, the child and adult care food program, and the summer food service program); the supplemental nutrition program for women, infants, and children (WIC); the nutritional program for the elderly; and the emergency food assistance program, among others. The food stamp program is the largest of these, with 20 million participants in 1998. Food stamps permit qualifying individuals to purchase food with special coupons, thus reducing their food costs and permitting them to purchase more food.

In addition to assisting low-income people and promoting better nutrition, one of the original objectives of these public food programs was to reduce agricultural surpluses by increasing the demand for food. These programs cost $33.7 billion in 1998. While the programs do increase the total demand for food, not all of the economic benefits of the programs flow to farmers and food marketing firms. By reducing the income required for food purchases, the programs allow participants to expand nonfood purchases on housing, medical care, and other items. The low-income elasticity of demand for farm products means that a 10 percent increase in food benefits will result in perhaps a 2 percent increase in food spending. And not all of that will filter back to the farmer. The programs also affect some farm products more than others. Low-income consumers spend a large share of their food dollar on meat, fish, and eggs. Since farmers receive a high share of the retail price of these products, these producers benefit most from the food stamp program.

SUMMARY

Food consumers' behavior and preferences condition, to a significant extent, all production and marketing decisions in the food industry. These preferences are complex attitudes relating to the physiological, sociopsychological, and economic values of food products. In high-income societies, the food product bundle consists of the raw farm product and an associated set of form, time, place, and possession utilities. Two key factors influencing the rate of change in the demand for food are population and income growth. Demographic trends such as urbanization, the changing nature of the family and household, working wives, educational levels, and the age distribution of the population are also influencing food consumption patterns. Rising consumer incomes result in a substitution of more expensive foods for staples, a declining share of income going for food, a broadening of the food product bundle, a greater demand for convenience foods, more eating away from home, and a greater consumer concern for the quality of foods and food markets.

KEY TERMS AND CONCEPTS

away-from-home food market
consumer food preference
convenience food
demographic trends
effective demand
Engel's law
foodways
income elasticity
product bundle of attributes
psychological satisfaction
superior and inferior foods

DISCUSSION QUESTIONS

1. Your friend argues, "Nutrition is the most important factor in buying food." Do you agree or disagree?

2. Why are American food preferences becoming more similar?

3. A famous psychologist stated that the economic factors, price and income, are not as important in explaining consumer behavior as culture, social influences, and psychological factors. Do you agree or disagree?

4. Comment on the following: "The United States should share its abundant food supply more generously with the low-income countries of the world."

5. A congressman introduced a bill to prevent consumers from purchasing "convenience foods" with public assistance money and food stamps. Comment on this proposal.

6. What are the demographic trends with the greatest potential for increasing farm product sales in the future?

7. What do you think the national diet will be like in the year 2010?

8. Suppose a food company develops a pill that can be taken daily, supplying all nutritional needs. However, it has no taste and leaves one feeling hungry. At what price could such a pill be sold for (assume it costs one cent to make each pill)? How would eating patterns change? Should the pill be developed?

SELECTED REFERENCES

Eating in the 20th Century (Jan.–April 2000). *Food Review.* Volume 23, Issue 1. Washington, D.C.: U.S. Department of Agriculture.

Farb, P., & G. Armelagos. (1980). *Consuming Passions: The Anthropology of Eating.* New York: Washington Square Press.

Food Consumption: Household Food Expenditures Briefing Room (January 2001). Economics Research Service: U.S. Department of Agriculture. http://www.ers.usda.gov./briefing/consumption/expenditures.htm

Food Consumption Briefing Room (January 2001). Economic Research Service: U.S. Department of Agriculture. www/ers.usda.gov/briefing/consumption/

Food Consumption: Food Supply and Use Briefing Room (January 2001). Economic Research Service: U.S. Department of Agriculture. www.ers.usda.gov/briefing/consumption/supply.htm

Food & Nutrition Assistance Policy Briefing Room (January 2001). Economic Research Service: U.S. Department of Agriculture. www.ers.usda.gov/Topics/view.asp?T=103816

Frazao, E. (Ed.) (1999). *America's Eating Habits: Changes and Consequences.* Agriculture Information Bulletin No. 750. Washington, D.C.: U.S. Department of Agriculture.

Frazao, E. (February 1995). *The American Diet: Health and Economic Consequences,* Agricultural Information Bulletin No. 711. Washington, D.C.: U.S. Department of Agriculture.

Harris, M. (1985). *The Sacred Cow and the Abominable Pig: Riddles of Food and Culture.* New York: Simon & Schuster.

Putnam, J. J., & J. E. Allshouse. (December 1994). *Food Consumption, Prices, and Expenditures, 1970–1997,* Washington, D.C.: U.S. Department of Agriculture. Statistical Bulletin No. 965.

Putnam, J. J., & J. E. Allshouse. (April 1999). *Food Consumption, Prices, and Expenditures, 1970–97.* Washington, D.C.: U.S. Department of Agriculture. Statistical Bulletin No. 965. http://www.ers.usda.gov./epubs/pdf/sb965/

Senauer, B., E. Asp, & J. Kinsey. (1991). *Food Trends and the Changing Consumer.* St. Paul, MN: Eagen Press.

What Do Americans Eat? (1998). *Agriculture Fact Book,* 1998. U.S. Department of Agriculture. pp. 1–13. http://www.usda.gov/news/pubs/fbook98/content.html.

Wilde, P. et al. (June 2000). *The Decline in Food Stamp Participation in the 1990's.* U.S. Department of Agriculture. Food Assistance and Nutrition Research Report No. 7.

Young, C. E. (May 1999). Moving Toward the Food Guide Pyramid: Implications for U.S. Agriculture. Washington, D.C.: U.S. Department of Agriculture. Agricultural Economic Report No. 779.

CHAPTER 5

FOOD PROCESSING AND MANUFACTURING

The Marketing Credo: Find Wants and Fill Them.

OBJECTIVES

After reading this chapter, you will be able to:

1. Explain how and why food processors add value to farm products.
2. Identify food processors' 4P marketing strategies and give examples of these.
3. Understand the reasons for some trends in food processing.
4. Appreciate the marketing and management problems of food processing companies.

Food manufacturers or processors are primarily responsible for adding form utility to raw farm products. Wheat is milled into flour, livestock is converted into meat products, fruits and vegetables are canned or frozen. These firms play a vital role in transforming bulky, perishable farm products into storable, concentrated, and more appealing food products. In so doing, food processors become involved in several other marketing functions, such as transportation, storage, and financing. Food processors occupy a strategic position in the food industry. Through the purchase of farm commodities, their activities are closely linked to farmers. As the source of many food product innovations and variations and as the major brand advertisers in the food industry, they are also in close contact with consumer markets. The relative importance of the various food processing sectors is shown in Table 5-1.

We will define food processing quite broadly. It may involve canning, freezing, or dehydrating farm products to make them more convenient food products. But it may also imply dissassembling of a raw farm product—for example, dividing livestock carcasses into separate meat cuts or crushing soybeans to separate the oil from

TABLE 5-1 Plant Numbers, Employment, and Value-Added of Food Manufacturing Industries, 1997.

Industry	*Number of Establishments*	*Value of Shipments ($ Mil.)*	*Production Workers (000)*	*Value-Added in Manufacturing ($ Mil.)*
Meat Products	3,077	102,103	370	23,518
Dairy Products	1,865	58,123	83	15,724
Preserved Fruits and Vegetables	2,029	51,708	169	24,506
Grain Mill Products	2,583	58,473	70	21,667
Bakery Products	3,508	32,741	136	20,712
Sugar and Confectionary Prods.	1,108	26,265	69	12,640
Fats and Oils	537	23,388	18	4,258
Beverages	2,398	68,387	72	35,210
Miscellaneous Foods	4,076	40,136	125	20,545
All Food and Kindred Products	21,233	421,265	1,113	178,783

SOURCE: *U.S. Census of Manufacturing, 1997.* U.S. Bureau of the Census.

the meal of the bean. Alternatively, food processors may combine different farm products to make something new, as when a processor combines meat, vegetables, fruits, and other items into a frozen prepared dinner. Some food manufacturers specialize in producing food ingredients for other firms to further process—for example, an egg-breaking plant that provides egg products for bakers and other firms. All of these food processing activities add value to farm products.

MARKETING MANAGEMENT IN FOOD MANUFACTURING

Food processors are quite adept at using all of the 4Ps in developing value-added products that will improve their competitive positions in the marketplace by better satisfying consumers' needs and wants. They practice market segmentation, target marketing, product differentiation, and positioning of their value-added, branded products. *Positioning* refers to the image in the consumers' mind that a firm's marketing strategy gives to its products to increase their value for customers. One firm may position its products as snack foods while another chooses to occupy the center-of-plate (main meal) position. Examples of food firms' positioning strategies include the *Pork, the Other White Meat* campaign and the *Orange Juice, It's Not Just for Breakfast* campaign.

Product Strategies

The goal of marketing management in food processing is to transform an undifferentiated, low-profit commodity into a differentiated, branded, high value-added, profitable food product. This is not magic, but neither is it an easy, inexpensive task. The marketing strategies employed by food processors are numerous. They are easily observed by reading any publication with food advertisements or by visiting the grocery store. In their product strategies, food processors attempt to incorporate all relevant aspects of the product bundle of attributes into their marketing strategies. Different foods and brands may emphasize quality, convenience, packaging, nutrition, or even price as the key marketing idea. Not every food product must—or even can—appeal to all consumers' tastes and preferences. There are mass-marketed foods that appeal to large numbers of consumers, but it is more common today for a new food to appeal to limited target markets or even very narrowly defined *niche* markets.

Branding is probably the most important product strategy of food processors. A *brand* is a name, term, symbol, or design that identifies the seller and differentiates the product from those of competitors. Branding permits the food manufacturer to certify the quality of products, transfer the goodwill of the firm to new products, and otherwise differentiate the product from competitors' offerings. A well-known and trusted brand can earn the food processor *brand loyalty* (a consumer franchise) from customers. This can be helpful in introducing new products, forestalling consumer substitutions of less expensive brands, and prolonging the product life cycle. It has been estimated that about 40 percent of U.S. food products are branded while the other 60 percent are sold as undifferentiated products.

There are other product strategies in this sector. Food processors in recent years have emphasized the development of convenience foods and stressed their "built-in maid service" aspects: ready-to-serve dinners, boil-in-the-bag foods, instant coffee, minute desserts, and brown-and-serve rolls. Processors have also led in the development of new processing techniques—dehydrating, irradiating, freeze-drying, aseptic processing—and in the use of new packaging materials such as foil, cellophane, polyethylene, and so on. The search for new food products continually spawns new industries. Sometimes they compete directly with older established industries; sometimes they complement or supplement them. The frozen food industry has grown rapidly. This in many ways competes directly with the canning industry. Freezing may eventually offer the meat packing industry a method of solving the many processing and distribution problems that stem from its very perishable product. The new processed potato products that have increased dramatically in recent years represent another example of a new processing industry.

Much of the innovation and new product development occurs in this sector of the food industry. An *innovation* is the discovery and application of a new idea. In developing their marketing strategies, these firms employ market researchers, food scientists, and advertising agencies to monitor the demand for and acceptance of new products. The success of food processors frequently hinges on scientific breakthroughs, such as freeze-dried coffee or soft margarine, minor changes in product composition or design, or even an advertising theme. Food processors introduce more than 15,000 new food products to the marketplace each year, with a failure rate of over 90 percent.

Three types of innovations have been important for food manufacturers: (1) new marketing methods and techniques—which often increase operational efficiency; (2) new products or services—which add more consumer value to products; and (3) new business organizations—such as the cooperative food processor, joint ventures between firms, or new market channels (e.g., the fast-food outlet). Frozen concentrated orange juice illustrates these forms of market innovation. This process was commercially developed in the 1940s and lowered marketing costs by concentrating the juice product. Many consumers preferred this product to fresh oranges or to single-strength juice, and a new food processing industry grew up in Florida. The single-serving, nonrefrigerated juice pack is a recent addition to this line of innovation.

New foods pass through a *product life cycle.* Early in their development they require substantial research and marketing costs. However, if the new product reaches the acceptance stage, it is frequently quite profitable for a pioneering firm. As the product moves to the mass market stage, it begins to attract imitators and loses its initial uniqueness; profits begin to wane. Price cutting, low profit margins, and widespread imitation characterize the market saturation stage. At this point, the firm hopes to have the next product innovation ready to introduce. Only a few brands and products—such as Crisco, Jello, Campbell's Tomato Soup—have escaped this product life cycle.

Pricing Strategies

Food processors may employ a number of pricing strategies. For example, one processor may use a gourmet strategy, with a high quality–high price mix, while another may use a value pricing strategy with a lower price and quality appeal. An important lesson in marketing is that not everyone wants the highest quality product and almost everyone is willing to sacrifice some quality for a lower price. Psychological pricing refers to a situation where a higher price, along with status advertising, encourages consumers to purchase products. Food manufactures also may package a product to a specific price point (e.g., $1.99) or use price discounts (cents-off, 2-for-1 sales) to attract consumers.

Distribution Strategies

Place or distribution strategies for food processors include selling through conventional foodstores, selling foods in nonfood stores, selling to the foodservice market, selling in vending machines, mail or catalog selling, home delivery, and even selling foods door-to-door by high school or scouting organizations. While most food manufacturers prefer the sales volume they get from mass outlets, they may also include more selective place strategies in their marketing mix. Again, marketing teaches that there is no one best strategy for reaching consumers, and multiple strategies are often preferred to a single approach. Food processors must also select a sales approach. Larger processors usually have their own sales offices, warehouses, and personal sales force. Smaller firms may sell their products through food brokers.

Processors may choose to market their products in many alternative ways. The three principal markets for food processors are industrial customers, foodservice firms, and consumer markets. These may involve local, regional, national, or international sales. Many food processors operate their own sales offices and wholesale operations but few are engaged in retailing directly to consumers. The large processor with a relatively full line of products will often operate its own warehouse and wholesaling system. If, on the other hand, the line of products is limited, the processor will often have the sales work done by a broker. Industries vary in the directions they have taken.

The meat packing industry, for example, has largely become its own wholesaler. It is handling a perishable commodity with wide fluctuations in volume. In order to keep such a product moving effectively, many packers consider it imperative that they have control of the distribution channel to the retailer and operate their own wholesaling establishments. Many dry goods manufacturers, on the other hand, rely on brokers to distribute their products to wholesalers and retailers.

Promotional Strategies

The promotional strategies of food processors are perhaps the most visible signs of their marketing efforts. Food manufacturers have many choices to make here in selecting the goal of the promotion (to remind, inform, or persuade); the theme or appeal (price, quality, etc.) of the promotion; the type of promotion (advertisement, sales promotion, etc.); which media (print, broadcast, direct mail, point-of-purchase, etc.) will carry the promotion; and who the promotion will be targeted to (the user, the buyer, or the influencer). Morever, the food manufacturer must fashion a combination of promotions to influence the buying decisions of both consumers and the retail distributors who will purchase and display the products. Increasingly, processors are finding it profitable to shift promotional dollars from direct consumer advertising (sometimes called pull promotion) to trade promotions (push promotion).

The major food processors differentiate their products through mass-media advertising, coupons, free samples, cents-off deals, promotional trade allowances, and point-of-purchase merchandising materials. These forms of competition have been criticized as cost and food price increasing. However, these promotional strategies can result in lower consumer prices and perhaps greater consumer satisfaction. Also, consumers can choose between the highly promoted manufactured foods and their private-label, lower-priced counterparts. Both forms of competition appear to fill a need in the marketplace.

Food processing firms are among the nation's leading advertisers, and food products are the most heavily advertised consumer products. Food products lead in expenditures for TV advertising, discount coupons, contests, and other forms of promotion. Food manufacturers' advertising expenditures average about 3 percent of their sales dollar, and in recent years these advertising costs have amounted to 1.5 percent of consumers' food-at-home expenditures. This advertising is important to the competition between food processors for the consumers' favor. For their part, consumers receive from this advertising and promotion some useful and some not-so-useful information, numerous persuasive messages that may influence their behavior, and perhaps some entertainment.

Overall, the food processing sector appears to perform rather well. Its value-adding functions can expand the markets for farm products and provide consumers with a range of food products to meet every taste and budget. Food processors provide consumers with choices from over 230,000 different packaged food products. They also contribute to a reliable flow of clean, safe, and nutritious products. Yet, the sector is not without its critics. Some argue that food processing and advertising have distorted consumer food preferences with adverse effects on the nation's health. Others suggest that innovation, product differentiation, and competitive tactics of food processors have raised food prices and profits.

THE STRUCTURE OF FOOD MANUFACTURING

Food processing is important to the U.S. economy, accounting for 1 percent of total U.S. employment and 12 percent of the value added by all manufacturing in 1998. Some 14,000 companies operated 21,000 food processing plants. Beef and pork processing is the largest food industry with soft drinks, poultry, and dairy processing next in size of industry. When measured by value-added (the difference between the sales price and costs of purchased supplies), the flavoring, breakfast cereal, bakery, and pasta food sectors are the largest.

As a result of economies of scale, mergers, and consolidations, the number of food processors and manufacturers has been declining, while the average size of plants has been increasing. Food manufacturing plants tend to have a higher sales volume than the average U.S. manufacturing plant. Also, a greater number of food firms are multiple-plant companies than is the case for all manufacturing plants.

There is considerable variation in the number and average size of food processing operations. Milk, bread, and canning plants tend to be more numerous but smaller in size. These size differences are the result of (1) economies of scale in food processing plants; (2) the perishability and cost of transporting the final product; and (3) the geographic concentration of raw farm products, which affects the ease and costs of assembling large quantities for processing at a central location.

Food processing plants tend to specialize by product lines. Flour millers are primarily engaged in milling flour; dairy processing and poultry plants are seldom used for processing other products. Sometimes related products are processed within a single plant, however. For example, several types of vegetables may be processed in the same plant, and complementary products, such as butter and cheese, may be processed jointly. Modern livestock packing plants, for the most part, specialize in either hogs, cattle, sheep, or poultry.

Mergers and acquistions of food processing companies in the past 25 years have significantly changed the structure of this industry. These often reduce the numbers of competing firms and increase their market dominance. Mergers may be motivated by the firm's desire to grow or expand into new markets, to diversify, to increase buying or selling market power, or to improve financial performance. Divestitures, or sell-offs of firms or subsidiaries, have also been common in food processing. Among the major food mergers of the 1980–1999 period were Philip Morris Company's purchases of Kraft, Inc. and General Foods Corporation; Grand Metropolitan PLC's purchase of Pillsbury Company; Kohlberg, Kravis, Roberts & Company's purchase of RJR Nabisco and Beatrice Companies; Sandoz AG's purchase of Gerber Products Company; and Nestle S.A.'s purchase of Carnation Company. In 1998, Cargill and Continental Grain, the largest and third-largest exporters of U.S. grains, agreed to merge. In 2000, Unilever, a British-Dutch company, acquired Best Foods.

Many of the larger food processors are highly diversified in their operations. Often these are multiplant, multiproduct, multimarket, and multinational companies. They purchase many different farm products and nonfarm commodities and may produce and sell hundreds of food products and brands. Many of these companies have diversified into nonfood industries, including transportation, publishing, clothing, toys, and other consumer goods. Some of the food manufacturers that diversified into nonfood lines in the 1970s reversed this diversification and have returned to their primary food businesses. Pepsico first moved into then out of the fast-food restaurant business.

Recent years have seen a globalization of the world food industry. America is a major exporter and importer of processed foods. The American food processing industry has had a positive trade balance in recent years. Many U.S. food processing firms also participate in world food markets through the ownership of foreign subsidiaries or through franchising, joint ventures, or licenses with foreign food companies. Building a plant overseas instead of exporting products from America might be seen as a way to reduce transportation costs. It also may allow better target marketing to local tastes and preferences of consumers. Foreign investment in the U.S. food processing industry has also grown rapidly in recent years.

The food processing industry today can be divided into two sectors: (1) a dominant core that consists of a few very large firms producing well-known brands and accounting for a significant share of industry sales, and (2) a competitive fringe that consists of a large number of smaller firms producing less well-known brands accounting for a small share of industry sales. The nature of these sectors varies for different products, but the dominant core firms are well known to most people. Some of the larger food manufacturing companies are listed in Table 5-2.

Selling competition between the dominant core and the competitive fringe of food processors centers on branding, advertising, and prices. The larger processors tend to emphasize heavily advertised brands, cents-off coupons, and new product development in their competitive strategies. Smaller processors often specialize in packing products under wholesaler- or retailer-controlled labels. These "private" labels are not extensively advertised and generally retail at lower prices than the nationally advertised brands. There may or may not be quality differences between these two products. This competition between advertised and nonadvertised brands

TABLE 5-2 The World's Largest Food Processing Companies, 1997

Sales Rank	*Company (Headquarters)*	*1997 Food and Beverage Sales ($ Bil)*
1	Nestle S.A. (Switzerland)	$38.8
2	Phillip Morris/Kraft Foods (New York, N.Y.)	33.4
3	Unilever (U.K., Netherlands)	26.7
4	ConAgra (Omaha, Nebraska)	24.8
5	Pepsi Co. (Purchase, New York)	19.1
6	Cargill (Minneapolis, Minnesota)	18.7
7	Coca Cola (Atlanta, Georgia)	18.0
8	Danone S.A. (France)	14.2
9	Archer Daniels Midland (Peoria, Illinois)	13.3
10	Mars, Inc. (New York)	13.0
11	Grand Metropolitan Plc (United Kingdom)	12.7
12	IBP (Iowa)	12.5
13	Kirin Brewery Co., Ltd. (Japan)	11.6
14	Anheuser-Busch Co., Inc. (St. Louis, Missouri)	10.3
15	CPC International, Inc.	9.8

SOURCE: U.S. Department of Agriculture, Economic Research Service.

is referred to as the *battle of the brands.* Generic, "no-frill" brands are also part of this competitive environment.

The increasing size and dominance of the large food manufacturers is a concern to many because it may influence competition of these firms in both their buying and selling markets. The 50 largest food manufacturers accounted for 35 percent of food manufacturing value added in 1967, 40 percent in 1977, 47 percent in 1987, and over 50 percent in 1997. The dominance of firms in an industry is measured by *market concentration*—the share of total industry sales accounted for by the largest companies. In general, concentration is higher in food processing than in other manufacturing sectors, and food processing concentration has been steadily rising over the years. However, while high market concentration levels exist in many food processing industries, low levels of concentration are also present. In addition, as shown in Figure 5-1, there are food processing sectors in which market concentration is increasing and other sectors in which it is decreasing.

The market concentration levels shown in Figure 5-1 are for the sales of food processors in national markets. These levels of concentration would be lower if measured for international sales of these companies, but higher when applied to the

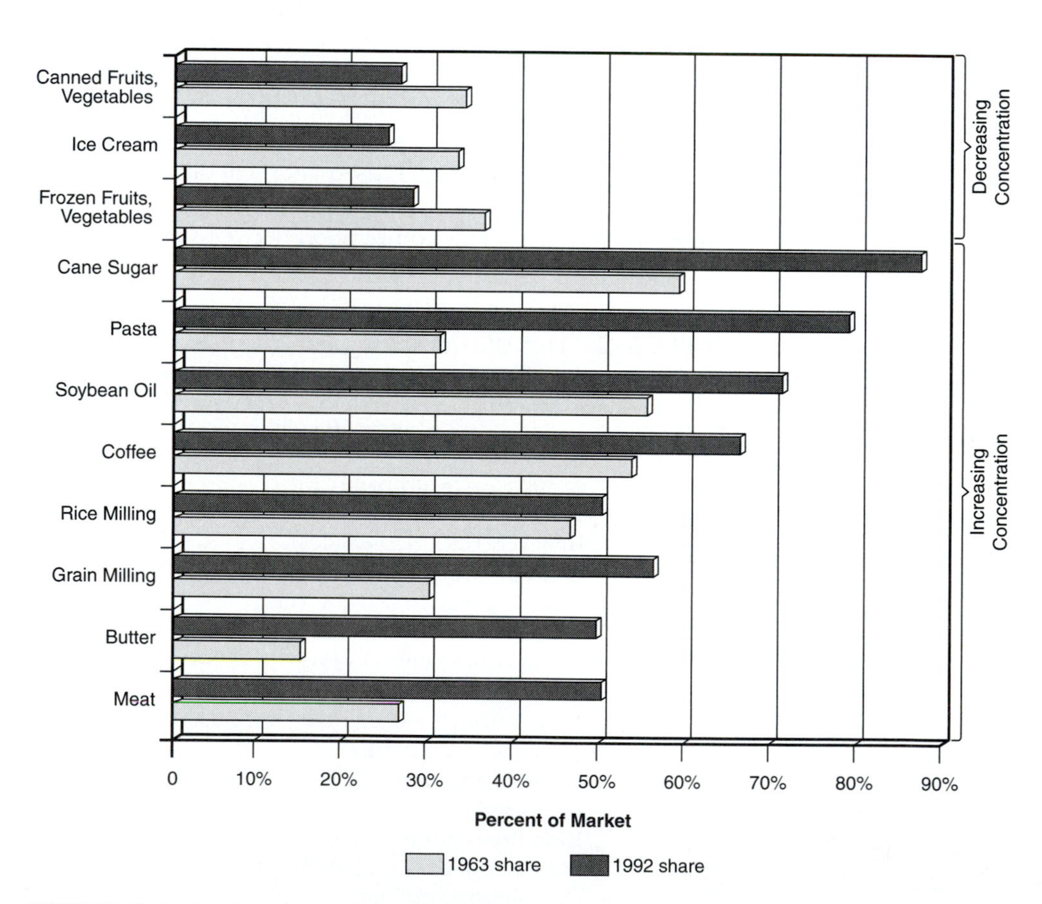

FIGURE 5–1 Market share of the four largest firms in selected food industries, 1963-1992. (U.S. Bureau of the Census.)

MINI CASE 5: MARKETING A NEW PRODUCT: SOYMILK

George Martin, general manager of SoyaFoods, was excited about the prospects for a product he has been selling for the past 10 years. Soymilk is a healthy beverage made from soya beans rather than from dairy milk. The nutritional benefits of soya foods have long been known to health conscious consumers, but the product has had only small sales and has been largely sold in health food stores. The U.S. Food and Drug Administration had just issued a report indicating that soyfoods have value in preventing heart disease. George has been getting phone calls from grocers all day long asking him to supply their stores with soymilk to meet the sudden demand. Most of the distributors have asked him to put their retailer label on the product, instead of the SoyaFoods brand. He is not sure he wants to do that because it means lower profits for his soymilk. However, SoyaFoods doesn't have the brand recognition of most food companies, and they lack the funds for a massive advertising campaign. He is also certain that many major food companies will soon enter the soymilk market with their own brands and other processing companies will supply the store-branded product if he doesn't.

1. Give your 4-P recommendations (price, product, promotion, and place) for how Mr. Martin should market his soymilk.

2. Should SoyaFoods put their own brand on the product or sell it under a grocer's brand label? Why or why not?

local markets in which many processors purchase their raw farm product supplies. Changes in the market share concentration of the dominant food processors results from new processing technologies, changes in the location of food manufacturing, the success of fringe processors in competing with core firms, mergers, and other dynamic market conditions.

THE LOCATION OF FOOD PROCESSING

The historical development of the United States has left its mark on the various food industries. As the country has expanded and agricultural production has moved into new geographic areas, many of the processing industries have also moved. This shift resulted in unused capacities and high costs in the firms that were left operating in old areas.

The grain milling industry is a good example of an industry that has undergone considerable movement. Milling first developed in the Atlantic coastal states. There, it was near both wheat supplies and the water power to operate the mills. Then, as new wheat areas farther west developed, Cincinnati and then St. Louis became important milling centers. With the growth of wheat production in the Northwest during the first years of the twentieth century, Minneapolis developed into the nation's leading milling center. Then, as the wheatlands of the Southwest were opened, Kansas City rose in importance. Finally, a change in the freight-rate structure in 1920, plus the growing opportunity to mill Canadian wheat, led to the growth of the Buffalo, New York, mills. As each of the gradual changes took place, much mill capacity was left in the declining areas.

The East Coast meat packing industry is to some degree the vestigial remains of the earlier years before the development of the livestock industry in the Midwest. At first much livestock was shipped to the East alive, for slaughter. The end meat product was so perishable that processing had to be done near the consumption areas. However, with the development of refrigerator cars, Chicago and other Midwestern cities became important packing centers. The packing industry in the past two decades has again followed production into the western part of the Cornbelt.

New refrigeration, transportation, and other technological developments will continue to change existing patterns. For example, new developments in whole milk transportation are tearing down the past barriers that restricted milk collection and distribution to relatively limited areas. Milk plants are consolidating and getting larger. At the same time, the area in which they can effectively sell their products is increasing. Therefore, the number of effective competitors in a given town may not be decreasing even though the independent dairy in that town has been forced out of business.

Overall, there has been a movement of food manufacturing from the northeastern and north central states to the southern and western states. This trend resulted from population migration, changes in the geographic location of agriculture, and changes in the way foods are marketed. Many other factors affect plant location decisions, including taxes, labor supply, community facilities, and personal preferences of managers. History, too, has left its mark on the location of food processing. The breakfast cereal industry is centered in Michigan because both W. K. Kellogg and C. W. Post were patients at the same Michigan health sanatorium at the turn of the century.

Should a food processor locate a plant in the farm production areas, near sources of raw farm products, or should the plant be located in the cities, near to consumers? The answer depends on the relative costs of transporting raw farm products to the plant and of shipping the processed product to consumers. Generally, if the farm product is quite bulky in comparison with the final product, and the unit value of the final product is higher than that of the farm product, the plant will probably be located near the source of farm products. This is the case for flour milling, meat packing, and butter and cheese manufacturing. On the other hand, if the final product is more perishable than the farm product, the plant will tend to be located nearer to consumers, as with baking, and milk and ice cream plants. The food processor will attempt to minimize the combined costs of raw product and finished product transportation. If raw products are cheaper to transport than final products, the plant location decision will be consumer-oriented. If raw products are more expensive to transport than final products, the location decision will be production-oriented. A change in either the cost of shipping farm products or final products might result in a shift to a lower-cost plant site.

THE LAW OF MARKET AREAS

Many food processors sell in only local or regional markets and others distribute foods nationwide. Usually, the national processors are multiplant companies with processing and distribution facilities located throughout the country. This raises the question of how processors decide the size of market territory to service from each

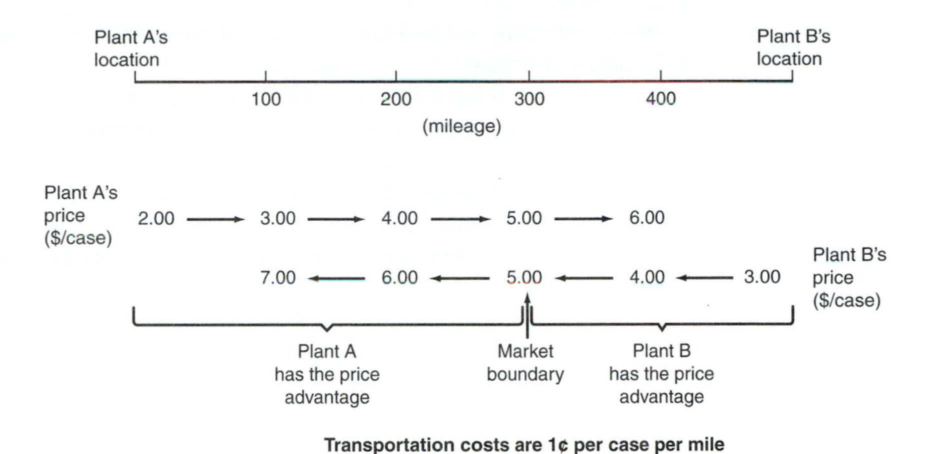

FIGURE 5–2 The law of market areas.

plant and distribution center. The law of market areas (LOMA) can help answer this question.

The *law of market areas* states that the territorial boundary between two or more markets or plants is a locus of points such that the final selling prices, including transportation costs, are equal for sellers in each market. The LOMA is illustrated in Figure 5-2. Plant A has a factory price of $2.00/case and plant B has a higher cost and factory price of $3.00/case. The two plants will divide the 500-mile market area between them in such a way that plant A will service a 300-mile range and plant B will service a 200-mile range. At this point or market boundary, each plant is charging the same price of $5.00. At all other points, either plant A or plant B has a competitive advantage. Note that a change in plant costs or prices will shift the market boundary.

FOOD SCIENCE AND MARKETING

Recent advances in food science and technology are impacting the entire food marketing system, from farmers to food processors to consumers. Whole new industries, products, and markets have been created by food science developments of recent years, such as frozen concentrated fruit juices, aseptic shelf-stable fruit juices, corn sweeteners, frozen baked goods, and dehydrated soups, to name a few. Food processors have developed fat substitutes and low calorie and low cholesterol foods to satisfy consumer demands. Food scientists are now able to produce food textures, colors, smells, flavors, and nutrients in almost any combination desired by processors.

By more closely controlling the characterisitics of foods, these technological developments allow plant and animal products to be tailored to specific processor and consumer wants. This creates incentives for processors to better coordinate the production activities of farmers through contracts and other arrangements. These "engineered" or "restructured" foods also provide processors with new opportunities to differentiate and target market their products by making entirely new product bundles of attributes for consumers.

Biotechnology also promises to change the nature of food production and processing. For example, a genetically altered tomato that has longer keeping quality is now being marketed. Biotechnology research has shown ways to produce more milk per cow, less fat per pig, and grain with more protein. Since biotechnology can alter such properties of foods as taste, organoleptic properties, and keeping quality, shall we also label biotechnology firms as food processors? What about firms that develop alternatives to traditional foods, such as fat, egg, and sugar substitutes?

The verdict is still out on the value and acceptance of these food science developments by consumers. Although they permit more carefully targeted food products, may lower costs of some foods, and contribute to the product differentiation objectives of food processors, the consumer will be the final judge of these technologies. While many "new foods" will succeed, it seems likely that Mother Nature's foods will continue to dominate our diets.

MANAGEMENT PROBLEMS OF FOOD PROCESSORS

Food processors experience problems and face challenges in two major areas: processing operations and procurement.

Processing Operations

Food processing involves significant investments in plant and equipment. In order to operate efficiently, these facilities should be used to full capacity year-round, every year. This is difficult to achieve when there are wide variations in farm product supplies from year to year and within seasons. Because of the short harvest season and perishability of many farm products, food processing plants may operate at above-capacity rates for a few months of the year and at below-capacity rates for the rest of the year. For example, livestock slaughter and processing plants typically operate at close to 90 percent of capacity utilization year-round while fruit and vegetable canning plants may average 60 percent capacity for the year. These variations can significantly influence food processing costs. Some progress has been made in improving the capacity utilization of food processing plants. Expanded storage facilities, contracts that coordinate farm supplies with processor needs, storage of the product at harvest for later processing, and extension of the processing season by manufacturing nonseasonal food items have contributed to the solution to this problem.

Processors face a dilemma when deciding on the number and size of plants to build. The operating efficiency gains of large, central processing facilities can be nullified by the costs of assembling large quantities of raw farm products and transporting final products long distances to consumers from centralized operations. Replacing a single, large plant with several smaller ones reduces some assembly and transport costs but may require sacrificing the operational efficiencies of large-scale, centralized plants.

A major thrust of food processors in recent years has been the replacement of human labor with machines and equipment. The continuous increase in labor costs has given added incentives for mechanizing and automating food processing operations. Batch operations are giving way to continuous assembly-line techniques. New machines and processes are continually being experimented with. Technical changes in recent years have been so widespread and rapid in some fields that it often is more economical to shut down an old plant and build a new one than to remodel the old one.

Investment in new plants, equipment, product development, and technologies is important to the progressiveness and productivity of an industry. Food processors spend a relatively small share of their budget for research and development. Yet, labor productivity in food manufacturing has grown more rapidly than in nonfood manufacturing industries. Many of the patents and technologies used by food processors have been adopted from other industries and from abroad.

Buying Operations

Food processors are major buyers of farm products. The value of farm commodities purchased represents 26 percent of the food processor's sales dollar. Their procurement decisions must consider how to handle many of the marketing functions, including storage, transportation, risk bearing, financing, and market intelligence. Because of variations in output and prices of farm products, the buying and pricing decisions of processors affect the division of risk shared between food producers and processors. Food processors have developed elaborate market information, financing, and grower assistance programs to better manage their purchasing function.

The buying power of food processors is generally greater in procurement markets than in their sales markets. This is because most food processors secure their raw product supplies from an area near their plants while they sell in larger regional and national markets. Not all processors will compete for the same farm product supplies, and in many local markets there may be only a few buyers. This lack of competition may result in marketing and price problems for farmers.

Food processors utilize a wide variety of ways to procure their raw product materials from farmers. Most processors do not produce their own raw product supplies but purchase these from farmers. Some purchase directly from producers at the farm, at the plant, or through country buying offices. Other food manufacturers use the services of such independent marketing agencies as livestock commission agents, grain elevators, or fruit and vegetable brokers to secure their supplies. For some commodities, the processor negotiates with farm cooperatives or bargaining associations for supplies and prices.

Processor contracting for farm product supplies is an increasing trend for many products. These contracts may contain a wide range of specifications, including the time and method of price determination, the delivery terms, the quality required, and even production practices. These processor contracts reallocate risk and financing between the producers and buyers of farm products. Most analysts agree that market contracts are quite different than the open market sales that they replaced, but there is considerable disagreement about their value for farmers.

Market orientation requires that processors more carefully coordinate their procurement activities with their processing and selling operations. This may involve actual ownership of the farm unit or contracting with producers for raw product supplies. The improved scheduling and coordination of such practices can reduce processors' supply and price risks, contribute to more successful marketing strategies, and improve profits.

SUMMARY

Food processors specialize in adding form utility to farm products. Their marketing activities include new product development, product alteration, packaging, labeling, promotion, and branding. Competitive rivalry between food processors for the customers' favor centers on products, prices, and promotional programs. The battle of the brands is being waged between nationally advertised and lesser-known, unadvertised food product labels. The food processing industry is composed of large and small firms, corporate and individually owned companies, single- and multiplant organizations, as well as specialized and diversified firms. Market concentration is increasing in some lines of food processing but decreasing in others. Food processors may locate in either production or consumption areas, depending on transportation costs and product perishability.

KEY TERMS AND CONCEPTS

battle of the brands
brand
competitive fringe
consumer franchise
dominant core
further processors
innovation
law of market areas
location decision
market concentration
niche products
plant capacity
positioning
private label
product innovation
product life cycle
quality control

DISCUSSION QUESTIONS

1. What would be the best marketing strategy for a small food processor just starting operations?

2. Give some reasons why many food companies are purchasing nonfood firms and nonfood firms are buying food processing operations.

3. Food engineers have suggested that the size, number, and location of food processing plants will depend on raw product transport costs, final transport costs, and the economics of plant size. How would these three factors affect those decisions?

4. Critics of the food industry sometimes allege that food processors manipulate consumer tastes and preferences through their advertising and merchandising campaigns. What does this mean? Do you agree or disagree?

5. Business theorists conclude that the law of market areas would result in hexagonal, honeycomb market territories when there are several competing firms. Why is it that actual selling territories do not follow this pattern?

6. Through cooperatives, such as the Sunkist Corporation or the Welch Grape Juice Company, farmers have become owners and operators of food processing facilities. Why have they done this? What are the pros and cons?

7. It has been suggested that food processors are too big and that they should be broken up into smaller firms. Do you agree? Why or why not?

SELECTED REFERENCES

Connor, J. M., & W. A. Schiek. (1997). *Food Processing: An Industrial Powerhouse in Transition 2nd ed.* New York: John Wiley & Sons.

Food Market Structures: *Food Processing Briefing Room* (January 2001). Economic Research Service, U.S. Department of Agriculture. www.ers.usda.gov/briefing/foodmarketstructures/processing.htm

CHAPTER 6

FOOD WHOLESALING AND RETAILING

The American supermarket has represented a revoluntionary step in the history of commerce, making a full range of goods available to a mass audience.
Ryan Mathews, Editor *Progressive Grocer*

OBJECTIVES

After reading this chapter, you will be able to:

1. Explain what functions food wholesalers and retailers perform in today's food industry.
2. Better understand the major economic and market forces shaping the food distribution sector and the structural changes affecting this sector.
3. Appreciate the key role food retailing plays in connecting producers to consumers and influencing producer and consumers' food prices.
4. Explain how food retailers attempt to gain competitive advantage and compete for the consumer's favor.

Food wholesaling and retailing represent the final stages of the food marketing channel. This is often called the "downstream" portion of the food industry, in contrast to farmers who are "upstream" in the system. Because of their proximity and high visibility to consumers, food retailers are at the front of the food marketing battlefield. They often are the first to feel a change in consumer buying patterns, a shift in food preferences, or the ire of dissatisfied customers. Through their purchasing and marketing strategies, they may have the power to influence consumer food demands and sales throughout the entire food system.

In an earlier period, it was felt that these market agents simply passed along processed or fresh foods to the consumer, with little change or added value. It is now clear, however, that the wholesaling and retailing marketing functions and operations are quite complex, and that these firms are not simply pass-through agents. The efficiency with which the functions are performed at these levels and the pricing-merchandising strategies of food retailers and wholesalers substantially affect all other players in the food industry.

PRINCIPAL TRENDS IN FOOD WHOLESALING AND RETAILING

The task of food wholesalers and retailers is to provide consumers with the right foods, at the right time, and at the right place *at a profit.* Think about the many places where you purchase food today. Food retailers include, in addition to the traditional supermarket, bakery shops, department stores, drugstores and discounters, gasoline filling stations, stop-and-shop convenience stores, the school cafeteria, fast-food shops, full-service restaurants, vending machines, and sport and entertainment facilities. Food is never far away.

Because of their close contact with consumers, food wholesaling and retailing firms are prime examples of how marketing agencies change in response to new demands and competitive conditions. These market agencies experienced continual change and adjustment in the twentieth century. Contrast the turn-of-the-century "ma and pa" corner grocery store with the modern supermarket. Gone for the most part are neighborhood markets, cracker barrels and other bulk-sales arrangements, personal service, credit, and home delivery. These were replaced with large supermarkets that handled a much wider variety of foods and nonfood items on a self-service, cash-and-carry basis. This trend reflected changing consumer preferences, competitive pressures, and changes in the relationships of food processors, wholesalers, and retailers.

There are other changing signs of food retailing. Consider the shopper who notices that many food products can now be purchased at K-Mart, WalMart, or a wholesale club. Consumers are purchasing prepared foods for take-out in convenience stores and supermarket delicatessens. Another busy family places their food order by phone and has their food delivered weekly. A group of students organizes an impromptu party and orders pizza to be delivered in 30 minutes. A food court in a shopping mall is shared by five different restaurants and a supermarket. A wedding dinner is served by a food catering firm. These alternative forms of "food retailing" have one thing in common: They find consumer wants and fill them.

The market agencies we are concerned with in this chapter are depicted in Figure 6–1. The wholesale assembly markets are generally located in food production areas. The job of these shipping point markets is to assemble large lots of product from diverse producers and prepare the products for shipment on to the processors or wholesalers located in the cities. Some farmers perform their own assembly market functions while others use the services of specialized firms.

A complex network of relationships between processors, wholesalers, and retailers has evolved over time. For example, most food retailers are affiliated in some way with food wholesalers, and many food retail firms also own processing facilities. By the same token, many food processors operate their own wholesaling operations. Through sales offices and branch warehouses, they deal directly with retail buyers.

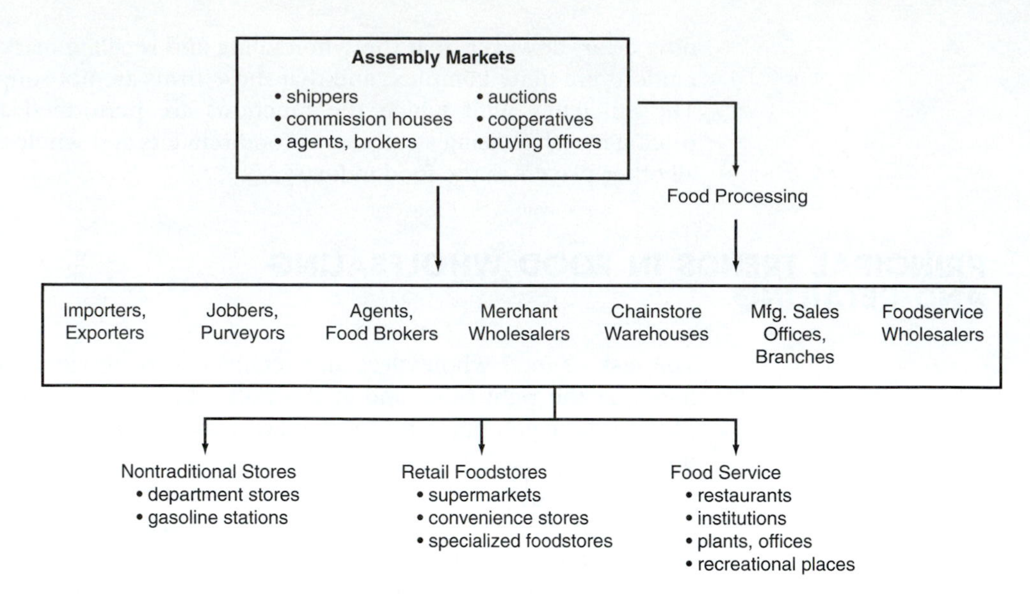

FIGURE 6–1 Structure of the wholesale-retail food distribution network.

Some food processing firms also compete indirectly with retail grocery stores through the ownership of restaurants. Thus, food processors, wholesalers, and retailers are expanding their activities beyond their traditional ranges of marketing functions.

FOOD WHOLESALING

The largest U.S. food wholesaling firms are shown in Table 6-1. Food wholesaling accounts for about 10 percent of the consumer's food bill. These firms specialize in adding spatial and possession utility to food products. They do little food manufacturing and generally do not sell directly to consumers. Their market success is largely determined by how efficiently they provide products and services to their suppliers and retail customers.

Food wholesaling is often a hidden but still critical step in food marketing. The job of the food wholesaler is to efficiently assemble various products in reasonable quantities from the shipping point firms and processors and sell them in smaller quantities to retailers. This is a valuable service. Food retailers stock relatively small amounts of literally thousands of very different items. Retailers could not possibly search out and deal with all of the producer and processor sources of their products. On the other hand, processors cannot, in many circumstances, profitably service the small-unit needs of retailers.

Types of Wholesalers

Three major types of food wholesaling firms are shown in Figure 6-2. *Merchant wholesalers* buy, sell, and store grocery products and perform numerous other marketing functions. About 75 percent of food wholesalers are classified as merchant

TABLE 6–1 The Largest U.S. Food Wholesaling Firms, 1999

Wholesale Firm and Headquarters	*1998 Sales (Million Dollars)*
SuperValue, Inc. (Minneapolis, MN.)	$11,845
Fleming Cos., Inc. (Oklahoma City, OK.)	11,476
McLane Co., Inc. (Temple, TX.)	10,350
C&S Wholesale Grocers, Inc. (Brattleboro, VT.)	5,100
Wakefern Food Corp. (Elizabeth, N.J.)	4,860
Nash-Finch Company (Minneapolis, MN.)	3,370
Topko Associates (Skokie, IL.)	3,330
Associated Wholesale Grocer's Inc. (Kansas City, KS.)	3,280
Richfood, Inc. (Richmond, VA.)[1]	3,010
Core-Mark International (San Francisco, CA.)	2,451

[1]Merger pending with SuperValue, June 1999.

SOURCE: *Chain Store Guide Information Services,* Tampa, Florida. Reprinted with permission.

wholesalers and they account for 69 percent of wholesale food sales. These firms may be either full-service or limited-function wholesalers. Full-service merchant wholesalers provide a wide array of services to their retail clients, such as inventory control, pricing, financial management and analysis, merchandising and advertising support, private brand label programs, credit and financing of new stores, and store site selection. These services have been especially important to small grocers in their competitive battle with the large chainstore organizations.

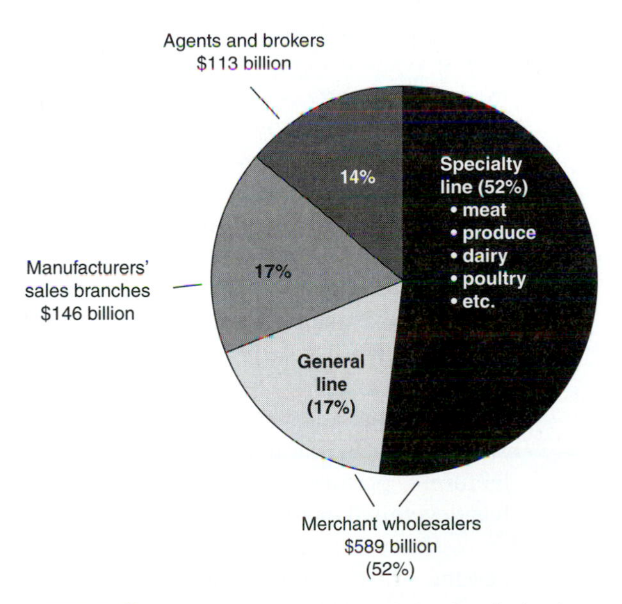

FIGURE 6–2 Types and sales of food wholesalers, 1997. (1997 Economic Census, U.S. Census Bureau.)

Merchant wholesalers also vary in the breadth of their product lines. General-line merchant wholesalers handle a wide variety of food and nonfood grocery products. Specialty wholesale houses carry a more limited product assortment, such as dairy or meat products. Limited-line food wholesalers are more numerous but smaller than general-line firms. The number of specialty-line food wholesalers has declined in recent years as these firms merged their operations and general-line wholesalers broadened their product lines.

General-line merchant wholesalers may be further classified according to their affiliation with food retailers. Historically, the independent wholesaler occupied a position of considerable influence in the food marketing channel. Small, independent retailers depended on these firms for help in stocking their stores and often for credit. Processors also depended on them to inventory and sell their goods to retailers.

The position of the independent food wholesaler began to decline with the advent of the food retailing chain. Very early in its development, the chain organization bypassed the independent wholesaler, set up warehousing facilities of its own, and began to buy directly from shipping points and processors. All of the large chainstores today operate their own distribution centers that perform the wholesale marketing functions for the parent chain. As a measure of self-preservation, independent wholesalers formed closer arrangements with independent retailers who also were fighting a battle with the chainstores. The wholesaler-retailer affiliations are known as voluntary or cooperative groups, and they have become a very important part of the retailing picture. Sales of retailer-affiliated wholesalers have grown much faster than the sales of independent wholesalers. The growth of affiliated retail-wholesale operations also put competitive pressures on the independent and specialty-line wholesalers. Many of these wholesalers responded by concentrating their sales efforts on the growing restaurant and institutional foodservice operations. Franchise restaurants may be supplied by either independent, contract, or in-house wholesalers.

Two new types of food merchant wholesaling organizations have emerged in recent years. Foodservice distributors do not sell to foodstores but rather specialize in supplying products and services to restaurants, fast-food outlets, airlines, and foodservice institutions. A second trend has been the growth of wholesale buying clubs that developed in the 1970s and grew rapidly in the 1980s and 1990s. These "wholesale" clubs are retailers today, but they originally sold in large lots to other businesses. They now require customers to be members, sell food products in multipacks or large containers, have a discount price appeal, and generally offer fewer services than supermarkets. Initially offering only packaged foods, these wholesale clubs have expanded their product offerings into fresh foods in their search for higher sales.

The entry of nontraditional department stores and discount retailers into food marketing has added these firms to the food wholesaling sector. Large retailers like WalMart, K-Mart, and Target often purchase their food supplies directly from manufacturers, performing both the wholesale and retail marketing functions. WalMart has developed its own private label foods.

Manufacturers' sales branches and sales offices are extensions of food processing firms' marketing activities to the wholesale level. Owned and operated by food manufacturers these perform the storage, selling, transportation, and intelligence marketing functions.

Wholesale agents and brokers perform a wide variety of sales-related marketing activities for their clients. They usually do not take title to products and are paid a commission for their services. Food brokers may handle a number of noncompetitive lines and are increasingly providing more merchandising services for their clients. While smaller food manufacturers are the principal users of brokers and agents, even large food manufacturers with their own sales forces will sometimes market through food brokers. As the increasing size of food processors permits more of them to have their own sales force and offices, the share of wholesale food sales handled by manufacturers' sales branches and offices has increased while the market share of agents and brokers has decreased.

Food Wholesaling Trends

The number of food wholesalers has been stable in recent times while their average size has grown significantly. This trend toward larger food wholesalers emphasizes the importance of volume for efficient wholesaling. In addition, there were many food wholesaling mergers in the 1980s that increased the sizes of wholesalers. Some very large regional wholesalers were created, such as Supervalu, which supplies more than 2000 retail stores in 26 states from 14 distribution centers.

These developments have increased concentration in both national and local wholesale markets. The 8 largest general-line merchant wholesalers increased their share of wholesale food sales from 22 percent to 44 percent between 1977 and 1992. Market concentration is also growing in foodservice wholesaling.

The profit levels of food wholesalers have generally exceeded those of other food marketing firms in recent years. Computerization, mechanization, electronic ordering and billing, and stable energy and labor costs have contributed to rising productivity, lower costs, and generally satisfactory profits in food wholesaling. The addition of nonfood lines and increased ownership of retail food stores have also contributed to food wholesalers' profits in recent years.

Like the rest of the food industry, wholesaling is experiencing a globalization of trade. There are Dutch, Canadian, German, and Japanese food wholesalers operating in America. U.S. companies, in turn, have made major investments in the Canadian, Mexican, and Latin American food wholesale markets.

The competitive positions of independent food wholesalers and brokers in the food marketing system have been threatened in recent years. Retail foodstore chains typically perform their own wholesaling operations and have less need for the services of wholesalers. At the same time, food brokers have lost market share as food processors have made more sales directly to chainstores, warehouse stores, and wholesale clubs.

FOOD RETAILING

Food retailing is the largest retail sector in the U.S. economy and the most expensive segment of the food marketing system. Consumer food expenditures account for about one-fourth of total U.S. retail sales and 11 percent of consumers' disposable income. Food retailing firms employ more than 80 percent of all workers in the food marketing system.

The retailer is located at the end of the food marketing chain, directly servicing the final consumer. Retailers perform many marketing functions. They must properly buy and merchandise literally thousands of different items. Modern supermarkets offer more than 18,000 items, and within a year several thousand new items are offered and many old items withdrawn. Some of these items move rapidly in large volume, whereas others have a very slow turnover. The store manager must continually monitor inventory so that all desirable items are in available supply. Some items are highly perishable and require elaborate storage equipment and continual sorting and discarding of unsalable material. Display space must be properly allocated. The retailer carries on the wholesaler's breakbulk job. For example, the wholesaler may purchase several bushels of apples or a carload lot of breakfast food. These items

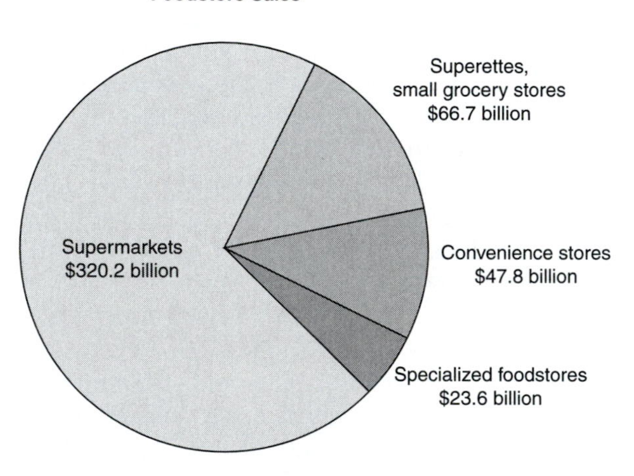

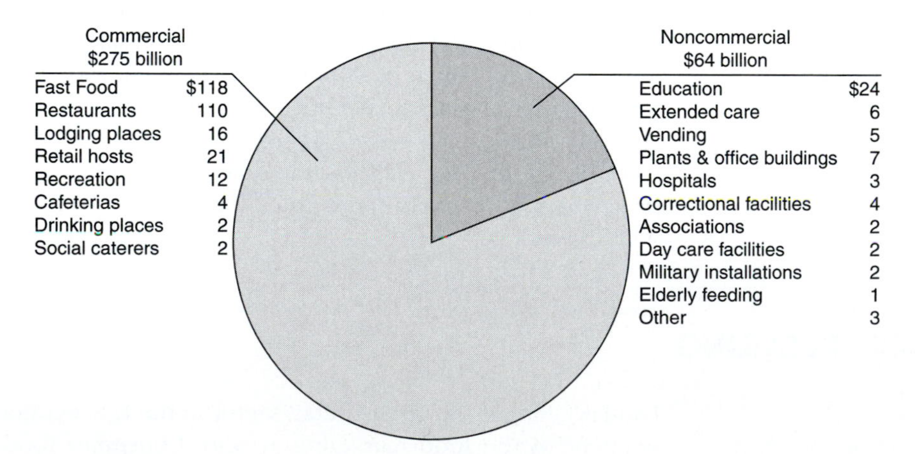

FIGURE 6–3 Retail food sales, by sectors, 1999. (U.S. Department of Agriculture. U.S. Bureau of the Census.)

then are sold in smaller lots to retailers, who then must package the apples for sale in 5-pound lots and display the individual boxes of cereal. Through advertising and point-of-purchase material, food retailers furnish information to customers about the prices and qualities of goods that are available. Finally, the retailer may carry the groceries to the shopper's car or perhaps even deliver them to the home.

There are two, quite distinct and often competitive, food retailing sectors, as shown in Figure 6-3. In 1999, consumers spent 53 percent of their food dollar in foodstores and 47 percent in foodservice establishments. The dividing line between these two sectors has grown less distinct as restaurants have added carryout foods to their operations and grocery stores have provided in-store delicatessens and eating opportunities for their customers.

The largest U.S. food retailers are shown in Table 6-2. Foodstores generally sell food for off-premise, home consumption. Grocery stores dominate foodstore sales, and supermarkets (variously defined as stores with $2 to 3 million annual sales and 10 to 40,000 square feet of space) accounted for 72 percent of total grocery store sales in 1996. This percentage is not all food as perhaps 20 percent of grocery store sales are for nonfood items. Convenience stores, specialty foodstores, such as bakeries and meat markets, and smaller grocery stores account for the other 28 percent of foodstore sales. A growing but undetermined amount of food is also purchased from nontraditional foodstores, such as department stores, drugstores, and gasoline filling stations.

Foodservice firms catering to the away-from-home food market account for the remainder of retail food sales. These include restaurants, cafeterias, institutional feeding operations (hospitals, schools, prisons, etc.), lodging and recreational places, and fast-food stores. In recent years foodservice sales have grown twice as fast as foodstore sales, and fast-food firms have experienced the most dramatic growth.

TABLE 6-2 The Largest U.S. Food Retailers, 1999

Retail Firm and Headquarters	*Number of Foodstores*	*1999 Sales (Million Dollars)*
The Kroger Co. (Cincinnati, OH.)	3,062	$45,481
Albertson's, Inc. (Boise, ID.)	1,647	36,772
Safeway, Inc. (Peasanton, CA.)	1,493	28,860
Ahold USA, Inc. (Atlanta, GA.)	1,033	21,300
Costco Companies, Inc. (Issaquah, WA.)	302	17,572
Wal-Mart Supercenters (Bentonville, AR.)	564	12,800
Winn-Dixie Stores, Inc. (Jacksonville, FL.)	1,180	13,618
Publix Super Markets, Inc. (Lakeland, FL.)	586	12,100
Delhaize America (Salisbury, NC.)	1,276	10,879
Great Atlantic & Pacific Tea Co. (Montvale, NJ.)	796	10,179
Sam's Club (Bentonville, AR.)	456	9,152
H.E. Butt Grocery Co. (San Antonio, TX.)	267	7,500
Supervalue, Inc. (Eden Prairie, MN.)	473	7,392
Meijer, Inc. (Grand Rapids, MI.)	128	4,605
Shaw's Supermarkets, Inc. (East Bridgewater, MA.)	169	3,700

SOURCE: *Directory of Supermarket, Grocery & Convenience Store Chains '00,* Business Guides, Inc., Tampa Florida. Reprinted with permission.

THE CHANGING NATURE OF FOODSTORES

The foodstore sector of the food marketing system has experienced considerable change over the years and is still evolving today. The rise of the chainstore, the development of supermarkets, the introduction of food discounters and convenience stores, the continual growth in variety of food products, and the battle between national and private food labels have all left their mark on food retailing. These developments have also affected the organization and competitive behavior of food retailing and have required continual adaptation to consumer demands and competitive pressures.

Table 6-3 traces the changing number and sizes of retail foodstores in the twentieth century. Between 1960 and 1997 the number of foodstores fell by 32 percent while average sales per store rose from $200,000 per year to $2.5 million. The number of foodstores increased during the 1980s and 1990s with the growth of smaller, neighborhood convenience stores. As a result of new product innovation and brand competition, the number of items stocked by food stores increased from 5,900 in 1960 to more than 18,000 in 1997.

How competitive is food retailing? The trend in retail foodstore market concentration is shown in Figure 6-4. The market share of the nation's four largest food retailers declined from a post-World War II high of 22 percent in 1958 to 16 percent in 1995. Much of this decline was attributed to a reduction in the A & P Company's market share from 11 percent in 1958 to 3 percent in 1980—evidence that even large food chains can experience market difficulties and competitive problems. Food retaining market concentration jumped in 1998 with the merger of the second- and fourth-largest grocers and the acquisition of the sixth largest firm by the largest U.S. food chainstore.

Market shares in local areas are frequently higher than the national figures implied by Figure 6-4. None of the national chainstores operates in every city, and the large chainstores often do not meet head-to-head in every market. Foodstore concentration is often higher in smaller cities. Market concentration in foodservice is generally lower than for foodstores.

TABLE 6-3 Number and Sales of Retail Foodstores, 1920-1997

Year	*Number of Foodstores (000)*	*Retail Foodstore Sales ($ Bil.)*	*Annual Average Sales per Foodstore*	*Number of Foods Items Carried*
1920	375	11.5	$30,667	700
1940	446	9.0	20,179	1,800
1960	260	51.7	198,846	5,900
1980	167	220.8	1,321,370	13,000
1992	180	369.2	2,043,165	16,000
1997	177	430.1	2,429,943	18,000

SOURCE: *U.S. Census of Business: Retail Trade.* U.S. Bureau of the Census; *Progressive Grocer.* Reprinted with permission.

Note: Foodstores include grocery stores, retail bakeries, meat markets, produce markets, and other outlets selling food for home consumption.

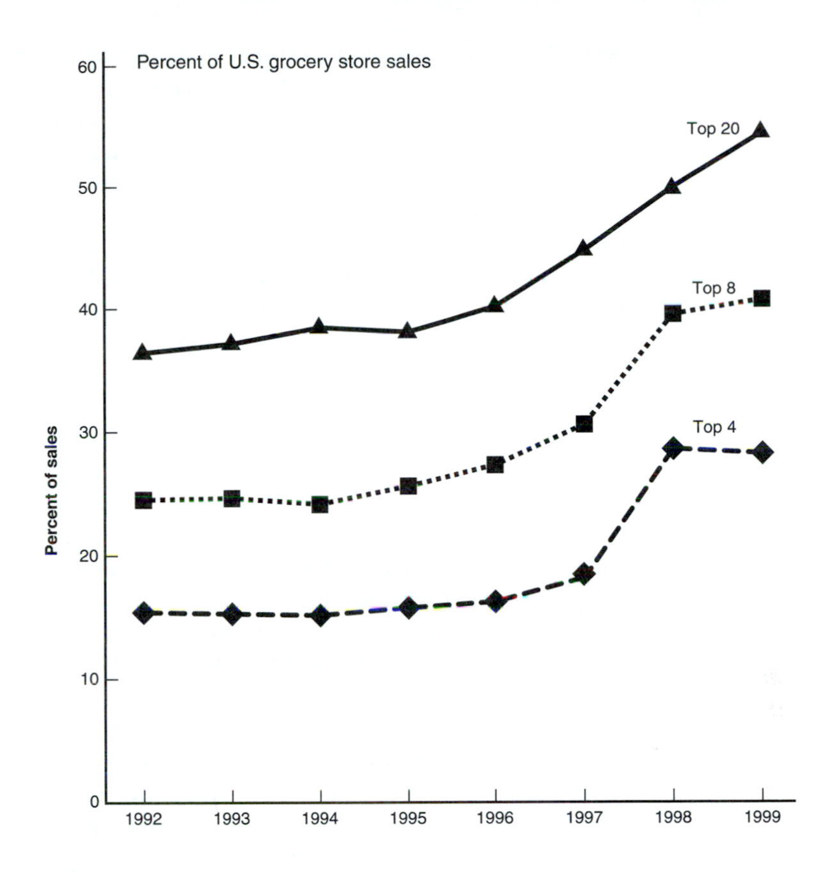

FIGURE 6–4 Percentage of U.S. grocery store sales made by the 4, 8, and 20 largest grocery chains, 1992–1999. (U.S. Bureau of the Census; U.S. Department of Agriculture.)

Mergers, firm restructurings, and foreign investments are continuing to reshape the retail food industry. Most mergers involve food firms, but nonfood firms sometimes merge with food retailers. German, Dutch, Canadian, and British firms own large stakes in some American food retail firms. In turn, U.S. food retailers have made major investments in Canadian, Latin American, and Mexican firms.

In order to fully understand the contemporary structure and competitive behavior of food retailing, we must review two revolutions in this industry: the chainstore movement and the supermarket movement.

The Food Chainstore Movement

Although the terms *chainstores* and *supermarkets* are used synonymously today, these two forms of food retailing developed quite independently and for somewhat different reasons. A grocery store is usually classified as a supermarket if its annual sales exceed $2 million; it emphasizes self-service and features dairy, meat, produce, and dry grocery departments. A chainstore operation is a group of 11 or more related grocery stores. Chainstores may or may not be supermarkets and supermarkets may or may not be part of a chainstore operation.

The modern food chainstore operation represents both a horizontal affiliation of retail stores and a vertical affiliation of food retailing, wholesaling, and, often, processing activities. The chainstore movement grew out of a search for operational efficiencies through large-scale buying and selling and a drive for competitive market advantage among food retailers. The origin of chain retailing is usually traced back to the establishment of the Great Atlantic and Pacific Tea Company in 1858. However, it was not until the decade of the 1920s that real growth occurred. By 1929, it was estimated that chainstores were doing about 38 percent of the total business of combination grocery and meat stores. At about this time, many food chains also reached their maximum number of outlets. For example, the A & P chain reached a peak of 15,737 stores in 1929; the Kroger Company reached its peak of 5,575 stores in that year.

Originally the chainstore organization collected several retail outlets under one management in order to secure the price advantage of large-volume buying from wholesalers and processors. At the retail level the chains adopted the policy of cash and carry. They aimed for low-cost, large-volume operations and competed with low prices instead of with services.

In time, however, the chains were not satisfied with horizontal expansion into more retail outlets. They began moving to expand vertically within the marketing system. They became their own wholesalers, often buying directly from growers in the production areas. To broaden their control over the marketing channel, they operated various processing facilities. Large chains acquired canning companies, cheese and butter firms, bakeries, and other miscellaneous food processors. These firms operated under the brand label and standards set by the chain. Chains also take large portions of the output of independent processors who package products under the chainstore's brand label.

Chainstore expansion was temporarily brought to a halt in the 1930s and 1940s by the general economic conditions, World War II, and the development of more vigorous competition from the voluntary and independent retailers' cooperative chain organizations. The number of chainstore units also fell as the small-service grocery gave way to the large supermarkets.

Chainstores' market share of grocery store sales rose from 50 percent in 1956 to 78 percent in 1996 (Figure 6-5). Despite this trend toward increased chainstore dominance, small chains and independent, single-unit grocery stores—many of them

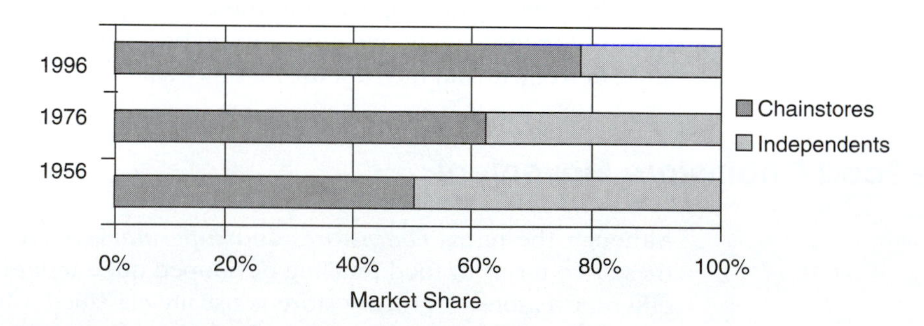

FIGURE 6–5 Chain vs. independent foodstore market shares.

family businesses—still have a share of the foodstore market. There are important marketing lessons to be learned from examining how these privately held, independent foodstores have survived in competition with the large, publicly owned corporate chainstores.

The food chainstore movement triggered competitive reactions on the part of those threatened by it. Because both the independent retailer and the service wholesaler had much at stake, it was logical for these two groups to develop some kind of joint action. This took the forms of retailer-owned cooperative wholesalers and the wholesaler-sponsored voluntary retail chains. In the former, independent grocers organized to purchase and operate their own wholesale facilities cooperatively. In the latter, a wholesaler brought together several retailers into a voluntary chain in which the wholesaler offered many services in return for the guaranteed business of the participating retailers. Both developments represented an effort to secure the low-price benefits of the corporate chain's buying and selling power and at the same time retain whatever advantages there might be from owner-operation of independent stores.

As a result of these competitive adjustments, food retailing is comprised of three segments today. The corporate chainstores (A & P, Kroger, American Stores, Winn-Dixie, Safeway, and so on) are the national and regional grocery wholesaling and retailing operations that are centrally managed. Affiliated independent chainstores (Super Valu, Red Owl, IGA, and so on) are independently owned grocery stores that have affiliated with food wholesaling operations in order to gain the economies of large-scale buying and to share the costs of retail services. There are also independent, unaffiliated foodstores not related in any formal way to a wholesaling operation. These grocers have experienced a long-term decline in market share.

The Supermarket Movement

The food chainstore movement was motivated by a competitive drive for greater retail operational efficiency. Chainstores lowered the costs of food marketing as compared with the independent, single-store "ma and pa" and high-cost wholesaling operations they displaced. The supermarket movement, in contrast, grew out of a competitive race to not only further lower food retailing costs but also to add greater utility to food shopping. The chainstore was economy-minded; the supermarket was market-oriented and consumer-driven.

The supermarket is best described as a full-line, departmentalized, cash-and-carry, self-service foodstore. This contrasts with its predecessors, the neighborhood, limited-line, full-service grocery store and the general store. Supermarkets can be thought of as food department stores. The supermarket gained operational economies over earlier stores. Self-service reduced labor expense by transferring clerking jobs to consumers—another illustration of the principle that marketing functions can be transferred but not eliminated. The larger volume of supermarket sales as well as specialized departmentalization of activities, also lowered retail costs.

Important to the evolution of the supermarket were the socioeconomic changes taking place in consumer markets in the 1940s and 1950s. These included suburbanization, the development of shopping centers, rising consumer incomes, the greater shopping mobility provided by the automobile, the growth in home freezer facilities, the increased use of mass-media advertising, and the increased tempo of new food product development. The one-stop shopping environment provided by the supermarket was nurtured by the new American lifestyle of the 1950s. Indeed, the supermarket became an American symbol of innovation, affluence, abundance, and efficiency.

The first self-service foodstore appeared in Memphis, Tennessee, in 1916. Henry Ford also applied mass production techniques to the Ford Commissary Stores in the 1920s in Detroit. Climate, land prices, and consumer mobility were particularly favorable to the early growth of supermarkets in the Los Angeles area. The supermarket was pioneered by independent grocers, not by the corporate chainstores. The chainstores with established investments in smaller, neighborhood stores were slower to adapt to the supermarket form of food retailing. However, the superiority of supermarkets soon became evident, and virtually all retailers converted to them in the 1950s.

Following the principles of the marketing mix, product differentiation, and market segmentation, retail foodstores have evolved in a variety of directions or formats in recent years, as shown in Figure 6-6. The *conventional supermarket*, with

Conventional supermarket	• Self-service • Highest percent of food versus nonfood items • Size range from 30,000 to 40,000 square feet
Combination food and drug store	• More product variety • Nonfood items 25 to 35 percent of sales • Nonprescription drugs and general merchandise • Prescription drugs • Size range from 35,000 to 45,000 square feet
Superstore	• Some prescription drugs carried • Generic and specialty product areas • Some self-serve bulk foods • Average size 35,000 square feet
Limited assortment store	• Limited number of product brands • Stock the most popular size products • Less than 10,000 square feet
Warehouse store	• Strong price appeal • Size ranges from 12,000 to 35,000 square feet Super warehouse stores range from 50,000 to 140,000 square feet • Primarily food, some health and beauty aids, but low general merchandise emphasis
Hypermarket	• Huge variety • Food and nonfood products • 100,000 to 200,000 square feet • Regional market
Gourmet foodstore	• Variable size • High service • Premium quality products, brands
Convenience store	• Neighborhood locations • Limited assortment • Convenience orientation

FIGURE 6–6 Alternative foodstore formats. (U.S. Department of Agriculture.)

30,000 to 40,000 square feet of selling space and up to 18,000 different items, is declining in numbers and market share. Despite its early success, it soon became apparent that no one store type could satisfy the needs of all consumers. *Combination stores* include a foodstore and drugstore under one roof. *Superstores* range up to 60,000 square feet, handle a wide range of general merchandise, feature service departments such as bakeries, delicatessens, and cheese shops, and emphasize low prices. The number of combination stores tripled between 1980 and 1985 and the number of superstores doubled. These stores add large units of capacity in food retailing and have often triggered heavy price competition in local markets.

The *limited assortment store* is a smaller foodstore, handling only about 500 nonperishable grocery products at deep price discounts. *Warehouse foodstores* eliminate services and frills and make a strong low-price appeal. Products are often displayed in their shipping cartons and shoppers may mark their own prices, bag their own purchases, and even supply their own grocery bags. A *hypermarket* is a super warehouse store ranging up to 200,000 square feet and providing one-stop purchases of groceries, sporting goods, auto supplies, and housewares. A *wholesale club* store requires customers to be a member and emphasizes deep discounts on large volume purchases. A *gourmet* foodstore handles a wide range of premium-quality and exotic foods and brands. A *convenience store* is a small, neighborhood foodstore featuring long hours, quick service, and a narrow assortment of fast-moving staple and impulse items. First opened in the 1960s, convenience stores account for about 15 percent of all foodstore sales.

These food retail developments demonstrate the creativity of the competitive market processes at work. Each of these retail store types has found a useful place in the market and satisfied a unique set of consumer wants. As a result, consumers face a wide variety of retail foodstore choices. From these, sovereign consumers can select their preferred product bundle of prices, variety, services, frills, and conveniences.

There have been several attempts to develop Internet food shopping sites. This form of grocery retailing has appeal for consumers who place a high value on their time and a low value on the store shopping experience. Some of these were startups of new firms to the grocery industry, such as Peapod, Netgrocer, Groceries Online, TeleGrocer, and Shoppers' Express. At the same time, many conventional retail grocery firms have begun selling food over the Internet. While there have been some successes at online grocery marketing, Internet grocery sales have encountered problems. Because of food quality differences, especially for meat and produce, shoppers are often reluctant to delegate in-store food selection to a third party. Some Internet grocery stores use personal shoppers filling customers' orders from local grocery stores. Others fill orders from a central warehouse. Groceries purchased on the Internet may be held for pick-up by the customer or delivered to the shopper's home. In either case, costs are generally higher than conventional shopping where the consumer performs the selection and delivery marketing functions. Internet food shopping will fill a niche for many busy and high-income consumers, but probably will not replace conventional foodstores in the near future.

Which of these store formats is most preferred by food shoppers? There is no easy answer to that. As shown in Table 6-4, prices, location, variety of selection, and numerous other factors are important to consumers in the selection of a foodstore. However, no one firm or store type seems capable of meeting the diverse needs of

TABLE 6-4 Factors Affecting Consumers' Foodstore Choices, 1997

Foodstore Characteristic	*Percent of Shoppers Indicating the Characteristic is Extremely Important in Their Foodstore Choices (multiple answers permitted)*
Low prices	68.1%
Good meat department	64.0
Good produce department	59.6
Convenient store location	52.7
Stocking of new items	38.3
In-store bakery	24.5
In-store pharmacy	24.5
Take-out foods	7.1

SOURCE: *Progressive Grocer,* October 1998, p. 138. Reprinted with permission. Copyright 1998 by *Progressive Grocer.*

today's food shopper. The same shopper may drive 100 miles to purchase a year's supply of some items at a superwarehouse store, shop weekly at a local supermarket, and pick up some milk from a neighborhoood convenience store from time to time. A consumer may spend hours cooking a large meal for a ceremonial family holiday and phone for a pizza delivery for the next meal.

COMPETITION AND PRICING IN FOOD RETAILING

The growing size of food chainstore organizations and their strategic position in the food industry raise many questions. Are food retailers too large and too powerful? Is there effective competition among food retailers? How do food retail pricing and merchandising strategies affect the sales and prices of others in the food industry?

Competition among food retailers today is certainly different from that of earlier periods, when neighborhood "ma and pa" grocery stores and general stores operated. The chainstore and supermarket movements have transformed food retailing in the direction of larger and more powerful firms. The small, independent grocers lost out in this process. Yet it does not automatically follow that competition in food retailing is less vigorous today than formerly. There is no indication that independent retailers are at a cost or merchandising disadvantage in competing with the national chains. On the contrary, in many local markets aggressive, well-managed independent retailers enjoy larger market shares than the national chains.

Moreover, attacked by new competitors from many sides, today's foodstores do not appear to have an entrenched, protected competitive position in the food industry. They must continually battle with nontraditional stores, the foodservice industry, and others for their share of the consumers' food dollar. A supermarket manager who is expanding a delicatessen operation to compete with fast-food stores, installing a food court in the store for the convenience of customers, offering customers no-frill brands to compete with nationally advertised brands, providing economy family packs to compete with food clubs, operating fast checkout lanes to imitate a convenience store, and providing in-store banking services is not behaving like a monopolist.

Consumer trends, such as increased mobility, rising income, and price conciousness have also added to the vigor of food retailing competition. Today's food shopper is no longer limited to the corner market that provides credit. Shoppers can choose from a number of food sources. Home gardening and preservation, health food stores, direct purchases at farmers' and roadside markets, and food cooperatives are small-volume but important alternative food sources. The away-from-home food market—and even dieting—must also be considered alternatives for the food consumer. It is by no means certain that competitive forces in retail food markets are less active today than in an earlier period.

The retail gross margin—the difference between what retailers pay for food and what they sell it for—is the "cost" of performing the food retail marketing functions. The gross margin includes all operating costs and net profits. Expressed as a percentage of sales, gross margin is widely regarded as a measure of trends in food retailing costs, efficiency, and profits. Food retail gross margins generally fell from 1932 to 1950, reflecting operational efficiencies of the chainstore and supermarket movements. Food retail gross margins rose from 1950 to 1965 as a result of increased promotional and labor expenses and profit rates. Retail gross margins were relatively stable from 1965 to 1975 as a result of efficiency gains that offset, to some extent, rising labor costs, competitive pressures, and the growth of food discounters. Foodstore gross margins rose again in the 1980s as the rising costs of labor-intensive services such as in-store bakeries and delicatessens exceeded the economies associated with larger stores, computerized checkouts, and precut, boxed beef. Rising debt levels also increased the 1980 gross margins. The food retail gross margin continued to rise in the 1990s.

This cyclical trend in food retail gross margins follows a familiar pattern called the *wheel of retailing.* Retailers tend to alternate between an emphasis on low service/low prices, and high service/high prices in attracting customers. The chainstore movement and the original supermarket movement of the 1930s represented the economy phase of the cycle. The high-service supermarket of the 1950s, with its trading stamps, games, and expensive facilities, represented the service phase of the cycle. The rise of food discounting in recent years is another turn in the cycle.

Food retail pricing and merchandising practices are of special concern to farmers and consumers. Through their pricing and promotional policies, food retailers can influence consumer demand, farm prices, and the movement of farm products. Undoubtedly modern food retailing is more than just a conduit for passing through foods and farm prices to consumers. Pricing foods at retail is a complex process, for two reasons. First, food retailers are multiproduct sellers with considerable discretion on how they price each item in the store. Second, they set prices to differentiate their stores from those of competitors. That is to say, retail food prices are a form of promotion. As a result, retail foodstore prices are sometimes influenced more by consumer demand and competitive objectives of grocers than by farm prices.

Food retailers price each food product as a component of a total mix of foods (and nonfoods) offered by the store. *Market basket pricing* provides the retailer considerable latitude in pricing any one food. Although total dollar sales must cover retail operating costs and the costs of purchased merchandise, there is no requirement that each retail food price must follow wholesale or farm prices. One store manager may elect to feature low-priced dry groceries and place a larger, compensating profit margin on meat or produce. Another store's competitive strategy may be to feature low meat prices and make up for low meat profits with higher canned goods profits. Consequently, within a community there are usually wide variations in retail prices for similar food products in different stores, even though the stores may have comparable market basket prices and pay similar wholesale prices.

Variable price merchandising is another retail pricing strategy that reduces the dependence of retail prices on farm or wholesale prices. Retailers use temporary, selective price cuts to differentiate their stores and attract consumers. The "loss leader" and weekend specials are examples of these pricing strategies designed to build store traffic. Trade estimates indicate that about 20 percent of grocery items are usually sold at or below cost. Chicken, coffee, bread, milk, and other popular items are frequently priced to attract the price-conscious shopper. This pricing strategy relies on the consumers' tendency toward one-stop shopping; therefore, low profits or losses on the featured items can be made up by purchases of the higher-profit items.

Still another food retail pricing strategy is to maintain low prices in the store on every item year-round. This is called *everyday low pricing* (ELP) and is the strategy of many of the newer foodstores, such as warehouse clubs, discount stores, and superstores.

These pricing strategies raise some important questions. Two criteria of pricing efficiency are that selling prices are related to purchase prices, and that farm and retail prices should move together in order to coordinate food supplies and demand. However, these retail pricing strategies break the link between retail and farm prices and may result in serious market distortions. Everyday low pricing may place downward price pressures on all suppliers in the food chain, from wholesalers to farmers.

Others have charged that these retail pricing strategies are designed to prevent consumers from making price comparisons among stores and food items. It is true that it is difficult for consumers to identify the lowest priced store in a market when there are so many items to price and prices change so frequently. However, the fluctuating or everyday low prices do provide consumers with real savings.

There is nothing illegal or conspiratorial in these retail food pricing practices. Indeed, they reflect competitive pressures and marketing strategies in food retailing, and it is probably unintentional when these strategies adversely affect farm prices. Before condemning retail food pricing strategies, it is well to remember that price features often have helped farmers move surplus food supplies. Most importantly, these alternative pricing strategies provide consumers with choices and opportunities to better satisfy their price and nonprice needs in the marketplace.

MARKETING IMPLICATIONS OF RETAILING DEVELOPMENTS

Because of their size, power, and strategic position, food retailers are referred to as the *gatekeepers* and *channel captains* of the food industry. This means they are making many of the major market decisions about quantity, quality, and services provided by the entire food industry. They serve as consumer purchasing agents, and they control one of the most important food marketing resources—the display space that is so vital to the success of farmers, food manufacturers, and food wholesalers. This provides retailers with potential market power in the food industry.

More than any other sector of the food industry, food retailers practice the marketing concept and orient the industry to satisfying consumer demands. Their pricing and merchandising decisions are directed toward either accommodating or influencing the consumer's purchasing behavior. Food chainstores are not indifferent to farm prices and agriculture, but they look forward to consumer markets for their profits, not backward in the marketing channel. They can adjust their merchandising strategies to assist farmers in selling seasonal gluts of production at reasonable prices. But most of the time today farmers and others in the channel are adjusting to the retailers' needs and strategies.

Food chainstores are reaching back into the marketing channel to control the timing, condition of delivery, and quality of foods—variables critical to the success of the retailers' merchandising strategies. This may lead to retail ownership of food processing operations, to purchasing by quality specifications, or to private label programs. Either way, food retailers have an increasingly greater voice in the production and delivery of food products.

The coordinating position that retailers now hold in the food industry was wrested from food manufacturers in the post-World War II period. Formerly, food processors were the focal point of the food industry. Food retailers gained this position through growth in size, the chainstore and supermarket movements, and the development of their own brands. These private and controlled labels weakened the dominance over the food industry that had been held by food processors through nationally advertised brands. Once again, this shows how competition and marketing strategy have altered the food industry.

MINI CASE 6: THE END OF THE FAMILY GROCERY STORE?

George Putnam, manager of his family's three grocery stores in Green Oaks, was worried. Over the years his family business had prospered by promoting a "buy local" theme, stocking special foods requested by his customers, and providing customer services and friendliness which other stores did not offer. However, George noticed lately that some of his customers were not shopping there regularly and were instead driving to a new shopping center in the next town. He also had just read that a huge, new Walmart combination food/nonfood would soon open outside of Green Oaks. In talking with his friends in the business, George was convinced that he could not compete with these new stores on the basis of price. In order to survive he had to find a new way to appeal to his customers.

What alternatives does George have in responding to this new competitive environment?

What do you think George did?

THE FOODSERVICE MARKET

As shown in Figure 6-3, America's more than 700,000 foodservice establishments range from the familiar fast food outlet to the university cafeteria to the five-star gourmet restaurant. These have little in common except that they provide food for eating away from home and generally involve more consumer services than the foodstore market. This market has grown more rapidly than the home food market in recent years. In a sense, foodservice represents another evolutionary stage in the separation of consumers from food production. The first stage occurred when people moved from the farm to the city; the second stage was marked by the transfer of some of the household food preparation and cooking activities to processing firms; the last stage is to move the kitchen and dining room outside the home. It's not a new idea. The ancient Chinese civilizations had central away-from-home kitchens and no doubt take-away foods.

The foodservice sector has experienced many of the same trends as foodstores. Traditionally made up of local, independent operators, franchised foodservice chains have been developed, such as McDonalds, Burger King, Kentucky Fried Chicken, and Pizza Hut. Some of the major restaurant chains are shown in Table 6-5. There are also full-service restaurant chains. Many of these have central wholesale commissaries and buying offices. Like the different foodstore formats, foodservice firms differentiate themselves by menus and store designs. There is also a foodservice wheel of retailing in evidence as firms have upgraded their services and diversified their menus in order to appeal to more consumers. It is estimated that more than 50 percent of all new restaurants fail in their first year.

Foodservice companies use a wide variety of competitive strategies focusing on the price/value equation, convenient locations, quality of food and presentation, menu specialization or diversity, a differentiated image, such promotional tools as advertising coupons and games, and special services like drive-through windows. Hamburger and pizza restaurants are the most popular, but ethnic foods are often featured in restaurants. Personal service, culinary art, atmosphere, and even "plate coverage" are important in white tablecloth restaurants.

Foodservice firms are influencing the demand and marketing of farm products. For example, fixed menus provide a constant, year-round demand for foods and require greater control over costs and supplies on the part of foodservice firms. The popularity of salad bars has greatly affected the demand for certain vegetables. Some foodservice firms use bid and specifications contracts to procure their food ingredi-

TABLE 6-5 The Largest U.S. Restaurant Chains, 1997

	Number of Stores:		
Company	*U.S. Stores*	*Stores Abroad*	*Sales (Billion $)*
McDonald's	14,204	8,928	$33.6
Subway	11,462	1,874	3.2
Pizza Hut	8,998	3,836	4.7
KFC	5,120	5,117	4.0
Burger King	7,584	2,060	9.8

SOURCE: *Food Review.* Economic Research Service, Sept.-Dec. 1998.

ents. McDonald's Corporation, the nation's largest foodservice chain, is the country's largest purchaser of beef. Because of its volume and quality specifications, McDonald's has revolutionized meat marketing and potato processing. The chicken, pizza, and seafood restaurants have similarly affected the markets for poultry, tomatoes, cheese, fish, and a host of other products widely used in the foodservice industry.

SUMMARY

Food wholesale and retail markets represent the final stages of the food marketing process. The chief activities at these market levels are assortment collection and dispersion, the addition of possession utility to food products, and facilitation of the consumer's choice of a desired food product assortment. The modern forms of food wholesaling and retailing have evolved in response to competitive market pressures, changes in the consumer market, and attempts by firms to improve operational efficiency and better satisfy consumer wants. The food retail chainstore movement was fueled by a desire to lower the costs of retailing through large volume purchasing. The supermarket movement was motivated by a desire to improve retailers' competitive position by increasing consumers' satisfaction with the retail shopping experience. Both chainstores and supermarkets now dominate the food retailing picture, but many independent wholesalers and retailers survive in the form of cooperative and wholesale-sponsored affiliated chainstores. A wide variety of foodstores has evolved to meet the different needs of shoppers. The techniques of food retail pricing have weakened the link between farm, wholesale, and retail food prices. Retailers have assumed the role of channel captains of the food industry. The food service industry provides a competitive alternative to the traditional home food market and is influencing the markets for many food products.

KEY TERMS AND CONCEPTS

affiliated retail chain store
chain store movement
channel captain
convenience store
cooperative groups
everyday low pricing
food brokers
gatekeeper
gross margin
hypermarket
loss leader
market basket pricing
private label
supermarket
superstore
variable price merchandising
voluntary groups
warehouse store
wheel of retailing

DISCUSSION QUESTIONS

1. Why do we have food retailers?
2. Do you see any parallel in the evolution of food retailing with other industries such as clothing, pharmaceuticals, gasoline, computers?

3. What is the significance of the fact that retailers do not price each food item based on its wholesale cost?

4. How did the chainstore and supermarket movements influence the market efficiency ratio?

5. Some food retailers are adding in-store restaurants and carryouts to their facilities. What are the reasons for this?

6. What are some reasons that food retailing passes through the stages known as the "wheel of retailing"? Where are we now in this cycle?

7. How do the following consumer shopping patterns and characteristics affect food retail market strategies: (a) one-stop shopping, (b) brand loyalty, (c) service orientation, (d) working wives, (e) rising incomes?

SELECTED REFERENCES

Epps, W. B. (March 1986). *Specialty Grocery Wholesaling: Structure and Performance,* Agricultural Economics Report No. 547. Washington, D.C.: U.S. Department of Agriculture.

Food Marketing Institute website (June, 2000). http://www.fmi.org/

Food Market Structures: Food Retailing and Foodservice Briefing Room (January 2001). Economic Research Service: U.S. Department of Agriculture. www.ers.usda.gov/briefing/foodmarketstructures/foodservice.

Food Market Structures: Food Wholesaling Briefing Room (January 2001). Economic Research Service: U.S. Department of Agriculture. www.ers.usda.gov/briefing/foodmarketstructures/wholesaling.htm

Kaufman, P. R., & C. R. Handy. (December 1989). *Supermarket Prices and Price Differences.* Technical Bulletin No. 1776. Washington, D.C.: U.S. Department of Agriculture.

Kaufman, P. R. (August 2000). *Consolidation in Food Retailing: Prospects for Consumers & Grocery Suppliers.* Agricultural Outlook. U.S. Department of Agriculture.

Price, C. C., & D. J. Newton. (November 1986). *U. S. Supermarkets: Characteristics and Services,* Agriculture Information Bulletin No. 502. Washington, D.C.: U.S. Department of Agriculture.

Stafford, T. H., & G. E. Grinnell. (September 1982). *Structure and Performance of Grocery Products Brokers,* Agricultural Economic Report No. 490. Washington, D.C.: U.S. Department of Agriculture.

CHAPTER 7

THE INTERNATIONAL FOOD MARKET

Today your competitor is not only the person across the street. It's also the guy across the ocean.

OBJECTIVES

After reading this chapter, you will be able to:

1. Explain why nations specialize and trade, even when they could be self-sufficient in food production.
2. Better understand the importance of international trade to the American farmer and food industry.
3. Distinguish between a free and a protectinist trade policy and indicate how and why nations often interfere with free trade.

World trade in agricultural products has grown rapidly, and the student of agriculture and marketing should be familiar with the theory, practice, and politics of trade. Although most food is eaten in the country where it is produced, some 15 percent of the world's food supply presently moves across international boundaries, compared with 2 percent in the 1950s. The United States is the world's largest exporter of agricultural commodities and also a major food importer. As shown in Table 7-1, about 22 percent of U.S. farm output—one of three acres—was exported in 1997, and the United States accounted for about 14 percent of world food exports. It has been estimated that U.S. farm exports sustain about 870,000 jobs.

American agricultural productive capacity and trade orientation have made important contributions to worldwide economic development and to the world food situation. The country's trading relationships have integrated U.S. agriculture with the

TABLE 7-1 Importance of Trade in the U.S. Food Industry, 1970-97

	Percent			
Item	*1970*	*1980*	*1990*	*1997*
Export share of farm production	14%	29	24	22
U.S. share of world agricultural exports	14	18	14	14
U.S. share of world agricultural imports	11	7	7	9
Agricultural exports as a share of all exports	16	18	10	9
Agricultural imports as a share of all imports	13	7	5	4

SOURCE: U.S. Department of Agriculture; Food and Agricultural Organization of the United Nations.

world agricultural economy. Today, the United States is part of an interdependent, global food economy. Many U.S. farm prices are now set in world markets, and it is difficult to understand today's food economy separately from the world food situation.

CONTRIBUTIONS OF AGRICULTURAL TRADE

Imports and exports of food have several consequences for American agriculture and the U.S. economy. First, exports represent a source of market expansion for U.S. farmers. Increased exports can raise farm prices and incomes by providing a growing market for U.S. farm output. The relationship of exports to farm prices during the 1970-1999 period is illustrated in Figure 7-1. Farm prices have generally risen during periods of expanding food exports.

Second, participation in world markets provides incentives for increasing the productivity and output of U.S. agriculture. Competition in world markets stimulates American farmers to increase their efficiency and output. Indeed, the ability to compete in world markets is one test of whether America should be an exporter or an importer of a commodity. Trade and international competition provide guides for farmers' production decisions. Because of these effects of trade on farmers' decisions and resource efficiency, a potential exists for increasing total output, and, consequently, the world's standard of living through increased trade.

Trade also influences the volume, variety, and price of the food supply for American consumers. Money earned from food exports can be used to purchase both food and nonfood imports desired by American consumers. At the same time, U.S. exports distribute the American food and industrial bounty to nations the world over. Trade-related commerce can also contribute to the economic development of low-income countries by providing them with needed imports and with purchasing power in world markets.

It is sometimes observed that trade plays a role in international diplomacy and foreign relations. It is often true that "two nations trading bushels are less likely to trade bullets." The improved political relationships and the opening of trade between America, the USSR, and the People's Republic of China in the early 1970s are examples of this proposition.

Agricultural exports make an important contribution to the U.S. *balance of trade,* the difference between the value of our exports and imports. The balance of

FIGURE 7–1 U.S. agricultural exports and farm prices, 1970–1999. (U.S. Department of Agriculture.)

trade is a measure of the country's ability to pay for its imports with exports and can influence the value of the dollar in world markets. From 1874 to the 1970s, the United States generally had a favorable balance of both agricultural and nonagricultural trade—exports exceeded imports. But the U.S. nonagricultural and total balance of trade turned negative in the 1970s as a result of growing consumer and industrial product imports. The agricultural trade surplus, however, as shown in Figure 7-2, offsets this total trade deficit to some extent. In a real sense, increased U.S. farm product exports help to pay for American imports of autos, steel, oil, shoes, and other products. After narrowing to $5 billion in 1986, the agricultural trade surplus averaged $19 billion in the 1990-1999 period.

Finally, increased trade in agricultural commodities has made the nations of the world more interdependent. International trade exposes U.S. farmers to world economic conditions. As a result of increased trade, American farm prices have become more variable and subject to climatic, economic, political, and social changes throughout the world. At the same time, U.S. food imports have tied U.S. food prices and supplies to world markets.

THE ECONOMIC REASONS FOR TRADE

If we were to observe the planet from a satellite, we would see a vast armada of ships, planes, trucks, trains, and barges moving food and other products from country to country. Copra moves from Asia to America; the return trip carries corn or

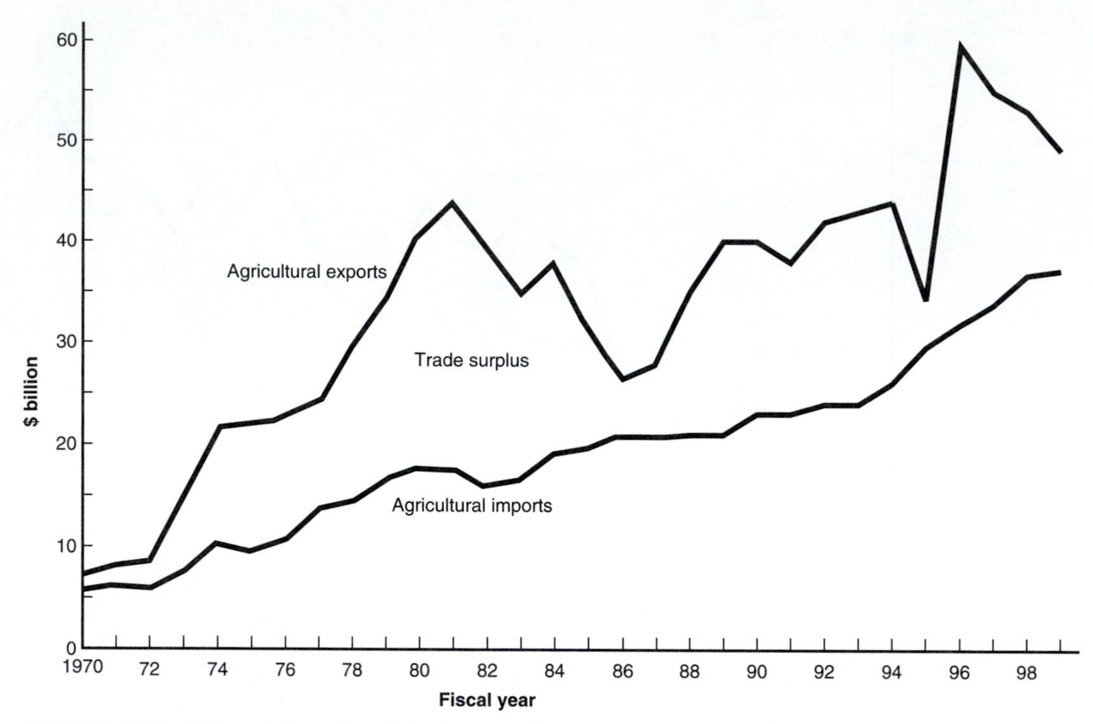

FIGURE 7–2 U.S. agricultural trade balance, 1970–1999. (U.S. Department of Agriculture.)

soybeans to India. Americans trade beef and pork for Japanese steel and TV sets. U.S. computers are shipped to Africa in exchange for cocoa and palm oil. In midocean, ships carrying U.S. feedgrains for European poultry pass ships carrying European-made autos bound for New York. Nothing seems to be in the right place!

If we focus in more closely, we will see some puzzling aspects of this international bazaar. The United States imports sugar from 32 nations despite a large domestic sugar-producing industry. America is the world's largest beef exporter and the second largest beef importer. America produces considerable rice, but Americans eat little rice and most of it is shipped overseas. American consumers are eating Polish ham, Scandinavian cheese, and Chilean fruits despite adequate supplies of domestic ham, cheese, and fruits.

Why Nations Trade

What is the reason for all this trade? Is it necessary? One explanation we can rule out is that countries only import products that they cannot produce at home. There are very few agricultural products that cannot be produced almost anywhere in the world, though often at a high cost. America and other countries import food products that are also produced at home.

A better explanation for trade is the profit incentive for countries to specialize in producing certain commodities and to trade these to countries specializing in other commodities. The United States might choose to trade wheat for bananas with a Central American country, not because bananas cannot be produced in the United States but because both countries gain economically by this trading of product specialties.

The incentive for specialization and trade lies in the economies of producing commodities that are, relatively speaking, best suited to a country's resource endowment. These resources include land, labor, climate, management skills, and other factors affecting the costs of production. Central American nations are endowed with a labor force and climate relatively better suited to banana production than to wheat production. The United States has a land and climate base relatively more suitable for wheat production than for banana production. The fundamental bases for food trade are that productive resources are unevenly distributed throughout the world, each commodity requires a somewhat different resource base, and consumers desire a varied diet.

But wouldn't the United States, with its diversified climate and resource base, be better off as a self-sufficient food producer rather than being dependent on other countries? The answer is, not if the high costs of domestic production of bananas, coffee, tea, and other products are considered. Moreover, even for commodities such as meat and dairy products, which are produced in America, the market provides the best answer to the question. In a free market, consumers decide whether foods will be imported through their purchase decisions.

There is more than just freedom of choice and diet variety, however, to support the case for international trade in food products. By their purchases, consumers indicate the appropriate products to be produced in each country. A consumer preference for Scandinavian dairy products rather than the American product is a consumer vote *for* the imports and *against* the domestic production. Through this competitive process, trade encourages farmers in each country to specialize in those products relatively best suited to their resource base. Put another way, competition in international markets discourages the production of high-cost products and encourages the production of low-cost products in each country. Trade therefore improves the efficiency of resource use throughout the world by encouraging the shift of resources to their most efficient uses. This resource efficiency gain can lead to greater output and/or reduced costs of production—both of which can contribute to an increased standard of living.

International price differences, not the government, will signal each country which products to produce, export, and import. Each country will find its most competitive commodity to be the one that uses its resource base relatively most efficiently. For each commodity, food costs and prices will vary between surplus-producing regions and deficit-producing regions. These price differentials will encourage exports from surplus areas to deficit areas. Each country will be specializing in different commodities, so each will be exporting and importing at the same time.

The Principle of Comparative Advantage

The above is a summary of the *principle of comparative advantage,* which was formally stated by the English economist David Ricardo in 1817. This principle holds that there are economic gains when, under free trade, nations produce and export those commodities that they can produce relatively most efficiently by virtue of their resource endowment and import those commodities that other nations can produce more efficiently.

Trading comparative advantage products for comparative disadvantage products is sometimes referred to as *indirect production.* If U.S. consumers wish to consume more meat or sugar or fruit or anything else, they can either produce more of these in America or they can produce other commodities, such as grains, and trade these for meat, sugar, or fruit. It is a make-versus-buy decision. The choice will depend on the relative efficiencies of producing alternative commodities here and abroad. The decision to produce directly or indirectly will influence world food production, trade, and pricing patterns.

The principle of comparative advantage goes counter to the common-sense notion that it is better to be self-sufficient than dependent on others. It also suggests that trade is not just a one-for-one exchange between countries. Specialization and trade can increase total world food output and, in turn, the standard of living of all trading nations. These ideas are widely recognized for specialization and trade within national boundaries. They are less widely accepted when products move across international boundries. But in fact, although there are some added political complexities involved in international trade not present in interstate trade, the principle of comparative advantage applies equally to both.

Every country will have a comparative advantage in some products and a comparative disadvantage in others. This is true even if one country produces all products more efficiently than all other countries. A nation's comparative advantage can change as a result of world climatic changes, new agricultural technologies, transportation costs, exchange rates, government policies, and other cost-affecting developments.

The Costs and Benefits of Trade

Despite the potential economic benefits from trade, there are some costs. The benefits of trade, for example, must be sufficient to offset the considerable costs of transportation involved in trade. The rising volume of world trade suggests this has been the case. Farmers sometimes feel they bear a heavy cost when imports displace domestically produced products or lower their prices. And consumers sometimes feel they bear a cost when farm exports raise domestic food prices. However, in trade analysis it is important to look at total costs and benefits for the nation as a whole rather than for any one group. The farmers' costs of imports must be viewed against farmers' export benefits, just as consumers' costs of exports must be viewed against their import benefits. The principle of comparative advantage suggests that the overall benefits from trade are greater than the costs for all groups, though this may not be true for every trader or trading country.

Another cost of trade is that associated with the frequent resource adjustments imposed by competitive world markets and changing comparative advantages. Some adjustments impose considerable costs on owners of land, labor, and capital and may be particularly severe in agriculture. For example, it is difficult for farmers to adjust their production to rapidly changing world market conditions. Another cost of trade and specialization is the loss of self-sufficiency and increased dependency on others. This may be simply a psychological cost or a very real cost, if other nations attempt to take advantage of this dependency by raising prices or threatening embargoes. For this reason, many nations choose to forego some gains from trade in strategic commodities, including agricultural products. The case for trade gains is also weakened when nations create artificial comparative advantages through subsidies or government policies that distort costs and prices.

U.S. FARM PRODUCT EXPORTS

U.S. exports of farm products have fluctuated over the years because of changes in world economic conditions, American trade policies, war and peace, and the changing worldwide demand for food. As Figure 7-1 shows, U.S. agricultural exports virtually exploded in the 1970s, reaching an all-time high of $44 billion in 1981. In that year the United States accounted for 38 percent of world agricultural trade tonnage. The U.S. tonnage and world share of this trade fell from 1981 to 1987 as global food production increased and trade competition intensified. U.S. farm product exports have recovered in recent years, again totaling $44 billion in 1994 and reaching a record high of $60 billion in 1995-1996. Agricultural exports then fell to $49 billion in 1999.

The importance of foreign sales varies widely among commodities. Figure 7-3 shows that crop products (feedgrains, foodgrains, and oilseeds) dominate U.S. farm exports. Grain and oilseeds have accounted for 48 percent of the value of U.S. farm product exports in the 1990s. These U.S. commodities have a broad world market, are appropriate for the country's resource base, and have made the United States a dominant farm product exporter. Recent years have seen an expansion in export sales of livestock, meat, and fruits and vegetables.

These traditional U.S. export crops, however, are low-value-per-unit, bulk commodities. *High value export products* (HVP) have a smaller share of natural resource cost and a larger share of marketing costs in their final value than traditional crops. These include unprocessed foods (eggs, fresh fruits and vegetables, nuts), semi-processed foods (flours, oils, fresh meat), and processed foods (processed meat, dairy, bakery products, and prepared foods). Until 1992, the value of U.S. bulk farm products exceeded that of HVPs. However, exports of HVP products have been expanding faster and now exceed that of the bulk commodities. Because of their higher added value, HVPs create more American national income and jobs than exports of the traditional commodities.

The export markets are more important for some crops and regions than for others. In the 1990s, the United States accounted for about one-third of world wheat exports, two-thirds of world corn exports, 60 percent of world soybean exports, 25 percent of world cotton exports, and 12 percent of world rice exports. As shown in Figure 7-4, more than one-third of the U.S. wheat, rice, tobacco, cotton, and soybean crops were exported in 1992. Eight states accounted for more than 50 percent of U.S. farm product exports: Illinois, Iowa, Texas, California, Kansas, Nebraska, Indiana, and Minnesota.

Who buys U.S. farm products? Trade patterns are shaped by population and purchasing power. Effective demand, not just need, determines the customer base for U.S. farm product exports. About one-half of American farm and food product exports are shipped to the higher-income coutries and one-half to the developing

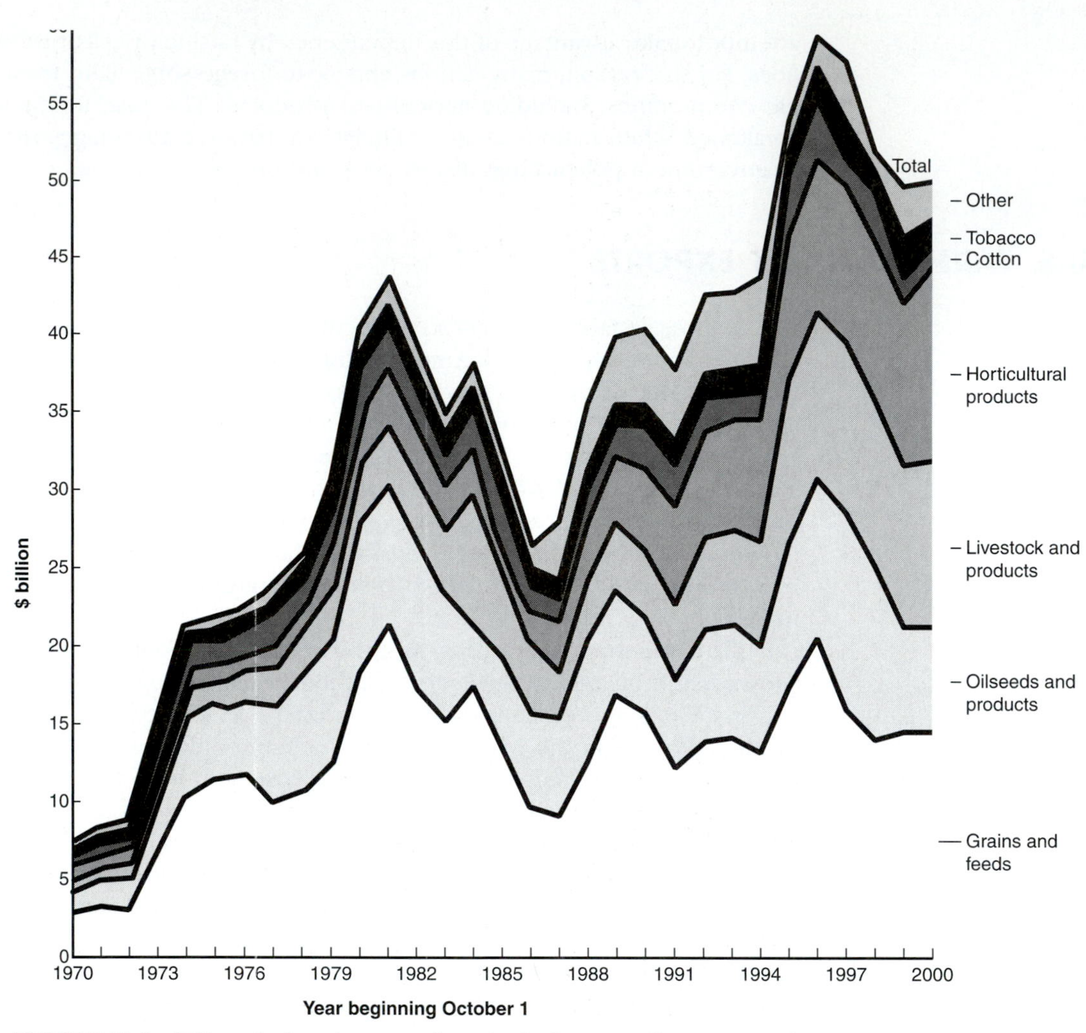

FIGURE 7–3 U.S. agricultural exports by principal commodity groups, 1970–2000. (U.S. Department of Agriculture.)

countries. As shown in Table 7-2, Japan, Canada, and Mexico are the largest buyers of U.S. farm products. The European Union is another large buyer, and the Far Eastern countries are rapidly expanding markets for U.S. agricultural exports. Developing countries are also a growing market for American food exports. In earlier years, much of American farm products exports was provided to low-income countries as aid. However, today most exports are for dollars.

U.S. FOOD IMPORTS

Despite its diversified agriculture and farm product trade surplus, the United States is among the world's largest importers of farm products. About 10 percent of the U.S. food supply is imported. During the 1990–1997 period, U.S. beef imports accounted

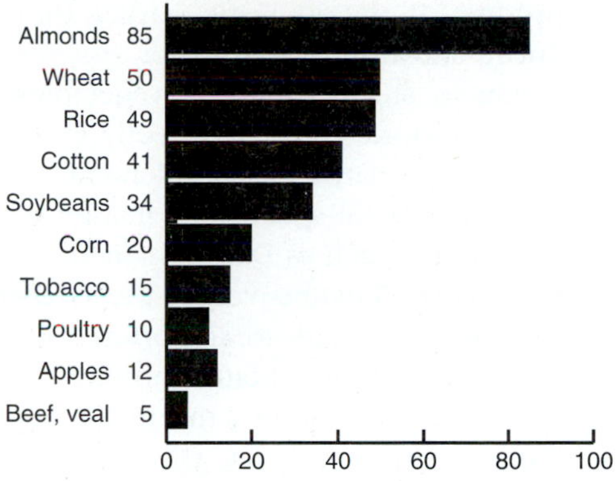

FIGURE 7–4 U.S. agricultural exports as a percentage of farm production, 1992-1996 average. (U.S. Department of Agriculture.)

for about 8 percent of domestic beef supplies, dairy product imports accounted for 2 percent of domestic dairy product supplies, tomato imports re presented 30 percent of domestic supplies, and sugar imports amounted to 25 percent of domestic supplies. There is considerable debate about the value and impacts of these imports. They contribute to a more varied and lower-cost diet for Americans, and they provide other countries with the purchasing power to buy American products. However, farmers frequently complain that food imports adversely affect their prices and incomes.

Not all agricultural imports are directly competitive with domestic production. *Complementary* or noncompetitive imports include bananas, cocoa beans, tea, coffee, spices, and the like. Although these imports do not directly displace U.S. farm

TABLE 7-2 Major Agricultural Product Trading Partners of the United States, 1999

Agricultural Exports From United States		*U.S. Agricultural Imports*	
Importing Country	*$ Billion*	*Exporting Country*	*$ Billion*
Japan	10.5	Canada	7.3
Canada	6.8	Mexico	3.9
Mexico	5.2	Indonesia	1.6
Republic of Korea	2.9	Brazil	1.5
Taiwan	2.6	Italy	1.4
Netherlands	1.9	Columbia	1.3
Russia	1.5	France	1.2
Hong Kong	1.7	Netherlands	1.2
Germany	1.3	Australia	0.9
United Kingdom	1.3	Thailand	0.9
Spain	1.1	New Zealand	0.8
		Germany	0.8

SOURCE: *Foreign Agricultural Trade of the U.S.* U.S. Department of Agriculture.

products in the domestic market, they do compete indirectly with domestically produced substitutes. The major *supplementary* or competitive imports include meat products, sugar, fruits and vegetables, wool, dairy products, and oilseed products. These imports compete directly with domestically produced food products and are increasing relative to complementary agricultural imports. Many of the supplementary imports fill specialized market needs. For example, meat imports include specialty items such as Danish ham and also boneless beef, an inexpensive ground beef product used extensively in the away-from-home food market. Some of the fruit and vegetable products are imported during the winter months and do not compete seasonally with United States supplies.

Table 7–2 shows the major exporters of food to the United States in 1997. Canada and Mexico are the largest exporters to the United States, accounting for more than 11 percent of U.S. farm product imports, including fruits, vegetables, livestock, and meat products. Brazil and Chile have become major exporters of tropical products and fruits to the United States. European exports to the United States consist of canned hams, cheeses, wines, and beers. Australia and New Zealand ship beef, cheese, and wool to America, and Africa exports coffee, cocoa beans, and sugar to the United States.

AMERICA'S COMPETITIVE POSITION IN WORLD FOOD TRADE

What is America's competitive position in worldwide food trade, and for which products does America have a comparative advantage? The answer is not as simple as declaring America the "breadbasket of the world." There is no central world organization that declares the comparative advantage products for each country, and in reality a nation's competitive trade position reflects government policies and economic structures as well as natural comparative advantage. Does American and world food trade follow the principle of comparative advantage?

The United States is a major world food exporter, it exports a large share of its farm output, and exports are more important in the food sector than in other sectors of the American economy. These facts suggest that the United States has a general comparative advantage in food production. Moreover, the fact that a larger share of U.S. farm ouptut is exported than imported demonstrates a favorable American competitive position in world markets.

It is obvious, however, that the United States does not have a comparative advantage in all farm and food products. The country's large expanse of fertile soils and weather patterns mean that its comparative advantage has historically been greatest in land-extensive field crops, such as corn, wheat, soybeans, and cotton. These are major U.S. farm product exports. Its exports of livestock and meat products are much less important, though growing. Also, note the complementarity of U.S. comparative advantage in these crops with its major trader, Japan, a country with limited land mass. The principle of comparative advantage states that the economic incentive to trade will be greatest for two nations with very dissimilar resource endowments. Further, as predicted by the principle, America imports a number of foods (coffee, bananas, spices, cocoa) that it could produce, but only at very high cost.

There is disagreement about the precise identity of U.S. comparative advantage products, but many observers would include grains, oilseeds, cotton, and tobacco

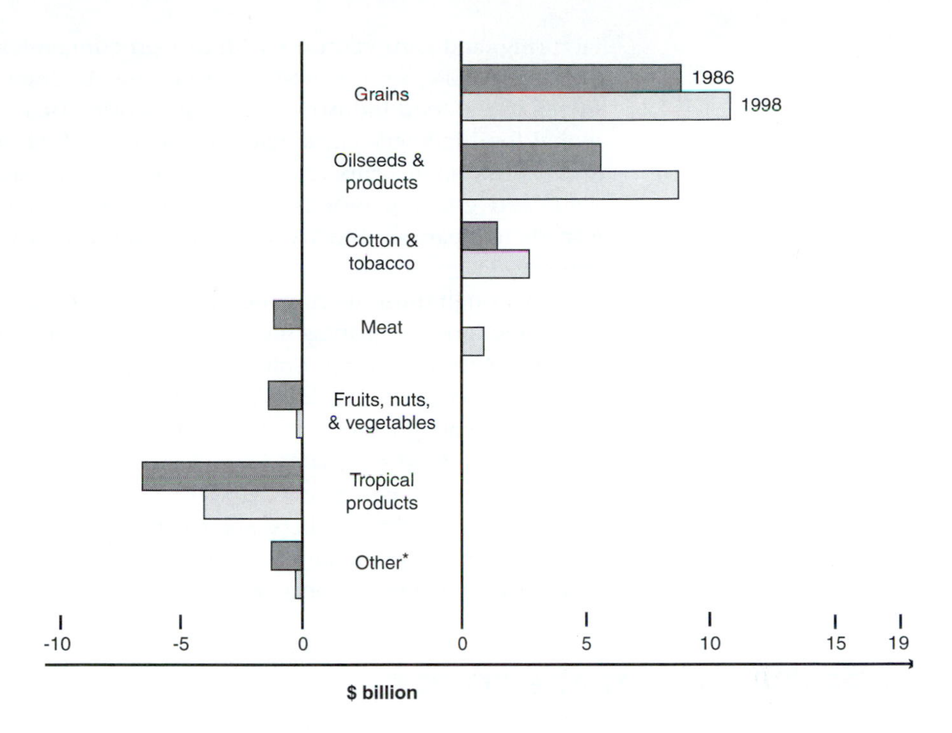

FIGURE 7–5 Export-import trade balances for selected U.S. farm products, 1986–1998. (U.S. Department of Agriculture.)

products in this category. Figure 7-5 shows the 1986–1998 trade balance (exports minus imports) of selected U.S. farm products. Products with a positive and growing trade surplus of exports over imports are more likely to be U.S. competitive products than products with negative or declining balances of trade.

In recent years the United States has been attempting to increase its exports of high value-added products in order to increase export sales and create more jobs in the food sector. There has been some success at this. Almost all of the growth in American food exports in the late 1980s and 1990s came from the value-added products, and high value product (HVP) exports accounted for 64 percent of U.S. agricultureal exports in 1999. This may signal a longer-term change in American food product competitive advantage, resulting perhaps from the country's leadership in food science and technology and expertise in marketing management. This has occurred at a time when production of lower value-added farm commodities has increased in the rest of the world.

FOREIGN INVESTMENT IN THE GLOBAL FOOD INDUSTRY

There are ways other than trade by which U.S. food marketing firms can participate in international markets. These alternatives to trade include direct investments in foreign food marketing companies as well as licensing, franchising, contract manufacturing, and joint ventures with foreign companies. U.S. companies such as Philip Morris, IBP, Pepsico, Coca-Cola, ConAgra, Kellogg, and Heinz have major investments in the food industry around the world. Most of the foreign investment in the global food industry takes place between the high income countries. Foreign affiliates of U.S. food firms are concentrated in Western Europe, Canada, Mexico, Taiwan, and growing rapidly in Latin America, China, and Central/Eastern Europe. Canadian, Japanese, and Western European firms are the major investors in U.S. food firms.

Through these worldwide activities, *multinational firms* facilitate rapid transfer of technology, management skills, and marketing strategies around the world. This involves exporting American management and marketing skills, rather than American products. While this may not sell more U.S. farm products, it generates profits and foreign exchange for the U.S. economy. The key marketing questions for international food companies are (1) whether to sell abroad, (2) which markets to enter, (3) how to enter new markets, and (4) what production and marketing strategies to use in foreign markets. Some firms use global marketing strategies—for example, a single brand marketed everywhere; others adapt their products, brands, and marketing strategies to local needs.

AGRICULTURAL TRADE POLICY

Because trade is so important to a nation's economy, all countries attempt to influence their trade in some fashion, either by promoting it or regulating it. The United States is no exception. How can U.S. farm product exports and the agricultural trade balance be increased? Why purchase from abroad when the products can be produced domestically? Why export American jobs and dollars to other countries? Is it in the national interest for America to become dependent on foreign food supplies? Shouldn't American farmers be protected from the cheap labor and unfair competitive practices of other countries? How do we compromise between (1) the farmers' wishes to expand export sales and limit imports, and (2) the consumers' wishes to limit exports and increase food imports? These questions are major issues in a nation's trade policy.

Protectionism versus Free Trade

Government trade policies vary between the two extremes of protection from trade for domestic industry to free trade. The case for free trade rests on the belief that unrestricted trade among all nations will result in more efficient use of the world's resources and a higher standard of living for all. However, many nations choose to forego these benefits in favor of protectionist trade policies. Throughout history there have been frequent shifts between these two policies.

Trade protectionism refers to market policies and devices that prevent free trade among nations. Among these protectionist devices are *tariffs* (taxes on imports), quotas, licenses, restrictions on imports, and government subsidies to industries that could not otherwise compete in world markets. In agriculture, trade protectionism is often the result of a country's internal domestic policies, for instance, helping farmers via price supports and subsidies. These well-intentioned programs have two effects: (1) They shelter domestic farmers from the rigors of world competition, thereby delaying domestic resource adjustments; and (2) they make the domestic market more attractive to imports.

There are other trade barriers that distort prices and interfere with international commerce. An *embargo* is a prohibition on exports of products to other coutries, usually as a punitive measure. *Most favored nation status* refers to a discriminatory trade preference granted to one country by another. *Dumping* refers to the practice of a country selling surplus products in foreign markets at below-cost prices. *Bilateral trade agreements* are special trade arrangements made between two countries that are not available to other nations. These devices demonstrate the political-economic bases of trade policy.

Since World War II, the United States and other nations have sought a freer trade policy in agriculture. However, barriers to agricultural trade still exist. In many cases there has been a substitution of nontariff trade barriers (quotas, licenses, voluntary restrictions) for tariff barriers.

In recent years many efforts have been made around the world to bring groups of countries together into *free trade areas, customs unions,* or *common markets.* The European Economic Community is one example. The North American Free Trade Area (NAFTA) is another. Though such groupings permit freer trade among the members, there is no assurance that the same attitude will be taken with outsiders. These common markets often protect and subsidize their own agriculture in order to seek agricultural self-sufficiency and may also subsidize exports.

The World Trade Organization (WTO) was established to assist member-nations in negotiating down international trade barriers for agricultural and other products. In 1995, this organizatin replaced the GATT (General Agreement on Tariffs and Trade), which held a number of world trade negotiations, the most recent being the Uruguay Round in 1986–1994. The WTO organized the Seattle Round in 1999. These trade negotiations have been somewhat successful in reducing world trade barriers for industrial products, but progress in reducing trade barriers has been slower for agricultural products.

NAFTA, the North America Free Trade Agreement (1993) was designed to promote freer trade between the United States, Canada, and Mexico. U.S. agricultural exports to Canada and Mexico increased from 9 to 13.2 billion dollars from 1993 to 1998 while U.S. agricultural imports from these countries grew from 7.4 to 12.5 billion dollars. Canada and Mexico purchase approximately one-fourth of U.S. agricultural exports and account for one-third of U.S. agricultural imports. NAFTA has also stimulated U.S. foreign investment in the Mexican and Canadian agriculture and food industries. This trade agreement and its impacts on U.S. business continue to be controversial, but there is little argument that it has increased trade within the North America nations.

The movement from a protectionist policy to a freer trade policy is a slow one. Trade negotiations involve numerous countries, each with different national interests. Even though it can be shown that a nation or the world will gain from free trade, some individuals, industries, and nations will be disadvantaged by a movement toward freer trade. It is difficult for governments to reconcile these micro and macro conflicts in trade policy. Moreover, it is always easier to grant protection than to withdraw it. The problem here is to distribute fairly the gains from trade among all parties—never an easy task. In addition, the benefits of protectionism to individuals and industries are quite significant in dollar value, whereas the costs of protectionism to consumers and the world's standard of living are diffused and not evident to most people.

Export Expansion Programs

To many people viewing the world population explosion, the plight of the hungry and malnourished millions in the world, and the tremendous U.S. agricultural capacity, it would seem to be easy to expand U.S. food exports. However, feeding the world with U.S. farm products encounters the problem of effective demand and competition from other major food exporters. Most U.S. agricultural exports go to the high-income countries. There are few untapped commercial markets that would allow rapid expansion of U.S. farm product exports. Until 1972, American farm exports grew slowly, along with increasing world population and economic development. An exceptional situation occurred in 1972 and 1973, when United States farmers gained access to the Russian and Chinese markets, which represent about a third of the world's population.

U.S. farm product exports are determined by the size of the international market for agricultural products and the share of this market that U.S. producers are able to secure in competition with other exporters. The overall global market for farm products depends primarily on population growth rates and income levels. U.S. producers' share of world agricultural trade, in turn, is influenced by agricultural output and trade policies of other countries and the prices of U.S. farm products relative to competitors' prices.

There are numerous U.S. government subsidy programs to expand U.S. farm product exports or improve the competitive position of American products in world markets. One approach is to reduce the cost of American commodities for foreign buyers. Recent farm bills have lowered farm price supports in order to improve the competitiveness of U.S. products. The Export Enhancement Program (EEP), begun in 1986, lowers the price of U.S. farm product exports by providing a bonus of commodities from government stocks to exporters. The largest export expansion program, the Export Credit Guarantee Program, guarantees credit for foreign buyers of U.S. farm products. The Targeted Export Assistance (TEA) Program provides U.S. producers of selected commodities assistance to counter the adverse effects of foreign subsidies, import quotas, or unfair trade practices on U.S. agricultural exports.

General economic and agricultural development assistance to low-income countries may also stimulate U.S. farm product exports in the long run. As economies develop and incomes rise, they become better customers for U.S. farmers. Mexico, India, South Korea, and Egypt are countries that formerly received U.S. food aid and are now major buyers of these products. The oldest U.S. food aid program is the 1954 Agricultural Trade Development and Assistance Act (Public Law 480), sometimes called the food for peace program. This provides long-term, low-interest credit for foreign buyers of American farm products and food donations and disaster relief.

MINI CASE 7: TO TRADE OR NOT TO TRADE

After their marketing class, Bill and Judy were discussing the recent increased imports of farm products into America. Bill protested, "America can and should be self-sufficient in food production. There is no reason why we should be importing things like meat, dairy products, fruits and vegetables and sugar, which can be produced right here at home. This just lowers the prices American farmers receive for their products." Judy disagreed saying, "It's true we can efficiently produce a large number of farm products right here at home. However, American consumers have a right to buy whatever foods they prefer, and they seem to view some imported foods as cheaper or a different quality than ours. What's more, trade is a two-way street. If we want to sell our farm products abroad, we have to buy something from other countries, many of which also want to export their farm products." And remember what the professor said about indirect production, "Sometimes it more economical to produce more of one product and sell it to buy another than to produce both products." "I don't see how that is possible," said Bill.

Who is right here? What does the principle of comparative advantage have to say about this? When would it make sense for a country to be self-sufficient in food?

The U.S. Department of Agriculture has responsibility for two other trade expansion programs. The Foreign Market Development Program (FMDP) provides government assistance to commodity groups in the form of foodstore promotions, nutritional information, trade servicing, technical assistance, and advertising. The Market Promotion Program (MPP) provides government subsidies for cooperatives, state departments of agriculture, and private U.S. companies selling branded products overseas.

U.S. products do not "sell themselves" in the highly competitive commercial foreign markets. Salesmanship, reliability as a source of supply, competitive pricing, acceptable product quality, and effective 4P marketing strategies are required to compete in world markets. In order to be successful in exporting, farmers and food marketing firms will have to analyze carefully market opportunities and customer needs. Specialists in foreign marketing problems will be necessary for our domestic firms.

Currency Exchange Rates

In the 1970s and 1980s, U.S. farm product exports and imports were significantly influenced by the value of the dollar. The *exchange rate* is the value of the dollar in relationship to other currencies, for example, the Japanese yen, the English pound, or the German mark. Fluctuating currency exchange rates alter the costs and prices of a country's imports and exports and can change the competitive position of an exporting country like the United States.

The dollar was devalued and generally fell against other currencies from 1970 to 1978, making U.S. agricultural exports less expensive to foreign buyers and stimulating U.S. exports. The dollar strengthened from 1979 to 1985, decreasing U.S. farm product exports and making U.S. imports less expensive. The value of the dollar generally declined in the early 1990s, making U.S. farm products more competitive in world markets. It then rose in the latter 1990s, reducing U.S. agricultural exports.

State Trading Programs

For the most part, food trade is carried out by private firms operating within the framework of national trade policies. In some countries, however, international food trade is centralized in trading monopolies. These monopoly agencies may be government organizations or producer groups granted trade monopolies by the government. Such agencies frequently become involved in domestic marketing programs, including storage, transportation, and pricing activities. The Canadian Wheat Board, a producer-controlled organization, has a monopoly position in Canadian wheat exports. Australia has similar export monopolies for wheat and barley. The Japanese Food Agency and Conasupo, the Mexican government's food and grain purchasing agency, regulate food imports into these countries. Prior to the recent market reforms, the central plan countries of Eastern Europe and Asia also traded as state monopolies. *Exportkhleb* was the Soviet Union's grain-trading monopoly. China's Grain Bureau is also a state trading monopoly. There is some concern that these state trading monopolies have a competitive advantage over countries that organize trade through the private, free-enterprise sector. The evidence for this is mixed.

SUMMARY

The United States is a major trader of agricultural commodities, and foreign trade in food products has a significant impact on farmers, middlemen, consumers, and the American economy. The principle of comparative advantage underlies the rationale for trade in food products. Trade has the potential for improving the allocation of worldwide resources, increasing economic output, and enhancing the world's standard of living. Over the long term, U.S. farm product imports and exports have been increasing. Agriculture's positive balance of trade has offset in part the overall U.S. negative balance of trade. Most U.S. farm product exports go to the higher-income countries of the world. Food imports that compete directly with domestically produced products are increasing more rapidly than noncompetitive imports such as coffee, tea, and bananas. There is growing foreign investment in the global food industry. Protectionist trade policies still exist in world agriculture, but some progress has been made in lowering trade barriers. The United States has a comparative advantage in producing several important agricultural commodities. The commercial export market for farm products is highly competitive and requires attention, and international marketing utilizes all of the 4Ps of marketing strategies.

KEY TERMS AND CONCEPTS

bilateral trade agreement
complementary imports
common market
customs unions
dumping
embargo
exchange rate
export market expansion
gains from trade
General Agreement on Tariffs and Trade
high value products
indirect production
most favored nation status
principle of comparative advantage
protectionism
resource endowment
state trading monopoly
supplementary imports
tariff
trade barriers
World Trade Organization

DISCUSSION QUESTIONS

1. How can an individual country benefit from trade? How could the world benefit? Farmers?

2. Why does a major food-producing country like the United States import any food at all?

3. If trade has so many advantages, why do so many governments follow protectionist policies?

4. A friend states: "The United States is the breadbasket of the world. U.S. agriculture is the most productive of all countries. We should strive to sell as much food as possible but limit imports so that we don't become dependent on others for our food supply." How would you respond to this?

5. Protectionist trade policies are frequently justified on the following grounds. Give your judgment of each.

- **a.** U.S. producers should not have to compete with cheap foreign labor.
- **b.** Taxes on imports are a good source of revenue for the U.S. Treasury.
- **c.** "Buy American" and keep the money in the country.
- **d.** Tariffs on imports increase the American standard of living by reducing imports that take American jobs.

6. Some countries have claimed that U.S. shipments under PL 480 were simply designed to dump U.S. food surpluses on foreign markets and that these supplies retarded the development of these countries' own agriculture. Explain how this could occur, and give your own recommendations for sharing, in a constructive manner, the U.S. food supply with developing nations.

7. Distinguish between food aid and food trade.

8. If the United States has a comparative advantage in farm products, why are so many programs needed to stimulate U.S. farm product exports?

9. How would a strong or weak dollar affect U.S. agricultural exports and imports?

10. Why is it so important for the U.S. to increase its exports of high value-added food products?

SELECTED REFERENCES

Ackerman, K. Z. and P. M. Dixit (1999). *An Introduction to State Trading in Agriculture.* Agricultural Economic Report No. 783. Washington, D.C.: U.S. Department of Agriculture.

AgExporter. Foreign Agricultural Service, U.S. Department of Agriculture. Monthly. http://www.fas.usda.gov:80/info/agexporter/agexport.html.

Ballenger, N., M. Bohman, and M. Gehlhar (April, 2000). Biotechnology: Implications for U.S. Corn and Soybean Trade. *Agricultural Outlook.* U.S. Department of Agriculture.

Bredahl, M. E., P. C. Abbott, & M. R. Reed. (1994). *Competitiveness In International Food Markets.* Boulder, CO: Westview Press.

Foreign Agricultural Trade of the United States—Briefing Room website, (June, 2000).http://www.ers.usda.gov./briefing/AgTrade/.

Henderson, D. R., C. R. Handy, and S. A. Neff, editors (1996). *Globalization of the Processed Foods Market.* Agricultural Economic Report No. 742. Washington, D.C.: U.S. Department of Agriculture.

Journal of International Food and Agribusiness Marketing. Quarterly. Binghamton, NY: Food Products Press.

Manchester, A. C. (September 1985). *Agriculture's Links with the U.S. and World Economies,* Agricultural Information Bulletin, No. 496. Washington, DC: U.S. Department of Agriculture.

Trade policy briefing room (January 2001). Economic Research Service, U.S. Department of Agriculture. www.ers.usda.gov/Topics/view.asp?T+103824

U.S. Department of Agriculture. (1986). *Embargoes, Surplus Disposal, and U.S. Agriculture: A Summary,* Agricultural Information Bulletin No. 503. Washington, DC: U.S. Department of Agriculture.

U.S. Department of Agriculture. (January 1990). *Exchange Rates and U.S. Agricultural Trade,* Agricultural Information Bulletin No. 585. Washington, DC: U.S. Department of Agriculture.

U.S. Department of Agriculture. (1993). Agribusiness spans national borders. *Agricultural Outlook,* May, 22–27.

Waino, John (August 1999). Agriculture and the Evolution of Tariff Bargaining. *Agricultural Outlook.* U.S. Department of Agriculture.

World Trade Organization website, (June 2000). http://www.wto.org/

PART III

PRICES AND MARKETING COSTS

CHAPTER 8

PRICE ANALYSIS AND THE EXCHANGE FUNCTION

Anyone can be an economist. All you must learn are two words, Supply and Demand.

OBJECTIVES

After reading this chapter, you will be able to:

1. Use supply and demand diagrams to locate equilibrium prices and quantities and predict price and quantity changes.
2. Explain some producer marketing problems using supply and demand curves.
3. Demonstrate how the law of one price connects all prices in a market.
4. Distinguish between price determination and price discovery.

The exchange functions of marketing, buying, and selling are the heart of marketing. As goods move through many hands before reaching the final user, title changes several times. Each time title changes, a price must be decided upon. Thus, pricing is an integral part of marketing. Price analysts claim that marketing is an important branch of the study of prices. Marketing people claim that pricing is a major branch of the study of marketing. Regardless of viewpoints, the study of marketing and that of prices go hand in hand. One really cannot understand marketing adequately without some grasp of the fundamentals of pricing. The basic economic theory of value and pricing is presented here.

ROLE OF PRICES IN THE COMPETITIVE ECONOMY

The food marketing process is a system of communication, conflict resolution, and coordination. The system must transmit to buyers and sellers information that will be useful to them in their decisions; it must achieve compromises between producers' and consumers' goals; and it must provide incentives to encourage efficient decision making. This is a large order, not easily accomplished. There are three ways for the marketing system to achieve these objectives: (1) custom or tradition; (2) central, authoritarian control; or (3) decentralized coordination via prices determined by competition.

Although custom and tradition do influence food industry decisions, they appear to be a poor choice for guiding decisions in a dynamic economy. For example, it is unlikely that we would have the supermarket, convenience foods, farmer co-ops, and numerous other market innovations if tradition were shaping the food industry. Nor are centralized, authoritarian decisions palatable to our economic philosophy. The alternative is to coordinate food industry decisions with prices determined in the market economy. This is a complex coordinating device. Basically, it requires that each firm and consumer make independent decisions, based on their own interests and guided by price signals. A key tenet of this market price system is that the firm's profit-seeking behavior will lead it to serve society's interest by allocating resources to their highest-valued use. Most of the U.S. food system, and much of the world food economy, is coordinated by this competitive price system. This reflects our philosophical bias toward decentralizing economic and political power, maximizing personal freedom of choice, and encouraging flexibility in decisions in order to foster efficiency, innovation, and technological change.

If a competitively functioning price system is directing the market, it has the advantage of being impartial. The idea of "fair treatment" is left to the composite judgment of the marketplace rather than to the decisions of individuals in positions of political power. Such a system of direction also has the advantage of being in continuous operation—there is a continuous adjustment to changing conditions. This is in contrast to the sluggish after-the-fact type of direction that usually occurs when market decisions are delegated to various public agencies.

In a competitive economy the pricing machinery is expected to transmit orders and directions. To firms, high and rising prices mean increased profits and more incentive to "go ahead." But to consumers, they mean "slow down" or perhaps "do without." The opposite directions would result from low or falling prices. Briefly, then, fluctuating competitive prices have the following three major jobs to perform:

1. They guide and regulate food producers' output and selling decisions.
2. They guide and regulate consumption decisions.
3. They guide and regulate marketing decisions over time, form, and space.

RELATIVE PRICES AND FOOD MARKETING DECISIONS

Profitable marketing and production decisions require careful selection from among alternatives. Substitution possibilities abound at every stage of the food system. Farmers can use the same resources to produce corn, soybeans, or wheat. They may in turn elect to market these crops or to transform them through feeding them to animals. Likewise, food processors may substitute corn oil, soybean oil, cottonseed oil, or palm oil in the preparation of cooking oils. Dairy producers can sell milk in the fluid state or as cheese or yogurt. Consumers have the greatest range of alternatives.

Beef, pork, chicken, and lamb are all substitute meat products, although consumers vary in the degree to which they view them as substitutes. Consumers also can obtain protein from vegetables rather than from livestock products. Peas can be substituted for carrots, national brands for private brands, A & P for Kroger stores, giant economy sizes for small packages, and restaurant meals for home-cooked meals.

How do farmers, marketing firms, and consumers choose among these alternatives? Tradition, custom, and personal preference play some role in market choices. But they seem inadequate to explain the frequent changes in market choices that are evident to everyone. Farmers do substitute one crop for another, and over a longer period they expand and contract farm enterprises. Food processors frequently change their formulas, add new products, and drop others. And consumers are notoriously fickle in substituting one food, one brand, or one store for another.

Price signals and the profit motive are a better explanation of these changing market choices. Farmers will plant more soybeans and less corn if prices and profit expectations favor soybeans rather than corn. Food processors also will respond to price and profit incentives, as when they substitute corn sugar for beet sugar in their processing formulas. And consumers can be encouraged to substitute chicken for pork, if prices favor such a substitution.

The *relative price* of substitute products influences what will be purchased, produced, and sold by farmers, marketing firms, and consumers. A relative price is simply a ratio stating the price of one substitute in terms of another. The relative price of soybeans to corn is 2:1 if the price of soybeans is $8.00 and the price of corn is $4.00. Put another way, this bushel of soybeans is twice as expensive as a bushel of corn. Producers and consumers will respond to changes in relative prices of substitutes in such a way as to improve their economic position. For producers, this means the relative price adjustment will be toward increased profits. For consumers, it means cost-minimizing and value-maximizing decisions.

Relative prices, not the absolute prices of individual products, serve as the signals for changing market choices. Relative prices change, of course, when there is a change in the price of either the numerator or the denominator of the price ratio. However, the relative price ratio will remain constant if both of the substitute prices move together. An equally important point is that a product's relative price may change, even if its absolute price does not. For example, a $2.00 corn price and a $6.00 soybean price will make these two crops as equally attractive to farmers as a $3.00 corn price and a $9.00 soybean price. The relative price is the same (3:1). But, if the price of soybeans rises from $6.00 to $8.00 and corn prices remain at $2.00, some farmers will respond to this relative price change by producing more soybeans and less corn. Knowledge of relative prices is useful if one is to understand and predict the market choices of farmers, marketing firms, and food consumers.

Farmers respond to relative price signals. For example, farmers usually transfer acreage from corn to soybeans when the soybean/corn price ratio is 2.5:1 or better, everything else being equal. On the other hand, corn will be substituted for soybean acreage when this relative price falls below 2.5:1. And, although wheat is normally used as a food grain, farmers frequently find it profitable to feed wheat to livestock when wheat prices are less than 40 cents above corn prices.

Consumers also are sensitive to changing relative food prices. One guideline that many consumers follow is that medium-size eggs are a better value than large eggs when the price differential between them is greater than 7 cents per dozen.

Another guide is that butter is a good buy when it is priced at less than twice the price of margarine. The point is that a product's price can only be considered "high" or "low" relative to the price of its substitute products.

SUPPLY AND DEMAND ANALYSIS

The heart of price formation in a competitive market economy is the supply and demand analysis. There is probably no more overworked and misunderstood phrase in economics than the *law of supply and demand.* To some it is a form of magic or divine guidance invoked to explain away any major problem or dilemma. To others it is something that can be used or ignored, depending on the desires of the moment. It is to these fundamental ideas of supply and demand that we shall now apply ourselves.

The Meaning of Demand

Demand is a schedule of different quantities of a commodity that buyers will purchase at different prices at a given time and place. The *law of demand* formalizes the relationship between quantities purchased and alternative prices. The lower the price, the more will be purchased. Conversely, the higher the price, the less will be purchased.

The law of demand results from consumers' attempts to maximize their satisfaction, given limited incomes, in a market where there are many product alternatives. The *principle of diminishing marginal utility* states that as a buyer consumes increasing amounts of a product in any time period the usefulness and desirability of each additional amount decreases. Even though we may desire steak at any meal, most of us would get more satisfaction from the first steak than from the second. We may even reach a saturation point where we would not pay anything for another unit of product. Thus, the price we are willing to pay for a product declines as the quantity we consume increases.

There are two other reasons for the law of demand. When a food price declines, say beef, its price relative to substitute products, such as pork and chicken also falls. As a result of this substitution effect, consumers will substitute the relatively cheaper food, beef, for the relatively more expensive foods (pork and chicken). The income effect of a price change, on the other hand, suggests that as the price of a food declines the consumer's real income (money income adjusted for the price change) increases. For most foods some of this increased income would then be spent to purchase more of the product whose price has fallen.

The response of quantity demanded to a change in price can occur for two different reasons. Consumers who are purchasing the product may choose to increase or decrease their rate of purchase. Alternatively, consumers who were not purchasing the product may decide to do so as price falls. The firm's marketing strategy is designed to stimulate purchases of both groups, but a different strategy may be employed for users and non-users.

Table 8-1 shows a hypothetical demand schedule of prices and quantities that might exist for corn at any given time and place. Of course, for any one time period

TABLE 8-1 Hypothetical Demand Schedule for Corn

Price Per Bushel ($)	*Amount Purchased (billion bushels)*
4.00	3.0
3.75	3.4
3.50	3.8
3.00	5.0
2.75	5.7
2.50	6.4
2.25	7.2

a definite amount of corn will be purchased at the stated price—say, 5 billion bushels at $3.00. But the demand schedule shows what amounts would have been purchased *if* the price had been different. Figure 8-1 is the graphic presentation of the demand schedule given in Table 8-1. This is the *demand curve* for corn. According to the law of demand, the demand curve will slope downward and to the right on a graph similar to Figure 8-1.

The demand curve in Figure 8-1 and the demand schedule in Table 8-1 both illustrate the nature of the relationship between quantity and price as it is established by prospective buyers. *If* the price were $4.00, 3 billion bushels would be purchased; if the price were $2.25, 7.2 billion bushels would be purchased, and so on for any other possible prices assuming that all other things remained unchanged. The demand schedule and curve do not indicate what *the* price and quantity are; they indicate only what effect different prices will have on the quantity purchased. *The* price that will exist has not yet been established, and demand alone cannot establish it.

Several important points must be kept in mind if the idea of demand is to be used correctly. First, it is a series, or schedule, of quantity—price relationships. The quantity of a product that consumers will purchase is not a fixed amount but depends

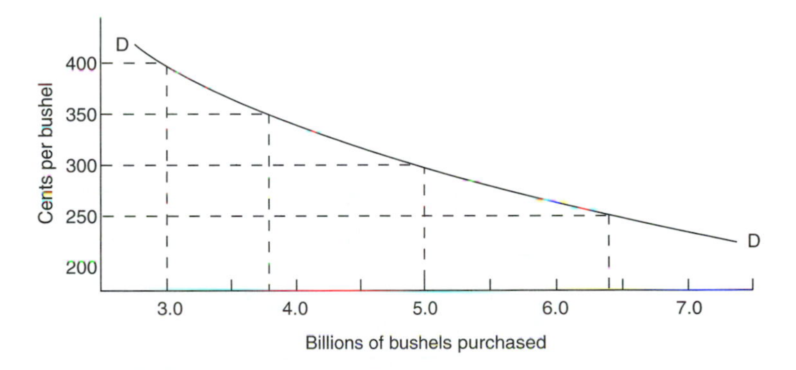

FIGURE 8–1 Hypothetical demand curve for corn.

on the price of the product. Care must be exercised in differentiating between demand—a schedule—and quantity demanded—a point on that schedule.

Second, demand indicates the differing amounts that will be purchased at differing prices and not simply the amounts needed by purchasers. The demand that is important in marketing is *effective demand.* Effective demand is the desire of the consumer for the commodity backed up by purchasing power. Low-income people both need and desire many things—more food, better clothing, better homes, and so on. However, their demand for these things is very limited because they do not have the purchasing power to make their needs and wants effective. The pertinent marketing question is always, How much will be bought at a price? and not, How much will be needed or desired?

We should add another important thought. In the example of the demand for corn, we have been concerned with the farm-level demand for this important product. However, consumers do not buy corn but rather the products made from corn, such as corn flakes and pork. Therefore, we say that the demand for corn is a *derived demand,* that is, its level is caused by the level of demand for the final products made from it. In Figure 8–2 the derived demands for retail pork chops are shown. There is a family of derived demand curves for each food product. The differences between the derived demand curves are the costs of adding time, form, space, and possession utility at each successive market level.

Finally, we need to distinguish between the total market demand schedule—all purchases at each price—and the firm's or consumer's market demand schedule. Consumer demand schedules are added horizontally to obtain the firm's demand schedule, and all firms' demand schedules can be added horizontally to obtain the total market demand schedule.

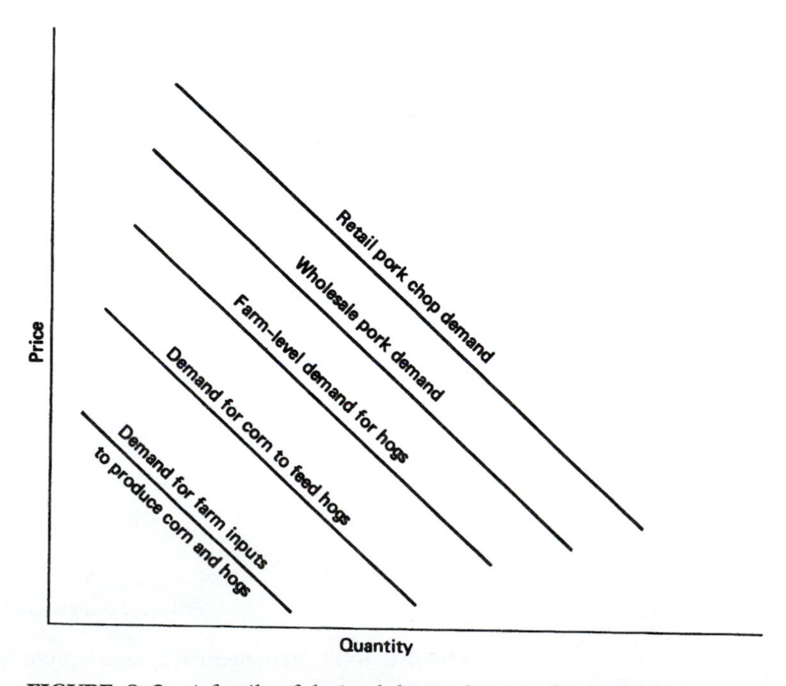

FIGURE 8–2 A family of derived demand curves for pork chops.

TABLE 8-2 Hypothetical Supply Schedule for Corn

Price Per Bushel ($)	*Amount Offered for Sale (billion bushels)*
4.00	5.6
3.75	5.5
3.50	5.4
3.00	5.0
2.75	4.7
2.50	4.4
2.25	4.0

The Meaning of Supply

Supply is a schedule of differing quantities that will be offered for sale at different prices at a given time and place. The *law of supply* is the relationship that exists between prices and quantity offered to the market. The higher the price, the more will be offered for sale; the lower the price, the less will be offered for sale. Whereas demand indicates the relationship between quantity and price from the buyers' viewpoint, supply indicates a similar relationship from the sellers' viewpoint.

As we did for demand, we can make a *supply schedule* of quantities and prices as illustrated in Table 8-2. According to the law of supply, the supply curve will always slope upward and to the right on a graph like that shown in Figure 8-3. Here, too, it must be remembered that at any one time and place only one point on the curve represents the actual situation. The curve represents what would be the effects of different prices on amounts offered.

As in the case of demand, the term *supply* should be used carefully. In economics, supply is a schedule of quantities and prices, not a point on that schedule. For example, the amount of hogs available today on the market is often referred to in the trade as the supply available when in fact it is the quantity supplied. To be realistic, we must accept this terminology confusion, but we should keep clearly in mind just what is meant. When supply is used in the economic sense, it always represents a series of price-quantity relationships.

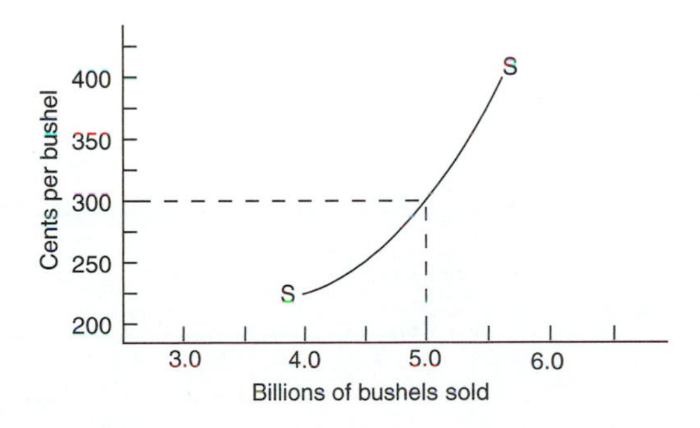

FIGURE 8–3 Hypothetical supply curve for corn.

TABLE 8-3 Hypothetical Demand and Supply Schedules for Corn

Amount Purchased (billion bushels)	*Price per Bushel $*	*Amount Offered for Sale (billion bushels)*
3.0	4.00	5.6
3.4	3.75	5.5
3.8	3.50	5.4
5.0	3.00	5.0
5.7	2.75	4.7
6.4	2.50	4.4
7.2	2.25	4.0

The Equilibrium Price

Supply and demand are often referred to as the two blades of a scissor. Just as it takes both blades to cut effectively, so must both demand and supply be considered in the determination of a price.

Table 8-3 shows the demand schedule of Table 8-1 and the supply schedule of Table 8-2. What will be the price of corn given these two sets of relationships? As we run down the list of prices we find that at $3.00 per bushel, buyers will take 5 billion bushels. At this same price sellers will produce and offer for sale 5 billion bushels. At $3.00, buyers will take all that sellers will offer. There will be no unsold corn; the market will be cleared. This, then, will be the price that will be established if the forces of competition are allowed to function. This is the *equilibrium price,* the point where demand and supply are equal. Figure 8-4 graphically presents the same picture. Where the demand (*DD*) and supply (*SS*) curves intersect, (*P*) is the equilibrium price. Correspondingly, 5 billion bushels is the equilibrium quantity.

It is the profit-motivated actions of buyers and sellers that make the price come to rest at $3.00 in Figure 8-4. Suppose prices were $3.50. At this price, sellers would be willing to sell more corn than buyers would want to purchase. The desire of these extra sellers to sell at some price above $3.00 would result in price conces-

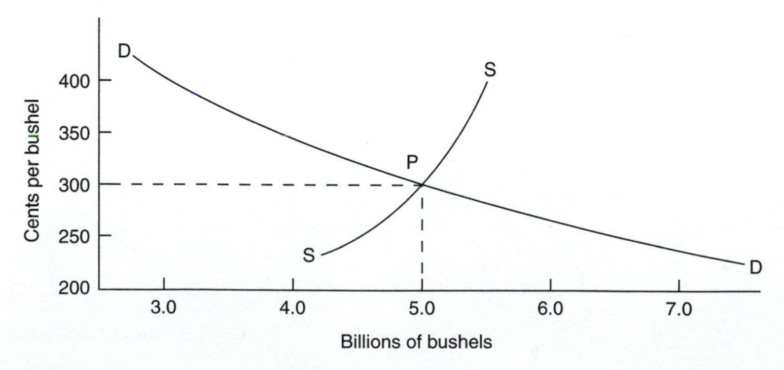

FIGURE 8-4 Hypothetical supply and demand curves for corn, illustrating the equilibrium price.

sions to attract buyers. Any price lower than $3.00 would find additional buyers but fewer sellers. In this situation, the desire of eager buyers to secure corn at what they considered advantageous prices would result in the bidding up of prices to $3.00. At the equilibrium price of $3.00, both the buyers and sellers would be satisfied, and 5 billion bushels would change hands.

The equilibrium price is a compromise between the desire of sellers for a higher price and the desire of buyers for a lower price. No other price is an acceptable compromise, although neither buyer nor seller is completely happy at the equilibrium price. Above the equilibrium price, it is a *buyers' market*—more is supplied than is demanded, and price cutting will occur. Below the equilibrium price, it is a *seller's market,* and eager buyers will bid up the price to the equilibrium level.

In the real world, the equilibrium price is not a point readily found or easily maintained. Rather, prices are always missing the mark in their search for the price that will clear the market. Price variations occur frequently as buyers and sellers search for the market-clearing price. Even though prices are seldom at the precise equilibrium level, it is useful to know that they will tend to move toward this market-clearing level. This fact allows us to use the supply and demand analysis in price forecasting, policy analysis, and in many other ways.

Two Food Economic Myths

This supply and demand analysis can be a powerful tool in understanding the price making process in the food industry. It can also be misused, however. One common mistake is to draw the supply and demand curves without slope, as shown in Figure 8-5A. The perfectly vertical demand curve would suggest that consumers have a fixed desire for food, regardless of its price. The horizontal supply curve would imply that we can produce unlimited quantities of food at a constant price. Both ideas are false. Consumers and food producers respond to prices, giving these curves their characteristic slopes.

A second error is to draw the supply and demand curves as shown in Figure 8-5B. Here, the quantity of food consumed is shown eventually overtaking the quantity of food produced, leading to food shortages and famine. The error in this graph

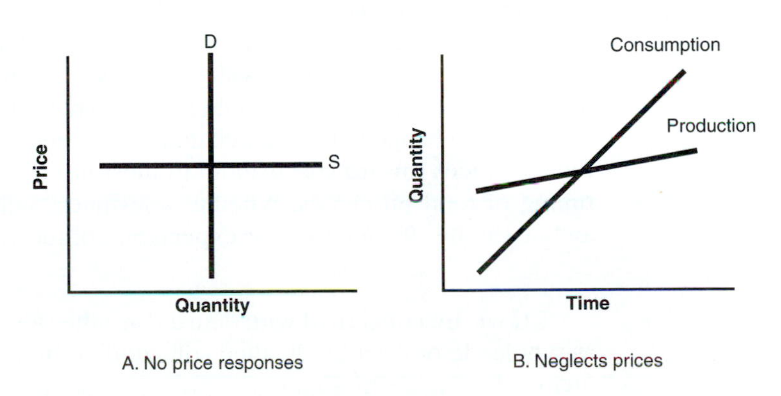

FIGURE 8–5 Two food industry myths.

is the total neglect of prices on supply and demand and their role in facilitating the matching of supplies with demand. If world population puts pressure on available food supplies, prices will rise and stimulate increased food production. This is precisely what happened in the 1960s and 1970s when anticipated world food shortages led to the increased food production in many developing countries associated with the Green Revolution. We should never take our food supply for granted and assume shortages are impossible. However, it is comforting to know that food supply and demand do adjust to prices, and one of those adjustments is increased food production.

Changes in Demand and Supply Curves

The demand and supply curves show the responses of buyers and sellers to changes in prices, everything else being equal. For example, when prices rise, producers are given a profit incentive to increase production and therefore the quantity supplied to the market. Consumers, on the other hand, receive a signal to reduce their consumption of that product. Thus, a change in price results in movements along the supply and demand curves and changes in the quantity supplied and demanded.

These curves can also shift in position. A shift in the supply or demand curve is quite different from a movement along the curves. If more of a commodity is produced or consumed at the same prices as previously, there has been a rightward shift in the supply or demand curve. If less is produced or consumed at the same prices as previously, there has been a leftward shift in the curves.

The factors that shift demand curves are assumed to be held constant when the demand curve is drawn. These include:

1. A change in the number of buyers. Such changes may occur from either population growth or an extension of the market served by the product, say by exporting.

2. A change in the incomes or purchasing power of people.

3. A change in the tastes and preferences for particular products. These may be a result of many factors, such as religion, habit, or personal desires. Demographic changes, such as age, urbanization, and household composition, may alter tastes and preferences for foods.

4. A change in the prices of related products. Most consumer goods have substitutes; and the price we will pay for a given amount of one product will depend to some extent on what we would have to pay for the possible substitutes.

5. A change in the expectations of buyers as to the future levels of prices and their attitudes toward speculation. In most instances there is some flexibility in the timing of food purchases. Whether consumers will purchase immediately or wait awhile can be affected by their expectation of future price levels.

If we are concerned with demand at other levels of the marketing system, such as wholesale or farm levels, then still another factor that can cause change must be included:

6. Marketing costs may change, shifting the retail or derived demand at the producer level.

The most important of these factors influencing the longer-run demand for food products as a group are changes in income levels and the population growth rate. The other factors above generally affect specific products and may be of a short-run nature.

The supply curve may also shift. If *more* of a product is offered for sale at the same or lower prices than before, supply has shifted to the right. Referring to the corn example, if 6 billion bushels are offered at $4.00, 5.7 billion at $3.75, and so forth, supply has shifted. If this new schedule were plotted on Figure 8-3, the new supply curve would fall to the right of the original. When *less* of a commodity is offered for sale at the same or higher prices than before, supply has shifted to the left. The curve for this supply shift would fall to the left of the original curve.

These factors are assumed held constant for a supply curve and shift the supply curve:

1. In the short-run, there may be a change that would induce sellers to offer their available stock of goods at a different schedule of prices. These would include such factors as costs of storage, the sellers' need for cash, and expectations about future prices.

2. In the intermediate and long-run periods, there may be a change in the costs of production of the commodity. This may be caused by changes in costs of farm supplies or in the technology of the production of the commodity itself. It may also be caused by changes in the costs of producing other commodities that compete for the same resources. For example, a technological change that increased the feed conversion rate for hogs and lowered their costs of production would shift the supply curve of hogs. At the same time, if there were no similar changes in cattle production, the supply of cattle would decrease, because hogs and cattle are farm enterprises that compete with each other for the same production resources.

It is important to remember that although supply and demand curves are constantly shifting, the demand shifters are totally different from the supply shifters. This makes it possible to analyze the effects of most developments by shifting only one of the curves and moving along the other. Changes in demand and supply have their real importance in their effect on the equilibrium price. Figure 8-6 shows the effect on equilibrium price of four possible supply and demand shifts. The changed curve is shown by the broken line and the new equilibrium price by P'. These examples can be expanded to show the results of all sorts of combinations of supply and demand changes. However, the results are often indeterminate if too many variables and shifts are analyzed at the same time.

Elasticity of Demand and Supply

The law of demand states that as prices go down, the quantities purchased will increase. The law of supply states that as prices go down, the quantities offered for sale will decrease. But buyers and sellers often want to know by *how much* will

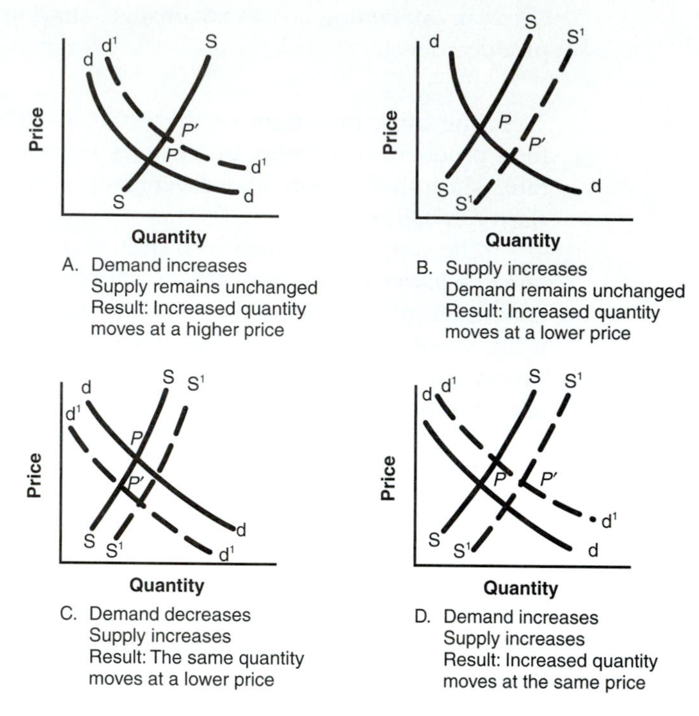

FIGURE 8–6 Situations illustrating some effects of changes in supply and demand.

quantity respond to changes in price? *Price elasticity* is a measure of the responsiveness of quantity supplied and demand to changes in prices. Price elasticity of demand indicates how much buyers' quantity demanded will change when prices change. Price elasticity of supply indicates how much sellers' quantity supplied will change when prices change.

Generally, demand curves are classified according to their elasticities into two broad groups: elastic and inelastic. The dividing point between these two classifications is unit elasticity.

Demand curves with unit elasticity are those in which the changes in quantity taken are proportional to the changes in price. An elastic demand curve is one for which the changes in quantity taken are proportionately *greater* than the changes in price. Inelastic demand curves are those in which the changes in quantity taken are proportionately *less* than the changes in price. Demand elasticity (E_p) can be approximated by the formula:

$$E_p = \frac{\text{\% change in quantity demanded}}{\text{\% change in price}}$$

The price elasticity of a unitary demand curve has an absolute value, ignoring sign, of 1. An elastic demand curve has an absolute value greater than 1, and an inelastic demand curve has an absolute value less than 1. For example, assume that price has decreased by 10 percent. The law of demand indicates that the quantity taken for commodities with demand curves of different elasticities would be as follows:

1. With unit elasticity, amount taken would increase by 10 percent.
2. With an elastic demand, amount taken would increase more than 10 percent—perhaps 12 or 15 percent.
3. With an inelastic demand, amount taken would increase less than 10 percent—perhaps 5 or 7 percent.

The price elasticity of demand for a product is primarily determined by the number of substitutes for it. Necessities with few substitutes, like salt, have highly inelastic demand curves. Marketing activities such as transportation, storage, and processing can alter the elasticity of demand for a food product by increasing or decreasing the number and availability of substitutes for consumers. The development of a new product may change the substitutability of one product for another, making the demand more or less elastic. Advertising may provide new information for consumers, making demand more elastic, or it may convince shoppers there are no good substitutes for a particular brand, making its demand less elastic.

On graphs of the same scale, the more inelastic the demand, the steeper is its plotted curve. However, the slope of a demand curve is not the same as its elasticity. Figure 8-7 shows how food products generally increase in price elasticity as they move toward the consumer. This is a result of the greater substitution possibilities for branded items at a local foodstore than for commodity items at the farm level.

Cross-elasticity of demand refers to the effect of a price change of one commodity on the quantity demanded of another product. It is useful to classify related food products as either complements or substitutes. Two foods are *substitutes*—for example, beef and pork—if they are alternatives and an increase in the price of one increases the consumption of the other. They are *complements* (such as ham and

FIGURE 8–7 Demand elasticity at various market levels.

eggs) if they are eaten together and if an increase in the price of one decreases the consumption of the other. Complementary products are often promoted together, as when the lettuce industry co-sponsors an advertisement with the salad dressing manufacturers. In contrast, producers of substitute foods are rivals for the consumer's dollar.

The most important implication of demand elasticity is its effect on consumer food expenditures and the amount of money received by food firms from selling different quantities at different prices. We can see this by studying the results of a 10 percent cut in prices on three hypothetical commodities of different elasticities. For a commodity with a demand of unit elasticity, the quantity purchased would increase 10 percent and the total revenue (price times quantity) would be exactly the *same* as before the price cut. For a commodity with an elastic demand, the quantity taken would increase more than 10 percent and the new total revenue from sales would be *greater* than before. However, for a commodity with an inelastic demand, something less than 10 percent more of the commodity would be taken and the total revenue would be *less.* This is shown in Figure 8–8, where the total revenues are indicated in parentheses. Note the effects of these price changes on total revenue.

Demand curves for the same product are often not of the same elasticity throughout the entire curve. Typically, elasticity is greater for small quantities and high prices of products, and the demand curve becomes less elastic at higher volumes and lower prices. Elasticity may also change over time as producers develop substitutes and consumers discover substitution possibilities.

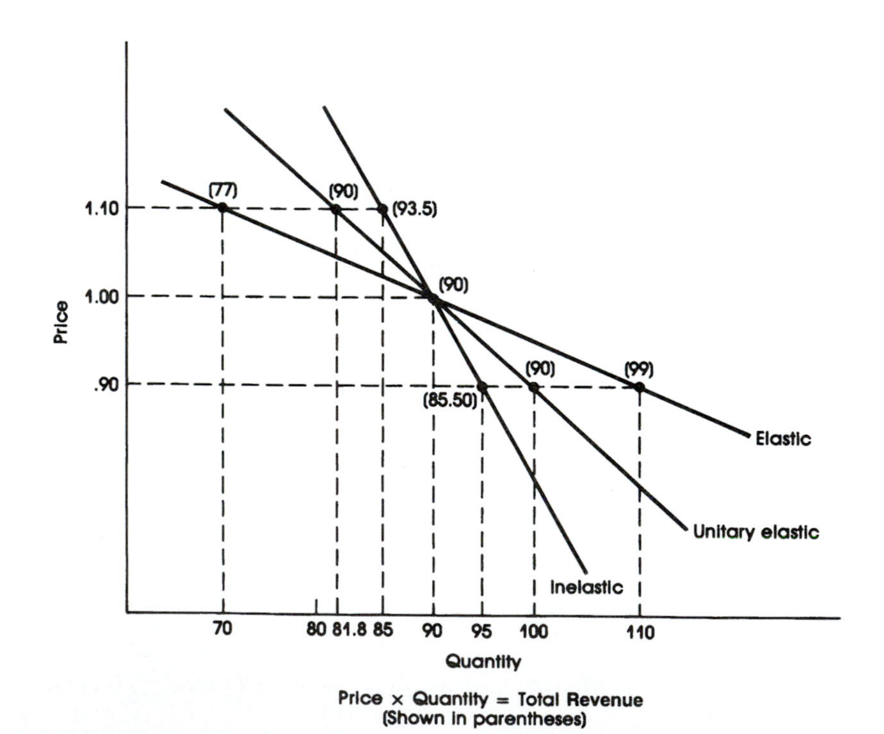

FIGURE 8–8 Impact of a 10 percent price change on total revenue: unitary, elastic, and inelastic demand curves.

Table 8-4 shows some estimated price elasticities of demand for selected food products. Notice the wide range in elasticities, that farm level price elasticity of demand is always lower than retail price elasticity, and that most farm commodities have an inelastic demand.

Will a large crop return more money to growers than a small one, even though prices are lower? What will be the effect of restricting output on the total returns from sales? Should a foodstore lower its prices to sell more products? Knowledge of demand elasticity and its relationship to total revenue is the key to answering questions such as these.

The same elasticity framework also can be applied to supply. Commodities that are very responsive to price changes have elastic supply curves. Those that respond relatively little to price changes have inelastic supply curves. On graphs of the same scale the more inelastic the supply the steeper the plotted curve.

Time is a very important consideration in supply analysis. As the time period under consideration lengthens, the supply curve tends to become more elastic. It is useful in price analysis to distinguish between the short run, the intermediate period, and the long run. The supply curves for each of these lengths of run are illustrated in Figure 8-9.

In the short run, the volume of product supplied to the market is fixed by the available harvest or quantity of products that must be marketed. In this case the supply curve is perfectly vertical, price adjusts to the quantity supplied, and the products will be marketed at any price above harvesting costs. Fresh fruits and vegetables

TABLE 8-4 Retail and Farm Price Elasticities for Selected Food Products

	Effect of a 1 Percent Change in Price on the Percentage Change in Quantity Demanded	
Commodity	*Farm Level*	*Retail Level*
Turkey	—	−1.56
Margarine	−.69	−.84
Chicken	−.60	−.78
Apples	−.68	−.72
Butter	−.46	−.65
Beef	−.42	−.64
Ice cream	—	−.52
Cheese	—	−.46
Pork	−.24	−.41
Fresh milk	−.32	−.34
Eggs	−.23	−.31
Potatoes	−.15	−.30
Breads, cereals	—	−.15

SOURCE: P. S. George, & G. A. King. (March 1971). *Consumer Demand for Food Commodities in the U.S. with Projections for 1980,* University of California, Giannini Foundation Monograph No. 26. Reprinted by permission.

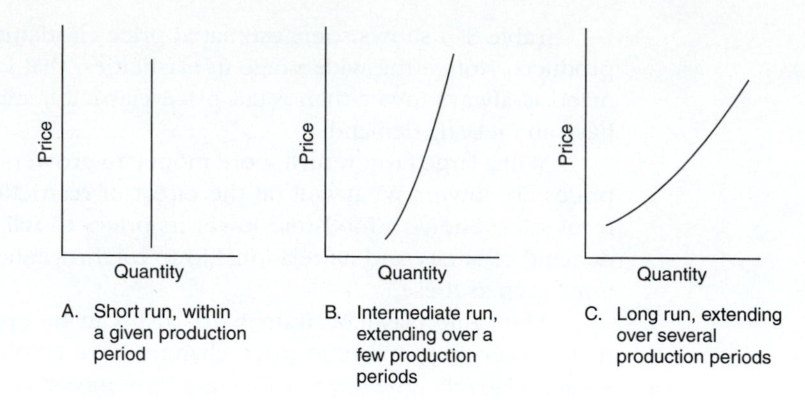

FIGURE 8–9 Supply curves illustrating the effect of time on elasticity.

at harvesttime illustrate this short-run pricing process. Because this stock of goods exists, costs that have been previously met in its production will not influence price.

In the long run, the food supply curve will have the familiar upward slope because there is sufficient time for producers to make investments in expanding productive capacity, which will increase output. Increased prices will increase output and vice versa. This may be many years, as in the case of fruit orchards that take several years to bear fruit and reach maturity. This period will be shorter for grains where land, equipment, and management could be shifted from one crop to another each year.

The intermediate period is that time during which products can be produced with the existing production facilities. The impact of food marketing decisions is most important in the intermediate period. This is the period when food producers and marketing firms can alter the quantity of products supplied to the market through their time, form, and place decisions. Again, the supply curve will be positively sloped as sellers increase supplies going to market when prices are higher and decrease supplies when prices are lower. For example, a livestock producer can increase or decrease the numbers of animals going to market depending on the price that is offered. A grain producer or marketing firm can sell additional grain out of storage if prices are attractive or keep the grain in storage if prices are judged too low.

The length of run is a highly useful tool in food price analysis. It is not a specific time period. Rather it is determined by the flexibility that sellers have in production and marketing. The appropriate length of run for price analysis will depend on the timing of the product's harvest, its urgency in marketing, and producers' marketing alternatives. A single-harvest, perishable crop, such as fresh strawberries, will have a short-run vertical supply curve. If the strawberries can be processed and sold later, the intermediate-run supply curve will be upward sloped. Grain producers are operating in the short run if they have no storage facilities and must sell at harvest. Storage facilities allow them to market in the intermediate run. Both grain and strawberry producers make long-run supply decisions over the years as they expand or contract their operations.

Time lags in food production and marketing can also complicate the analysis of food prices. Some observers who notice that large receipts on today's market move at lower prices, or that a large crop brings lower average prices, conclude that the law of supply is not valid. However, these observations often fail to recognize the time lags between the price stimulus and the quantity response. High livestock prices on the market today will result in larger quantities marketed one and two days later because of the time necessary to move the shipments from the farm to market. The correct supply schedule in this instance would be one that related today's prices with receipts two days later. Similarly, the quantity of hogs going to market during the fall months is a response to prices that existed during the previous year. The supply schedule in this case would relate current prices with the level of production forthcoming twelve months later.

APPLICATIONS OF SUPPLY AND DEMAND ANALYSIS

It is not sufficient simply to be aware that supply and demand influence food prices. The student must understand the mechanics of supply and demand shifts in order to be able to understand, analyze, and predict food price changes. Some selected applications of supply and demand analysis will be helpful in building these skills.

The Instability of Farm Prices

Farm prices tend to be more variable than nonfood prices. The reasons for this lie in the inelastic supply and demand curves for agriculture and the unpredictable changes in food supplies as a result of weather conditions, disease, and other factors. Because of the inelastic curves, shifts in either demand or supply will result in proportionately larger price changes. In turn, instability of farm incomes is directly related to unstable farm prices. Because the demand for most farm products is inelastic, falling prices will reduce total revenue and rising prices will increase revenue. This price and income instability complicates the farmer's investment market planning processes.

Supply Control in Agriculture

Inelastic demand curves provide farmers with a powerful incentive to restrict output by shifting supply leftward and raising gross farm income, as shown in Figure 8-10. This incentive would not exist if farm demand curves were elastic. However, it has generally been difficult for farmers to achieve the levels of supply control necessary to raise their total revenues. This is because each farmer tends to increase output in response to higher prices resulting from supply control programs. This shifts the aggregate farm supply curve to the right, eroding any price increases resulting from the supply control. There are also free rider problems in supply control programs. Consequently, most successful agricultural supply control programs require government authority to prevent supply expansion from eroding higher prices.

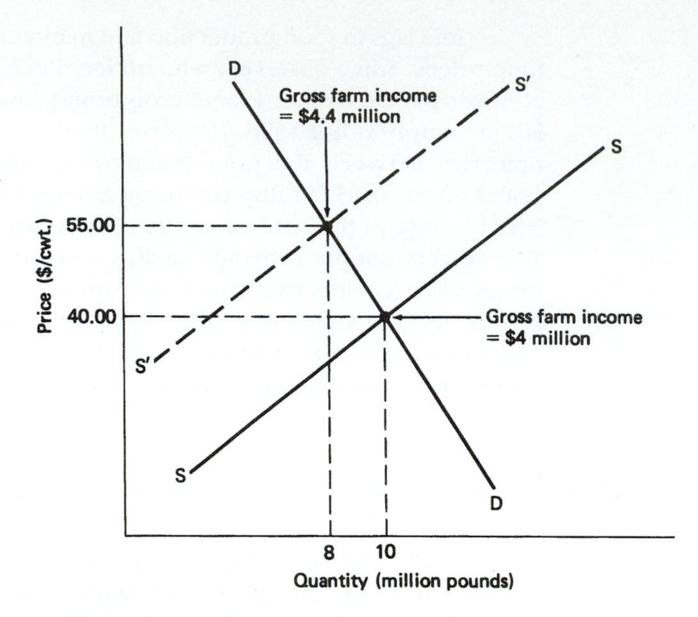

FIGURE 8–10 Effects of supply restriction on gross farm income.

Large Crop Penalty, Small Crop Reward

It is sometimes observed that farmers' returns are higher when poor weather reduces yields than when good weather produces high yields. Why are farmers punished in the marketplace when they succeed in producing good yields and high output and rewarded when output is lower? The explanation lies in the inelastic demand for most farm products.

The agricultural supply curve is drawn for normal, expected yields. An unexpected decrease in yields shifts the supply curve leftward. An unexpected yield increase shifts supply rightward. The leftward shift results in an increase in farm income because of the inelastic demand. The rightward supply shift lowers farm prices and, to add insult to injury, this movement along an inelastic demand curve reduces gross farm income. The farmer is penalized for an unexpectedly large crop! What is a farmer to do? Most farmers learn to average their incomes over the good and poor crops. A less-than-altruistic farmer might hope that other farmers suffer poor yields but that he or she reaps bumper crops! A large crop at high prices is the best of all possible worlds.

WHO BENEFITS FROM COST-REDUCING FARM TECHNOLOGY?

It is sometimes observed that farmers receive less benefit from new, output-increasing agricultural technologies than consumers. Why would that be?

New farming techniques and technologies shift the farm supply curve rightward. As prices fall along an inelastic demand curve, gross farm sales also fall. This means that the benefits of cost-reducing agricultural technology are passed on to the consumer in the form of lower relative food prices in the long run. This has been called the *agricultural treadmill* because of the farmers' constant need to adopt new technology that benefits food consumers more than producers.

Does this mean that farmers should resist the development and adoption of new farm machinery and higher-yielding crops? Not necessarily. Since many of these techniques are developed in the farm input sector or in universities, farmers do not control their development. In any case, there is good reason for farmers to adopt new technologies and lower their costs in the short run, even though this simply moves all farmers to a new level of the cost-price squeeze in the long run. Farmers who are the first to adopt new technology—sometimes called *early innovators*—gain a profit advantage over those who adopt later, before the full supply shift is completed.

How Does Trade Affect Food Prices?

Food exports are a demand shifter for a country, while food imports are a supply shifter. Increased exports of a commodity—for example, U.S. wheat—shift the American demand for wheat to the right and move quantity supplied up the American supply curve of wheat. This increases farm wheat prices and wheat farmers' total revenue as well as consumer prices for wheat products. The wheat prices will be greater in the short run than in the long run as wheat producers increase output in response to the higher prices.

American food imports shift the U.S. food supply curve to the right, reducing farm and retail food prices. The resulting movement along the inelastic food demand curve reduces farm receipts and consumer food expenditures. The aggregate effects are about the same for both complementary and supplementary food imports. No wonder farmers generally resent food imports and are big promoters of farm exports.

U.S. farm exports can also be a source of farm price and income variability since they vary from year to year with food production conditions in other countries. A drop in export demand for U.S. farm products may cause a severe decline in farm prices, just as a surge in export demand will cause farm prices to spike upward in the short run.

Trade can also affect the elasticity of demand for a farm product. Because of the large number of countries competing in foreign trade, the export demand for a farm product is usually more elastic than the U.S. home demand curve. Therefore, as more of a farm product is exported, its elasticity of demand increases. Many analysts now believe that the demand for U.S. farm products that have large export markets has moved from inelastic to elastic.

What Is the Effect of Food Price Ceilings and Floors?

Although there is a tendency for prices to go to market-clearing levels, it is possible for government to prevent this. A *price ceiling* is a legally set price below the equilibrium price; a *price floor* is a legally imposed price above the equilibrium level.

These devices prevent market prices from falling below floors and from rising above ceilings. Price floors are sometimes used by governments to "support" farm prices and incomes. Price ceilings are sometimes used by governments to prevent inflation, although there is some question about their value in doing so.

Whatever the intentions behind them, price floors and ceilings distort markets and prices by preventing buyers and sellers from reaching a market-clearing price. As shown in Figure 8-11, a price floor generates a surplus of supply over demand ($Q_1 - Q_0$). This may be stored, dumped, or sold in noncompeting markets. Price ceilings, on the other hand, result in shortages, black markets, rationing, and out-of-stock problems.

Much agricultural legislation has dealt with how to raise farm prices and incomes using price floors. These supports may help farmers in the short run. But farm price floors can cause significant resource allocation problems in the long run. For example, price supports artifically stimulate increased productive capacity and output in the long run. They also attract imports and can reduce the competitiveness of U.S. farm products in world markets. While food surpluses may seem like a valuable asset in a hungry world, they are expensive, not only to store but in their use of resources that might be employed elsewhere.

During wartime and other periods of high inflation, the government and the public often see the problem as how to keep prices from rising too fast or too high. The Office of Price Administration (OPA) during World War II and the Office of Price Stabilization (OPS) during the Korean War were direct attempts to control rising prices through price ceilings.

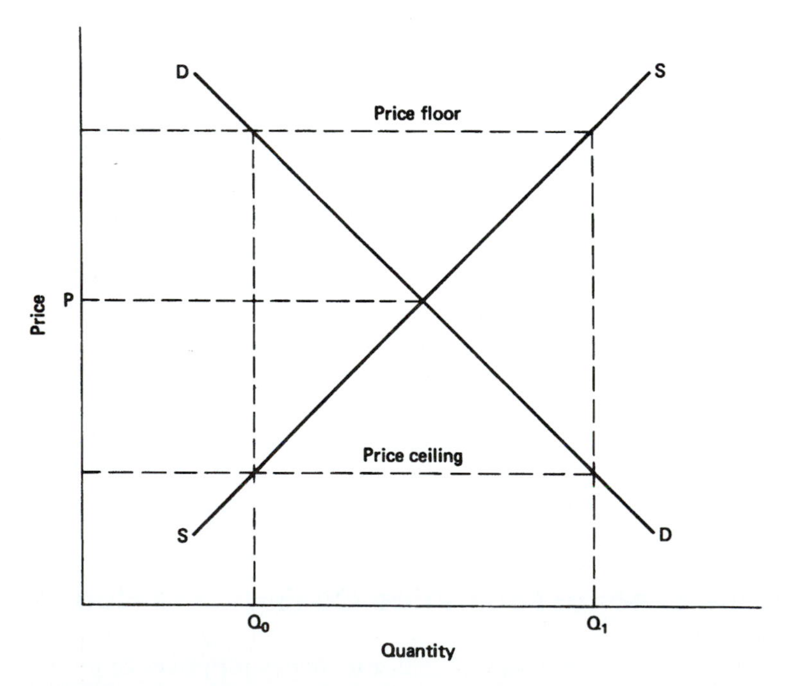

FIGURE 8–11 Food price floors and ceilings.

In initiating a price control program, prices are "frozen" as of a given date by government order. From these freeze points, adjustments up or down are made as needed. The enforcement technique is simply to make it illegal to charge higher prices. The simplicity of the technique, however, belies the difficulty of the job. Effective price control is immeasurably complex. Food prices are especially difficult to control. With large annual and seasonal variations in production, differences in qualities, and numerous and small producers, price ceilings always seem to be out of adjustment at some place or time. A rationing and subsidy system has to be devised.

Price Rationing

Do prices determine the quantities supplied and demanded or do these quantities determine prices? It is a chicken and egg situation. Equilibrium prices and quantities are simultaneously determined. However, we normally think of price changes influencing output and selling decisions. The term, *price rationing*, refers to the process by which prices rise and fall in an attempt to match the available supply with current demand. However, in the short run, the cause and effect is often just the opposite. Output and selling decisions determine prices. This is because many food products are highly perishable and must be marketed, regardless of price, giving them a perfectly vertical supply curve. In this situation, quantity supplied, in combination with the demand curve, is determining prices.

MINI CASE 8: ARE FARM AND FOOD PRICES DIFFERENT FROM OTHER PRODUCT PRICES?

After their marketing class, Billy and his new friend Sue were discussing what the professor said about the differences and similarities between food prices and the prices of other products which people need and buy. Billy thought that food prices are different because of biology, perishability, and the necessity of food. "I think food prices are too important to be left to the impersonal forces of supply and demand," said Billy. "Farmers need a fair price for their products and so do food consumers" he said. "Wait!," asked Sue, "who would determine that price?" "Food has an upward-sloping supply curve and a downward-sloping demand curve," she said, "just like any other product." "And the market-clearing price for farm and food products is determined where the supply and demand curves intersect. If you pay farmers less than this," Sue lectured, "there will be shortages and food shoppers will be standing in line to buy groceries." "And," Sue continued, "consumers will have to pay farmers more if they want more food produced." "Supply and demand are important in food prices," she said. Billy agreed, but he didn't like it when she seemed to understand food marketing better than he did.

Do you agree or disagree that food and nonfood prices should both be determined by the forces of supply and demand? Why or why not?

This short-run price rationing is very important when the food supply falls as a result of a disaster, such as a flood or drought. This temporary reduction in food supplies will increase prices as a way of rationing the available supply among buyers. Understandably, consumers are not happy with these higher prices and their related reductions in quantity demanded. Often they blame food middlemen for raising prices and exploiting the situation. However, these price rises have some beneficial market effects. First, a rising price of one food encourages consumers to switch to other foods whose relative prices have fallen. This reduces the upward pressures on prices. Secondly, in the long run, these higher prices stimulate increased production which alleviates the food shortage and lowers food prices.

A Food Price Dilemma

The fact that food consumers and producers react to food prices in opposite ways creates a dilemma for those who wish to help one group without regard to the welfare of the other. For example, governments sometimes attempt to raise farmers' prices through price supports above market-clearing levels. This helps farmers at the expense of higher prices for consumers. Similarly, price ceilings help food consumers at the expense of farmers. The short-run rationing effect of prices also can conflict with the long-run production incentive of prices. Take, for example, U.S. food aid shipments to other countries. These free or low-cost imports may provide necessary supplies for those suffering from food shortages and dietary deficiencies. However, in increasing food supplies and reducing local food prices they also reduce incentives to increase food production in recipient countries. It is sometimes difficult to give short-term food aid to other countries without distorting longer-term price incentives there. Of course, this shouldn't be used as an excuse not to assist hungry people. But it might influence the form of the aid. Perhaps it is better to "teach a man to fish" than to "give a man a fish."

The Relationship of Prices to Costs in Agriculture

Many observers have noted the tendency for farm prices to approximate costs of production over the long run. Why would that be true? The reasons for this are related to the competitive structure of agriculture and the long-run adjustments of farm output to profits and prices.

In the short run there is no necessary relationship between farm prices and costs. Prices are established by the intersection of a vertical supply curve with a downward sloping demand curve. This may occur at a profitable point where prices are above average costs or at an unprofitable point where prices are below costs. In reality, farm prices often do fall below costs of production.

In the longer run, however, firms will expand their output in response to profits and contract output when prices are below costs of production. This gives an upward slope to the supply curve so that profits will bring forth increased output and losses will reduce output. This means that in the long run production will adjust such that average prices will just tend to equal average costs. That is to say, the re-

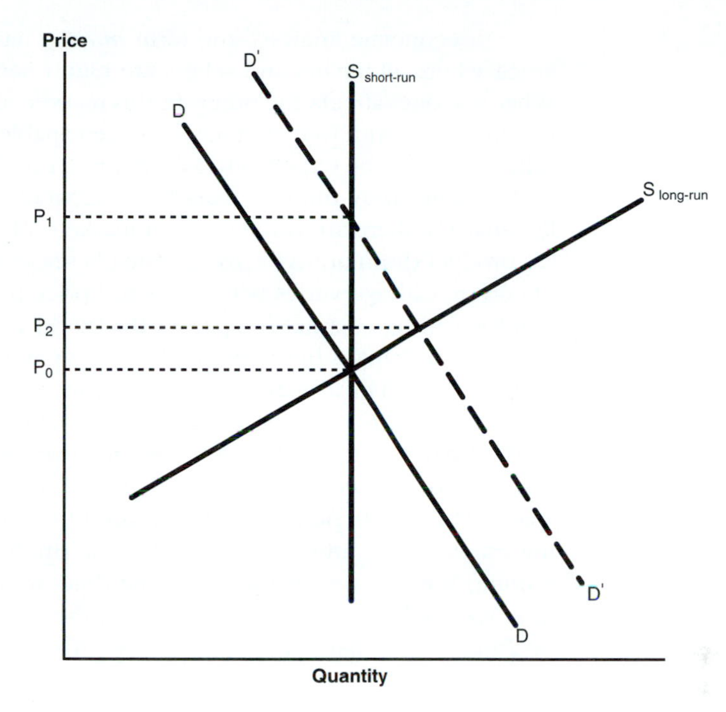

FIGURE 8–12 Short- and long-run price and output responses.

turns to capital, land, labor, and management will be no more, or no less, than they would receive if they were used in other forms of production.

This is shown in Figure 8–12. A rightward shift in the demand curve to D′D′ increases prices in the short run from P_0 to P_1, raising prices above costs of production and creating profits. In the long run, however, these profits stimulate producers to increase output, rotating the supply curve to the right and reducing prices to, say, P_2. The short-run profits are eroded by the long-run supply response, and profits return to a more normal level.

Because of these long-run adjustments, it is sometimes said that low prices and profits in farming "cure" themselves. Sustained low prices and profits below costs will lead to reductions in output and eventual higher prices and profits. Sustained prices and profits above costs will lead to an expansion in output and eventual lower prices and profits. The response period for these changes is determined by the ease of resource adjustment.

MARKETING AND THE LAW OF ONE PRICE

A market has been defined as an arena for organizing business activity. We have also seen that the business of the food marketing system is to add time, form, place, and possession utilities to farm commodities. In the market economy, prices and the profit motive play a key role in coordinating this marketing activity. The way in which this occurs bears closer examination.

In economic analysis, the term *market* has a special significance—that of an arena where all buyers and sellers are highly sensitive to each other's transactions. What one does affects the other. In this market, all buyers and sellers must be able to communicate with one another, must be capable of exchanging products with each other, and must be exposed to similar price signals.

We normally think of markets as geographic areas. But in the context of market analysis, there are three types of markets: (1) geographic (place utility) markets, (2) product (form utility) markets, and (3) seasonal (time utility) markets. If the price of corn in Chicago varies with the world price of corn (which it does), we may infer that the Chicago market is a part of the world corn market. Or if the price of beef is sensitive to changes in the price of pork, we would suspect that both are in a single market, the red meat market. Or, if the June 1999 price of soybeans fluctuates with the October 1998 price of soybeans, we would say that both months are in the same market, namely the 1998–1999 soybean market. Market boundaries, then, are defined by the degree of interdependence of buyers and sellers over time, form, and space. This interdependence is measured by price sensitivity. These boundaries are not fixed. They could be compared to the ripples caused by dropping a pebble into a pond. We may find it useful, at one time, to identify a "beef market"; at another time we might refer to beef as being in the "red meat market"; at still other times, beef might be considered a part of the "all-food market." A market is a flexible tool of analysis.

Geographic markets may be local, regional, national, or international in scope. Bulky products such as hay have a small geographic market area because high transportation costs prevent distant buyers and sellers from trading with each other. Milk and livestock can be transported more economically so they tend to have regional or national markets. Wheat and cotton are easily transported and have international markets. With improved communications and transportation facilities, market areas usually increase in size.

There is a special relationship among prices in space, time, and form markets. It is known as the *law of one price* (LOOP). This law states that under competitive market conditions all prices within a market are uniform, after taking into account the costs of adding place, time, and form utility to products within the market. The law can be very useful in determining the size of a market, predicting price changes within a market, and evaluating the pricing efficiency of a market.

Figure 8–13 illustrates the law of one price in the international corn market. This map shows the prices and transportation costs for four shipping points (Farnhamville, Seneca, Champaign, and Columbus), three export terminals (Chicago, New Orleans, and Norfolk), and the major port of entry into the European Economic Community, Rotterdam. Grain tends to flow from the surplus-producing shipping-point areas, through the U.S. ports, and on to the deficit-producing European countries. The prices in Figure 8–13 are aligned according to the law of one price. They differ only in the cost of transportation—the cost of adding place utility to the corn. Put another way, any price can be divided into two parts: (1) a uniform price of corn throughout the market, and (2) the costs of adding place utility to the corn.

Under competitive market conditions, there is a tendency for prices within a market to remain aligned according to the law of one price. To illustrate: If the price European consumers are willing to pay for corn were to increase by 10 cents per bushel, there also would be an increase in the price of corn at the three U.S. ports

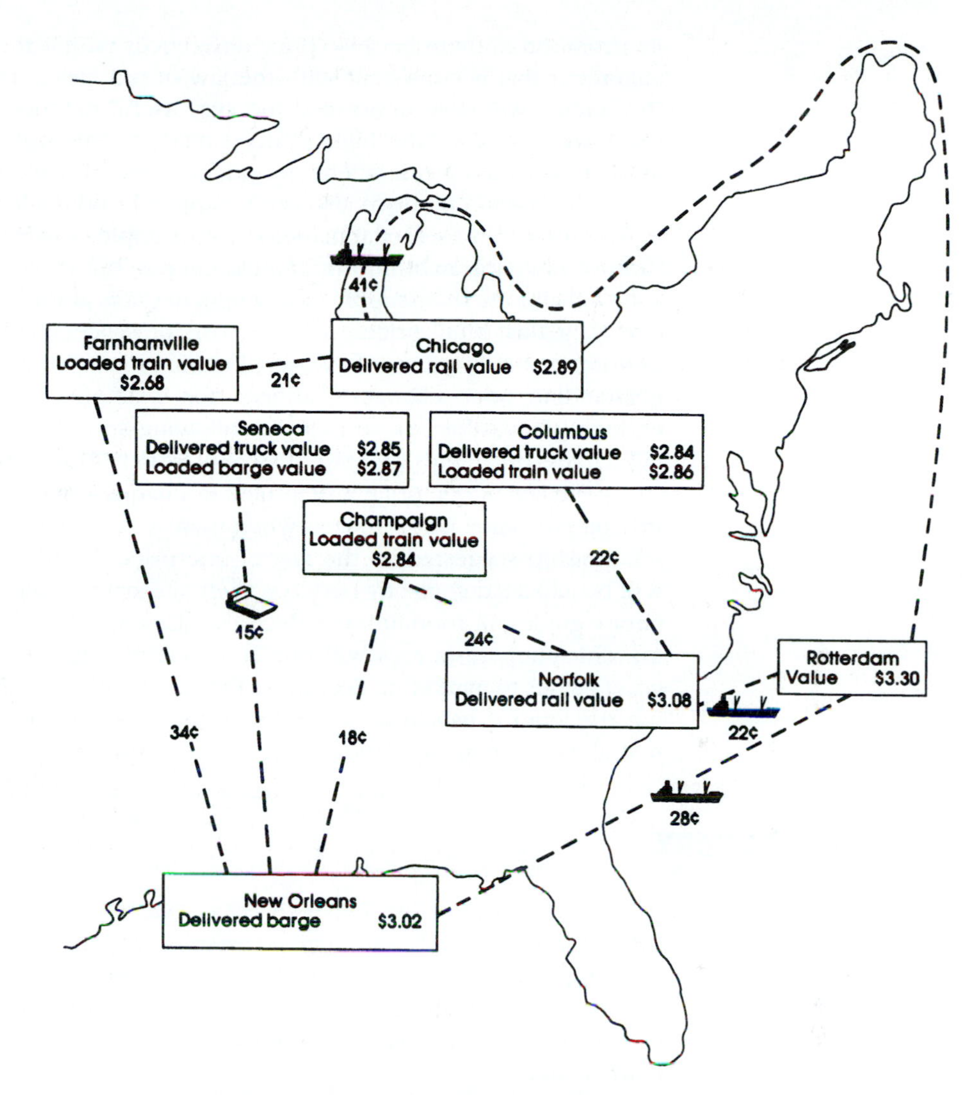

FIGURE 8–13 Prices in the international corn market.

and, ultimately, at the farm level. Within the market area, these prices are tied together and will move together.

The law of one price results from the profit-seeking behavior of food marketing firms and traders. If the Rotterdam corn price increases by 10 cents per bushel, grain traders will see that extra profits can be made by increasing the flow of corn from the United States to Europe, because the price difference exceeds the transportation cost. As traders respond to this profit incentive, they will bid up the U.S. price of corn and reduce the Rotterdam price. After all adjustments are made, the United States–Rotterdam price difference will return to the level of the cost of transportation. The U.S. farm price and the Rotterdam price will be higher than originally. This means that the traders' profit-seeking behavior will reduce the extra profits promised to them because price differences within the market will return to the alignment that is consistent with the law of one price. In the meantime, however, the traders will have performed the very useful function of shifting the corn from the lower-valued to the higher-valued market. This process of buying low in one market to sell at a profit in a higher priced market is called *arbitrage.*

The law of one price also can be applied to the addition of form and time utility to product and seasonal markets. For example, a dairy processor can change the form of raw farm milk into fluid milk, cheese, butter, ice cream, or yogurt. Processors will allocate their supply of raw milk to these alternative products in such a way that their final retail prices, less the costs of processing each product, will equal a uniform raw-milk price. For seasonal commodities, storage costs are the key to understanding price changes. Farmers, for example, would be expected to store enough grain at harvest so that the following spring's price would be equal to the fall harvest price, plus the cost of storage from fall to spring.

The law of one price is helpful in evaluating pricing efficiency within a market. Pricing efficiency is maximized when there is a tendency for prices to maintain the relationship suggested by the law of one price. Under these conditions, resources will be allocated correctly between their alternative uses and prices will serve as accurate guides for food industry decisions. Conversely, with less than maximum pricing efficiency, resources will not be efficiently allocated, prices will not serve as good guides to marketing decisions, and there will be some loss in output. Poor pricing efficiency is usually associated with the presence of large, dominant firms, trade restraints, price manipulation, poor market information, or other barriers to trade.

PRICE DISCOVERY

It is useful to distinguish between *price determination*—the process by which the broad forces of supply and demand establish a general, market-clearing, equilibrium price for a commodity—and *price discovery*—the process by which buyers and sellers arrive at a specific price for a given lot of produce in a given location. The equilibrium price target must be "discovered" and applied to each transaction in the marketplace.

Price discovery is a human process, marked by errors in judgment and fact and subject to the relative bargaining powers of buyers and sellers. There is no guarantee that buyers and sellers will always or immediately discover the equilibrium price. In real life, variations in the price-discovery process make it profitable for buyers and sellers to shop around among alternatives and to bargain on prices and other terms of trade.

This distinction gives us a two-stage price-discovery process for farm products. Stage one consists of evaluating the supply and demand forces and estimating the market-clearing price. Stage two is the application of this estimated price to a specific trade, taking into consideration grade premiums, quality, discounts, buyer and seller services, and bargaining power. Pricing errors can arise at both stages of the pricing process, but these errors can be minimized, and price discovery improved, by good market information.

Five systems of price discovery for farm products have been identified: (1) individual, decentralized negotiation; (2) organized, central market trading; (3) formula pricing; (4) bargained prices; and (5) administered prices. These alternative pricing systems vary in pricing efficiency, in the amount of price information they generate, in the freedom allowed buyers and sellers in arriving at prices, and in the relative bargaining strength conferred on buyers and sellers.

In *individual, decentralized negotiations,* each farmer bargains separately with buyers of farm products until a price is established. Private treaty negotiations are quite common in agriculture. The resulting fairness of prices depends on the information, trading skills, and relative bargaining power of buyers and sellers. Consequently, prices discovered in this way tend to vary widely for different transactions. Moreover, the time and energy costs of this form of price discovery are rather high compared with the alternatives.

Organized, central markets shift the locus of price discovery from the farm gate to a central, often public, marketplace. All buyers and sellers and their supplies and demands are represented in the central market. Terminal markets and auctions are examples. These markets generate considerably more information than private treaty markets and probably also reduce some costs of price discovery. Also, prices for individual transactions are likely to be more uniform. Finally, because of the open outcry and public nature of these markets, they are said to be more *transparent* than other price discovery points. This means the pricing machinery is more open and exposed to the participants.

Formula pricing systems evolved in attempts to secure the benefits of central-market price discovery without physically routing all produce through central markets. Egg producers, for example, frequently are paid a formula price, which is related to the market price reported by the U.S. Department of Agriculture Market News Service or the Urner-Barry report (a private market newsletter). These prices are adjusted, again by formula, for transport costs and quality differences. Meat frequently is purchased in advance to be priced "at the yellow sheet," a private market newsletter. Formula pricing can reduce transaction and bargaining costs. However, formulas can become obsolete; moreover, they require at least one "correct" price on which to base other prices.

Bargained prices are common for fruits, vegetables, nuts, and milk. Bargaining implies collective pricing on the part of farmers. The collective bargaining process used in labor is frequently cited as the model for farmers to follow in order to discover farm prices. However, there are differences in labor and farm products. Bargained price discovery probably works best when there are relatively few producers in a concentrated geographic area selling a product that can be stored or withheld from the market.

Administered pricing systems are those in which the government becomes a third party in the price discovery process. Price supports, price ceilings, and supply control programs are the techniques of administered pricing. These have been used for a number of agricultural products.

SUMMARY

Prices are the key signals that direct and coordinate the decisions of producers, consumers, and food marketing firms in the market economy. These prices are the result of supply and demand forces operating within the framework of a competitive marketplace. Supply and demand analysis can be useful in understanding pricing forces, predicting the effects of market changes on prices, and generating a better understanding of the problems that farmers face in the market. There is a tendency for farm prices to return to the market-clearing equilibrium point when disturbed. Shifts in the supply and demand curves result in new equilibrium prices. In the short run, farm and food products tend to have an inelastic demand and supply. As a consequence, farm prices and incomes tend to be highly variable, farmers can increase their gross incomes if they can restrict production and supplies, and the benefits of cost-reducing agricultural technology are passed on to consumers in the long run. Discovering market-clearing farm and food prices is a difficult task for farmers and food marketing firms. There are several alternative ways to discover prices in agricultural markets.

KEY TERMS AND CONCEPTS

administered prices
agricultural treadmill
arbitrage
bargained prices
buyers', sellers' markets
complementary products
cost-price squeeze
cross elasticity
derived demand
diminishing marginal utility
early innovators
equilibrium price
formula prices
law of demand, supply
law of one price
price ceiling, floor
price discovery
price elasticity of demand
price rationing
price system
relative price
short, long run
substitute products
transparency

DISCUSSION QUESTIONS

1. You wish to help a businessman determine the elasticity of demand for his product. He indicates that sales revenues behave as follows:

Price Change	Sales Change	Elasticity
\$4.00 to \$3.00	Increase	_______
\$3.00 to \$2.00	No change	_______
\$2.00 to \$1.00	Decrease	_______

Fill in the elasticity of demand for each price range.

2. Why are some foods more price-elastic than others? How does this affect the producers of these products?

3. What is the effect on farm prices and incomes when there are inelastic demands for most farm products?

4. Suppose beef is selling at \$40/cwt at the farm level, and the farm-price elasticity of demand is –.40. What would be the market-clearing price of beef if there was suddenly a 2 percent increase in beef supply? (Hint: Assume supply is perfectly inelastic and shifts rightward by 2 percent.)

5. What impact would the following have on the equilibrium price?

a. A successful consumer boycott of meat.
b. Increased unemployment.
c. A disease that reduces the hog crop by 10 percent.
d. A new combine that reduces harvest costs by 5 percent.
e. Consumer complaints that food prices are too high.
f. An increase in wages of food marketing workers.

6. How does price discovery differ from price determination? Why are there so many variations in agricultural price discovery systems?

7. What is meant by this statement: "Prices coordinate resource allocation in agricultural production and marketing."?

8. Apply the law of one price to the seasonal increase in grain prices from fall harvest to the following spring.

SELECTED REFERENCES

Bressler, R. G., & R. A. King. (1970). *Markets, Prices, and Interregional Trade* (Chapter 4). New York: John Wiley and Sons.

Tomek, W. G., & K. L. Robinson. (1981). *Agricultural Product Prices* (Second edition). Ithaca, NY: Cornell University Press.

Waugh, F. V. (1964). *Demand and Price Analysis: Some Examples from Agriculture,* Technical Bulletin No. 1316. Washington, D.C.: U.S. Department of Agriculture.

CHAPTER 9

COMPETITION IN FOOD MARKETS

When products compete, they get better.

OBJECTIVES

After reading this chapter, you will be able to:

1. Explain why competition is so important in food and agricultural markets.
2. Distinguish between the many forms of competition found in food markets.
3. Appreciate the unique competitive situation which many farmers face.

Competition plays a number of important roles in our market economy. Competitive behavior helps organize economic activity and answer the basic economic questions: What to produce? How much to produce? How to produce? And how to distribute goods and services? Competitive activity also produces a set of prices that influence buyers' and sellers' behavior and resource allocation throughout the market. Further, competition is the mechanism by which we attempt to harness the profit seeking of business firms to the public interest. Finally, competitive activity is an important agent for change in food markets as it stimulates low-cost organization and the development of new technologies, products, and marketing strategies.

Freedom is at the heart of our concern for competitive food markets: (1) freedom of consumers to choose what they wish to eat; (2) freedom of firms to develop new products and market them; (3) freedom of new firms to enter the food industry; (4) freedom of farmers to make decisions about what to produce, how to produce it, and where and when to sell it; and (5) freedom of buyers and sellers to bargain together and arrive at mutually advantageous exchanges. These freedoms are vital to our decentralized, private enterprise economy. When competition breaks down, there is no longer assurance that private profit-seeking behavior is serving the public good. Some form of market intervention is often the result.

TYPES OF COMPETITION

Competition in the food industry takes several forms. *Product competition* refers to rivalry between alternative or substitute products, such as beef and pork, for the consumers' dollar. *Firm competition* concerns the rivalry between sellers of similar products. Firm rivalry frequently focuses on *brand competition*—rivalry between competing brands within a product class, for example, Del Monte versus Green Giant canned peas. *Interregional competition* is illustrated by the rivalry between California and Florida oranges, or Maine and Idaho potatoes. *International competition* takes place between nations, and *institutional competition* relates to the rivalry between competing market institutions, for example, grocery stores, fast-food restaurants, and vending machines. *Functional competition* arises when two or more firms vie to determine who will perform a particular marketing function, such as storage, financing, or transportation.

There are other ways to view market competition. *Horizontal competition* involves rivalry between firms at the same market level—processors or wholesalers or retailers. *Vertical competition* is concerned with the bargaining relationships between buyers and sellers of food and how the consumers' food dollar is divided. It is also frequently useful to distinguish between *price* and *nonprice competition*, which emphasizes the other 4Ps in a firm's marketing strategy.

The behavior of a firm is influenced by the environment and structure of the industry in which it operates. The structure of an industry is described by the number of firms, the similarity of their offerings, and the ease by which new firms can enter and leave the industry. The pricing and output behavior of firms in these industry structures will differ as will the industry's performance. Economists have classified industries into four basic structures, as shown in Table 9-1.

TABLE 9-1 Comparisons of Industry Market Structures

Industry Characteristics	*Perfect Competition*	*Monopolistic Competition*	*Oligopoly*	*Monopoly*
Number of sellers	Very large number	Many	Few	One
Product similarity	Identical for all firms	Differentiated, variations	Similar to different	
Ease of entry for new firms	Easy, no great obstacles	Relatively easy	Difficult, some obstacles	Blocked entry
Firm's influence on price	None for individual firm	Some, limited by substitutes	Substantial, but limited by rivals' prices	Little restraint, unless regulated
Examples	Some farmers, futures markets	Small foodstores, restaurants, farm supply stores	Large chainstores, food processors, wholesalers	Public utilities

PERFECT COMPETITION

Perfect competition is also often referred to as pure or atomistic competition. A market is said to be perfectly competitive if it satisfies the following conditions.

1. There are large numbers of buyers and sellers, no one of which is large enough to influence price through its actions alone.

2. There is no product differentiation. Buyers do not have a preference for one seller's product over another's. This does not mean there are no differences in quality. However, products of like quality will not be differentiated by brand name or advertising.

3. Firms are free to enter or leave the market without significant technological, legal, financial, or other obstacles.

4. All buyers and sellers have perfect and equal knowledge of all prices and the factors that affect market conditions. In addition, they will utilize this information in an economically rational manner so as to maximize their own individual gain.

5. Prices are free from restrictions and firms cannot collude to fix prices.

In perfect competition, the individual firm can sell all of its production at the going market price. If its price is slightly above the market price, it can sell nothing. If it is slightly below, it will be swamped with buyers. Its individual output is such a small part of the total that it can have no effect on price. In addition, because its products are exactly the same as the products of every other firm, it cannot establish preferential demand for its output. Under conditions of perfect competition, then, the individual firm has no price decisions to make.

With the selling side of operations thus simplified, the firm devotes its time to adjusting its production. The firm will adjust its output to that point at which total profits are maximized. In other words, under conditions of perfect competition, the place of the individual business is to accept the prevailing market conditions and adapt itself to them.

A situation where farmers are in perfect competition is illustrated in Figure 9-1. Each farmer produces such a small portion of the total farm supply that he or she "sees" only a point on the downward-sloping market demand curve. To the individual farmer, it may appear that he or she can produce and sell as much as he or she wants at the $3 price determined by the market supply and demand curves. The farmer's demand curve (*dd*), then, is perfectly horizontal at the $3 price established by all farmers' total supply and market demand curve.

This is a key feature of many farm markets. The individual farmer feels he or she can produce and sell as much as desired at the going market price without influencing that price. However, if all farmers increase or reduce output, the market supply curve will shift and the market price will change.

Some of the consequences of farmers being in nearly perfect competition include:

1. Individual farmers are price takers. They receive whatever price is dictated by the total market demand and supply curves. Consequently, they do not

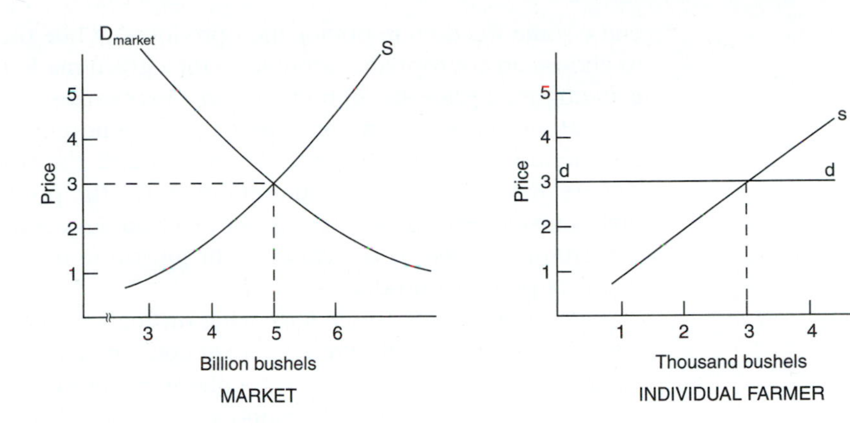

FIGURE 9–1 Supply and demand curves for the market and the farmer under perfect competition.

"price" their products as other firms do. Their major decisions are how much to produce and how to market their output.

2. Because individual farmers cannot affect their prices, there is a strong incentive for them to increase their profits by lowering their costs and by improving operational efficiency in agriculture.

3. Farmers are subject to a perpetual cost-price squeeze. Any short-term increase in profits will stimulate a long-term increase in production, which then forces market prices back to "normal" profit levels.

4. Farmers do not view each other as rivals. They each make their production decisions independently.

5. The principal marketing decisions for farmers are the timing of sale, the place of sale, and perhaps the form in which the commodity is sold. Individual farmers do not attempt to influence the demand for their products through advertising and other marketing techniques.

Considering these consequences of being in nearly perfectly competitive markets, it is not surprising that farmers would attempt to make their markets more imperfectly competitive. This requires that farmers change the conditions that put them in their competitive state. Among the ways they have done this have been farm growth, attempts to consolidate the marketing of independent farmers through cooperatives, efforts to jointly control the output and flow of products to markets, and programs to brand farm products. There have been some successes here, but many farmers still remain in a near perfectly competitive state.

MONOPOLY

The extreme opposites of perfect competition are a monopoly and a monopsony. A *monopoly* is a single-seller market; a *monopsony* is a single-buyer market. Monopolists are said to be price setters because the individual firm's demand curve *is* the market demand curve. Hence, monopolists have some protection from rivals and enjoy some freedom in pricing their products. While the perfect competitor's task is to choose an appropriate output level for a given market price, the monopolist's task is to choose a price and output level that maximizes profit.

However, we must not assume that a monopolist has complete freedom to set production and prices. On the demand side, if the price is pushed too high, consumers may seek out a substitute product. On the production side, if profits are too high, outside firms may enter into production. Freedom to extract monopoly profits is therefore limited by the closeness of substitute products and the barriers to entry that face potential producers.

Monopolies can come into being through the control of the sole source of an important raw material, through patent control of a very important process, because of large economies of scale, or through the actions of government. Like perfect competition, examples of perfect monopoly are rare. Where they have existed for any length of time, it is likely that they have been established or protected by the power of government.

There are near-monopolies involved in food marketing: telephone and energy companies, and other public utilities. However, these monopolies are regulated by government, and their profit seeking, in most cases, is limited also by the availability of alternatives. For example, natural gas, electricity, propane, and even solar energy may all be used as energy sources in the food industry.

OLIGOPOLY

In an *oligopoly* (few sellers) or an *oligopsony* (few buyers), the control of sales is in the hands of a few large firms. This market concentration is so great that the leading firms can influence the market price through their output decisions. Each firm, in making production and price decisions, must consider the effects of its actions on the market price and how rival firms will react. In such a situation of interdependence of firms, the oligopolist must always consider the potential retaliation of business rivals.

In a true oligopolistic situation, the real business challenge to the individual firm is how to live with its competitors and not openly fight them. A fight among giants may end by crippling all concerned, with none of them benefiting. Such a situation results in a stability of prices and production that may not represent profit maximizing for any one firm but rather a situation in which all can find tolerable existence. Price wars sometimes break out among oligopolists and upset this stability.

An oligopolistic industry is likely to develop a hierarchy of leaders and followers. When the leader changes its price, the followers do likewise almost immediately. Such a price situation does not mean there will be no rivalry among firms. The rivalry, however, will take the form of nonprice efforts, such as product differentiation and innovation, value-added services, and advertising and merchandising efforts.

Because of their interdependence, oligopolistic firms sometimes attempt to form associations—*cartels*—to agree on prices, level of output, and division of profits. Cartels reduce competition and make life easier for oligopolists. Many farm products are traded in international markets by government-sponsored cartels. Generally, in the United States cartels or cartel-like behavior is illegal unless sanctioned by government.

MONOPOLISTIC COMPETITION

Monopolistic competition lies between perfect competition and oligopoly. There are several firms in the market—not enough to meet the "no effect on others" criteria of perfect competition, but more than would make them as interdependent as oligopolists. Each firm, however, seeks to make its product or service unique or different from that of every other firm. In a sense, then, each firm is a little monopoly. But the monopoly is of very limited power because in the consumer's eyes the products of competitors are very close, though not perfect, substitutes.

The demand facing each firm is no longer the horizontal, perfectly elastic curve that faces the firm under perfect competition. Because of product differentiation the demand has some downward slope, but it is still very elastic because of the closeness of substitutes. Prices under such conditions will also be nearly alike among firms, not because of fear of retaliation of rival firms, but because of the likely loss of consumer markets if prices are out of line.

In the monopolistic competition situation, prices are not as "sticky" as in the oligopoly situation. As in oligopoly, however, competition will emphasize such marketing strategies as product innovation and differentiation, packaging, branding, advertising, credit and discount policies, field service, and a host of other nonprice weapons to secure business. The successful marketing manager must use all of the 4P marketing strategies to continually persuade the consuming public that his or her company's products are a better buy than those of rivals. The market acceptance of a differentiated product or service is likely to be quite dynamic, and no firm can afford to lag behind its competitors.

MARKET STRUCTURE AND PERFORMANCE

The economic performance of these competitive structures differs significantly. Although they are not found in pure form in the real world, perfect competition and monopoly provide useful standards for evaluating the effects of imperfect competition on prices, output, and efficiency.

Except for perfectly competitive markets, the profit-seeking behavior of the firm may not lead to the most efficient allocation of resources. Firms in a perfectly competitive industry will produce that output and aggregate price level that result in the greatest operational efficiency (lowest possible cost) and pricing efficiency (the best allocation of resources among competing uses). By comparison, monopolists will produce a lower output, resulting in higher prices and profits. Economists consider the reductions in operational and pricing efficiency associated with monopoly to be a misallocation of resources.

The performance of monopolistically competitive industries falls between these two extremes. Because of the existence of product substitutes and the possibility of new firm entry, the price, output, and efficiency levels of monopolistic competitors may be reasonably close to the perfectly competitive levels. Many would consider these rather modest losses in efficiency a small price to pay for the product variety offered by monopolistic competitors. Few consumers would prefer the homogeneous products of perfect competition even if this competitive state were attainable.

The economic performance of oligopoly is the most difficult to evaluate. Oligopolists may behave in a wide variety of ways that affect their performance. Whether oligopolistic market structures result in price and output levels approximating those of perfect competitors or monopolists depends on their reactions to each other's behavior, their ability to enforce a group decision, and the threat of firm entry. If oligopolists form illegal conspiracies or *cartels,* their performance will resemble that of a monopoly. Oligopolistic price leadership and followership or administered pricing arrangements can also produce significant departures from the perfectly competitive performance levels. The oligopolistic misallocation of resources increases as the products become more differentiated and entry becomes more difficult.

Economists have expressed other objections to oligopolistically and monopolistically competitive markets. Many contend that firms in these markets are plagued with chronic excess capacity, may permit their costs to rise unnecessarily, and may not aggressively pursue new cost-reducing technologies. Other economists express concerns that these firms engage in excessive, sometimes manipulative, forms of nonprice competition in an attempt to win consumers and avoid price competition.

Students of marketing must bear in mind the different perspectives that economists and business practitioners have in evaluating economic and market performance. The economist uses an idealistic, perfectly competitive model to evaluate the allocation of resources among firms and industries. The business manager has a narrower focus on increasing sales and profits of the firm in a marketplace with often threatening and successful, if not numerous, competitors. While sharing the economist's concern with keeping costs down, the marketing manager has no hesitancy in attempting to influence consumer demand through both price and nonprice promotional techniques. The manager believes that consumers will be the ultimate judge of market prices and performance through their buy or no-buy decisions. Both perspectives are helpful in understanding and evaluating market performance and behavior in the food industry.

COMPETITIVE CONDITIONS IN FOOD MARKETS

Agricultural producers have long been used as examples of nearly perfectly competitive firms. While there is not completely free entry or perfect knowledge in farm markets, these markets do come close to the perfectly competitive conditions of large numbers of sellers and undifferentiated products. Consequently, farm markets display many of the characteristics—and farmers experience many of the consequences—of perfect competition.

This situation is changing, however, as the number of producers declines, their sizes increase, and capital and management requirements in agriculture rise. Moreover, farmers are actively seeking to escape the confines of perfect competition through cooperative and joint marketing activities, farmer promotional programs, and other means.

In their imperfectly competitive markets, food marketing firms utilize many different types of competitive weapons. Probably the most important is product differentiation. Through the selection of marketing strategies, sellers differentiate their products in order to increase their appeal to buyers, to reduce the substitutability for their products, and to increase their latitude in pricing. All of these can enhance firm profits. Typical differentiation techniques include branding, packaging, sales promotion, and advertising.

Product differentiation occurs at all levels of the food system. Grocers attempt to differentiate their stores through price differences, location, store design and layout, product assortments, shopper services, advertising, and even the friendliness of employees. Food processors rely on new products, packaging, and brand advertising to differentiate their offerings. Food assemblers and wholesalers emphasize prices and services in their marketing efforts.

What are we to make of all this product differentiation? An evaluation of product differentation is not so simple. It likely increases the range of consumer choices in the marketplace and sometimes provides bundles of attributes that better satisfy the preferences of diverse consumers. However, product differentiation can also complicate consumer choices by substituting nonprice for price competition. Differentiation may also increase prices and firm profits, provide the illusion of product differences where none exists, and generally insulate sellers from the discipline of price competition.

Market concentration, the proportion of industry sales made by the largest firms, is another source of imperfect competition. Successful competitors frequently eliminate their rivals or discourage new firm entry, contributing to more concentrated markets. While farming is largely unconcentrated, higher levels of market concentration are often found in food processing, wholesaling, and retailing. In general, the higher the level of market concentration, the less perfectly competitive the market.

Table 9-2 suggests generally increasing but a wide range of market concentration in food sectors. Food processing and wholesaling tend to be more concentrated than food retailing and eating places. However, these figures are for national market concentration, and local concentration can be much higher for grocery stores and restaurants. Within food processing, there is a wide range of market concentration. Moreover, market concentration is increasing more rapidly in some industries than others. Higher levels of market concentration are generally related to economies of scale and high levels of product differentiation. A guideline is that strongly oligopolistic industries have a four-firm concentration ratio of at least 50 percent; a 33-50 percent ratio denotes a weak oligopoly; and unconcentrated industries have ratios of 33 percent or less. All of these competitive structures are represented in the food industry.

Barriers to new firm entry also exist in food marketing and may contribute to imperfectly competitive behavior and market structures. An existing firm may enjoy a cost advantage over potential rivals through command of financial resources, access to raw materials, technological know-how, or the existence of economies of scale. These conditions may make it very difficult for a new, small firm to enter the industry. Product differentiation and brand loyalty that come from years of heavy brand promotion may also serve as barriers to entry in the food industry.

TABLE 9-2 Concentration Trends in Food Marketing, 1967–1997

		Percent of Total Industry Sales						
Sector	*Number of Firms*	*1967*	*1972*	*1977*	*1982*	*1987*	*1992*	*1997*
Food Chainstores	Top 4	19%	18	17	16	17	16	18
	Top 20	34%	35	35	35	37	37	44
Eating Places	Top 4	na	4	4	5	8	8	na
	Top 20	na	9	12	15	17	17	na
	Top 50	na	13	18	20	22	23	na
Wholesalers, General	Top 4	na	na	15	17	26	35	28
	Top 20	na	na	37	43	54	59	55
	Top 50	na	48	57	64	71	76	71
Food Processors	Top 50	35	38	40	43	47	na	na
Meat Packing	Top 4	26	22	19	29	32	50	na
	Top 8	38	na	na	43	50	66	na
Butter	Top 4	15	45	49	41	40	49	na
	Top 8	22	na	na	61	63	78	na
Canned Fruits	Top 4	22	20	22	21	29	27	na
and Vegetables	Top 8	52	na	na	35	40	42	na
Fluid Milk	Top 4	22	18	18	16	21	22	na
	Top 8	30	na	na	27	32	30	na
Breakfast Cereals	Top 4	88	na	89	86	87	85	na
	Top 8	97	na	na	na	99	98	na
Bread, Bakery Products	Top 4	26	na	33	34	34	34	na
	Top 8	38	na	na	47	47	49	na
Cookies and	Top 4	59	na	na	59	58	56	na
Crackers	Top 8	70	na	na	71	73	70	na
Soybean Oil Mills	Top 4	55	54	54	61	71	71	na
	Top 8	76	na	na	83	91	91	na
Soft Drinks,	Top 4	13	na	na	14	30	37	na
Carbonated Water	Top 8	20	na	na	23	40	48	na
Roasted Coffee	Top 4	53	na	na	65	66	66	na
	Top 8	71	na	na	76	78	75	na
Macaroni, Spaghetti	Top 4	31	na	na	42	73	78	na
	Top 8	48	na	na	66	82	85	na

SOURCES: U.S. Census of Business, 1967–1997; Economic Research Service, U.S. Department of Agriculture.

Mergers have also changed the market structure of the food industry and often lead to higher levels of market concentration. Figure 9-2 shows the trend in numbers of food marketing mergers for the 1982–1996 period. When two similar food companies merge or consolidate their operations, it is referred to as a *horizontal merger.* When two firms who buy and sell from each other merge, it is a *vertical merger.* When a food and nonfood company merge, it is termed a *conglomerate merger.* All three forms of food market mergers have been high in recent years. This causes concern about the competitive nature of food marketing and the future of the industry.

Location is a source of competitive advantage in many sectors of the food industry. Food stores and restaurants may have a limited spatial monopoly for nearby

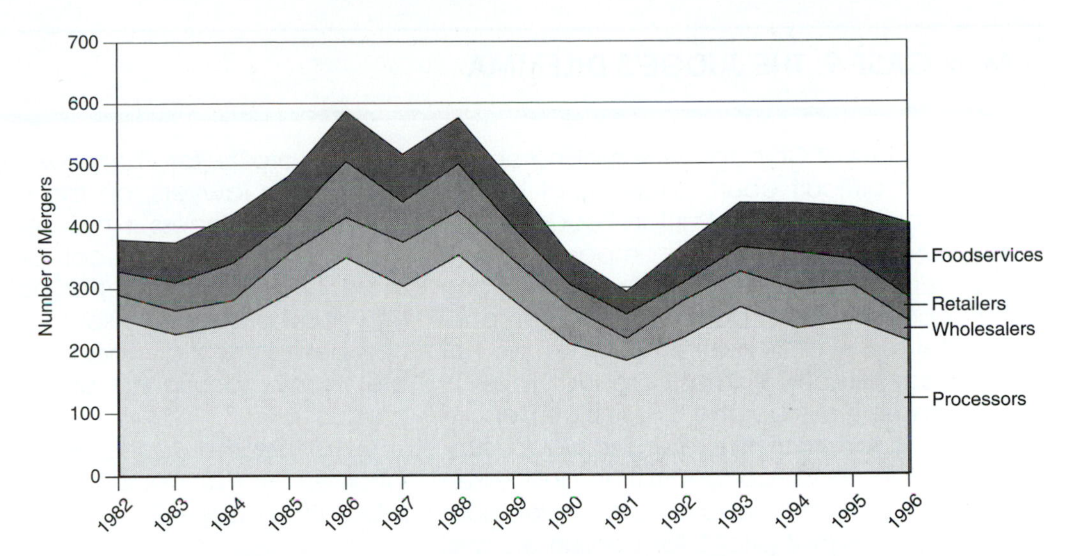

FIGURE 9–2 Numbers of food marketing mergers, 1982-1996. (Economic Research Service, U.S. Department of Agriculture.)

customers. Only a few buyers may compete for farm supplies in local markets. The strength of market power in these situations is influenced by buyers' and sellers' knowledge and costs of more distant market alternatives.

Information is sometimes overlooked as a source of imperfect competiton. Deceptive advertising, of course, distorts and corrupts the competitive process. But beyond this deception, buyers and sellers never have the complete market information required for perfect competition. Consumers do not know all their purchase alternatives, have difficulty making all the complex quality and value judgments they face, and certainly cannot be aware of all food price changes. Buyers and sellers may sometimes be able to take advantage of incomplete market information in their transactions with others.

Studies have investigated the impacts of imperfect competition in the food industry. They have generally documented the existence of excess capacity, the substitution of nonprice for price competition, varying degree of product differentiation and brand loyalty, and numerous instances of anticompetitive behavior such as price fixing, market sharing, and predatory practices. However, the food marketing system seems in general to be as competitive as other industries. Its overall economic performance is not perfect, but neither is it seriously deficient. It is not at all clear that consumers would prefer a food industry more closely resembling perfect competition or that society in general would be better off.

WORKABLE COMPETITION

Competition in the food system exists and is a force, even though it is not perfect. We must have some criteria for judging when competition is effective. How can we decide when competition is "adequate" or "effective"? The following guideposts

MINI CASE 9: THE JUDGE'S DILEMMA

Judge Jones of Chicago was making one of the most difficult court decisions of his career. Major Foods and Puritan Foods, two of the largest U.S. food companies were proposing a merger of the two firms. Each firm currently held dominant market positions for many of its brands, but they did not compete directly in most product lines. In opposing the merger the U.S. Justice Department argued that the merger would substantially increase concentration and lessen competition in the food industry, eventually leading to higher prices for consumers and greater profits for the new company. The company's lawyers, on the other hand, argued that "bigness is not necessarily bad," that the new, larger company would be more efficient and better able to develop new food products to satisfy consumers and compete in world markets, and that there are many competing brands for all of its products.

What are the economic and social issues involved in this decision? How would you rule on this merger?

have been developed for judging an industry to be workable or reasonably competitive.

1. There must be an appreciable number of buyers and sellers. They do not need to be so numerous as to have no individual market influence, but the number must be great enough to provide market alternatives for buyers and sellers.
2. No firm must be powerful enough to be able to coerce its rivals or others in the marketing system.
3. Firms must be responsive to incentives of profits and loss—they must not be so large that they can ignore market incentives over long periods of time.
4. There must be no agreements or collusion among rivals.
5. Entry and exit must be free enough for rivals to contest each others' markets.
6. There must be free access of buyers with sellers, with no substantial preferential treatment of any particular trader or group.

These criteria, admittedly difficult to apply in practice, accept the real-life propositions that wide differentiation in products does exist, that both price and nonprice competition are valued by consumers, and that large firms and high market concentration are inherent in the modern economy.

Because of the public interest involved, government has a role to play in preventing monopoly and maintaining effective competition. Our antitrust laws are designed to make certain types of business behavior illegal and to promote fair competition. Other laws deal with the regulation of businesses, such as public utilities, in which there are monopolistic tendencies. Certain laws, such as those dealing with farmer cooperatives and marketing orders, attempt to strengthen the weaker participants in the market bargaining process. Society has also sanctioned the actions of government that attempt to create the general conditions for more effective competition.

SUMMARY

Competition plays a key role in harnessing the rivalry and the profit-seeking of the marketplace in order that it will serve the public interest. The competitive state of food markets disciplines firms' behavior and practices, encourages new technologies and products, and regulates prices and profit levels. Farm markets tend to be nearly perfectly competitive, making farmers price takers. Food marketing firms are closer to the oligopolistic and monopolistic types of competition, and these firms can influence prices by output and marketing strategies. Product and firm differentiation, market concentration, barriers to entry, location, information, and trade barriers are important reasons for competitive imperfections in food marketing. The standards of workable competition represent a compromise between unattainable perfectly competitive market conditions and the real world of imperfect competition. Public policies can influence the state of competition in food markets.

KEY TERMS AND CONCEPTS

cartel
market concentration
monopolistic competition
monopoly
oligopoly/oligopsony
perfect competition
price takers/price setters
price war
product differentiation
workable competition

DISCUSSION QUESTIONS

1. What does it mean to say that farmers are price takers and food marketing firms are price setters?

2. What is the difference between a true monopolist and monopolistic competition?

3. If as the U.S. Attorney General you were charged with maintaining competitive food markets, how would you decide which sectors to investigate first? What information would be helpful in evaluating competition in these sectors?

4. Your senator charges that monopolistic food companies are cheating consumers out of millions of dollars through advertising, merchandising, cents-off campaigns, and other nonprice gimmicks. How do you respond?

5. What are the principal differences between how businessmen view competition and the economic view of competition?

6. Evaluate the following statement: "We wouldn't need competition in food markets if businessmen were honest."

7. Given their choice, do you think farmers would choose perfect competition over imperfect competition in their markets? What can farmers do to make their markets less perfectly competitive?

SELECTED REFERENCES

Armstrong, G. and P. Kotler (2000). *Marketing: An Introduction.* Upper Saddle River, NJ: Prentice-Hall.

Bucklin, L. P. (1972). *Competition and Evolution in the Distributive Trades.* Englewood Cliffs, NJ: Prentice-Hall.

Business Week (Weekly). New York: McGraw-Hill.

Clark, J. M. (1961). *Competition as a Dynamic Process.* Washington, D.C.: The Brookings Institution.

Food Market Structures: Industrial Organization of Food Markets Briefing Room (January 2001). Economic Research Service, U.S. Department of Agriculture. www.ers.usda.gov/briefing/foodmarketstructures/industryorg.htm

Fortune (Monthly). New York: Forbes, Inc.

Marion, B. W. (1986). *The Organization and Performance of the U.S. Food System* (Chapters 4, 5, and 6). Lexington, MA: Lexington Books.

National Commission on Food Marketing. (June 1966). Overview and appraisal. In *Food from Farmer to Consumer* (Chapter 12). Washington, D.C.: U.S. Government Printing Office.

Porter, M. E. (1985). *Competitive Advantage.* New York: The Free Press.

Wall Street Journal (Daily). New York: Dow Jones & Company.

CHAPTER 10

FARM AND FOOD PRICES

. . . aggregate demand for farm products in the United States will in certain years expand relative to aggregate supply, and in certain other years lag behind aggregate supply.

Willard Cochrane, *The Development of American Agriculture,* 1981, p. 412

OBJECTIVES

After reading this chapter, you will be able to:

1. Discuss the major factors affecting farm and retail food prices.
2. Better understand the relationships between farm, nonfarm, and retail food prices.
3. Explain how the business cycle, agricultural production cycles, and seasonal variations affect farm prices.

Prices and incomes play key roles in farm and food marketing decisions. In the long run, prices, incomes, and profits are expected to allocate farm and marketing resources efficiently. In the short run, these economic signals and incentives motivate food producers and marketing firms to shift resources from low-valued to high-valued products and markets. Fluctuating food prices are necessary to allocate resources efficiently, yet they can present problems for farmers who must adjust to the present and plan for the future.

FORCES INFLUENCING FARM PRICES

The forces influencing farm prices can be grouped into four broad categories. Supply conditions affecting farm and food prices include farm production decisions, weather, disease, harvested acreage, and food imports. The demand conditions include income, prices, tastes and preferences, population, and exports. The food marketing sector influences farm prices through its value-adding activities, price and cost behavior, and procurement strategies. Finally, government may influence farm prices through price supports, supply controls, trade policies, or policies influencing domestic demand for food.

U.S. agriculture has often been viewed as isolated and insulated from the national and global economies. By this interpretation, farm prices and incomes are influenced primarily by such internal factors as producer decisions, agricultural technology, weather, and the demand for food. This picture is no longer a valid one of agriculture, if it ever was. Today's farm prices and incomes are significantly influenced by powerful forces outside of the food system: government agricultural policy, international trade policy and currency exchange rates; government monetary policies and interest rates; government taxing and spending decisions; inflation and unemployment rates; and the economic policies of foreign governments. The student of farm and food prices today must understand these macroeconomic forces and their impacts on farmers and the food industry.

There is sometimes a debate over whether farm prices are determined at the farm level, in the marketing system, or by consumers at the retail level. Often it appears that farm prices are "set" first—related in some way to costs of production—and then marketing costs are added to arrive at the retail prices. This impression is reinforced when a change in retail prices follows in time a change in farm prices.

To others, it appears that the buyers of farm products—food processors, wholesalers, or retailers—"set" the price of food and that this "determines" farm prices. This impression is created when price-taking farmers sell to less than perfectly competitive buyers of farm products. The fact that prices are more easily *discovered* at points of product concentration in the marketing system also leads many to conclude that prices are *determined* at those points.

Actually, prices are determined jointly—by consumer demand, farm supply, and the food marketing system. No one of these is any more important than the other in the determination of farm prices. A change in any one usually results in adjustments of the other two. Therefore, it is rather pointless to argue about precisely where farm prices are determined.

GENERAL FOOD PRICE TRENDS AND RELATIONSHIPS

Food prices are one of the most controversial aspects of food marketing. Are food prices too high or too low? Do food prices cause, or moderate, inflationary trends? How well do farm and retail food prices move together? Examining trends and relationships in food prices can provide a better understanding of the food marketing process.

Figure 10-1 illustrates the trends in wholesale food and nonfood prices between 1960 and 1999. Although farm and industrial wholesale prices moved together broadly over the long run, there were significant short-run differences in these prices. First, farm prices fluctuated more than nonfarm prices. Second, farm

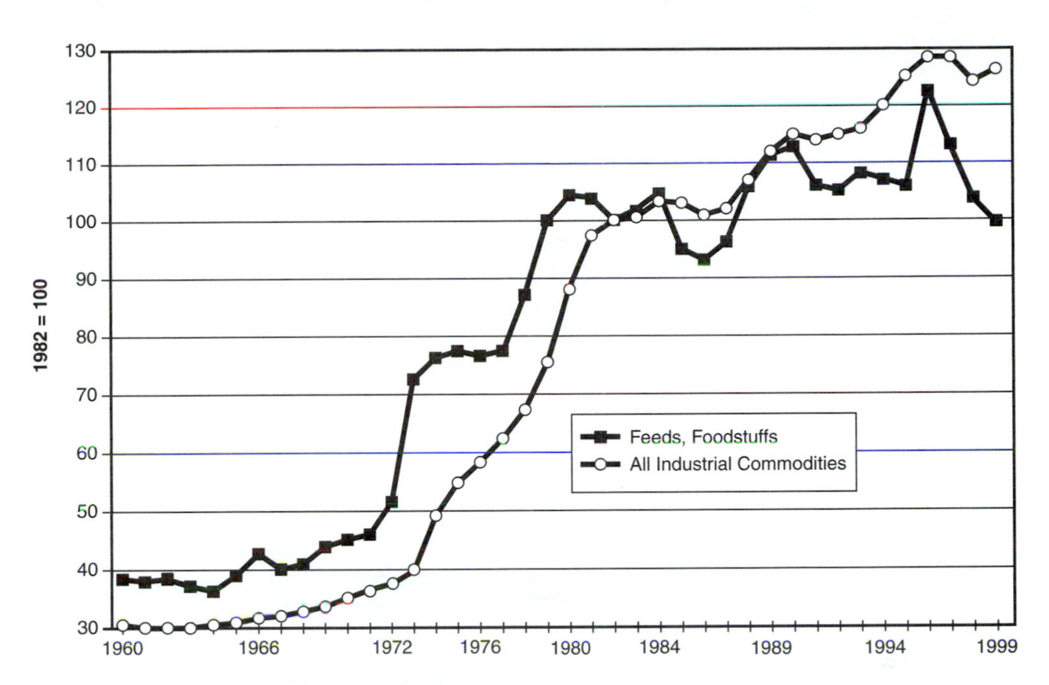

FIGURE 10–1 Producer price index, farm products and industrial commodities, 1960–1999. (U.S. Bureau of Labor Statistics.)

prices were flexible both upward and downward, whereas nonfarm prices tended to rise steadily. A major reason for this difference in year-to-year price movements is the greater control over output that is possible for nonfood products. Annual variability in food prices reflects the difficulty of tailoring supply to demand when the product is susceptible to weather, disease, and other unpredictable elements. Factory production of nonfood goods is much more controllable and stable.

Annual retail food and nonfood prices have also moved together over the long run, as shown in Figure 10–2. Between 1970 and 1999, retail food prices rose 297 percent while all other retail prices increased by 270 percent. There are three reasons for these prices moving together over the long run: (1) 80 percent of retail food prices is determined by food marketing costs, such as energy, labor, packaging, and freight, which affect, equally, both food and nonfood prices; (2) farmers' purchases of supplies from the industrial sector tend to build these costs into food prices; and (3) over the long run, the same economic forces affect both food and nonfood prices.

How closely do farm, wholesale, and retail food prices move together? How long does it take a change in price at the farm level to reach the consumer? These are questions of pricing efficiency. There are several reasons why farm and retail food prices might not move together on a day-to-day basis. First, there are lags in the time necessary to move the product from the farm to the grocery shelf. Fresh foods move through the marketing channel more rapidly than highly processed food, but all products spend some time in transit and in inventory. Therefore, a snapshot of today's farm and retail prices would not record exactly the same commodity at each market level. Changes in marketing costs and profits also reduce the relationship between farm and retail food prices. In addition, many consumer products often are made from an individual farm commodity. For example, a steer is used for meat,

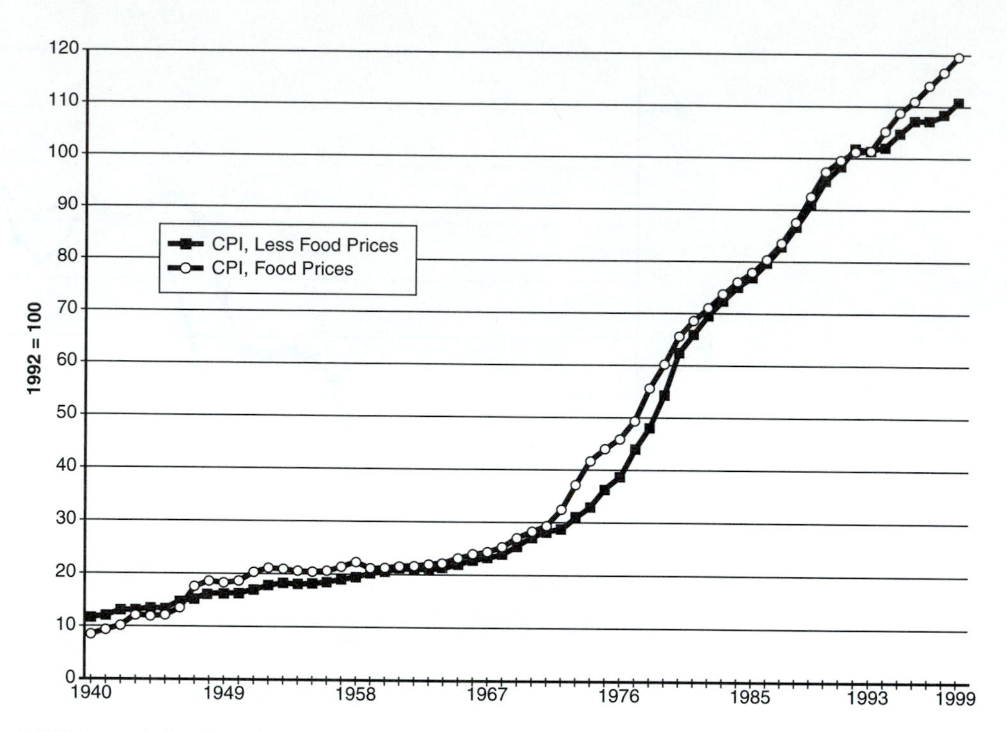

FIGURE 10–2 Trends in retail food and nonfood prices, 1940–1999. (U.S. Bureau of Labor Statistics.)

hide, tallow, and other by-products. Soybeans are crushed to produce oil and meal. Because each of these final products is subject to different economic conditions, the farm price of the commodity may not correlate precisely with any one of the retail prices. Merchandising strategies, such as weekend specials and loss leaders, may also alter the day-to-day relationship between farm and retail food prices.

Figure 10-3 indicates the relationship of annual farm and retail food prices during the 1960–1999 period. Generally, farm and retail food prices move up and down together from year-to-year. However, farm prices are more variable than retail food prices, both when prices are rising and when they are falling. The correlation of farm and retail prices is greater for fresh foods than for highly processed foods, and is more pronounced for annual price trends than for monthly, weekly, or daily prices.

There is continuing concern about short-term pricing efficiency in food markets. In particular, many farmers feel that farm price reductions are not fully and quickly transmitted to consumers. If true, this would be a deficiency in pricing efficiency. The evidence is somewhat mixed on this aspect of food pricing efficiency. However, most studies have reported that:

1. Farm and retail food prices do move together, although the degree of association varies widely by the amount of processing involved.

2. The greatest impact of a change in farm prices on retail prices occurs in the first month of the change, with the full impact spread out over a number of months.

3. Retail and wholesale food prices respond equally to both rising and falling farm prices.

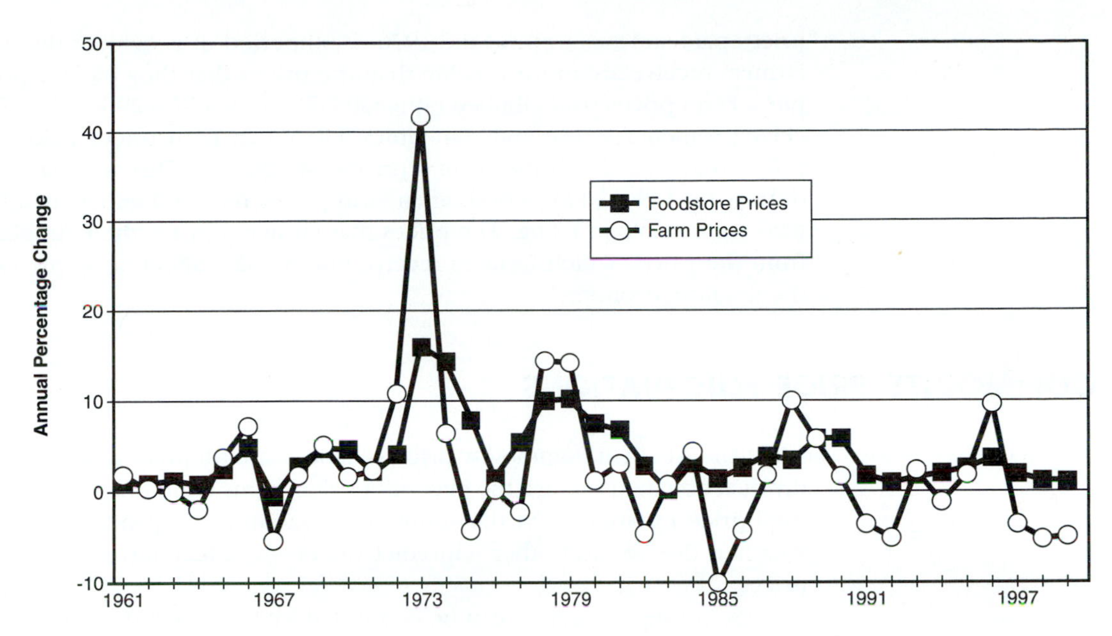

FIGURE 10–3 Farm and retail food price changes, 1960–1999. (U.S. Departments of Agriculture and Labor.)

4. Grocery store food prices reflect short-term changes in farm prices more closely than do prices for food eaten away from home.

Farmers' net incomes are influenced by both the prices at which they sell their products and the prices that they pay for such items as seeds, machinery, feeder livestock, fertilizer, fuel, and other inputs. Figure 10–4 shows the relationship of farm

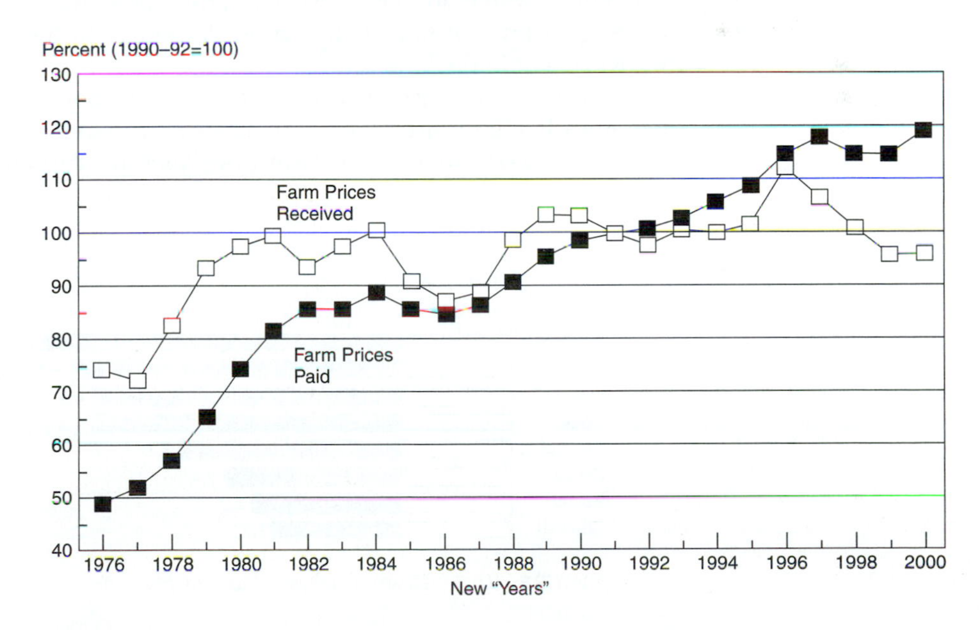

FIGURE 10–4 Farm prices paid and received, 1976–2000.

prices paid and received for the 1976–2000 period. It is evident that the prices that farmers receive are more variable than the prices that they pay for production supplies. Farm prices paid climbed rather steadily from 1976–2000 while farm prices received sometimes rose and sometimes fell. When farm prices paid rise faster than prices received, the farmer is in a price-cost squeeze. This was particularly evident during the 1996–2000 period when farm prices received generally fell while prices paid were stable to rising. The prices that farmers pay for their supplies are derived from the prices which farmers receive but are also affected by prices and costs in the nonfarm economy.

COMMODITY PRICE FLUCTUATIONS

Wide and frequent commodity price variations are the rule, while stable prices of individual commodities are the exception. Numerous conditions contribute to agricultural price instability. On the supply side, variations in producer output decisions, weather, disease, and other unpredictable events affect acreage, yields, output, and prices.

Some supply factors can be controlled by farmers and some cannot, and farmers' efforts to tailor supply to demand can be frustrated by uncontrollable events or the elements. Producers may respond to price predictions or expectations by changing acreage, but the yield may be lower than average. The results of these supply changes are illustrated in Figure 10–5. These *price flexibilities*—the percentage change in price caused by a percentage change in quantity supplied—are approximately equal to the reciprocal of the elasticity of demand. They are used to predict the effect on prices of changes in agricultural output.

Farmers' decisions to expand or contract output are often a source of farm price variations. A high price will stimulate increased production or marketings of a commodity, which in turn lowers prices. There may be biological lags in this process that prevent immediate supply responses to high or low farm prices, but they will occur in the long run.

Demand factors also contribute to price variations in agriculture. International trade links U.S. farm markets with consumers throughout the world, adding world demand variations to the U.S. farm price picture. Particularly destabilizing is the

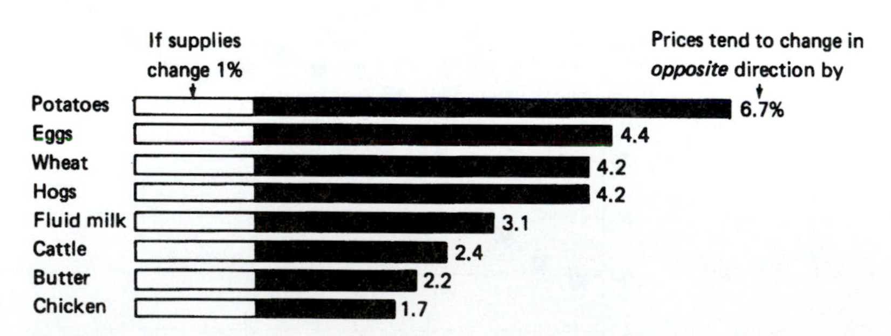

FIGURE 10–5 How a change in supply affects prices received by farmers.

TABLE 10-1 Average Annual Real Prices Received by U.S. Farmers for Selected Commodities in 1970 Dollars, 1929-1998

Year	*Wheat $/bu.*	*Corn $/bu.*	*Hay $/ton*	*Rice $/100 lbs.*	*Cotton cents/lb.*	*Soybeans $/bu.*	*Peanuts $/100 lbs.*	*Beef $/100 lbs.*	*Chicken $/lb.*
1929	2.34	1.72	32.02	5.03	38.0	4.22	8.46	21.47	—
1940	1.86	1.66	27.09	4.99	27.4	2.49	9.22	20.94	.47
1950	3.23	2.45	34.77	8.21	64.6	3.98	17.58	37.48	.44
1960	2.28	1.31	26.70	5.97	39.6	2.80	13.12	26.77	.22
1970	1.33	1.33	24.20	5.17	22.0	2.85	12.80	27.10	.14
1980	1.87	1.47	33.44	6.03	35.2	3.59	12.40	29.48	.13
1990	.77	.68	23.91	1.99	19.9	1.70	10.30	22.58	.08
1998	.64	.46	20.71	2.02	15.28	1.27	6.12	14.19	.09

SOURCE: *Agricultural Prices,* U.S. Department of Agriculture, Annual. Actual prices divided by the Consumer Price Index, 1970 = 100.

practice of some large countries that make substantial purchases of U.S. farm products one year but not the next. Some countries view the United States as the world's residual or standby food supplier, and this causes year-to-year price variability.

Domestic changes in consumer incomes, employment, and business conditions influence the demand for food and its price. In the short run the food demand curve shifts up and down a relatively inelastic supply curve, and most of the market adjustment must come in prices.

Table 10-1 shows the level and variation of selected "real" farm prices during the 1929-1998 period. These prices have been adjusted to eliminate the effects of inflation on the level of prices. These price variations make it difficult for farmers to make production, marketing, and investment decisions. They result in rather wide swings in net farm income and profits; they require frequent and sudden farm resource adjustments, and these are difficult to make without significant farmer losses or change in farm resource prices. Farm price and income variations also have affected farm marketing practices such as storage, contracting, and timing of sales.

Food marketing firms must adjust to agricultural price variations. Many processors operate their own farming operations to supply needs when prices are high, and many have become involved in farm contracting to assure both quality and prices. These firms also can use the commodity futures market to protect themselves and to profit from changing farm prices.

We will now review three sources of farm and food price variability: the business cycle, agricultural production cycles, and seasonal production patterns.

FARM PRICES AND COSTS OVER THE BUSINESS CYCLE

Farm prices change with shifts in the aggregate supply and demand curves for farm products, as shown in Table 10-2. If demand shifts rightward faster than supply—as in 1914-1920, 1941-1951, and 1970-1981—farm prices will rise. If supply shifts

TABLE 10-2 Changes in Aggregate Demand, Supply, and Farm Prices, 1914-1990

Period	*Aggregate Demand and Supply Shifts*	*Farm Price Trends*
1914-1920	World War I demand shifts rightward faster than food supply	Farm prices rise 120%
1929-1932	Demand falls with income and employment, little change in supply	Farm prices fall 47%, returning to 1900 levels in 1932
1941-1951	War-related demand expands faster than food supply	Farm prices rise 222%
1951-1970	Technological change in agriculture shifts the supply curve rightward faster than increases in domestic demand; government price supports cushion downward price pressures	Farm prices fall 9%
1970-1981	World food demand shifts rightward faster than world supplies	Farm prices rise 132%
1982-1990	World food supply shifts rightward faster than world demand	Farm prices fall 12%

SOURCE: U.S. Department of Agriculture. Adapted from W. W. Cochrane (1958). *Farm Prices, Myth and Reality* (Chapter 3). Minneapolis: University of Minnesota Press. Reprinted by permission.

rightward faster than demand, or demand shifts leftward—as in 1929-1932, 1951-1970, and 1982-1987—farm prices will fall. What causes these shifts in farm supply and demand?

Over the years, the American economy has followed a cyclical pattern of boom and bust—prosperity, then depression or recession, inflation, then deflation. These changes in real growth of economic output and price levels are referred to as the *business cycle.* There have been twenty-one U.S. business cycles in the twentieth century, ten since World War II. The cycles are caused by wars, the psychology of consumers and investors, and a host of other factors.

Business cycles affect not only the general economy but also farm prices, costs, and incomes. During prosperous periods, rising employment and incomes contribute to a strong demand for food and to rising farm prices. Similarly, during depressions or recessions, declining consumer demand weakens farm prices. For example, during the 1990-1991 economic recession real foodstore sales declined as consumers purchased less expensive foods and brands.

These increases and decreases in farm prices would not necessarily affect net farm incomes if the costs of farm production also rose and fell with the business cycle. A decline in the price of cattle from $60/cwt. to $50/cwt. would not be disastrous for farmers if the costs of producing beef fell proportionally. However, the prices of farm inputs purchased from the nonfarm sector—which account for two-thirds of all farm costs—do not respond to the business cycle in the same way as farm prices.

Machinery, hardware, chemicals, and other supplies purchased by farmers are produced by firms that adjust production in response to changes in the business cycle. These firms shift their supply curve leftward in response to declining demand by reducing plant capacity, cutting back on output, and laying off workers. This moderates the downward pressure on industrial prices during contracting phases of the business cycle.

Farmers react differently to the business cycle. They maintain, or even increase, output regardless of the business cycle. For example, agricultural output increased 15 percent during the 1930s, a period when the total output of the U.S. economy declined. Agriculture, then, takes the impact of depressions and recessions in its prices without making price-moderating output adjustments as shown in Figure 10–6. An economic recession is illustrated by a leftward shift of the demand curves for industrial and agricultural products. The industrial sector cushions the price-reducing effects of this shift by reducing employment, closing factories, and limiting output, which is represented by the leftward shift in the industrial supply curve. Farmers do not limit output like this in a recession. As a result, farm prices fall more during a recession than industrial prices, including the costs of farm supplies.

The differences in the supply response of agriculture and industry to the business cycle are the result of the competitive nature and cost structure of these two sectors. In the industrial sector, there are few enough firms of sufficient size that each can affect price through output decisions. These monopolistic competitors and oligopolists reduce production during business contractions in order to cushion price and profit impacts. Nearly perfectly competitive farmers cannot, individually, prevent falling prices by reducing production. In addition, new technologies may actually shift the farm supply curve *rightward* during a business contraction, because these technologies originate outside of agriculture. Moreover, farmers generally have a larger part of their costs in the fixed category than do manufacturers; thus, prices must fall further in order to trigger output responses.

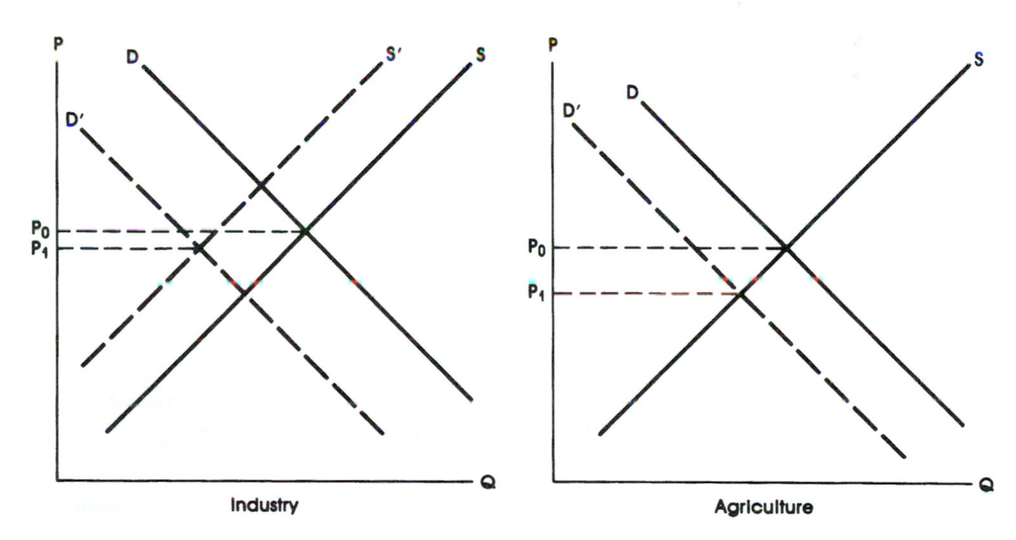

FIGURE 10–6 The effects of the business cycle on the farm and nonfarm economy.

MINI CASE 10: AMERICAN FARMERS AND THE MACROECONOMY

Mary and Bill Williams were discussing their farm costs and profits over breakfast one morning. "It is really tough to make it in farming today," said Bill. "I can match anyone on yields, farm management, and marketing savvy," he said. "But it seems like our farm prices and profits today are determined more and more by things outside of my control and even outside of the farming community," he complained. "I agree," said Mary, "interest rates, inflation, employment, the federal budget, government spending, and lots of other things are part of farming today." "We need to learn more about how the general economy is affecting us," said Bill, "and what we can do to live in this new world."

How does the general economy (economic growth, inflation, and employment) affect farmers today? Why is this so important to modern farmers? What can farmers do as citizens, spenders, savers, and earners to survive in today's marketplace?

AGRICULTURAL PRICE CYCLES

The business cycle affects farm prices through periodic shifts in the aggregate supply and demand of food. Agricultural price cyles, in contrast, are regular fluctuations in prices owing to periodic expansions and contractions in the production of individual agricultural products. Some agricultural price cycles are illustrated in Figure 10-7. The

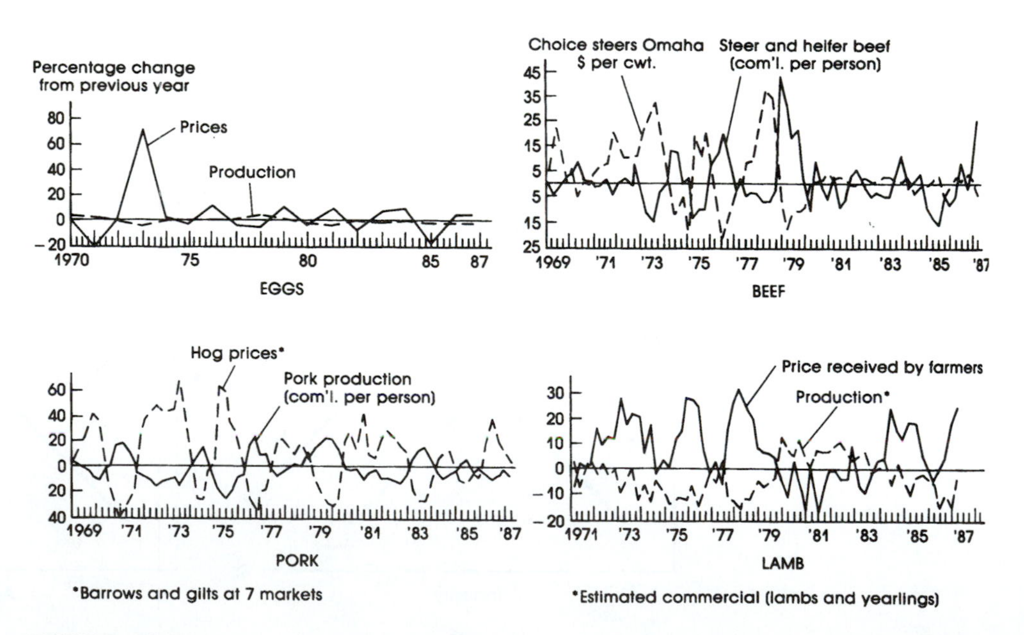

FIGURE 10–7 Price and production cycles for four commodities. (U.S. Department of Agriculture.)

price cycles run counter to the production cycles. When supplies increase, prices fall, and when supplies decrease, prices rise. The price swings are much greater in magnitude than the production swings because of the inelasticity of farm demand.

Agricultural price cycles are caused by the tendency of farmers to base future production plans on current prices and profits, rather than on future prices. For example, let us suppose that hog production is relatively low and hog prices are high. Hog producers look at their favorable earnings of the past year and decide to expand their hog enterprises. Others who previously had left the business decide to reenter. But to expand hog production means that more female pigs must be withheld from market and bred, raising price further. Time must elapse before pigs can be born, fattened, and sent to the market. All in all, some two to three years will elapse before the full, intended expansion may result in additional hog supplies on the market. By that time increasing supplies will be driving prices down. Producers will be appraising the situation as unprofitable and decide to raise fewer hogs, pushing prices down further. The faint of heart may liquidate their hog enterprises altogether. The cycle will then reverse itself; production will decline and prices will increase.

This pattern of cyclical ouput and prices is shown in Figure 10-8. High prices at P_0 eventually stimulate output expansion to Q_1. Output Q_1 can only be sold at the lower price, P_1. These lower prices eventually lead to output reductions to point Q_0. After the output falls to Q_0, prices rise to P_0. The cycle then begins again. This is sometimes referred to as a cobweb cycle since the price and output points trace a cobweb-like design around the supply and demand curves.

The length of agricultural price cycles (peak to peak) depends on the psychological and biological lags involved in producing the commodity. The *biological lag* is the time period between when farmers decide to expand or contract production and when market supplies actually change. There is a longer biological lag for cattle than for chicken, because it takes longer to expand the cattle herd than it does to in-

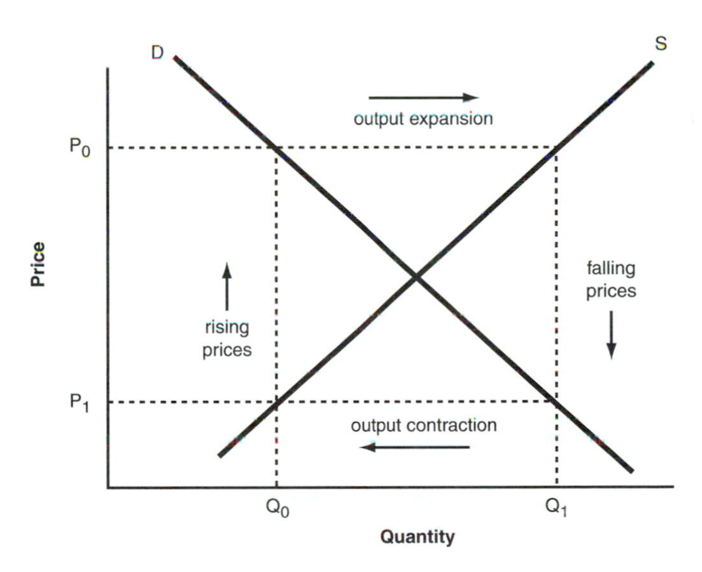

FIGURE 10–8 Agricultural price and output cycles: the cobweb diagram.

crease chicken production. Most agricultural price cycles are longer than the biological lag, however. The hog cycle, for example, is three to four years, whereas the cattle cycle runs eight to twelve years. The *psychological lag* accounts for the remainder of the cycle. This is the period of time when prices must be high or low in order to convince farmers that production plans should be changed. Producers do not and could not change output with every daily or weekly price change.

Several conditions are necessary for the development of identifiable price cycles. First, there must be a time lapse between a change in price and producers' response to that change. The lags provide this. Second, producers must target production decisions to current prices rather than to expected future prices. Third, producers must have reasonably good control over output. There are no well-defined price cycles in crops, as there are in livestock, because producer decisions are only partially responsible for crop size. Finally, price cycles are more likely to develop where there are large numbers of producers and each believes that the individual output decision will not influence prices.

Knowledge of agricultural price cycles can be helpful in decision making for farmers, marketing firms, and consumers. Price analysts and forecasters can also benefit from a knowledge of price cycles. However, it should be noted that the cycles do not follow perfectly predictable patterns. The magnitude and length of the cycles vary over time. Extraordinary conditions—such as a sharp change in food supplies, a sudden surge of export demand, or yield-affecting weather—may prolong the expansionary phase of a cycle or cut short a contracting phase. At best, the cycles are ten-

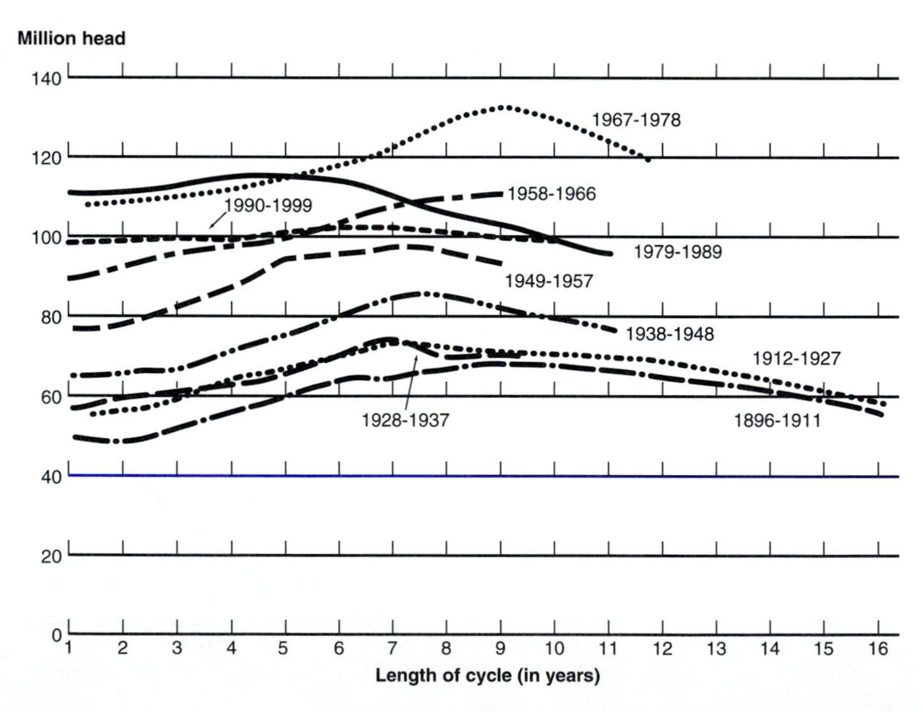

FIGURE 10–9 Cattle production cycles, 1896–1999. (U.S. Department of Agriculture.)

dencies. In many cases, turning points are evident only after they have passed. The cattle production cycles shown in Figure 10-9 illustrate the regularity yet variation in cycles.

Farmers adjust to agricultural price cycles in three ways. Most farmers contribute to cycles by expanding during periods of high prices and contracting during periods of low prices. Other farmers opt to produce at a steady rate over the long haul, regardless of the cycle, averaging the high prices with the low prices. Still others attempt to gear their production counter to the cycle—expand when others are contracting and contract when others are expanding.

These cycles contribute to farm income variability and make it difficult for producers to plan and finance their operations. Why don't farmers stop the livestock production and price cycles? This is easier said than done. The cycle is caused by rational producers' adjustments to changing farm prices. The cyclical problem reflects the near-perfectly competitive nature of farming and the biological and psychological lags in agriculture. These are difficult to change. Many people thought the livestock cycles would disappear as production was concentrated into the hands of fewer farmers, but the cycles have persisted.

SEASONAL PRICE VARIATIONS

Seasonal price variations are the more or less regular patterns of price changes occurring within a crop or marketing year. These variations are the result of seasonality of demand, production, and marketing patterns. Fall turkey prices are a classic example of a demand-induced seasonal price variation. The seasonal pattern of grain prices from harvest to harvest illustrates a supply-induced seasonal price change.

Retail food prices follow a regular seasonal pattern, as illustrated in Figure 10-10, with some deviation from year to year. This pattern primarily reflects meat

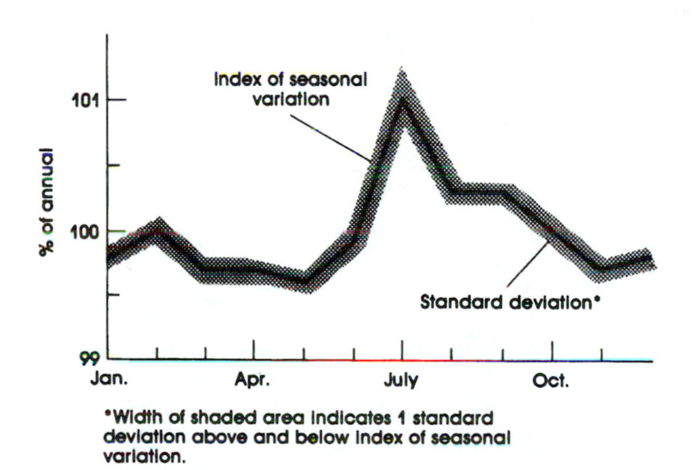

FIGURE 10–10 Retail food price seasonality. (U.S. Department of Agriculture.)

prices, which are lowest in the spring and fall and rise to a summer peak. Crop prices tend to be more stable over the year, with rising prices of some commodities matched by falling prices of others.

Variations in both supply and demand are responsible for the livestock seasonal price patterns. Cattle sales are generally highest in the fall at the end of the grazing season. Farmers avoid winter hog farrowings (births), so the pig crop is at a seasonal low in the December–February period. The summer increase in red meat demand usually raises prices at that time. These seasonal price patterns are illustrated in Figure 10–11.

The seasonality of crop prices depends on the suddenness of the harvest, the potential for storing the crop over the year, and the cost of storage. For most crops, prices reach their low point during harvest when the supply available for sale greatly exceeds the day-to-day demand. For example, July is normally the low wheat price, November for corn, as shown in Figure 10–11.

The seasonal crop price rise from harvest until just prior to next harvest is influenced by the cost of storage. The market must pay the storage costs of firms who

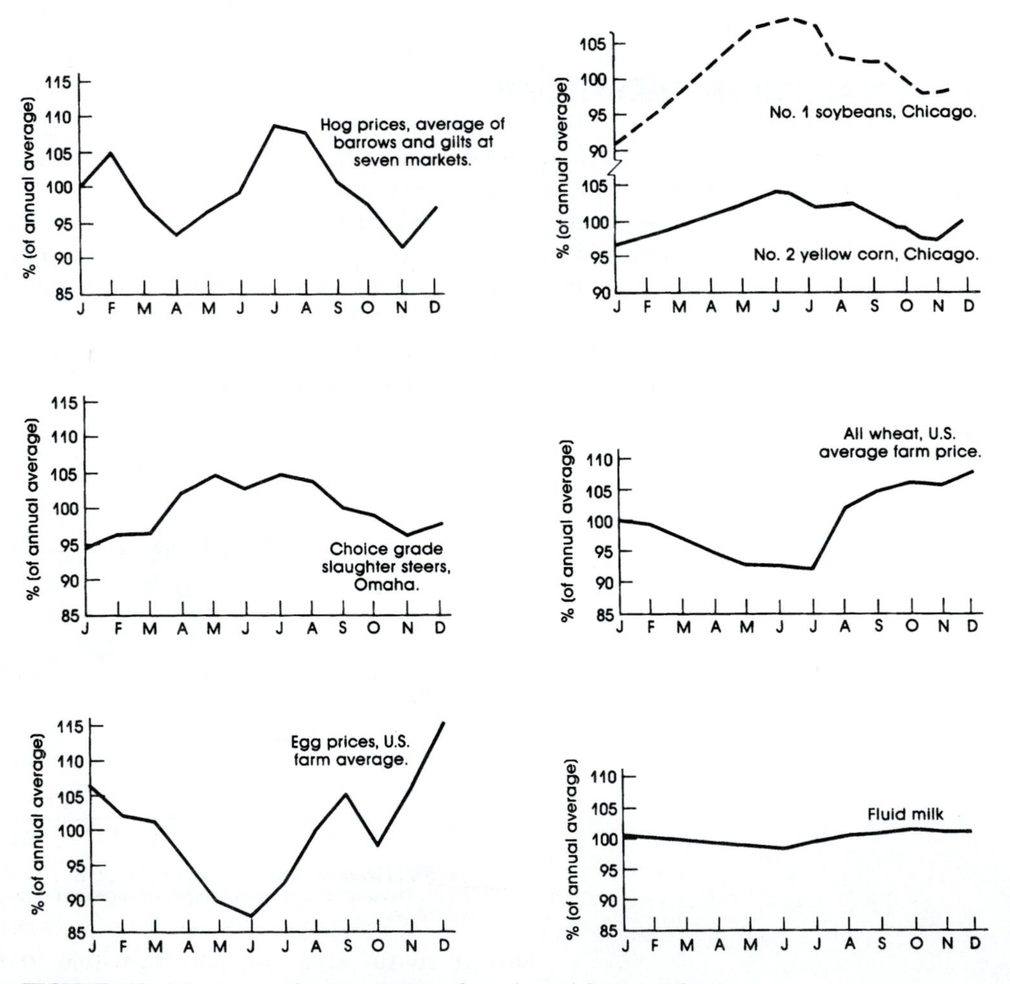

FIGURE 10–11 Seasonal price patterns for selected farm products.

defer selling at harvest. Thus, the seasonal price rise will tend to be greater for commodities with more expensive storage costs.

Changes in seasonal production practices, demand patterns, storage capacity, or storage costs will change the seasonal price variations of farm products. For example, year-round egg production and hog feeding have reduced the seasonal price variation for eggs and pork. Food storage and processing technologies have altered the seasonal price patterns of highly perishable commodities. Development of new products has to some extent reduced seasonal turkey prices.

As with price cycles, knowledge of seasonal food price movements can be valuable to farmers, marketing firms, and consumers. Again, however, seasonal price variations are not perfectly predictable in any year, nor wholly reliable from year to year. The high and low price periods shift unpredictably, and there are years when the seasonal price rise will be much greater or much less than the cost of storage.

FARM INCOME AND PRICES

Net farm income is influenced by three factors: (1) the volume of farm products sold; (2) the price of farm products; and (3) farm costs of producing and marketing these products. A change in any one of these will change net farm income. Moreover, these components do not move independently. A change in farm prices affects the quantity of farm products produced, just as the price of farm products influences the costs of farm inputs.

Figure 10–12 shows the trends in gross and net farm income and production expenses during the 1965-1999 years. Net farm income was relatively stable from 1965 to 1972 as gross farm income rose at about the same rate as production expenses. But

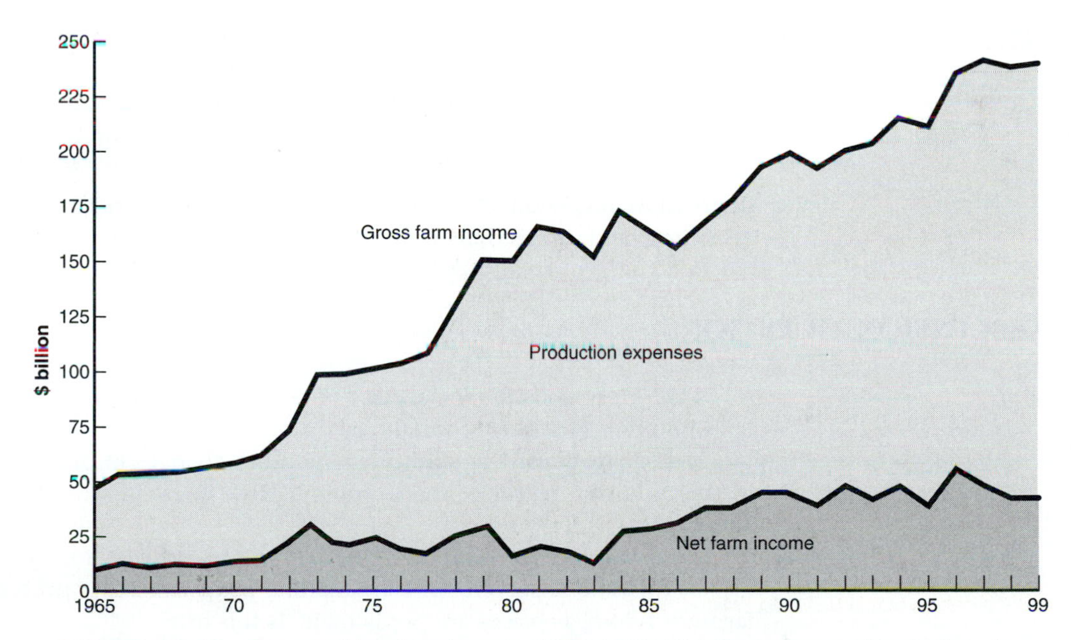

FIGURE 10–12 Gross and net farm income, 1965-1999. (U.S. Department of Agriculture.)

over the long term, and especially since 1972, farm production expenses have been driving a larger and larger wedge between gross and net farm incomes.

How have these farm income trends affected farmers? Wide fluctuations in income affect different farmers in different ways. Young farmers, who usually have large debts, may be more adversely affected than older, more well-established farmers with considerable equity ownership. The latter can average the highs and lows of farm income over longer periods of time. Farmers also have adjusted to the level and variability of farm incomes by increasing their nonfarm income.

SUMMARY

Farm prices and incomes play an important role in allocating resources in agriculture and in rewarding efficient producers. These prices and incomes are increasingly influenced by macroeconomic and global factors, many of which lie outside the food and agriculture system. Whereas commodity prices often appear to be set by food marketing firms, they are in reality jointly determined by farm supply, consumer demand, and the food marketing agencies. There are wider variations in farm prices and incomes than in nonfarm prices and incomes because of the greater inelasticity of demand for food, farmers' failure to adjust output to prices when aggregate demand falls, and the greater difficulty in controlling the output of a biological commodity. Three important sources of farm price variations are the business cycle, agricultural production cycles, and seasonal changes in production and supplies. The livestock price and production cycles are caused by a biological and psychological lag in farmers' response to changing prices and profit opportunities. Net farm income is influenced by farm prices, the quantity of farm products sold, and farm costs of production.

KEY TERMS AND CONCEPTS

biological, psychological lag
business cycle
cobweb
depression, recession
gross and net farm income
price cycle
price flexibility
real prices
seasonal prices

DISCUSSION QUESTIONS

1. How does the inelasticity of demand for farm products contribute to agricultural price and income variations?
2. Why doesn't agriculture respond to changes in prices in the same manner as the industrial sector of the economy? How does this result in a farm cost-price squeeze?
3. What steps could be taken to stop the cattle or hog cycle?
4. Some observers have argued that price instability presents no problem for farmers as long as prices are predictable. Is this true?
5. Comment on the statement, "Inflation is good for farmers."

6. How can consumers benefit from knowledge of cyclical and seasonal farm price variations?

SELECTED REFERENCES

Agricultural Graphics (July 2000). National Agricultural Statistics Service, U.S. Department of Agriculture. http://www.usda.gov/nass/aggraphs/graphics.htm

Agricultural Outlook, Monthly (July 2000). U.S. Department of Agriculture. http://www.ers.usda.gov./epubs/pdf/agout/ao.htm

Agricultural Prices Paid and Received (July 2000). National Agricultural Statistics Service, U.S. Department of Agriculture. http://www.usda.gov:80/agency/nass/aggraphs/agprices.htm

Agricultural Prices, Monthly (July 2000). National Agricultural Statistics Service, U.S. Department of Agriculture. http://www.usda/gov/nass/pubs/rptscal.htm

Consumer and Producer Price Indexes, Monthly and Annual (July 2000). U.S. Bureau of Labor Statistics, U.S. Department of Labor. http://stats.bls.gov/proghome.htm#OPLC

CHAPTER 11

FOOD MARKETING COSTS

Food marketing doesn't cost, it pays.

OBJECTIVES

After reading this chapter, you will be able to:

1. Describe the food marketing margin and explain how it affects farm and consumer food prices.
2. Explain the reasons for the rising food marketing bill and evaluate the role of profits and wages on the marketing bill.
3. Better understand the reasons why the farmers' share is so low overall and the reasons why the share varies for individual products.

Excessive profits, inefficiency, unnecessary services, and high marketing costs are often cited as responsible for high retail food prices and low farm prices. As a result, few areas have been studied as intensively as the costs of food marketing. The U.S. Department of Agriculture has a long history of research into marketing costs. Scarcely a Congressional session passes that does not see legislation requiring investigation of food marketing costs. In 1921, the Congress directed a Joint Commission of Agricultural Inquiry to investigate ". . . the cause of the difference between the prices of agriculture products paid to the producer and the ultimate cost to the consumers." In 1935, the Congress gave the Federal Trade Commission the responsibility of analyzing ". . . the distribution of the consumer's dollar paid for farm products between the farmer, processor, and distributor." In 1966, the National Commission on Food Marketing was established to study, among other things, the reasons for the difference between farm and retail prices and the reasons why these differences were widening. Each of these studies has added to our knowledge of food marketing costs and their impact on farm and retail food prices.

The central questions relating to food marketing costs are: Does food marketing cost too much? Why are marketing costs so high and rising? How do changes in marketing costs affect farm and retail food prices? How could food marketing costs be reduced? And, are food marketing profits excessive? In our study of food marketing costs, it is important not to overgeneralize. As we shall see, the costs and profits of food marketing vary considerably over time and for different products. In addition, as in all lines of business there are profitable and unprofitable firms in food marketing, although this is frequently hidden by industry averages.

THE FOOD MARKETING MARGIN

Consumer food expenditures can be broken down into their constituent marketing and farm components. Changes in these marketing and farm "shares" are watched carefully because they are indicators of trends in costs, profits, and services provided by farmers and food marketing firms. The portion of the consumer's food dollar that goes to food marketing firms is referred to as the *marketing margin.* This is the difference between what the consumer pays for food and what the farmer receives. In a sense, the marketing margin is the price of all utility-adding activities and functions performed by food marketing firms. This price includes the expenses of performing marketing functions and also food marketing firms' profits.

The allocation of consumer's food dollars between farmers and food marketing firms is one of the most controversial aspects of food marketing. Consumers do not earmark part of their expenditures for farm production and another part for marketing services. This division of the consumer's dollar is determined by competition and bargaining between these two sectors of the food industry. In effect, consumers face two prices for food: the farm price and the marketing "price" or margin. These prices reflect the costs of producing farm products, the costs of marketing services, as well as consumers' desires for these two "products." Consumers influence farm prices by substituting one food for another, just as they influence the marketing margin by substituting high-service foods (for example, prepared dinners) for lower-service foods (such as fresh produce at a roadside market).

The food marketing margin is labelled AB in Figure 11-1. Notice that for this diagram, the demand curve is drawn for consumers at the retail foodstore while the

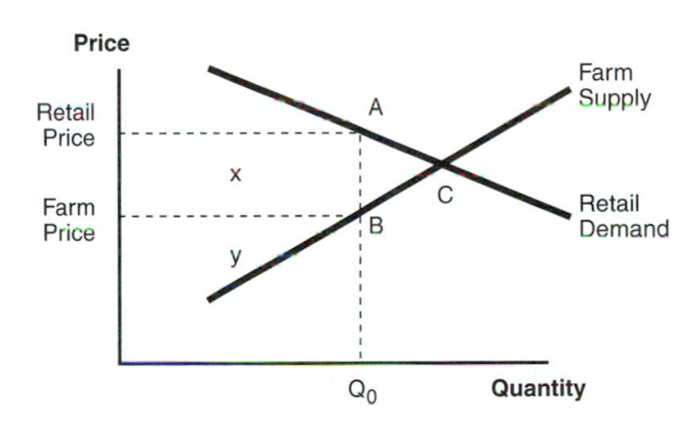

FIGURE 11–1 The food marketing margin.

supply curve is drawn at the farm level. If there were no marketing costs, the retail and farm price would be the same, at point C, and the farmer would receive 100 percent of the consumers' food dollar. With marketing costs, the margin is the difference between the retail and farm price of food. Area x is the payment to marketing firms for their services and area y is the payment to the farmer.

There are some widely held misconceptions about the food marketing margin. Many believe that a small margin denotes greater marketing efficiency, and that this is more desirable than a large margin. If this were true, farmers' markets—where the marketing margin is zero and the farmer receives all of the consumer's food dollar—would represent the most efficient method of food marketing. In fact, although direct marketing by farmers is becoming more prevalent, it is difficult to envision marketing our entire food supply in this direct-from-farmer-to-consumer way. Food marketing functions may or may not be performed efficiently, but efficiency cannot be judged solely by the size of the marketing margin.

Another widely held belief is that the large marketing margin reflects "too many" middlemen, and that the margin could be reduced by eliminating middlemen. The rule that middlemen may be eliminated but not their marketing functions and costs applies here. It is more correct to say that the size of the marketing margin depends on the number and costs of marketing functions performed rather than the *number* of middlemen. The division of labor resulting from the addition of more and highly specialized middlemen might well increase rather than decrease marketing efficiency.

Another misconception is that a large marketing margin "causes" low farm prices, and that an increase in the margin must necessarily lower the farmer's price. Here it is important to remember that marketing functions add both value and costs to raw farm products. Thus, an increased marketing margin can also increase the retail value and price of food. So it is quite possible that the farm price and the marketing margin will rise together as retail food prices rise. We should also remember that some of the marketing activities, such as advertising and merchandising, are designed to increase the demand for food, and this can lead to increased farm sales and higher farm prices.

Finally, the size of the food marketing margin is sometimes taken as a measure of the profits to be gained by farmers and consumers as a result of performing additional marketing functions. Both farm and consumer cooperatives have been justified as a means of lowering the food marketing margin and of capturing marketing profits for their patrons. However, the marketing margin is composed of both *costs* and profits. There is no guarantee that farmers or consumers will perform marketing functions as efficiently as middlemen and thus capture food marketing profits.

THE FOOD MARKETING BILL

The food marketing bill is the difference between total consumer expenditures for all domestically produced food products and what farmers receive for equivalent farm products. The marketing bill is calculated annually and serves as one measure of the food marketing margin. In 1999, consumers spent $618 billion for food at retail, including away-from-home purchases. Eighty percent ($498 billion) of this $618 billion went to food marketing agencies and the rest (20 percent) represented the farm value of food marketed. By this measure, then, consumers were paying four times as high a "price" for food marketing services as for basic farm products. That is to say, farmers received 20 percent of consumer food expenditures, whereas food marketing firms received 80 percent.

It is difficult for most people to view with detachment this 80:20 division of consumer food expenditures. To many, these shares seem both unfair and an underestimation of the contribution that agriculture makes to society. After all, don't farmers produce the raw ingredients without which there would be neither middlemen nor food on our tables? Yes, but we should be careful in making quick value judgments about this. The marketing bill tells us nothing about the *level* of farm prices or profits. Nor is it an indicator of costs or efficiency in either sector. Much more information is needed to evaluate the marketing bill in order to determine whether it is "fair."

Figure 11-2 shows the food marketing bill increasing steadily and more rapidly than the farm value of food during the 1958-1999 period. Three factors are responsible for this rising food marketing bill. First, as a result of population growth, the physical quantity of food that is marketed has increased, raising the total expenses of

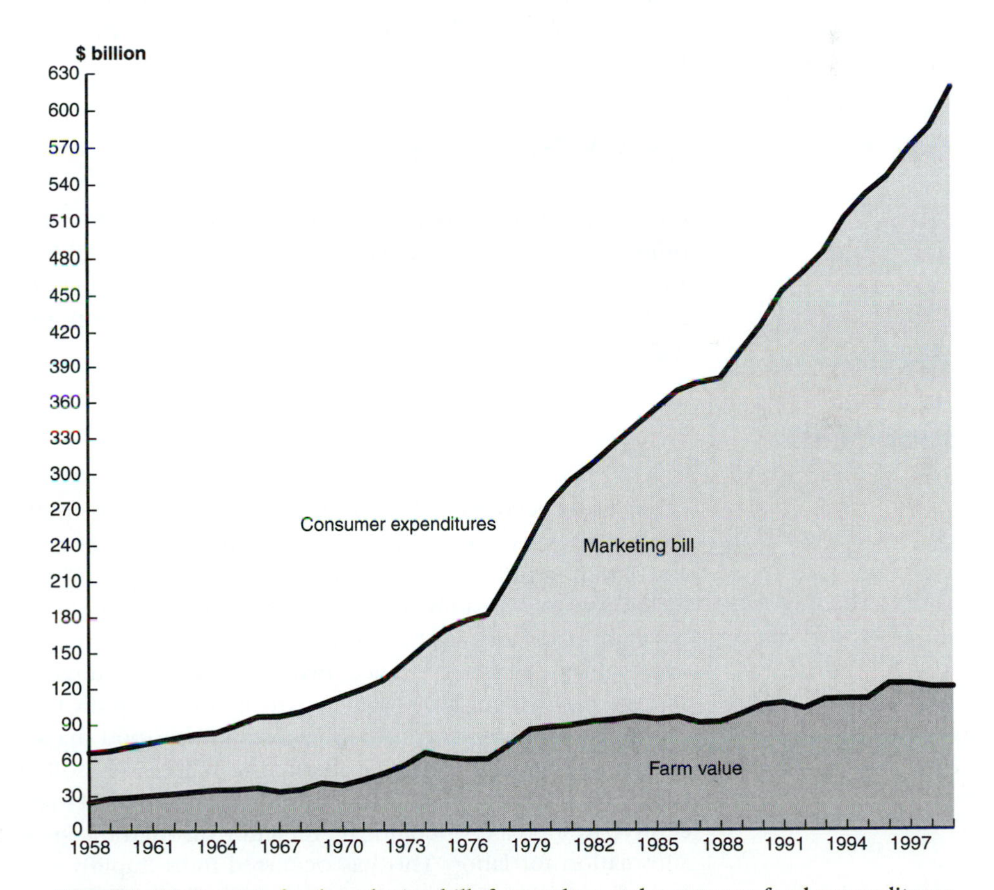

FIGURE 11–2 U.S. food marketing bill, farm value, and consumer food expenditures, 1958-1999. (U.S. Department of Agriculture.)

marketing food. Second, the costs of most food marketing inputs, especially labor and energy, have added to the rising cost of marketing food. Finally, consumer desires for additional food marketing services, such as represented by convenience foods, have further increased the food marketing bill. Notice that the rising food marketing bill has not depressed farm value, which has also risen. It is simply that the bill is rising faster than the farm value. If food marketing middlemen were trying to exploit farmers and force them out of business, the farm value should have fallen over the years.

Although each of these factors contributed to the rise in the bill shown in Figure 11-2, their importance has varied over the years. In the 1950s and 1960s, the trend toward more convenience foods had a greater impact on the bill than it had in more recent years, when increased costs of marketing services and inflation accounted for most of the rise in the bill. In recent years, the rise in demand for convenience foods has slowed as consumers returned to eating more fresh foods.

The fact that the food marketing bill is rising more rapidly than the farm value of food should not be surprising. Food marketing services are more income-responsive than are raw farm products. The same thing is happening elsewhere in the economy. As incomes rise, people buy more convenience foods and services, not just more products. Hence, as consumer incomes rise, we would expect the growth in demand for marketing services to outpace the rise in demand for farm foods.

COST COMPONENTS OF THE MARKETING BILL

Food marketing firms incur a number of costs when performing marketing functions. The cost components of the food marketing bill for 1999 are shown in Figure 11-3. Labor is the most important food marketing expense. Other significant components are packaging materials, taxes, transportation, profits, energy, and advertising. Let's look at these costs in more detail.

Labor Costs

Labor costs accounted for 48 percent of the food marketing bill in 1999. These include wages, salaries, employee health and welfare benefits, earnings of proprietors and family workers, and gratuities for foodservice. Wage rates in food marketing have increased steadily, but not dramatically, in recent years, accounting for about one-half of the rise in the marketing bill. Employee benefits, such as paid vacations and holidays, health insurance, private pensions, and payroll taxes for social security and unemployment have been increasing more rapidly than hourly wage rates.

The predominance of labor costs in the food marketing bill has three important consequences. First, the marketing bill closely follows the rate of increase in labor costs. Second, rising labor costs have given food marketing firms a powerful incentive to increase operational efficiency through the substitution of machinery and automation for labor. This has occurred more rapidly for food processors than for food distributors. However, all food marketing activities are constantly being scrutinized for potential labor-saving improvements. Third, the dominance of labor costs

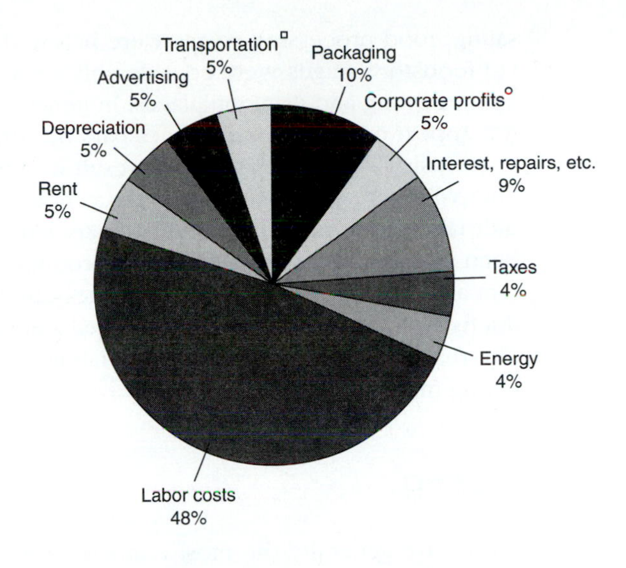

FIGURE 11–3 Components of the bill for marketing farm foods, 1999. (U.S. Department of Agriculture.)

in the food marketing bill introduces a downward rigidity in the marketing margin. Wages do not adjust to downward food price pressures.

Are wage rates too high in food marketing? Table 11-1 provides a comparison of wage rates in food marketing with wages paid in related fields. For the 1980–1997 period, wages in grocery wholesaling were about equal to those in nonfood whole-

TABLE 11-1 Average Hourly Earnings in the Food Industry, 1980–1997

	Average Hourly Earnings of Nonsupervisory Employees		
Sector	***1980***	***1990***	***1997***
Manufacturing			
Food processing	$6.85	$9.62	$11.49
All manufacturing	7.27	10.83	13.17
Wholesaling			
Food wholesalers	6.96	10.45	12.87
All wholesalers	6.96	10.79	13.44
Retailing			
Retail foodstores	6.24	7.20	8.59
Eating, drinking places	3.69	4.97	6.05
All retailers	4.88	6.75	8.34

SOURCE: U.S. Department of Labor.

saling, food processing wages were below the all-manufacturing wage rate, and retail foodstore wages were considerably above the all-retail wage rate and the wages paid in eating and drinking places. In general, wage rates in food marketing firms do not appear out of line with wages paid by comparable nonfood firms.

Rising wage rates need not result in higher food marketing costs. If labor productivity increased at the same rate as hourly wages, there would be no increase in unit labor costs and thus no inflationary labor cost pressures. However, wages have been rising more rapidly than labor productivity in much of the food marketing system as wage gains have followed wages elsewhere in the economy while labor productivity in food marketing has lagged behind productivity gains in other sectors. The resulting steady rise in per unit labor costs has contributed to a rising food marketing margin.

Profits in Food Marketing

Profits are generally the most controversial component of the food marketing bill. This is because people have different conceptions of "profit." To the accountant, profit is what is "left over" after all expenses are paid. On this basis, some conclude that profits are an unnecessary residual that can be reduced or eliminated without serious consequences. To the businessperson, profits are a reward for efficient behavior, and profit-seeking is a vital force that encourages lower costs and improved products. To eliminate these profits would destroy the economic incentives that have contributed so much to the American standard of living. The economist, on the other hand, views profits as another *cost* of doing business. Profits are the cost of attracting capital for investing in the growth and efficiency of the food marketing system. This is not to say that all profits are justified and therefore immune from criticism. Where profits result from anticompetitive behavior or market conditions, corrective action is justified. Although profits themselves are not evil, then, there is room for argument as to whether a particular level of profit is justified.

Comparing profit rates between firms and industries is hazardous. Differences in accounting methods and reporting techniques are one reason for this. In addition, there is no logical basis for arguing that profit rates should be equal in all firms or industries. Profits will vary depending on the riskiness of the business, the competitive nature of its markets, and a host of other factors. Nevertheless, comparisons of profit rates between firms and industries frequently are made.

Profit ratios are usually reported in two ways. Net profits as a percent of sales are calculated by dividing dollar net profits by total sales. This profit measure is useful to show the share of the consumer's dollar going for profits. However, it is difficult to use this measure to compare profits among firms and industries because of their different turnover rates and sales volumes. Many financial analysts prefer to calculate dollar profits as a percent of the investment in the firm—usually measured by stockholders' equity. This facilitates comparing a firm's returns to invested dollars with other investment alternatives. For both of these measures, profits may be taken before or after taxes.

Table 11-2 indicates that food industry profit rates are lower when expressed as a percent of sales than when measured as a return to stockholders' equity. Moreover, food marketing firms' profits as a percent of sales are generally lower than non-

TABLE 11-2 Average Profit Ratios of Food and Nonfood Firms, 1975-1999

	Manufacturing		*Retailing*	
Period	*Food Manufacturing*	*All Manufacturing*	*Food Chains*	*All Retailing*
	Percent return on stockholder equity (after taxes)			
1975-1980	14.3	14.2	11.8	16.4
1981-1986	14.1	11.0	14.0	13.8
1987-1992	17.5	10.4	12.6	9.9
1993-1999	18.6	14.9	16.1	11.6
	Percent return on sales (after taxes)			
1975-1980	3.3	5.2	0.9	2.2
1981-1986	3.5	4.1	1.1	2.2
1987-1992	4.6	3.8	0.7	1.5
1993-1999	5.3	5.4	1.3	1.9

SOURCE: U.S. Census Bureau, U.S. Department of Agriculture.

food firms' profits even though food manufacturers' and retailers' profits as a percent of equity are generally greater. Not surprisingly, then, food industry executives emphasize returns on stockholders' equity when they are attempting to attract investors, and they point to their profits as a percent of sales when they are attempting to document their "low" profits to consumers and government agencies.

Profits in food manufacturing have increased over the long term, as shown in Figure 11-4. These increases have been attributed to the development of new, highly differentiated food products, the diversification of food companies into more profitable nonfood lines, and improvements in plant operational efficiencies. However, profits both rise and fall. As shown in Figure 11-4, food manufacturers' profits declined from their recent record high of 21 percent in 1988 to 12 percent in 1993. This was a result of stagnant food industry sales and costly nonprice competition for retail shelf space.

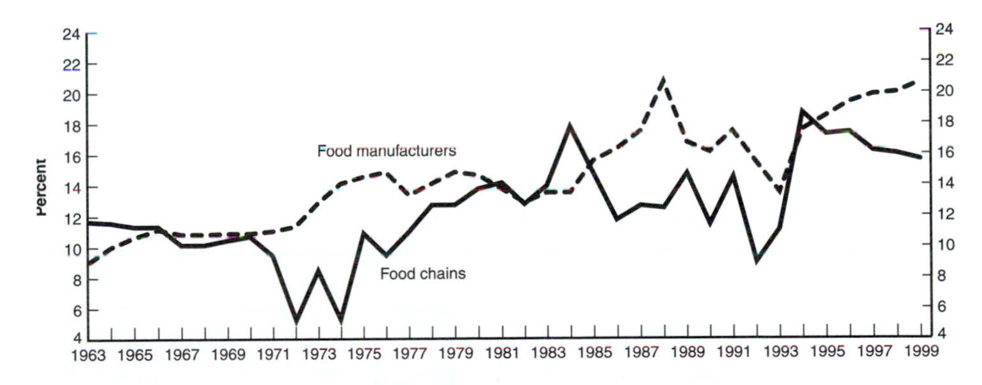

FIGURE 11–4 Rate of profit on stockholders' equity by food chains and manufacturers, 1963-1999. (Department of Commerce.)

Food chain profits were relatively stable at the 10 to 11 percent levels during the 1963–1977 period, with the exception of the 1972–1975 period when farm prices rose rapidly and the major chain stores were engaged in a price war. However, chain store profits generally rose from 1975 to 1994, reaching a high of 19 percent in 1994. Then, foodstore profits declined in the 1990s as a result of slow sales growth and increased competition from nontraditional outlets selling grocery products.

Profit rates vary for different food industries, depending on their competitive structure and degree of product differentiation. Historically, food industry profit rates have been lowest for meat, sugar, edible oils, milk, and grain mill processors. These are relatively homogeneous products. Differentiated products such as frozen foods, bakery products, breakfast cereals, dairy products, and beverages have typically had above-average profit rates in the food industry.

Are food marketing profits excessive? And are they responsible for high marketing costs? The answer is not so simple. First, profits account for 5 percent of the marketing bill, a significant but not dominant share of the bill. A large increase in food marketing profits would be necessary to raise the food marketing bill appreciably. Secondly, food marketing profits appear to be about comparable with profits elsewhere in the economy—neither higher nor lower when viewed as a return on stockholders' investment. Profit rates on the order of a 10 to 15 percent return on investment are considered satisfactory but not spectacular by financial analysts. On the other hand, food marketing profits are certainly not unduly low, except when measured by return on sales. Considering the relatively low risk of the food industry in general, profits appear to be adequate to attract capital to the industry and to reward investors sufficiently.

There is still another complication in evaluating food industry profits. Profits are not a simple indicator of market performance. High or rising profits may reflect superior management or efficient operations, but they may also result from such market imperfections as concentration, product differentiation, and barriers to entry. While profit levels in food marketing do not appear excessive, we cannot conclude from this that the food marketing industry is perfectly competitive or optimally efficient. Imperfectly competitive market conditions frequently lead to higher

MINI CASE 11: MARKETING COSTS AND THE FAMILY FOOD BILL

Melinda, a junior at State College, was home for vacation and was helping her mother put away the groceries they had just purchased. As her mother put away each item, she sighed and commented, "Food seems so much more expensive today. Why are food costs so high?" Seeing an opportunity to show her mother how much she had learned at school, Melinda said, "We studied that in our food marketing course, and I got an A." "Well," her mother exclaimed, "Tell me Miss Expert, and while you're at it help me lower my grocery bill so that we can afford your tuition next semester."

Outline the major points which Melinda should cover in her explanation of food prices and marketing costs.

costs and prices—hence, lower profits—rather than to extra-normal profits. Conversely, operational inefficiencies may be disguised by high price levels and not result in lower profits.

THE MARKETING COST INDEX

The food marketing bill provides one measure of food marketing costs. The U.S. Department of Agriculture's food marketing cost index provides another measure. This index shows the annual changes in variable operating costs incurred in processing, wholesaling, and retailing foods. The cost index is computed as a ratio of the current costs of a given set of food marketing inputs to the cost of these inputs in a base year. The index reflects changes in the costs of food marketing labor, packaging, materials, transportation, advertising, energy, rent, maintenance, and interest. It does not reflect changes in marketing productivity or profits.

FARM-RETAIL PRICE SPREADS

The marketing bill provides an aggregate view of the division of consumer food expenditures between farmers and food marketing firms. Farm-retail price spreads allow a more detailed view of this division for individual food products. The marketing bill is concerned with expenditure margins for all foods as a class, whereas the farm-retail spreads are concerned with price margins for individual foods.

The *farm-retail spread* is another measure of the marketing margin. It represents payments, including profits, for all marketing functions performed in assembling, processing, transporting, and retailing food after it leaves the farm. Changes in farm-retail spreads reflect changes in marketing costs, profits, or both.

During the 1970s the farm value, retail price, and farm-retail spread for a market basket of foods rose at about the same rates as shown in Figure 11-5. However, in the 1980s farm values did not rise as rapidly as retail food costs, so the farm-retail spread widened. This means that a rising farm-retail spread was contributing more to the increases in consumer food prices than were changes in the farm value of foods.

Farm-retail price spreads are not simply the difference between farm and retail food prices. Instead, the spread is the difference between the retail price per unit and the farm value of an *equivalent amount* of food sold by farmers. For example, a 1,000-pound steer produces about 417 pounds of retail beef cuts. Thus, it requires 2.4 pounds of live steer to "produce" a pound of retail beef. Consequently, in 2000 the farm-retail price spread for beef was $1.57 per pound—or the difference between a pound of choice beef at retail ($3.06 average all cuts) and 2.4 pounds of live steer (65.4 cents/lb. × 2.4 = $1.57).

Figure 11-6 shows how farm-retail spreads can be utilized to trace changes in marketing costs over time. The retail price of milk is composed of five margins: the retail margin, the wholesale margin, the processing margin, the assembly margin, and the farm value or "farmer's margin." The figure suggests that over the 1974-1992 period, all of these margins increased, but the retail functions were primarily responsible for the rising retail costs of milk in the 1988-1992 period.

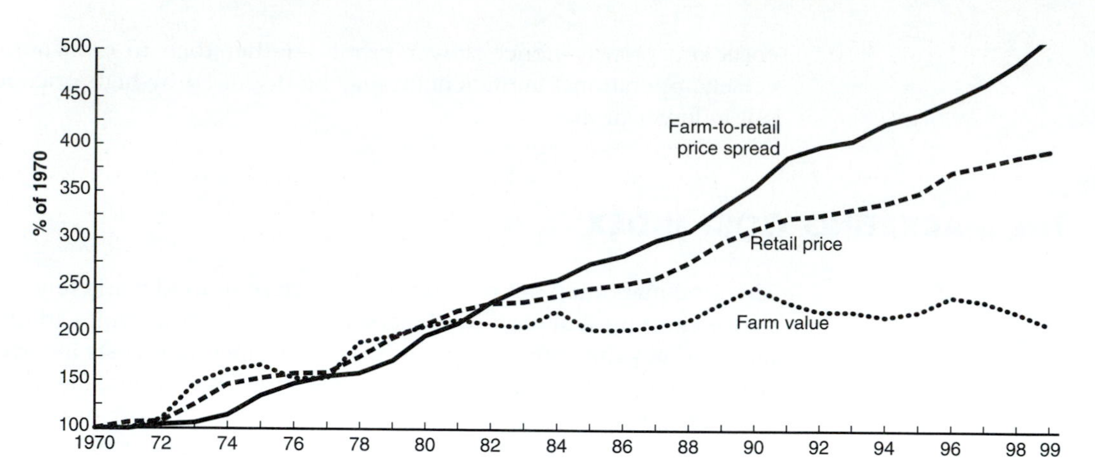

FIGURE 11–5 Retail price, farm value, and price spread for food, 1970–1999. (U.S. Department of Agriculture.)

THE FARMER'S SHARE

The difference between the retail price of food and the marketing margin is referred to as the *farmer's share.* This is the portion of the consumer's food dollar that farmers receive, expressed as a percentage of the consumer's food dollar. There are two ways to measure the farmer's share, and these give somewhat different estimates of this widely quoted statistic.

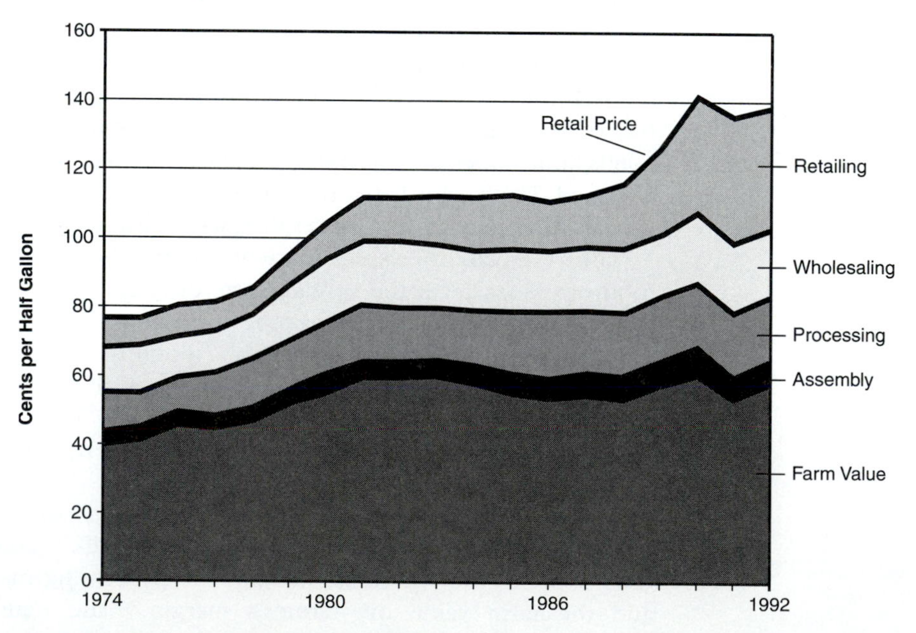

FIGURE 11–6 Distribution of the consumer's milk dollar, by agency, 1974-1992. (U.S. Department of Agriculture.)

Alternative Measures of the Farmer's Share

The marketing bill approach calculates the farmer's share as a ratio of the farm value of all domestically produced farm foods to the dollar value of consumers' food expenditures. For example, consumers spent $618 billion for food originating on U.S. farms in 1999. This food had a farm value of $120 billion, so the 1999 farmer's share was 20 percent ($120 billion divided by $618 billion). Conversely, the share of consumer food expenditures going for marketing activities was 80 percent, since the farmer's share and the "marketing share" must total 100 percent.

Alternatively, the market basket approach to calculating the farmer's share takes the ratio of the farm value of a market basket of domestically produced food items to their retail foodstore value.

These two measures of the farmer's share are compared in Table 11-3. The market basket approach generally yields a higher farmer's share than the marketing bill approach. However, both measures tend to move together over time. The major reason for the difference in the two series is that the marketing bill approach prices food eaten at home and away from home while the market basket approach uses only grocery store prices. Since foods eaten away from home have a smaller farmer's share, the marketing bill approach gives a smaller farmer's share.

Which is the correct measure of the farmer's share? It depends on what you want to measure. Changes in the farmer's share of the consumer's food dollar over time can occur for two reasons. One, there may be a change in the basic relationship between farm and retail food prices, as a result of changes in supply and demand at the farm or retail levels or a change in marketing costs. Two, the farmer's share can change with consumption patterns, as for example when consumers substitute more highly processed, convenience foods (with lower farmer shares) for fresh foods (with higher farmer shares). Since it prices the same market basket over time, the market

TABLE 11-3 Alternative Approaches to Measuring the Farmer's Share, 1960–1999

	Farmer's Share of the Consumer's Food Dollar	
Years	*Market Basket Approach*[1]	*Marketing Bill Approach*[2]
1960–64	38%	33%
1965–69	39	34
1970–74	41	34
1975–79	38	32
1980–84	35	28
1985–89	31	25
1990–99	25	21

SOURCE: U.S. Department of Agriculture.

[1]The farm value of 74 domestically produced food items as a percent of their retail value in food stores.

[2]The farm value of all domestically produced farm foods as a percent of consumer food expenditures for food eaten at home and away from home.

basket approach is most useful for monitoring changes in the pure farm–retail price relationship. Changes in the farmer's share as measured by the marketing bill approach reflect both these changing price relationships and changing consumption patterns. Because it prices a larger number of foods, many feel the marketing bill statistic is more representative of what consumers really purchase than the market basket. Nevertheless, the market basket statistic is widely used since it is calculated monthly and provides detailed farmer's shares for individual foods. The marketing bill statistic is only calculated annually for eight categories of food products.

During the 1945–1980 period, the farmer's share of the consumer's food dollar tended to hover around the 40 percent level for the market basket approach and around the 33 percent level for the marketing bill approach. Periods of falling farmer's share (e.g., 1958–1964 and 1969–1971) were balanced with periods of rising farmer's share (e.g., 1964–1966 and 1971–1973). This previous relative stability in farmer's shares suggests similar trends in farm and retail food prices and consistent contributions of farmers and food marketing firms to retail food prices over the long run. However, the farmer's share declined in the 1980s and 1990s as inflation increased the marketing margin relative to farm prices.

The Farmer's Share by Commodities

The general farmer shares shown in Table 11–3 are averages for several different commodities. As shown in Table 11–4, some foods have a farm share much higher than the average while others have a relatively small share.

Different commodity farmer shares reflect variations in time, form, and place utilities added by farmers and food marketing firms. Commodities for which the farmer provides most of the value-added utilities have a higher farmer's share. Commodities for which the marketing agencies provide a relatively large share of the utilities have smaller farmer's shares. These relative contributions to final product value are influenced by the following product characteristics:

1. The degree of processing.
2. Perishability of the product.
3. Product seasonality.
4. Transportation costs.
5. Bulkiness in relation to product value.

It is evident from Table 11–4 that the degree of processing influences the farmer's share. Farm products marketed to consumers in relatively unprocessed, fresh form have higher farm shares than processed foods. For this reason, animal products tend to have higher farmer shares than crops. The farmer's share declines as the degree of processing increases for both animal and crop products.

Transportation costs—as influenced by perishability, geographic concentration of production, seasonality, and product bulkiness—also affect the farmer's share. The relatively high costs of long-distance shipments, storage, protective services such as refrigeration, and spoilage result in lower farmer shares for fresh fruits and vegetables than for other fresh products. For example, lettuce production is geographically concentrated in the West, and its transportation costs alone exceed its

TABLE 11-4 Variations in Farmer's Share among Food Products, 1999

Product	*Farmer's Share*	*Product*	*Farmer's Share*
Chicken	49%	Pork	25%
Beef	49	Fresh apples	21
Eggs	47	Potatoes	19
Milk	39	Flour	19
Cheese	32	Lettuce	18
Sugar	31	Shortening	18
Fresh California oranges	28	Frozen french-fried potatoes	10
Peanut butter	26	Bread	4

SOURCE: U.S. Department of Agriculture.

farm value. This means that marketing costs will rise even if consumers substitute fresh foods for processed foods. In general, then, the farmer's share of the consumer's food dollar is larger for bulky, perishable, fresh-marketed commodities and lower for highly processed, concentrated food products.

The Meaning of the Farmer's Share

Caution is advised in interpreting and evaluating the size of the food marketing margin and changes in the farmer's share. A large marketing margin or a declining farm share are not necessarily indicative of the level of farm prices, farm income, marketing efficiency, or the value of food to consumers.

An example will point out the hazards of making judgments about farm prices and income from the farm share. Table 11-5 illustrates the prices, marketing margins, and farm shares of fresh and frozen potatoes. Notice that producers of fresh potatoes receive a higher farmer's share than farmers whose potatoes are processed and sold in the frozen form. Yet, because of the difference in retail prices, the grower's dollar return is larger for processed than for fresh potatoes. A small share of a large number can easily exceed a large share of a small number.

We should remember that the farmer banks dollars, not percentages. This example suggests that a falling farm share does not necessarily indicate falling farm prices or returns. Suppose that more and more potatoes were consumed in their

TABLE 11-5 Retail Price, Marketing Margin, Grower Return, and Farm Share for Potatoes and Potato Products, 1999

Product	*Retail Price*	*Marketing Margin*	*Farm Value*	*Farmer's Share*
Fresh potatoes	$.39/pound	$.31	$.08	19%
Frozen french fried potatoes	$1.02/pound	$.92	$.10	10%

SOURCE: Calculated from U.S. Department of Agriculture data.

processed and frozen forms. The total aggregate marketing costs for potatoes would increase and the farmer's share of the consumer's potato dollar would decrease. But if through this change, consumers could be encouraged to buy more potatoes at higher prices reflecting the increased processing costs, the net dollar returns to potato producers might actually be increased. This could represent a situation in which higher marketing costs were desirable from both the consumers' and producers' point of view.

The size or trend of the farmer's share are not reliable measures of marketing efficiency. Rather, in most cases the share merely reflects the complexity of the job that must be done in marketing the product. In fact, there is reason to believe that in very prosperous times when the share is relatively large, marketing may be less efficient than in depressed times when the share is small. Prosperous periods often encourage poor management and organization, because large profits come easily. Hard times provide incentives to operate as efficiently as possible.

It is doubtful that the statistics of the farmer's share merit the attention they receive. The important thing is not the size of the share, but the total return received by agricultural producers from the sale of their products. Higher marketing costs and a more prosperous agriculture are compatible ideas. As incomes rise, increased demands for more processing and marketing services increase marketing costs. However, this need not lower farm prices. Higher marketing costs can shift the consumer and farm demand curve to the right, raising farm prices.

In some instances, maybe enough is not being spent to market the product to its best advantage. This situation could be true regardless of the size of the farmer's share. Firms are continually attempting to discover new ways of attracting and influencing the consumer. In a sense, they are always asking the question, Does marketing cost enough? It is not a matter of low dollar costs alone but rather of getting the marketing job done with the best combination of resources. The end product of the marketing job is the movement of goods into consumption with top priority given to consumer satisfaction.

INTERRELATIONSHIPS OF THE MARKETING MARGIN AND FOOD PRICES

Farmers often feel that rising marketing costs depress farm prices. Consumers frequently complain that falling farm prices are not readily passed on to them in the form of reduced retail prices. Both criticisms raise questions about the relationships between the marketing margin and farm and retail food prices. Mathematically, the marketing margin is always equal to the difference between the retail price and the farm price. If one of these changes, the others must adjust in order to maintain the equality. However, this does not tell us which determines which, or how changes in one specifically influence the others.

There are two views of the price-margin relationship. The "cost-plus" theory is that the retail price of food is "built up" by adding the marketing margin to the farm price. Thus, changes in farm prices or marketing costs are simply passed through to consumers in the form of higher retail food prices. It follows that the consumer—not the farmer—would bear the cost of a rising food marketing bill or higher farm prices. In contrast, the "derived demand" theory of prices and margins suggests that the farm price is what is left over from the retail price after all marketing costs are paid. According to this view, an increase in marketing costs would reduce farm prices, unless retail prices also increased.

The cost-plus theory appears to be the correct one over the long run when the consumer's food dollar must cover all farm and marketing costs. However, in the shorter run, the derived demand theory of farm prices appears to be a more accurate view of the real world. Rising marketing costs reduce farm prices. The fact that there are frequently periods when farm prices are below costs of production also supports the derived demand view of prices.

Rising marketing costs typically raise the food marketing margin after a time lag which varies by products. The specific effects of a rising food marketing margin on consumers' and farmers' prices depend on the relative elasticities of demand and supply. In the short run, the consumer demand curve is more elastic than the farm supply curve, and an increase in marketing costs is felt primarily by farmers in the form of lower prices. In the longer run, as the farm supply curve becomes more elastic, the rising marketing costs are passed on to consumers.

There is also some controversy about whether the food marketing margin remains constant as prices change. In the short term (day-to-day, month-to-month), the marketing margin seems to be relatively stable in dollar value. This has important consequences for pricing efficiency, the stability of farm prices, and the farmer's share. The tendency for the marketing margin to remain constant in dollars, with short-term variations in farm and retail prices, is referred to as the "sticky" or ***inflexible marketing margin.*** This inflexibility causes the farmer's share to rise during periods of increasing food prices and fall during periods of declining food prices.

Figure 11-7 illustrates the effects of an inflexible dollar marketing margin on changing retail and farm prices. With a sticky marketing margin, any change in the retail price is immediately transmitted to the farm level, and changing farm prices are immediately reflected at the retail level. This results in farm price variability and rather wide swings of the farmer's share in the short run. If, on the other hand, the dollar marketing margin adjusted simultaneously with a changing retail or farm price, farm prices and the farmer's share would be more stable.

Several reasons are given for the stickiness of the dollar marketing margin in the short-run period. First, most of the costs of performing marketing functions are related to the physical volume of food market rather than to the price of food. It costs the same to grade, store, transport, and process a bushel of wheat regardless of its price. There is no reason why such marketing costs should adjust to food prices. The dominance of labor costs in the food marketing bill also contributes to the sticky marketing margin. Wages simply do not adjust to price changes in the short run. Increased unionization of food marketing workers further adds to the inflexibility of the marketing margin. Finally, imperfect competition in the food marketing industries contributes to margin inflexibility. The market power of imperfectly competitive middlemen makes them more margin setters than price setters.

In addition to the tendency of margins to be sticky, it is sometimes observed that the marketing margin rises when farm prices fall and declines when farm prices rise. The explanation for this observation is that farm prices fall when supplies of farm products are rising. This increase in farm product supplies shifts the demand for marketing services (transportation, storage, processing, etc.) rightward. This

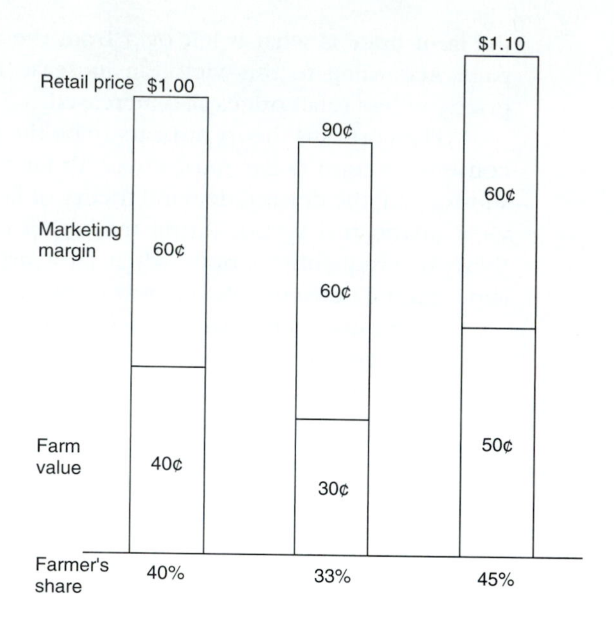

FIGURE 11–7 Effects of a fixed dollar marketing margin on farm and retail prices.

shift, in turn, increases the price of marketing services, that is, the marketing margin. The opposite effect occurs when farm supplies are falling.

THE FUTURE OF FOOD MARKETING COSTS

Are there any unexploited opportunities to reduce food marketing costs, increase farm prices, or reduce retail food prices? What can we reasonably expect the future trend to be in marketing costs and in the food marketing margin?

Continued Large Marketing Margin

Even with a highly efficient marketing system that functions well, the costs of marketing food will continue to be high and to increase. The food marketing task grows more complex and more expensive with urbanization, geographic specialization of agriculture, the affluence of consumers, and increasing population.

Cost inflation and consumer demand for marketing services hold the key to the rate of future increases in the marketing margin. Most marketing costs are influenced by general economic forces outside of the food economy, especially labor, transportation, packaging, and energy costs. These rising costs will maintain their pressures on the rising food marketing bill, and government regulations, affecting such areas as occupational safety, plant sanitation, energy sources and uses, and environmental protection, also will add costs. Consumer demand trends will further contribute to the rising food marketing bill in coming years. The demand for most processing and convenience services has already contributed to the rising marketing bill. This demand is closely linked to trends in consumer incomes, which are expected to continue to rise in the long run.

But couldn't we reduce marketing costs by returning to a more simple food supply? It depends on what is meant by "simple." A return to self-sufficient household food production would certainly lower food marketing costs, but this seems infeasible for many people today. Consumers can and do purchase foods direct from farmers, and this can lower marketing costs. But it is unlikely that the entire food supply can be marketed this way. The sale of more fresh foods in place of processed foods may increase marketing costs if the products are produced in specialized areas around the world, increasing transporation costs. Bulk, unpackaged sales of food products may increase spoilage and merchandising costs. It is not so easy to lower food marketing costs today.

Opportunities for Reducing Costs

In 1966, the National Commission on Food Marketing made a comprehensive evaluation of food marketing costs. The conclusions of this study, which still seem valid, were that the rising cost of labor was chiefly responsible for the growth in the farm–retail spread; that few unnecessary or wholly wasteful marketing functions were performed in the food industry, but that some selling activities, especially advertising, could be reduced without loss of consumer value; and that most food marketing functions were performed efficiently but that there were places where efficiency could be improved.

As we have seen, profits are not abnormally high or distressingly low *on the average* in food marketing, and in any event profits do not account for a very large share of the food marketing bill. Therefore, while profits in the food industry should be monitored, reducing profits is probably not a very good approach to lowering food marketing costs.

Labor is the most obvious area in which to cut costs, but food marketing wages could not be reduced significantly without afffecting the ability of the industry to attract and hold qualified labor. Therefore, the logical approach to reducing labor costs is through improved productivity resulting from mechanization and automation. Progress will be slow, however, because productivity gains in food marketing have lagged behind labor efficiency gains elsewhere in the economy. The efficiencies gained by such new marketing techniques as improved packaging and transportation methods, computerized ordering and inventory systems, and electronic foodstore checkouts have been more than offset by the higher costs associated with new consumer services, product and brand proliferation, and promotional costs.

Some observers have suggested the elimination of "unnecessary" marketing functions as a cost-reducing measure. There are value judgments here. Moreover, the rule that it is difficult to eliminate marketing functions without changing the value of food products to consumers applies here. Integration of marketing functions into a single firm promises some economies, as has occurred in food retailing and wholesaling, but in general, those who seek to eliminate the middleman often simply change their identity.

Attempts to reduce other food marketing costs or functions in order to improve operational efficiency often reduce consumer satisfaction, pricing efficiency, or consumer freedom of choice. Some food industry critics have argued, for example, that many food industry advertising, packaging, and promotional practices could be eliminated without loss. This viewpoint, however, presumes to judge what is a necessary and an unnecessary cost—a judgment that many feel should be left to the consumer and to the competitive market processes.

The National Commission on Food Marketing cited the shift of the food industry from a commodity orientation to a marketing/merchandising orientation as a major reason for the increasing farm–retail spread. This shift probably sacrificed some pricing and operational efficiency for nonprice competitive advantages and perhaps consumer satisfaction. Food manufacturers' and retailers' success increasingly hinges on the differentiating value of cost-increasing services such as advertising, coupons, games, and elaborate merchandising. These activities may or may not represent "true" consumer desires, but there is no doubt that they have added significantly to the cost of food marketing.

Market researchers often report that the costs of food marketing could be reduced if the large number of firms were reduced to an "optimal" number, or if all firms would distribute their products optimally among geographic markets. These studies often neglect the competitive impacts of improvements in operational efficiency. Should we reduce the number of food marketing companies to gain economies of size, even at the expense of weakening competition in the marketplace?

Thus, the goal of lowering food marketing costs must always be reconciled with the consumer's freedom of choice, the firm's freedom of behavior, pricing efficiency, and consumer satisfaction. The result is a compromise, which is easy to improve upon only if one goal is viewed independently of the others.

SUMMARY

The food marketing margin is the price of all the utilities added in the marketing process. The size, composition, and behavior of this marketing margin are controversial aspects of food marketing. The bill for marketing food continues to rise each year because of the increased volume of food marketed, higher costs of marketing services, and consumer demands for more marketing services. Marketing utilities are more income-responsive than the demand for farm commodities, so the marketing bill is about twice the value of farm foods. Food marketing labor costs and profits do not appear to be excessive, but neither are profits and wages abnormally low in the industry. The farmer's share varies widely for different food products. The dollar marketing margin tends to be sticky or inflexible in the short run, and this contributes to instability in farm prices and variability in the farmer's share. Although some progress is being made in improving operational efficiency and lowering unit costs of food marketing, rising marketing costs are contributing to an increasing marketing margin. There does not appear to be any potential for dramatically reducing food marketing costs in the near future. Considering the contribution that marketing activities can make to increased consumer satisfaction and demand, farmers and consumers might well ask, Is enough being spent to market the nation's food supply?

KEY TERMS AND CONCEPTS

cost components
farm-retail spread
farmer's share
market basket
marketing bill
marketing margin
marketing profits
profit rates
sticky margin

DISCUSSION QUESTIONS

1. You have been asked to explain to a group of laypersons why the farmer's share for beef was 49 percent in 1993 while the share for bread was only 4 percent. Outline the major points you would make in your presentation.

2. What are the principal efficiency conflicts preventing rapid reductions in food marketing costs?

3. Using your own figures, demonstrate that farm prices can rise even when the farmer's share is falling. What are you assuming about the dollar and the percentage marketing margin?

4. Prove the following: If the marketing margin is a constant percentage, percentage changes in retail prices will exactly equal percentage changes in farm prices. Whereas, if the margin is a constant dollar amount, percentage changes in farm prices will exceed percentage changes in retail prices.

5. Evaluate the statement, "Food marketing costs too much."

6. What do you think would be a "fair" division of the consumer's food dollar between farmers and middlemen?

SELECTED REFERENCES

Elitzak, Howard (1999). *Food Cost Review, 1950–97.* Agricultural Economics Report No. 780. Economic Research Service, U.S. Department of Agriculture. This data is periodically updated and made available at: http://www.ers.usda.gov/briefig/FoodMark/

Food Marketing and Price Spreads Briefing Room (January 2001). Economic Research Service, U.S. Department of Agriculture. www.ers.usda.gov/briefing/foodpricespreads/

PART IV

FUNCTIONAL AND ORGANIZATIONAL ISSUES

CHAPTER 12

THE CHANGING ORGANIZATION OF FOOD MARKETS

Nothing is changing in food marketing. Except everything.

OBJECTIVES

After reading this chapter, you will be able to:

1. Explain how specialization, diversification, decentralization, and integration have changed the organization of the food system.
2. Evaluate how these changes in market organization have affected farmers, food marketing firms, and consumers.
3. Better understand some of the public policy issues concerning the future organization of the food system.

The food industry is a dynamic one. It is in a constant state of flux. The food marketing machinery continues to make adjustments to changes in agricultural production patterns, new marketing technologies, and trends in food consumption. These adjustments alter the market organization of the food industry—the kinds of firms involved, the relationships of firms within the industry, the allocation of marketing functions among firms, and the nature of food product flows. Three important market organization trends in the food industry are (1) specialization and diversification, (2) decentralization, and (3) integration.

VERTICAL COORDINATION IN FOOD MARKETS

The food market consists of communication, physical distribution, and exchange systems. It might be thought of as an orchestra. Each level of the market contributes its utility to the final product. The numerous decisions about what to produce, how much to produce, where, how, and when to sell must somehow be coordinated. *Vertical market coordination* refers to the process of directing and harmonizing the several interrelated and sequential decisions involved in efficiently producing and marketing the nation's food supply.

Simple economies do not have vertical coordination problems because there are limited marketing activities, and often, many production and marketing functions are performed by the household or the same firm. In advanced market economies, more marketing functions are performed, the marketing channels are longer, and firms specialize in performing separate but related marketing functions. This presents coordination problems.

Vertical market coordination is particularly critical in the food industry because of the length of the marketing channel, the large number of specialized firms involved, the inherent uncertainty of prices, supplies, and qualities of farm products, and the urgency of marketing perishable products. For these reasons, errors in food production and marketing decisions are quite costly to both the firms in the industry and to the efficiency of the system.

Vertical market coordination in the food industry may take the form of either competition or cooperation between buyers and sellers. While there is a natural rivalry between buyers and sellers in a distribution channel, there is also a community of interests in producing the correct products for the correct markets at the correct time, place, and price. Farmers rely on marketing firms to provide the utilities that consumers desire just as marketing firms are dependent upon farmers for their raw product supplies. Because of this interdependence of all parties in the modern food distribution channel, cooperative market behavior in coordinating the various stages of food production and marketing is as common as competitive behavior.

There are two ways to vertically coordinate food industry decisions. Prices can serve to communicate the needs of each firm to others in the marketing channel. Alternatively, other coordinating devices—for example, contracts—can be used to communicate information between firms, such as price, quality, timing, and other market considerations. The search for improved vertical market coordination has resulted in a change in food market organization in recent years. Two important changes in vertical market coordination are, first, that the major decisions—what to produce and how much to produce—have in large part passed from the farmer to food marketing firms. Food marketing firms are directing and coordinating the food economy today, including many farm decisions. Agricultural production often takes place on order from food marketing firms. Second, there is a trend toward replacing open market prices with administered coordination techniques, such as contracts, bargaining associations, and the like. Both trends are affecting the farmer's position in the food industry.

Evolving Market Organization

In the colonial period of our history, the food marketing channel was a simple one. Much of the food was consumed on the farm, and the remainder was marketed directly to nearby consumers. As population centers developed, country storekeepers came on the scene to assemble farm products for local sale or shipment to distant markets.

During the nineteenth century, food marketing channels continued to lengthen as the distance between farmers and consumers increased. Direct farmer sales to consumers declined as specialized assemblers of farm products and wholesalers developed to bridge the gap between farmers and consumers. Public central terminal markets were established in the major population centers where crops and livestock products were collected and sold to processors, exporters, and retailers. The Chicago Union Stockyard, established in 1865, was an important livestock terminal market. Wholesale terminal markets for fresh fruits and vegetables, eggs, butter, and other products developed near railroad yards and port facilities in the larger cities. Food processors were often located adjacent to these terminal markets.

Each level of the food marketing channel was made up of substantially independent firms. Country buyers, such as cream stations, elevators, and poultry and egg buyers were independent businesses buying from farmers and selling to processors and wholesalers in the cities. Most processors were engaged only in processing activities. Wholesalers performed a few specialized marketing functions, as did retailers.

All of this trading was coordinated by prices determined in open market competition among buyers and sellers. There were few contractual arrangements and little in the way of formal bargaining relationships. Prices were discovered in the central terminal markets where all buyers and sellers met and could physically examine the produce. Although not without problems, such markets may have approximated the perfectly competitive conditions.

If the nineteenth century was a period of open, lengthening food marketing channels and increased specialization of marketing functions, the twentieth century has been a period of closing of these channels, diversification of marketing functions by firms, and a substitution of administrative arrangements for price coordination.

SPECIALIZATION AND DIVERSIFICATION IN FOOD MARKETS

Specialization and diversification are two pervasive characteristics of food markets. Farmers usually specialize in a few, or in one, commodity or enterprise. Food processors operate specialized plants or company divisions that provide a wide variety of brands and products. The supermarket can be thought of as a collection of specialized departments (dry groceries, meat, produce, and nonfoods) selling a diversified assortment of products. Which is more efficient, specialization or diversification? And why are they occurring together in the food industry?

Specialization in the Food Industry

Specialization and division of labor are two of the universal characteristics of markets. In the food industry we observe product, functional, and institutional specialization. Battle Creek, Michigan, is the breakfast cereal capital; citrus is a southern crop; food processors specialize in adding form utility to farm products; and livestock production takes place in and around the corn belt. Specialization is so common in the food industry that it is taken for granted. Farmers and food marketing firms usually find it profitable to specialize because it can improve operational efficiency and increase profits.

Specialization and division of labor have three important consequences for the food marketing system: (1) increased interdependency of food producers, marketing firms, and consumers; (2) increased volume of exchange and thereby the importance of marketing activity; and (3) a tendency toward larger firms. The latter bears further explanation.

Typically, food marketing plants' costs of production decline with rising output. Table 12-1 provides some examples. These economies of size result from the specialization of labor, machinery, and management that is possible in large plants. Not all costs rise with output, and average costs of firms fall as these fixed costs are spread over more volume of output. Thus, there is a profit incentive for food marketing firms to increase in size. However, there are limits to this profitability. Beyond a particular volume, the average costs of food marketing firms stabilize and eventually may even rise as the plant grows in size and experiences diseconomies such as bureaucratic costs or the transportation costs to and from the plant increase. A typical food firm's average costs are shown in Figure 12-1. There is a range of output (*AB*) where costs fall with size of plant as operational efficiency is improved. The output range, *BC*, is a level of stable average costs; and beyond *C*, plant costs rise with further growth.

Knowledge of food marketing firms' average costs can be helpful in evaluating the trend toward larger food marketing firms. The cost savings of firm growth from *A* to *B* in Figure 12-1 represent potential food price reductions for consumers and possible price increases for farmers. However, as these operational efficiencies are gained by the firm and there are fewer, larger firms, pricing efficiency may be reduced. Therefore, there is no guarantee that the cost savings will be passed on to farmers or to consumers. This is the *pricing-operational efficiency dilemma.*

TABLE 12-1 Illustrations of Economies of Size in Food Processing

Industry	*Plant Size*	*Average Costs*
Dairy processing (1962-69)	6,000 qts. per day	6.7 cents/qt.
	100,000 qts. per day	3.4
	800,000 qts. per day	2.4
Bread baking (1965)	250,000 lbs. per day	3.87 cents/lb.
	500,000 lbs. per day	3.59
Canned tomatoes (1969)	100 cases per hour	$3.60/case
	800 cases per hour	3.03
	1,500 cases per hour	2.85
Soybean processing (1952)	25 tons per day	63.5 cents/ton
	200 tons per day	37.9
	1,000 tons per day	35.1

SOURCES: *Market Structure of the Food Industry.* U.S. Department of Agriculture Marketing Research Report No. 971. September 1972, p. 23; *Organization and Competition in the Milling and Baking Industries.* National Commission on Food Marketing, June 1966, pp. 132-35; G. A. Mathia et al. *An Economic Analysis of Whole Tomato Canning Opportunities in the South.* North Carolina State University, Economic Information Report No. 17, May 1970; *Size of Soybean Oil Mills and Returns to Growers.* U.S. Department of Agriculture Marketing Research Report No. 121, 1953, p. 24.

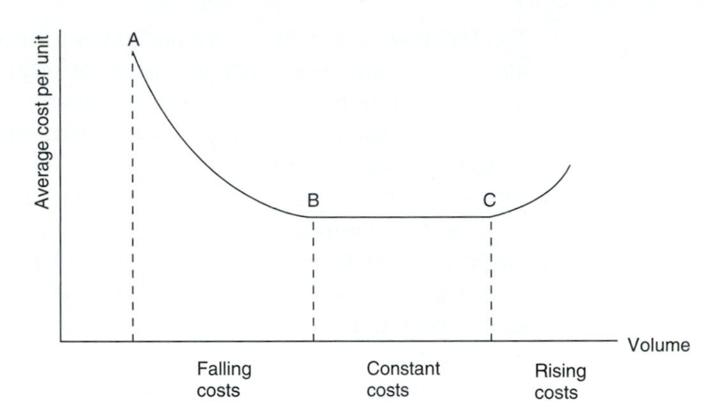

FIGURE 12–1 A hypothetical food marketing plant's average cost curve.

Other uses can be made of these cost relations. The falling average cost is usually steeper for food processors than for wholesalers and retailers as a result of their mix of fixed and variable costs. Thus, there may be more justification for firm growth in processing than in the distributive trades. The constant cost-volume range presents a public policy issue. Is there any reason to encourage firms of size *C* when they are no more operationally efficient than firms of size *B?*

Numerous studies have been made of these economies of size in food marketing. The following generalizations have emerged: (1) Larger plants usually have lower costs than smaller plants, and no firms operate in the rising cost range; (2) the economies of size are greater for highly capitalized food industries (such as dairy processing, flour milling) than for lower-capital industries (such as fruit and vegetable packing); and (3) many food processing plants are much larger than is justified by economies of size. The National Commission on Food Marketing found that the smallest firms in the food industry suffered handicaps because of inefficient size; medium-sized plants were operationally as efficient as larger plants; and the larger firms experienced economies of size in advertising and sales promotion. Some researchers have concluded that economies of size are insignificant in food manufacturing, and that maximum operational efficiency can be achieved by relatively small plants.

Diversification in the Food Industry

In view of these economies of specialization and size it may seem contradictory that many food marketing firms are diversifying their product offerings, their range of marketing functions, and the institutional levels of the market in which they participate. Many examples may be cited. Farmers perform storage and transportation functions, and some farmer cooperatives operate food processing plants. Food processors have farming operations and are merging with nonfood firms. Many retail chain organizations operate food processing plants and are diversifying their product offerings into nonfood areas. Are these trends consistent with specialization? Do they require a sacrifice in specialization economies?

There are a number of explanations for this diversification. For one thing, specialization is not necessarily inconsistent with diversification. Food manufacturers can operate multiple plants, each specializing in separate products and lines. Another reason for diversified growth is that many large food companies have exhausted the specialization economies or are prevented from further growth by antitrust regulations. Growth by diversification may be their only alternative.

There also appear to be considerable economies to market and product diversification in advertising, promotion, and merchandising. A broad product line can be an asset in the battle for retail shelf space and consumer attention. Diversification also can protect the firm from the risks of price changes and market losses for a single product.

Functional diversification in the food industry is motivated principally by market coordination economies. Food chains and processors often reach back into the food marketing system to influence the quality and timing of products consistent with their scheduling and merchandising programs. For example, a food processor may be able to lower costs by contracting for an assured quality or delivery schedule for farm products. These coordination economies reflect the mutual interdependency of all firms in the food industry.

We are likely to see continued specialization and diversification in the food industry, both in the United States and worldwide.

DECENTRALIZATION OF FOOD MARKETS

Decentralization is a major structural change that has occurred in food marketing since the early twentieth century. Decentralization means that farm products move from farms and into the hands of processors and wholesalers without utilizing the services of the traditional central markets. The buying agents of processors, wholesalers, and retail firms contact producers and take title to the products in the farm production area.

Centralized markets are compared to decentralized markets in Figure 12-2. In a centralized marketing process, farm products are delivered to urban *terminal markets* (so named because they are often located at the end of a rail line) for the performance of such marketing functions as exchange, standardization, and market information. There is a physical concentration of buyers and sellers and other market agencies in these public markets. In decentralized marketing, buyers and sellers move into the production area to operate at widely separated locations. Instead of products coming to the processor, wholesaler, or retail buyers, these buyers go to the products at country markets. Thus, decentralization constitutes a rearrangement of the marketing functions among firms—not an elimination of these functions—and a displacement of the central wholesale terminal market by direct sales from shipping-point markets to wholesalers, retailers, and processors.

Decentralization involves more than just bypassing the terminal market and rearranging marketing functions. The trend has altered significantly the nature of food production and farm marketing patterns. Farmers often act as their own sales agents in decentralized markets, whereas in centralized marketing, they were more likely to employ the services of a wholesaler or broker as a sales agent. This means that in decentralized markets the farmer not only is more involved with marketing products,

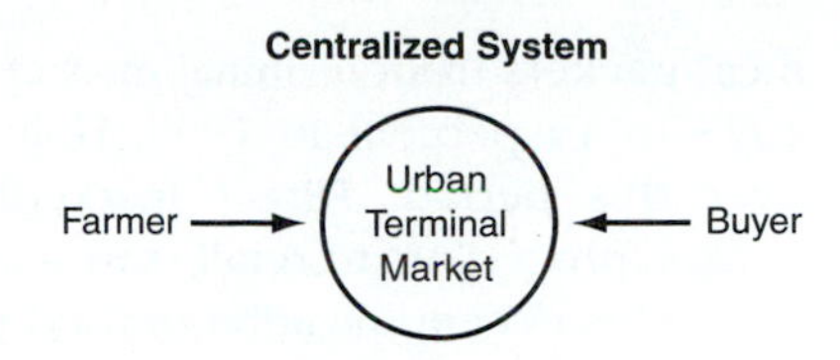

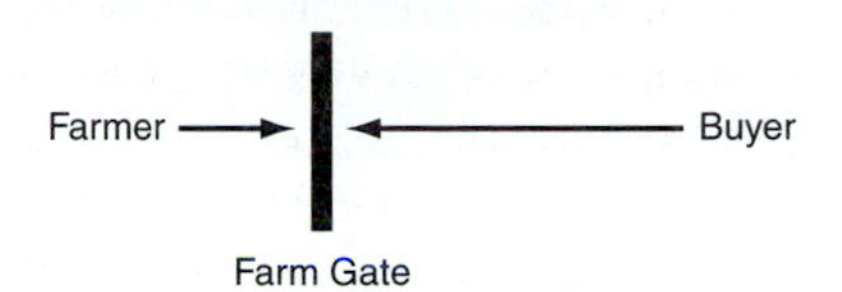

Figure 12–2 Centralized versus decentralized markets.

but also is dealing directly with the large buyers of products. Moreover, there is no "perfect" market buffer between the farmer and farm product buyers. The price discovery process shifts to shipping-point markets. Finally, buyers of farm products exert greater control and coordination over farm decisions in decentralized markets than is possible in central markets.

Extent of Decentralization

Decentralization has proceeded somewhat more rapidly for livestock products than for crops, although fresh fruit and vegetable markets are highly decentralized today. The trend in livestock decentralization is shown in Table 12-2. Cattle purchases by

TABLE 12-2 Decentralization in the Purchase of Livestock and Eggs, 1923-1998

	Percent of Packer or Processor Purchases, By Market Outlet							
	Cattle			*Hogs*			*Eggs*	
Year	*Terminal Markets*	*Auctions*	*Direct Purchases*	*Terminal Markets*	*Auctions*	*Direct Purchases*	*Terminal Markets*	*Auctions, Direct Purchases*
1923	90	—	10	76	—	24	77	23
1940	76	—	24	47	—	53	NA	NA
1950	75	—	25	40	—	60	40	60
1960	46	16	38	30	9	61	30	70
1970	18	16	66	17	14	69	17	83
1980	8	15	77	14	9	77	NA	NA
1990	17	17	83	10	10	90	NA	NA
1998	14	14	86	3	3	97	NA	NA

SOURCE: *Packers and Stockyards Resume,* U.S. Department of Agriculture.

meat packers from terminal markets declined from 90 percent of total purchases in 1923 to 14 percent in 1998. Hog purchases by packers followed a similar trend over this period. Direct marketing of shell eggs, broilers, and turkeys from packer/processors to retail stores also increased.

This decentralization of food marketing channels occurred in two steps. First, the buyers of farm products bypassed the central terminal markets and began purchasing directly from shipping-point markets or from farmers. The following step was the transfer of processing facilities from the urban terminal markets to the producing areas. This movement of physical facilities permitted some new firms to enter the packing industry, and the new facilities resulted in an increase in packing plant efficiency.

Reasons for Decentralization

Perhaps the best way to understand the reasons for this decentralization trend is to examine the reasons for the initial development of the centralized market system and then look at the changes that have occurred in these factors.

Factors Favoring the Development of a Centralized Market System	*Factors Favoring Development of Decentralized Markets*
1. Limited transportation system with reliance on railroad. Only a few advantageous points for product concentration.	1. Development of flexible truck and highway system.
2. Poor communication meant that buyers and sellers meet physically to make exchanges.	2. Improved communications permits buyers and sellers to trade without coming face to face.
3. High perishability and poor standardization make physical inspection of products necessary before buying.	3. Improved refrigeration, storage, and grades and standards allow products to be traded by sample or description.
4. Small, unspecialized farmers make it expensive for buyers to deal directly with producers.	4. Fewer and larger production units make direct purchases feasible.
5. Variations in regional consumer preferences and small retailers prohibit mass marketing by large-scale distributors.	5. Development of large-scale retailing and mass markets makes direct buying possible.

It is not possible to state categorically which of the preceding changes was the most important in fostering decentralization. Probably developments in transportation and communication first started the process. However, in recent years, decentralization has received its impetus from product, farm production, and consumer developments. Through *direct buying,* decentralized markets permit food marketing firms to better control and coordinate the flow and quality of farm products than was possible under centralized marketing. Quality control is improved through *specification buying*—the process of buyers stating market requirements and securing product offers from alternative suppliers.

Implications of Decentralization

The principal concerns with decentralization relate to its effects on pricing and operational efficiency as compared with central markets. Pricing efficiency considerations seem to favor centralized markets, whereas operational efficiency is often better served by decentralized markets.

Decentralized markets are often more economical to operate. For example, shipping farm products directly to buyers, without routing them through central terminal markets, results in handling, transport, and shrinkage savings. Freight savings are particularly evident where slaughter and processing facilities have moved out to the production areas. Many of the older terminal market facilities were also congested and inefficient, contributing to the cost savings of decentralized markets.

The greater concern is with the effect of decentralization on pricing efficiency in food markets. Terminal markets were close to perfectly competitive in the large numbers of buyers and sellers involved, in the great amount of public market information they generated, and in the homogeneous nature of graded farm products. Consequently, price discovery was considered quite good in these markets, and terminal market prices were believed to fairly represent market conditions and product values. By contrast, in decentralized markets there are fewer buyers and sellers at each shipping-point market; it is more difficult to collect adequate market information from the widely scattered, private transactions; and specification buying makes it more difficult to compare market prices.

Price discovery, then, is more complex in decentralized markets, and decentralized markets *appear* to be less perfectly competitive than central markets. Farmers face large-volume buyers in these markets, and price discovery is a more subjective process, open to the relative bargaining power of buyers and sellers. Consequently, there is less assurance that the law of one price will align all market prices competitively.

Do farmers receive less when they market directly than they would if they sold through terminal markets? A simple answer cannot be given. Some farmers prefer decentralized marketing; others prefer terminal markets. Obviously their experiences with the two markets influence their preferences. Farmers are likely to become more involved with marketing their products in decentralized markets than in centralized markets, where commission men and brokers perform more of the marketing functions. Some farmers feel, too, that they have more bargaining power in decentralized markets because they are pricing the product before releasing it to the market. In terminal market sales, commodities are often shipped to market prior to pricing them, and farmers have no alternatives except to sell them at the going market price. Larger farmers have generally favored direct sales while smaller producers have continued to support the terminal market outlets.

Another concern is the continuing practice of using terminal market price quotations to price direct, decentralized sales. With decentralization, many terminal markets have become *thin markets;* that is, they handle a small and declining volume of product. Many observers feel that the prices discovered in these thin markets do not always accurately represent true market conditions and should not be used as guides in pricing direct sales.

Electronic Marketing

With improved communications and computer technologies, buyers and sellers can now negotiate the terms of trade for a product from their offices using teletypes, conference calls, video screens, and the Internet. This electronic marketing makes it unnecessary to concentrate buyers and sellers and products in a physical place. Electronic marketing can improve the performance of widely dispersed decentralized markets. Product shipments are made after the price has been determined in the electronic marketplace. These new electronic developments can then combine the pricing efficiency of centralized marketing systems with the operational efficiency of decentralized markets.

The first electronic marketing system was developed in 1962 to market slaughter hogs in Virginia. In the 1970s, Missouri feeder pigs were auctioned by telephone conference call, a computerized network (TELCOT) was developed to auction Texas cotton, and an electronic egg market clearinghouse was established. There are currently many additional experiments with electronic marketing. The National Electronic Marketing Association (NEMA) is a computerized lamb, hog, and feeder cattle system; CATTLEX, is an electronic feeder cattle market; and Computer Aided Marketing Programs (CAMP) is an electronic fruit and vegetable market. Farmers can now buy and sell farm supplies and products at a variety of World Wide Web sites.

INTEGRATION OF FOOD MARKETS

Another market organization trend in the food industry has been the tendency toward mergers and *integration.* This process refers to expansion of firms by consolidating additional marketing functions and activities under a single management. Examples are food retailers who establish wholesaling facilities, one milk processor purchasing another's plant and routes, or the joining together of a meat packer with a bus company. In each case, there is a centralization of decision making into the hands of a single management.

Varieties of Integration

There are three basic kinds of integration. *Vertical integration* occurs when a firm combines activities unlike those it currently performs but that are related to them in the sequence of marketing activities. Such integration is illustrated by meat packers who reach both backward toward the producer and operate their own livestock buying points in the countryside and forward toward the consumer and operate their own meat wholesaling establishments.

Horizontal integration occurs when a firm gains control over other firms performing similar activities at the same level in the marketing channel. The development of line elevators, in which many individual elevators are brought under one management, is one example; the merger of two food processors is another.

Firms often expand both vertically and horizontally. Modern retail grocery chains are a good example of this type of growth. They have expanded horizontally by adding additional retail food stores; and they have grown vertically by operating their own wholesale establishments, sometimes owning their own canning factories or operating their own country buying points.

There is still another type of organizational expansion, sometimes called *conglomeration.* Here other agencies or activities that do not have any direct relation to the core business of the individual firm are brought under a unified management. This is a form of diversification. During the 1960s and 1970s, many dairy, meat, canning, and tobacco companies diversified into other industries and products. This process was slowed and even reversed in the 1980s and 1990s as many firms returned to their core business.

Another way to view integration in the food industry is by studying the extent of the transfer of decisions among integrated firms. *Ownership integration* or merger occurs when all the decisions and assets of one firm are completely assumed by another, as, for example, when a food processor buys a food wholesale firm or sets up its own wholesaling facilities. In contrast, *contract integration* involves an agreement between two firms on certain decisions, but each firm retains its separate identity. For example, a vegetable producer may sign a contract with a food processor specifying variety, delivery date, or price of raw products.

All of these integration efforts are attempts to organize or coordinate the marketing process to obtain increased operating efficiency or more power over the selling and/or buying process. Like decentralization, integration of the marketing channel may have both advantageous and disadvantageous results. Integration has the effect of shortening and closing the marketing chain. Integrated supplies do not enter the open market for changes in ownership and pricing. Commodities flow between integrated firms on an administered basis rather than on an open market, competitive basis. Integration and decentralization have proceeded hand in hand in the food industry. Both contribute to closer coordination of product flows and quality control.

Three methods of vertically coordinating the food system are illustrated in Figure 12-3. In the traditional, open market system each independent firm exchanges products freely with others in the system, and prices are discovered at each transaction point. In a contractual system, some of the exchanges are governed by privately negotiated legal documents, which may specify prices, markets, quantities and quali-

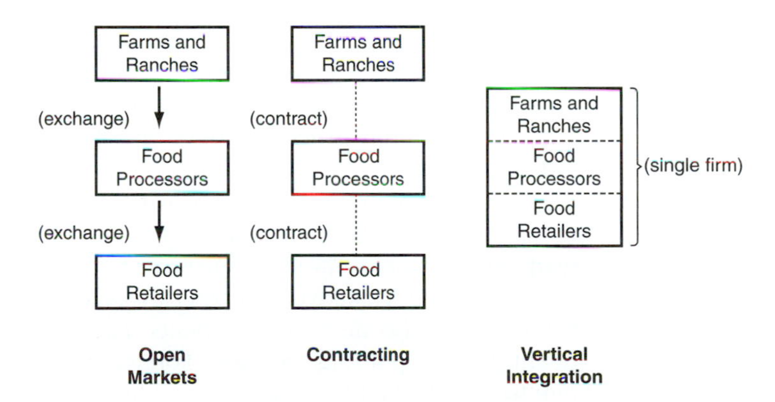

FIGURE 12–3 Alternative forms of vertical coordination in the food system.

ties in advance of sale. Once the contracts are signed, buyers and sellers have less independence in marketing, and there is less open market buying and selling. In a vertically integrated system two or more market stages are combined into a single firm, and products move from one stage to the next by managerial decisions rather than by open market trading. These alternative forms of vertical coordination often coexist in the same markets.

Reasons for Integration

The motives for integration in the food industry vary widely among firms and products. A number of reasons for vertical integration of food firms and functions have been suggested, including the profit potential from assuming additional functions; risk reduction through improved market coordination; improved bargaining power and the prospect of influencing prices; and lower costs through gaining operational efficiencies. Fundamentally, vertical integration represents a substitution of an administered coordinating system for price coordination of markets. To the extent that prices fail adequately to coordinate these markets, firms find the market a poor guide for their decisions and so substitute integrated arrangements.

Other motives stimulate horizontal integration in the food industry. Buying out a competitor is a time-honored way to reduce competition, gain a larger share of the market, and perhaps improve profits. Gaining economies of size and specialization is perhaps a more noble motive for horizontal integration. Because of its potential impact on competiton, the antitrust laws limit horizontal integration in the food industry.

The motives underlying conglomerate integration in the food industry appear to be related to risk reduction through diversification; the acquisition of financial leverage; and the empire-building urge. Conglomerates may also have been blocked by law in attempts to grow horizontally. Food conglomerates allegedly practice cross-subsidization—the use of nonfood profits to support aggressive price competition in food markets.

FOOD INDUSTRY MERGERS AND ALLIANCES

The U.S. food marketing system has undergone a major restructuring in recent years. Integrative mergers and acquisitions, consolidations, divestitures (selling off parts of firms), corporate takeovers, vertical alliances, and leveraged buyouts (purchasing a company using the assets or debt of the firm) have significantly reorganized the food industry.

Mergers are quite prevalent in the food industry. The current merger movement dates from 1977 and differs from earlier movements in several respects. First, many of the mergers are accompanied by significant divestiture activity. The rise of the "unfriendly" takeover and the leveraged buyout by corporate management or "raiders" are also new developments. Finally, food industry mergers are more frequently crossing foreign borders. Foreign investment in the U.S. food marketing system has grown rapidly in recent years. At the same time, American investment in foreign food firms has also increased. Often two firms in different countries will form a *joint venture* or *strategic alliance* that pledges them to cooperate in mutual production and marketing programs.

What are the effects of these mergers on food market performance? They have changed the ownership of many companies and brands, increased the debt levels of several food companies, and contributed to the internationalization of the U.S. food industry. The mergers have also resulted in larger firms. Beyond these changes it becomes more difficult to judge the effects of mergers. Whether it is easier or more difficult for a new firm to enter the market, whether the mergers have facilitated capital investment in the food industry, and whether on balance the mergers increase or decrease efficiency in the food marketing sector are subject to debate.

Food firms are also forming vertical alliances with their purchasers and suppliers. In this case, buyers and sellers mainitain their separate operations, but they coordinate their activities in some way. For example, a fast food company may purchase food supplies exclusively from only one vendor. Each firm maintains its independent identity, but there is a close relationship between them. These strategic vertical alliances may reduce risk and transactions costs between buyers and sellers.

INTEGRATION INTO FARMING

American farmers are not strangers to integration. In recent years, the declining number of farms, and their increasing size, have been the result of horizontal integration of farmers. Through cooperatives, many farmers are vertically integrated into the food marketing channel. And, to the extent that farmers invest in nonfarm securities (stocks, bonds, and so on) and exercise their ownership rights, farmers are conglomerate integrators.

There is increasing concern with vertical and conglomerate integration that crosses over the farm gate and provides nonfarm firms some influence on farm management decisions. Does "corporate agriculture" (excluding the family farm corporation) threaten the independence of farmers? How will a fully integrated food industry perform? Will agribusiness corporations use their control over farming to lower farm prices, raise retail food prices, and improve their profits? Integration into agriculture has become a key public policy issue.

Extent of Agricultural Integration

We noted in Chapter 3 that the family farm still dominates U.S. agriculture, and that most "corporate farms" are family-owned. This, however, is not an adequate measure of agricultural integration or a measure of the influence of off-farm firms on farm managment decisions. Contract integration of agriculture is more common than ownership integration. In 1999, about 10 percent of all farms had contracts accounting for about 50 percent of all farm products marketed.

Table 12-3 shows the importance of contractual and ownership vertical integration into agriculture. There has been a long-term increase in agricultural integration since 1950 and a corresponding decline in open-market sales. Integration is still growing for some products, notably fresh market vegetables, hogs, and eggs. However, integration of the major field crops did not grow significantly, and even fell in some cases, over the 1970-1990 period. Many farm products are approaching 100 percent integration, including citrus, sugar, and turkeys. Overall, perhaps 35 percent

TABLE 12-3 Production Contracting and Vertical Integration at the Producer-First Handler Level, Selected Commodities, 1960-1994

	Percent of Output					
	Production Contracts[1]			*Vertical Ownership*		
Products	*1960*	*1980*	*1994*	*1960*	*1980*	*1994*
Feed grains	0.1	1.2	1.2	0.4	0.5	0.5
Food grains	1.0	1.0	0.1	0.3	0.5	0.5
Soybeans	1.0	1.0	0	0.4	0.5	0.4
Cotton	5.0	1.0	0.1	3.0	1.0	1.0
Fresh vegetables	20.0	18.0	25.0	25.0	35.0	40.0
Processing vegetables	67.0	88.1	87.9	8.0	10.0	6.0
Potatoes	40.0	60.0	55.0	30.0	35.0	40.0
Hogs	0.7	1.5	10.8	0.7	1.5	8.0
Eggs	7.0	43.0	25.0	5.5	45.0	70.0
Broilers (chickens)	90.0	91.0	92.0	5.4	8.0	8.0
Total Farm Output	8.3	11.5	10.7	4.4	6.2	7.6

SOURCE: *Food Marketing Review, 1994-95.* U.S. Department of Agriculture, Agricultural Economics Report No. 743, September 1996, p. 12.
[1]Excludes marketing contracts.

of farm production is now contractually or ownership-integrated with food marketing firms. Contracts are used more by larger than by smaller farmers.

The sectors of the food industry that have integrated into agriculture have varied for different commodities. Input suppliers in the feed business have integrated the broiler and egg industries. Food processors and manufacturers have integrated the processed fruit and vegetable industry. Farmers integrated the sugar beet, citrus, and milk industries through cooperatives and bargaining associations. Food retailers and wholesalers have not integrated into agriculture to any great extent, although they have integrated various other stages of the marketing system.

Types of Farm Contracts

Several types of contracts are used in vertically integrating farmers with their input suppliers and buyers of farm products. These differ in the number of decisions influenced by the contract, the sharing of costs and risks, and the specificity of contract terms.

Market-specification terms specify some of the product characteristics that will be acceptable to the integrator and usually establish the basis of payment to the producer. A grain farmer, for example, may sign a forward price contract to deliver the crop to a buyer at harvest. The producer receives minimal financial or technical help. Little or none of the producer's price or income risk is assumed by the integrator, as returns are still fundamentally tied to the open market. The contracts guarantee the producer a buyer if the specifications are met. Such contracts cause very little integration of the two parties in the sense of any centralized management control.

Resource-providing terms often specify certain production resources to be used and the place of their purchase. The integrator usually provides the producer with financing, ranging from operational to fixed investment financing, and a degree of managerial help and supervision. Product prices are usually based on the open market, and income guarantees to the producer are minimal. In such contracts the integrators may influence the technology and size of operations of the producer in order to increase and stabilize the market for their own products.

Management- and income-guaranteeing contracts often include the marketing and production stipulations of the above two types of contracts. In addition, they provide for the transferring of part or all of the market price and income risks from the producer to the integrator. This is usually done by paying the producer a prearranged return per unit of product or by guaranteeing against market-oriented financial loss. In these contracts the integrator assumes a substantial part of the managerial responsibility of the producer. These contracts come closest to obtaining the managerial and financial control and risk that occurs when the integration is effected through complete ownership.

In 1998 market-specification contracts covered about 22 percent of the value of farm output while resource-providing and management contracts covered about 12 percent of output.

Reasons for Farm Production Integration

There are substantial reasons why marketing firms may want to have some control over the farm source of their raw materials, or in the case of those selling to farms, such as feed companies, control over their market. One of the big costs to processors and handlers is the great fluctuation—yearly, seasonal, and day to day—of both the amount and quality of their raw products. In recent years this problem has taken on added significance as the development of large-scale, self-service retailing has increased the demand for an orderly flow of a large volume of uniform products. Moreover, food processors have found it necessary to exert more control over the attributes and qualities of farm products in order to better target their products to specific consumer markets. In a sense, integrated coordination replaces price coordination.

Formerly the difficulties of effectively managing any large-scale agricultural production by nonresident managers were so great that the costs of such efforts were greater than the potential gain. This barrier, however, has been increasingly reduced as scientific management practices have been developed in agricultural production. More and more, if certain practices are followed under adequate supervision, farm products can be produced in a specified form and on a predetermined schedule of output. The term *industrialization* has increasing validity for many agricultural enterprises and, as in a factory, management and supervision can be separated successfully from the actual production activities.

In addition, this move toward scientific management is often accompanied by increased needs on the part of farmers for the capital necessary to put the new technology into practice. Also, the larger and more specialized farming becomes, the greater is the vulnerability to unpredictable prices. The advent of "predictable production" has made farm and off-farm integration feasible. The farmer's need for financing and price insurance have furnished the avenue by which integrators could secure the farmer's cooperation.

The Future of Farm Production Integration

The growth of contract farming in recent years has led many observers to predict that U.S. agriculture would soon be fully integrated with food marketing firms. This integration has occurred for some commodities but by no means for all of them. Farm contract integration is widespread today, but large segments of agriculture, particularly the livestock and grain sectors, remain largely unintegrated.

Farm leaders early expressed dismay and gloom concerning a development that would make a "hired hand" of the once proudly independent farmer. This idea was apparently a fear more of farm leaders than farmers themselves. Many farmers do not see a conflict between their independence and their position as contract growers. The fact that many farmers voluntarily sign contracts when they have market alternatives suggests that contracts can be beneficial to both farmers and integrators. Many prefer the income protection the contract gives them and feel they have enough to say about the management of their operations.

On the other side, not all integrated arrangements are highly profitable for the integrators. Analysis of different types of livestock and poultry contracts finds both profitable and unprofitable ventures. The importance of the integrator's ability underscores the obvious point that if management is to be "lifted" from the farmer to the integrator, the integrator must be able to supply the necessary managerial skills for the larger and more complex operation.

The major limiting factor to the use of contract integration between farm and nonfarm firms rests in the production processes of the enterprise itself. Only if the production activities have become standardized and the technology quite scientific can important management direction be provided in absentia by the integrator. If successful operation still takes close, personal supervision and skill, it is difficult to transfer management to the integrator.

If the production process permits feasible integration, will integration necessarily take place? Answers to this rest on the mutual gains from integration for farmers and potential integrators. Does the producer need the extra production know-how and financing that the integrator might provide? Is the present market mechanism operating poorly in providing the integrator the appropriate kind of product in adequate amounts at the needed time? Will the market absorb the additional production, which integration usually encourages in its initial stages, at satisfactory prices?

Some farm enterprises meet many of the preceding criteria for integrated operations, and production under contract is expanding rapidly. Other farm enterprises do not, and the extent of contract operation is less prevelant.

Implications of Producer Integration

There seems to be little doubt that integrated production and marketing can result in improved operational efficiencies. However, as with decentralization, there is considerable debate over its effect on pricing efficiency. In one sense, integration tends to hasten the further decentralization of the market channel, in that it effectively contracts for the output of each individual producer. As integration progresses, less and less of the production moves through the "open market" for pricing

MINI CASE 12: TO CONTRACT OR NOT TO CONTRACT?

Bill Murray and Jim Watson were discussing their farm prices and returns over coffee one morning. "I made a good decision this year when I signed that contract to sell my crops to my local buyer at a fixed price prior to harvest," said Jim. "Without that contract I would have lost money on this year's bumper crop," he said. "Well . . . contracting can be helpful sometimes," cautioned Bill, "but I prefer to wait and see what the price is at harvest." "Last year when the drought pushed up prices I made a lot more than you guys who had signed contracts to deliver at lower prices," he boasted." "That sounds smart but risky," added the waitress who poured them both another cup of coffee. "Farming is a risky business," said Jim.

What are the pros and cons of farmers using contracts to establish prices and delivery of their product in advance of harvest?

purposes. In the mid-1960s, for example, the reporting of live broiler prices by the federal market news service was discontinued in several southern states. With most of the broilers being produced and sold under contract, there were not enough being sold alive to establish a realistic live price. As the use of contract arrangements becomes a major part of the market structure of a commodity, it becomes increasingly difficult to maintain an effective open-market channel for those who do not wish to accept contracts. For the same reason, price discovery becomes more and more difficult with the integration of food markets.

The operation of an integrated channel theoretically need not reduce effective competition as long as there are many firms competing for the contracts of capable, informed producers. In fact, in situations where the existing market news, grading procedures, and pricing facilities operate poorly, the use of integration arrangements can improve the operating situation. However, in practice, the growth of integration has sharpened the debate concerning the relative bargaining power of farm producers and marketing agencies. Contract details are complex and often are not public information. As with labor contracts, the negotiation of the contractual terms becomes the basic "pricing" activity between the parties to the contract. The fairness of this process and how to police it are continuing concerns.

FUTURE ORGANIZATION OF FOOD MARKETS

The organization of food markets is being shaped by numerous trends, including: (1) the industrialization of agriculture, (2) specialization and diversification of firms, (3) technological change in food marketing, (4) decentralization, (5) integration, (6) globalization, (7) government regulation, (8) the information revolution, and (9) the shift from a commodity-oriented to a merchandising-oriented food industry. These trends will continue to shape and mold the industry.

Efforts to more closely vertically coordinate the several market levels of the food industry will continue, and the results will determine who controls the food industry and its market performance. Farmers in particular are expected to press their efforts to be more than just a raw-material supplier, residual-income claimant, and price-taker in the food system.

The direct marketing and integration trends will proceed at different rates for various commodities. Some parts of agriculture will follow the broiler and sugar beet industries and will achieve full integration of most production and marketing stages. The remainder of agriculture will follow the traditional concept of independent farming, limited integration, and open-market sales. Both sectors will require new price discovery techniques to substitute for the terminal markets.

When will the Henry Ford of the food industry integrate all sectors of the food production and marketing chain? There are companies with plans to fully integrate from farm to fork, and some are well along in this goal. This may be the next step in the organization of food markets. However, there are also companies withdrawing from unprofitable integrated arrangements in the food industry.

SUMMARY

Specialization, diversification, decentralization, and integration are the key trends shaping the modern organization of food markets. Powerful economic incentives are propelling these trends, and they raise several public policy issues. These changes in market organization also affect the status and behavior of farmers in the food industry. Specialization and diversification in food marketing are leading to larger, more complex firms. Efforts to vertically coordinate farm production and food marketing decisions have fostered both ownership and contractual integration of food marketing firms into agriculture. Horizontal and conglomerate integration of food marketing firms are also pronounced trends. Decentralization and integration of food markets affect the operational and pricing efficiency of these markets.

KEY TERMS AND CONCEPTS

conglomerates
contract integration
cross-subsidization
decentralization
direct buying
diversification
electronic marketing
horizontal integration
industrialization
joint venture
mergers
specification buying
strategic alliance
terminal market
thin markets
vertical coordination
vertical alliances
vertical integration

DISCUSSION QUESTIONS

1. Explain the problem of vertical coordination in the food industry. Why can't price be relied on to serve this coordinating role?

2. What are some reasons for the variations in contract and ownership integration among the commodities shown in Table 12-3?

3. Compare the farmer's marketing task for a highly integrated commodity, such as broiler-chickens, to that for a less integrated commodity, such as corn.

4. What kinds of competitive regulations would you recommend for the following kinds of integration?

- **a.** Foreign nationals purchase U.S. farms.
- **b.** A steel firm purchases and operates U.S. farmland.
- **c.** A vegetable canner owns company farms.
- **d.** A vegetable canner signs contracts with farmers for their supplies.
- **e.** Farmers own a vegetable cannery.
- **f.** Farmers purchase more farmland.

5. What do you think is the possibility, or feasibility, of one company fully integrating the food industry, from the farm supply sector all the way to food retailing?

6. A farm leader states, "Farmers should resist integration into agriculture, and there should be laws to prevent it." How do you respond?

7. Many observers feel that food marketing firms are taking the initiative in vertically coordinating the food industry. What have farmers done and what more can they do to participate more actively in the vertical coordination process?

8. How do you explain the rather slow acceptance of electronic marketing in the food industry?

9. How is a contract sale of farm products different from an open-market sale?

10. How have consumers been affected by the organizational trends discussed in this chapter?

SELECTED REFERENCES

Banker, D. and J. Perry (January–February 1999). More Farmers Contracting to Manage Risk. *Agricultural Outlook.* U.S. Department of Agriculture.

Contracting In Agriculture: An Overview of the Issues (December 2000). U.S. Department of Agriculture, Economic Research Service, http://www.ers.usda.gov/whatsnew/issues/contracting/

Heifner, R. et. al (1999). *Managing Risk in Farming: Concepts, Research, and Analysis.* Agricultural Economic Report No. 774. Washington, D.C.: U.S. Department of Agriculture.

Martinez, S. W., & A. Reed. (June 1996). *From Farmers to Consumers: Vertical Coordination in the Food Industry,* Agricultural Information Bulletin No. 720. Washington, D.C.: U.S. Department of Agriculture.

Perry, J. et. al, (May 1997). Contracting–A Business Option for Many Farmers. *Agricultural Outlook,* U.S. Department of Agriculture.

Reed, A.J. and J.S. Clark (2000). *Structural Change and Competition In Seven U.S. Food Markets.* Technical Bulletin No. 1881. Washington, D.C.: U.S. Department of Agriculture.

CHAPTER 13

COOPERATIVES IN THE FOOD INDUSTRY

Either we all stand together or we all hang separately.

OBJECTIVES

After reading this chapter, you will be able to:

1. Explain the purposes of agricultural coooperatives.
2. Distinguish between the cooperative and other forms of business organization.
3. Describe the problems and potentials of agricultural cooperatives.

Agricultural history is full of examples of the continuing battle of the farmer against the abuses, either real or imaginary, of the marketing middleman. Farmers have continually complained about having to sell cheap as a producer and buy high as a consumer. They have also been concerned about their relative bargaining power. Cooperative organization has been adopted by farmers as one possible solution to these problems.

Mutual action by farmers to help solve large tasks has long been a part of agricultural life. Barn raisings, husking bees, and threshing rings all required that many act together with a common purpose. Cooperative enterprises of today are similar in nature but have been established as formal business organizations.

Most cooperatives in the United States are farmer cooperatives. They are sometimes called the off-farm arm of farmers. They have been organized to provide a wide variety of services, to help sell the farmers' products, and to help farmers purchase their needed supplies. Many look upon cooperation as a way by which the many, independent, small farm units can effectively compete in a business world composed of larger, more powerful firms.

Cooperatives are important in today's farm economy. More than 3800 farmer cooperatives reported a business volume of $106 billion in 1997. Membership in farmer cooperatives totals about 3.2 million, indicating that many farmers belong to more than one cooperative.

WHAT IS A COOPERATIVE?

There are many different definitions of cooperatives. H. E. Babcock, an early cooperative leader, said that cooperatives are a legal, practical means by which a group of self-selected, selfish capitalists seek to improve their individual economic position in a competitive society. Other leaders in cooperative thought have defined a cooperative as "a business voluntarily owned and controlled by its member–patrons and operated for them on a nonprofit or cost basis." Two aspects of these definitions deserve attention. First, a cooperative is a legal, institutionalized entity that permits group action that can compete within the framework of other types of business organization. Second, cooperatives are voluntary organizations set up to serve and benefit those who are going to use them.

These definitions have been incorporated into the body of law that sanctions American cooperation. The passage of the Capper-Volstead Act by Congress in 1922 helped codify this legal concept of agricultural cooperation, although the details of laws vary among the several states. Three fundamental concepts differentiate a cooperative from other forms of business enterprises. These concepts must be incorporated into the organizational and operating pattern of an enterprise in order for it to qualify as a cooperative.

The first of these distinctive concepts is that the *ownership and control of the enterprise must be in the hands of those who utilize its services.* The control is exercised by the owners as the patrons of the business rather than by the owners as investors in the business. In no other form of business enterprise is there a comparable patron–owner relationship. Such a relationship means that the primary objective of the cooperative enterprise is to do the job assigned to it at a minimum cost and with maximum satisfaction for its owner–patrons. In contrast, the primary objective of nonpatron firms is to maximize returns over costs for the benefit of the owner–investors. In order to assure the effectiveness of this concept, some provision is often made in the bylaws of cooperatives to limit the amount of business that can be transacted with nonmembers. In addition, the voting control of the business is restricted in various ways to help ensure that the user is dominant over the investor orientation. Traditionally, control has been on a one-person, one-vote basis regardless of the amount any individual has invested. Several states now permit some variation from this view and allow proportional voting rights according to the extent of either ownership or patronage.

The second distinctive cooperative concept is that the *business operations shall be conducted so as to approach a cost basis,* and any returns above cost shall be returned to patrons. From this concept arises the common practice of referring to cooperatives as nonprofit business concerns. The *patronage refund* of cooperatives is the device used to return to the owner–patrons the overcharges or underpayments that have resulted in earnings above cost. Cooperatives return more than $1 billion to their farmer–members each year. In noncooperative businesses, earnings or profits belong to the business for distribution or use as the business sees fit. In cooperatives such earnings are a liability owed to the patron–owners.

The third distinctive cooperative concept is that the *return on the owner's invested capital shall be limited.* The capital requirements of a cooperative may be no different from those of any other type of business organization engaged in similar activities. However, the relationship of the investor to the business is quite different. In a cooperative the patron-owners invest their money primarily so that the organization may provide desired services for them. Their decision to enter or remain as patron-owners of the cooperative is made largely on the basis of their opportunity to benefit as a patron-user. In noncooperative forms of business, investors offer their money in expectation of a profitable return on it. The need for capital may be as urgent for a cooperative as for any other kind of business, but the methods of capital accumulation must acknowledge the fact that returns on the capital are limited.

These distinctive differences give rise to the principal unique quality of a cooperative. The *point of view* guiding its activities is that of the owners of the business who are *also* its customers and users. Cooperatives seek to undertake profitable ventures like any other business. However, these profits accrue to the owners through their own use of the organization instead of to owners as investors. These distinctive differences also give rise to several operational differences between cooperative corporations and noncooperative corporations. These differences are summarized in Table 13-1.

KINDS OF COOPERATIVE BUSINESS

When viewed according to their tasks performed, cooperatives fall into four broad categories—marketing, purchasing, service, and processing associations.

TABLE 13-1 Similarities and Differences of Three Types of Business Organizations

Condition	*Individual or Partnership*	*Corporation*	*Cooperative*
Operated for profit?	Yes	Yes	Yes
How are earnings distributed?	To owners or partners	To owners on basis of shareholdings	Largely to patrons on patronage basis
Who controls firms, selects management, performs other management duties.	Individuals or partners	Board of directors elected by stockholders	Board of directors elected by patron-owners
How is voting done?	None or by agreement	Usually 1 vote per share	Usually 1 member, 1 vote, or proportional to business
What is owners' liability?	All property of owners	Assets of corporation	Assets of cooperative
Who are the customers?	General public	General public	Chiefly members, but sometimes nonmembers

Marketing Cooperatives

Marketing cooperatives sell farmers' products. They are an example of farmers vertically integrating into the food marketing channel. These cooperatives may collect members' products for sale, grade, package, and perform other functions. Cooperative livestock commission organizations, producers' milk associations, and cooperative elevators are examples of cooperatives acting as marketing agents. The objective of such organizations is to secure the greatest possible amount for the products of their farmer-owners. Some associations act solely as commission agents. Some associations act as bargaining agents and do not actually handle the products. Others will actually buy the commodity from the farmer for resale. Some cooperatives specialize in handling only a single commodity; others handle multiple commodities.

Marketing cooperatives accounted for approximately 30 percent of all farm cash receipts in 1998, as shown in Table 13-2. The proportion of commodities marketed cooperatively varies from commodity to commodity, as can be seen in Table 13-2. The cooperative market share is highest for milk and lower for livestock and livestock products.

Purchasing Cooperatives

Purchasing cooperatives sell supplies to farmers. Some purchasing cooperatives are engaged only in retailing and wholesaling. In other instances, such as in fertilizer and petroleum, they manufacture the products they sell and acquire the sources of raw materials. Most states have large statewide associations that are examples of this type of cooperative. The objective of such organizations is to provide savings for the farmer on purchases. The principal source of such savings will usually come from lower prices or from higher-quality and better-adapted supplies and equipment.

TABLE 13-2 Farm Product Marketing Cooperatives, Number and Share of Market, 1970-1998

	Number of Cooperatives		*Percent of Cash Farm Receipts*	
Product Marketed	*1970*	*1995*	*1970*	*1998*
Milk	971	241	73	86
Cotton, Cottonseed	554	16	26	43
Grains, Oilseeds	2,539	1,104	32	40
Fruits and Vegetables	499	283	27	19
Livestock and Products	546	109	11	14
Total[1]	5,415	2,085	26	30

SOURCE: *Rural Cooperatives.* (February 2000). U.S. Department of Agriculture, Rural Business-Cooperative Service.

[1]Includes "other".

The sale of various farm supplies accounts for most of the business volume of these cooperatives, although increasing amounts of items for farm household use are being handled. Purchasing cooperatives sold $25 billion of farm supplies in 1998, accounting for 29 percent of the value of farm production expenditures, compared to 15 percent in 1970. The importance of purchasing cooperatives varies by production items. They sold 50 percent of petroleum sales to farmers in 1998, 45 percent of fertilizer, 21 percent of feed, and 10 percent of seed.

Service Cooperatives

Service cooperatives were organized in the 1930s and 1940s to provide their members with improved services or with services they could not otherwise obtain. Today the services provided may include credit, insurance, electric power, telephone, irrigation and drainage, hospitals, and mortuaries. Membership may be of rural or urban people or a combination of the two. Farmers obtain substantial amounts of their farm credit from the Production Credit Associations and Federal Land Bank Associations. Rural Electric Associations furnish electricity to a large number of rural people. Cooperatives provide electric and telephone service to more than 45 million rural customers. Over half of farmers' fire insurance is carried by their mutual insurance companies.

In recent years, many marketing and purchasing cooperatives have added customer services to their offerings. These may take the forms of farm management advice, custom chemical application, marketing advisories and assistance, and other services valued by farmers. These assist the cooperatives in meeting the needs of customers and differentiating themselves from competitors.

Processing Cooperatives

The processing cooperative engages in the packing or processing of the farmer's products. Cheese- and butter-manufacturing, sugar-refining, fruit-packing, and vegetable-canning associations are examples of this type of cooperative. This is another form of vertical integration by which farmers attempt to add value to their products and capture a larger share of the consumer's food dollar. Sunkist oranges, Ocean Spray cranberries, Sun-Maid raisins, Land O' Lakes dairy products, Sunsweet prunes, Welch grape juice, and Diamond walnuts are successful cooperative brands. In recent years a new generation of cooperatives, so-called new generation cooperatives (NGC), has evolved. These have moved farmers further downstream in the food marketing channel by specializing in adding more value to farmers' products.

TYPES OF COOPERATIVE ORGANIZATIONS

Cooperatives can also be classified on the basis of membership affiliation and control. From this viewpoint, cooperatives are usually grouped as independent local associations, federated associations, centralized associations, or a combination of these types of organizations.

Independent Local Associations

The simplest type of cooperative is the independent local association in which people hold direct membership and are able to participate in the affairs of the cooperative. The relatively small area of coverage and number of people involved mean that the opinion and action of each member can have influence. Because of the size limitation, however, such cooperatives are often limited in what they can do. These local cooperatives often join to form larger organizations to conduct mass marketing, purchasing, or manufacturing operations.

Federated Associations

The federated association is a cooperative composed of several local associations that operate together as an integrated unit. Farmers are members of the local association, and the local is a member of the regional association. Usually the motive for banding together is to secure greater business power and efficiency. In such an association the basic channel of control is from the local up to the overall organization. The local associations have a considerable degree of autonomy, and any powers that are not expressly granted to the central organization are retained by the locals. Savings made from the operations of the overall association are allocated back to the member local associations. The local then adds them to whatever savings have accrued from its own operations, and this total amount is then distributed to the patron-member.

Centralized Cooperative Associations

Centralized cooperatives are those in which the patron is a direct member of the central organization and exercises control through delegates sent from the different areas to the annual meeting of the central organization. The central organization in turn controls the local branch cooperatives that serve the members. This plan has the advantages of centralized control that makes possible prompt and uniform action by all the local outlets, but it lacks the direct membership participation possible in federated cooperatives. The central association itself is dominant and delegates certain powers to the local branches. In such an association the local units have a limited amount of autonomy. Savings are distributed directly from the central association to the members.

Mixed Associations

Many large cooperative organizations today are neither totally centralized nor totally federated but are a mixture of the two. Associations that are basically federated in nature often undertake new operations that are organized on a centralized basis. Through stronger bargaining power or other methods, the overall association may essentially gain control of the member local cooperatives. On the other hand, cooperatives that were originally organized on a centralized basis sometimes find it practicable to establish at the local level a committee of farmers to suggest operating procedures for the local units. Theoretically, these committees do not have any absolute authority, but often in practice they control the policies of the local units.

Most states have statewide organizations combining both the marketing and purchasing operations. These state associations, in turn, often combine into regional and national associations. This may be done in order either to operate manufacturing enterprises or to enhance bargaining power with noncooperative manufacturing concerns. In some instances, the regional or national associations may acquire raw materials for their manufacturing enterprises. Savings made from the operations of these regional and national associations are apportioned among their member state associations. The way in which the state associations handle these savings depends on their individual type of organization.

HISTORY AND STATUS OF AMERICAN COOPERATION

Informal cooperative enterprises have been undertaken in this country since the early colonial days. Throughout the nineteenth century, cooperatives were attempted for most commodities. A brief review of cooperative history follows.

The Active Period, 1910–1930

After the turn of the century, serious attention was given to the organization of all kinds of cooperative businesses. The movement gained momentum during World War I and reached its peak in the postwar depression of the 1920s. The number of marketing and purchasing cooperatives doubled in the fifteen-year period 1915–1930, as shown in Table 13-3.

TABLE 13-3 Trends in the Number, Membership, and Sales Volume of Farm Marketing and Supply Cooperatives, 1915–1997

		Membership		*Sales Volume*	
Year	*Number of Cooperatives*	*Total (000)*	*Per Cooperative*	*$ Million*	*Per Cooperative ($000)*
1915	5,424	651	118	635	117
1920	7,374	NA	NA	1,256	70
1930	11,950	3,000	251	2,400	201
1940	10,600	3,400	320	2,280	215
1950	10,035	6,584	705	8,726	810
1960	9,345	7,273	797	12,036	1,354
1970	7,790	6,355	816	19,080	2,449
1980	6,293	5,379	854	66,254	10,528
1990	4,663	4,119	883	77,266	16,570
1997	3,791	3,239	695	106,474	28,085

SOURCE: *Statistics of Farmer Cooperatives,* Annual, U.S. Department of Agriculture, Rural Business-Cooperative Service.

Several factors favored the expansion of cooperation in this period. First, rapidly falling prices in the period immediately following World War I led many co-operatives to be organized in order to stabilize or raise prices. In addition, there were complaints that private suppliers of necessary farm supplies, such as fertilizer and feeds, were taking exorbitant margins. Farmers further complained bitterly that middlemen were not interested in improving either quality or service. There was just enough truth in many of these complaints that Congress, through the passage of the Capper-Volstead Act in 1922, officially sanctioned cooperatives as a way of "self-help" and as a way of restoring and maintaining reasonable competition in the marketing and purchasing of agricultural products and supplies.

The Consolidation Period, 1930–1950

In the flush of early enthusiasm it was inevitable that many cooperatives would be started that were unsound in their business structure and practices. Also, it was soon evident that cooperation could not be depended on to stabilize general price fluctuations and automatically cure the many ills of the general economic situation. These things, plus the depression of the 1930s, contributed to the short life of many cooperatives.

In order to gain economic strength, many small independent cooperatives consolidated into the large federated associations. Several regional supply cooperatives were organized between 1925 and 1935. The number of cooperatives tended to decline throughout most of this period. However, this decrease was caused largely by the decline in numbers of marketing cooperatives. Purchasing cooperatives continued to grow both in numbers and memberships.

Cooperation expanded into many fields. The depression years of the 1930s gave impetus to governmental sponsorship of both credit and electrification cooperatives. The Production Credit Associations, the National Farm Loan Associations, and the Rural Electric Associations were originally established during this period. The government also assisted in the formation of rural telephone associations. There was also an expansion in insurance cooperatives, irrigation cooperatives, and other miscellaneous associations to perform special services for rural areas.

The Period of Growth, 1950–2000

While the number of farm cooperatives and total membership declined over the 1950–1999 period with falling numbers of farmers, the size of cooperatives has expanded along with farm output. As shown in Table 13-3, from 1970 to 1997 cooperative membership was halved as sales per cooperative rose five-fold.

Cooperative sales more than tripled during the decade of the 1970s. This was a time of cooperative restructuring, mergers, and international growth. Throughout this period cooperatives continued to grow both in membership and dollar volume of business. Cooperatives' market share of farm cash receipts grew from 25 percent in 1970 to 30 percent in 1997. Cooperatives have taken their place among the nation's largest businesses.

Cooperatives are continuing to evolve and find new places in the market. Since the late 1950s there has been increased interest in the potential role of cooperatives in bargaining for price and contract terms with processors and handlers. The increased use of state and federal marketing orders often fosters the cooperative organization of growers. The American Farm Bureau has established an affiliated organization, the American Agricultural Marketing Association, whose purpose is to help establish and coordinate the activities of various bargaining associations across the country. Recent years have seen the formation of many strategic alliances among cooperatives themselves and with noncooperative firms. In addition, there has been growing involvement of marketing cooperatives in owning and operating processing facilities and in export marketing. Cooperatives that have moved into higher value-added marketing activities are often referred to as new generation co-ops. Some of the newer cooperatives challenge the older principles of cooperation in their closed memberships, investor-orientation, and equity capital sources.

COOPERATION BY REGIONS AND COMMODITIES

As shown in Table 13-4, Iowa, Minnesota, California, and Wisconsin are the leading states in cooperative business volume. Minnesota leads all states in the number of cooperatives and co-op membership. The North Central states account for over one-half of total cooperative business. The importance of farm cooperatives varies by commodities. In general, continuous, large-volume production by specialized farm units favors cooperation much more than scattered production by nonspecialized farm units. As shown in Figure 13-1, grains, oilseeds, and dairy products accounted for more than 60 percent of cooperative marketing associations' sales in 1997. Similarly, the impor-

TABLE 13-4 States with the Largest Farmer Cooperative Sales, 1997

State	*1997 Sales ($ Billion)*
Iowa	10.9
Minnesota	9.8
California	9.1
Wisconsin	6.5
Illinois	6.0
Nebraska	5.1
Kansas	4.0
Texas	3.7
Washington	3.3
North Dakota	3.3
Missouri	3.2
Ohio	3.0
New York	2.6
Indiana	2.5
Michigan	2.2
Pennsylvania	2.0

SOURCE: U.S. Department of Agriculture, Rural Business-Cooperative Service.

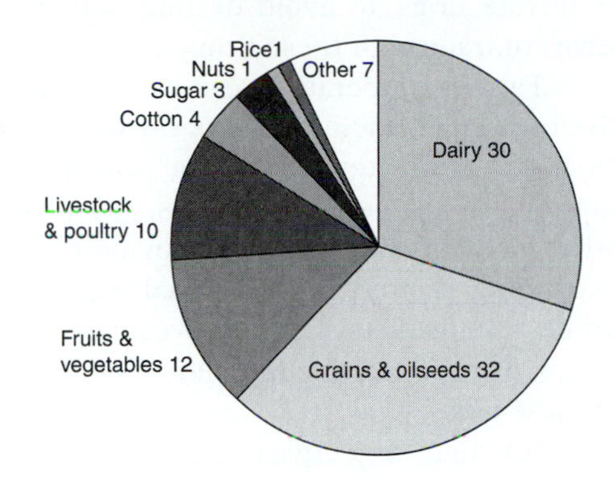

Total 100% based on net marketing business of $77.6 billion

Supplies handled by cooperatives

Percent

Seed 3

Other 13

Petroleum 27

Farm chemicals 12

Fertilizer 21

Feed 24

Total 100% based on farm supply business of $19.2 billion

FIGURE 13–1 Distribution of cooperative sales, by products, 1997. (U.S. Department of Agriculture.)

tance of cooperatives in farm input markets varies. In 1997 feed, fertilizer, and petroleum products accounted for 72 percent of purchasing cooperatives' sales volume.

Farmer marketing cooperatives have been more active in farm supply and assembly markets than in processing and retail markets. Cooperatives play only a small role in food processing and distribution.

PURPOSES OF COOPERATIVES

Cooperatives enable farmers to do collectively what they cannot do independently as small-unit firms. They alter the competitive structure of markets. Consolidating the purchases and sales of numerous farmers into a single agency, cooperatives permit farmers to operate as a single large firm while maintaining their separate identities. Farmers may form and support cooperatives to assume the marketing functions of private firms, to avoid dealing with proprietary businesses, or to improve their terms of trade with these firms.

Farmer cooperatives can assist farmers in achieving a number of objectives, including (1) enhancing returns through increased efficiency, improved market coordination, or greater bargaining power; (2) reducing farmers' costs of purchasing supplies or marketing products; (3) providing farmers with products or services otherwise not available or improving product and service quality; (4) stabilizing and expanding markets; and (5) enabling farmers to move into supply, assembly, and processing markets. Cooperatives may also function as a "competitive yardstick" by which noncooperative firms are evaluated. Most cooperatives contribute to several of these goals.

All these objectives can improve the economic well-being of the individual members. Many studies have shown that, whatever other benefits there may be from cooperation, the economic one is of the foremost importance. It is not enough for a cooperative to do a job as well as it is done by other agencies. The cooperative must do it better from the viewpoint of its patron–owner. A good cooperative should be the pacesetter for the industry with which it is associated.

Cooperatives are not magic. There are things even a successful cooperative cannot do for its members. A clear understanding of these limitations is necessary for successful operation. The cooperative should recognize that it cannot set prices unless it controls supply. This means that it cannot guarantee cost of production to its members. Cooperative associations cannot eliminate the marketing functions performed by other middlemen, though they may be able to perform them more efficiently than other agents. Cooperative membership is voluntary, and coops are not guaranteed membership or sales unless they provide quality products and services. Cooperatives are circumscribed by the same set of economic restrictions as any other form of business organization. Their success depends not on their uniqueness, but rather on their ability to operate profitably and satisfy their patron–owners.

The fundamental premise guiding the establishment of a successful cooperative association is that there be an economic need for the venture. There must be the opportunity to do a job better than is being done. This can take the form of more advantageous prices or better quality products and service. Whether there is such a need that can be met by a cooperative must be determined by an objective study of the existing facts. Far too many cooperatives have been wished into existence only to fail when it was discovered there was no real additional service they could perform.

After the need has been established, it must be determined whether the factors necessary for a successful business operation are available. The importance of these factors is fundamentally the same as for any other business undertaking:

1. Can an adequate volume of business be secured and maintained? The economies of large-scale operation are just as important to cooperatives as to private corporations.

2. Can adequate and reasonable financing be secured? To build an efficient plant takes capital—to build less than an efficient plant invites failure.

3. Is efficient management available and will the association pay its price? In management, as in other things, high quality demands a high price. Successful cooperatives need as high a level of managerial ability as other businesses.

4. Is the membership prepared to meet competitive challenges?

The most common reasons for cooperative failure are (1) lack of sufficient capital, (2) inadequate membership support, and (3) ineffective management.

Do cooperatives increase farmers' prices and incomes? It is difficult to give a single answer. Cooperatives operate in a competitive environment, and they have no more magical powers to command higher prices than do private firms. They cannot repeal the laws of supply and demand, and their lack of production controls limits their ability to influence prices. However, cooperatives are a farmer tool that can influence the returns of their members through value-adding activities, more efficient operations, and the channeling of marketing profits back to farmers. Not all cooperatives achieve this potential, but successful ones do. Research has shown that cooperatives can be as efficient as and operate at cost levels as low or lower than their proprietary counterparts.

PROBLEMS OF MODERN COOPERATIVES

Cooperative business units have been increasing in size and complexity. Merger activity has been common among cooperative businesses as well as among noncooperative corporations. There is also a tendency toward more vertical integration in both marketing and supply organizations. This growth in size of cooperatives has been necessary both to take advantage of modern management and technology and to compete effectively in the marketplace.

Such growth, however, has sharply focused on a developing weakness of one of the cooperative operating principles. Close control of the membership so as to obtain a business oriented to user rather than investor needs was one of the important premises of early cooperation. This was assumed to be obtained by the one-person, one-vote provision in most cooperative bylaws.

Producers themselves are often dissatisfied with the one-person, one-vote principle. There is a tendency for the production units to become fewer, larger, and more specialized. This means that a given cooperative may depend on relatively few large patrons for most of its business volume and that a much larger number of smaller patrons actually furnish very little volume. These few large patrons are not satisfied to be outvoted by the more numerous, smaller patrons. In many instances the one-person, one-vote idea has been set aside in favor of cumulative voting on the basis of business volume.

Modern cooperatives must solve the problem of maintaining user control and orientation in the current world of complexity and bigness of business organization. This will probably have to take some form other than the dependence on the one-person, one-vote principle to elect the controlling board from among the patrons. Somehow, cooperatives must find a solution that will keep the larger patrons loyal and satisfied, obtain the services of a knowledgeable board of directors, and still retain the user-benefit orientation of operation.

Financing

Cooperatives require the same amount of capital as noncooperatives to perform similar functions. But with limitations placed on voting rights, share transferability, and the returns paid on invested capital, cooperatives cannot utilize the noncooperative method of selling additional shares to the investing public in order to secure additional funds.

When cooperative enterprises were relatively simple, only limited amounts of capital were needed. However, the desire of cooperatives to expand has focused attention on the techniques of securing growth capital. In general, cooperatives have relied on members to provide much of the financing needed, although other sources also have been used. Many cooperatives borrowed heavily in the 1970s to finance their growth. Much of this capital was provided by the Farm Credit System. Those cooperatives with high debts suffered severe financial strains in the late 1970s and the 1980s as a result of high interest rates and low farm income.

How does the cooperative obtain its equity capital for operations, growth, and expansion? The sale of common or preferred stock provides capital for some cooperatives. The market for such stock must be limited mainly to cooperative members. Common stock has voting rights, usually one vote per member, but no guaranteed income. If nonmembers hold common stock, they usually are not permitted to vote. Preferred stock usually carries a fixed dividend and no voting rights. These sometimes can attract investors other than cooperative members.

In the early years all earnings of the cooperatives were returned to patrons in the form of cash refunds. Most cooperatives have discontinued this practice in favor of retaining some part of the earnings in the assocation for capital purposes. Of the equity capital of cooperatives, about half is carried in the form of these retained earnings that are either covered by being allocated to the patron in the company books or by certificates of equity issued to the patron. This procedure has the advantage of spreading the financial burden among the users of a cooperative. Because the earnings allocations are based on the amount of business done with a cooperative, ownership is divided in direct proportion to the use each patron makes of the association.

The methods under which a member may withdraw an investment from a cooperative will vary from one association to another. Generally, when a member ceases to be a farmer or moves out of the trade area, the investment is returned. If a member dies, the investment is usually returned to the estate. Usually the member may also sell stock or earning certificates to any buyer, subject to approval of the cooperative directors, or hold them until redeemed by the cooperative. Some cooperatives have a fixed time for redeeming these investments; others do not. If the cooperative regularly redeems its certificates after a fixed time, such practice is generally referred to as the use of a revolving fund. These funds may receive interest, but usually they do not.

In the present setting of increasing size and complexity of modern manufacturing and distribution, huge amounts of capital are necessary to make effective and competitive entry into new fields and markets. Securing these large amounts of needed capital is a major problem of the modern cooperative. The past process of obtaining capital either internally from the business itself or from traditional borrowing is not likely to be adequate. Some innovative way of obtaining capital that will not change the patron–owner orientation of the cooperative over to the owner–financier orientation of the regular corporation needs to be found.

Management

Three groups of people are involved in the management of a cooperative—the members, the board of directors, and the hired managers. The members exercise their control through their elected directors. The directors have the responsibility of formulating general operating policies and of obtaining a manager to carry out these policies and report the results to the members. Managers are charged with the operating responsibility of the cooperative enterprise. They put into actual practice the policies laid down by the board.

The problem of securing adequate hired management for a cooperative is essentially no different from that of any other business. To obtain excellent management, cooperatives must be willing to pay a price competitive with other businesses. They must also be willing to finance a staff that will furnish them with needed research and information.

The problem of the board of directors is quite different. Corporations staff their boards with whatever talent they need and can find from the business and financial world. Cooperatives, however, usually require that their boards be selected from the patron-members. Excellent farmers do not necessarily make excellent managers and advisers to a large and complex manufacturing or distributing enterprise. How to secure such talent and yet retain the patron-user viewpoint is another cooperative problem.

Membership Relations

The nature of cooperative business makes it imperative that good relationships be maintained between the members and their association. In many ways this is more important to cooperatives than stockholder relations are to noncooperative corporations. The stockholders' relationship to the corporation is that of investors, and when they are dissatisfied with the returns they may dispose of their stock through sales to the public. Members of a cooperative, however, are at the same time both owner-investors and owner-users of the business itself. They cannot as readily divorce themselves from the responsibility of the business by selling their holdings.

Many of the problems of membership relations are fundamentally by-products of the increased size and complexity of the associations. In large associations members often feel it is "*the* co-op" instead of "*my* co-op." Often only a small minority of the member-owners takes an active interest in the problems and management of their association.

Other problems of members stem from the changing structure of farming itself. Cooperatives, historically, have prided themselves on their free and voluntary membership and that each member is treated equally both in matters of voting control and rates of patronage refunds. This system was adequate when farming was largely made up of small general farmers and the cooperative enterprise was rather simple. However, today's farm structure is increasingly made up of a relatively few, very large specialized producers, on the one hand, and a large number of very small producers, on the other. These two diverse groups do not have equal interests in the cooperative organization and they are not of equal importance from the viewpoint of the success of the cooperative. Many cooperatives are having to face the issue of treating some members differently from others if they wish to retain their patronage. Some cooperatives have established cumulative voting, differential rates of payment, a limitation on eligibility for membership, and a contractual obligation between members and the cooperative.

A good member is an informed member. Farmers who support cooperatives are those who know what cooperatives have done and what they can and cannot do. Successful cooperatives rank this task of keeping an informed and participating membership high on their list of priorities.

Relations with the General Public

In the early years, with few exceptions, the general public was disposed favorably toward cooperation. This attitude took concrete form in the passage of legislation favorable to cooperatives by both the state and federal governments. The Capper-Volstead Act, although not exempting cooperatives from antitrust prosecution, for all practical purposes made prosecution improbable. The federal government provided public funds for the Farm Credit Administration to aid cooperatives with research and other services. It has actively participated in the formation of some cooperatives. In the U.S. Department of Agriculture there is a special division, the Rural Business-Cooperative Service, that provides research services for cooperatives.

In recent years—probably as a result of the growing size and success of some cooperatives—the public image of the cooperative has been a mixed one. Some observers believe that farmer cooperatives have evolved into something that was never intended and that cooperatives are no different from other businesses. As a matter of official agricultural policy, the cooperative is still encouraged as a major tool for helping farmers secure equitable treatment in the marketplace. On the other hand, income-tax procedures have been changed to remove some of the automatic advantages that cooperatives previously enjoyed. In addition, courts appear more willing to find large cooperatives in violation of antitrust laws.

CONSUMER FOOD COOPERATIVES

Just as farmers frequently view cooperatives as a way to improve their market position, consumers also have formed food-buying cooperatives to lower food costs and participate more fully in food retail decisions. These buying cooperatives range from informal groups of neighbors who purchase in large volume to full-fledged, integrated wholesale and retail food operations owned and managed by consumer–patrons.

Consumer food co-ops are a growing but still small part of the retail food industry. They probably account for less than one half of 1 percent of retail grocery sales. Food co-ops are more important in other countries, such as Great Britain and Sweden.

Food co-ops are most prevalent in the midwestern states of Ohio, Michigan, Wisconsin, Minnesota, and the Dakotas. The largest co-ops are found in the major cities: Berkeley Consumers' Cooperative of California, Greenbelt Consumers' Cooperative of Washington, DC, the Common Market Food Co-op of Madison, Wisconsin, Chicago's Hyde Park Co-op, and the New Haven, Connecticut and Palo Alto, California Cooperatives.

In contrast to these traditional consumer cooperatives that compete directly with corporate supermarkets, new wave consumer cooperatives are experimenting with new food retailing techniques. In the 1970s and 1980s, several food-buying clubs were organized as cooperatives. These pre-order food cooperatives consist of a group of consumers who jointly place food orders with wholesalers and distribute the food to members. Another type of new wave cooperative is a small limited-line grocery store concentrating on dairy products, especially cheeses and yogurts, bulk whole grains, fresh fruits and vegetables, so-called organic foods, and a few nonfood items.

Consumer food co-ops face many of the same problems as farm cooperatives—the need to provide unique services, the difficulties of hiring and keeping good managers, and the difficulties of securing adequate capital. Their cooperative nature does not guarantee them automatic success. Properly financed and managed, they provide a viable alternative to, but not a replacement for, the traditional and specialty retail food stores.

SUMMARY

Cooperatives are a unique and important marketing tool for farmers. They are owned by and managed for the benefit of their farmer–patrons. They provide farmers an alternative to private or publicly owned marketing channels. Agricultural cooperatives are important in the areas of farm supplies, farm services (credit, electricity, telephones, and the like), farm marketing, and food processing. Agricultural cooperatives may be independent local associations, federated associations, or centralized associations. Although the number of agricultural cooperatives is declining, the dollar volume of farm cooperative business has grown. To be successful, farm cooperatives must either increase member returns, decrease costs, or render a service not provided by the noncooperative sector. Cooperatives face challenges in the areas of member relations, control of decisions, financing, management, and public relations. To date, consumers' retail food cooperatives have not been as successful as farmer cooperatives.

KEY TERMS AND CONCEPTS

Capper-Volstead Act
cooperative
cooperative patron
Farmer Cooperative Service
independent, federated, and centralized cooperative associations
marketing, processing, service, and supply cooperatives
new generation coop
patronage refund

DISCUSSION QUESTIONS

1. Most cooperatives follow the Rochdale Principles, named for a pioneering cooperative in Rochdale, England. What are these principles and how faithfully do modern cooperatives adhere to them?

2. It is sometimes said that to compete successfully with established firms, co-ops must copy their management and operating procedures. Does this confict with the uniqueness that is so important to the purpose of co-ops?

3. Why do you think cooperatives are so much more important in the marketing of dairy products than in livestock marketing?

4. What are the differences between federated and centralized cooperative associations?

5. Why haven't consumer food buying cooperatives been as successful as farmer cooperatives?

6. Discuss the most important factors affecting the success of an agricultural cooperative.

7. List the pros and cons of the present financing arrangements for farmer cooperatives.

8. It has been observed that farmer cooperatives have had their greatest success in performing marketing activities closely associated with farming. Why have farm cooperatives been less successful in moving to other levels of food marketing such as food processing, wholesaling, and retailing?

SELECTED REFERENCES

Cobia, D. W. (1989). *Cooperatives in Agriculture.* Englewood Cliffs, NJ: Prentice Hall.

Cook, M. et. al (Eds.) (1997). *Cooperatives: Their Importance In the Future Food and Agricultural System.* Las Vagas, Nevada: Food and Agricultural Marketing Consortium.

Cotterill, R. W., (Ed.) (1994). *Competitive Strategy for Agricultural Marketing Cooperatives.* Boulder Colorado: Westview Press.

Roy, E. P. (1976). *Cooperatives: Development, Principles, and Management.* Danville, IL: The Interstate Press.

U.S. Department of Agriculture, Rural Business-Cooperative Service Homepage (2000). http://www/rurdev.usda.gov/rbs/index.html.

CHAPTER 14

MARKET DEVELOPMENT AND DEMAND EXPANSION

Advertising doesn't cost, it pays.

OBJECTIVES

After reading this chapter, you will be able to:

1. Distinguish between the various kinds of food market development activities.
2. Explain the purposes of food product promotion and how it affects the sales of farm products.
3. Discuss the pros and cons of farmer advertising programs.

Advertising and promotion are important elements of the food marketing picture. Advertising contributes to possession utility by encouraging the flow of products from low- to high-value markets. In this way advertising and promotion also contribute to the exchange and facilitating marketing functions.

Total U.S. advertising expenditures amounted to $215 billion in 1998—about 2 percent of the gross domestic product, or $1,200 for each man, woman, and child. The food industry is one of the largest advertisers. Food marketing firms are among the largest users of such national advertising media as TV, radio, and magazines. In 1998, food marketing firms spent an estimated $12 billion on food-related advertising. About 4 percent of each retail food dollar goes to pay for food advertising.

Advertising and promotion are among the more controversial aspects of food marketing. Do these activities contribute to or interfere with consumer sovereignty and freedom of choice? What value do firms and consumers get for their advertising dollar? What effect does advertising have on competition in the food industry? And, should farmers advertise and promote food products?

MARKET DEVELOPMENT IN THE FOOD INDUSTRY

Food market development refers to a wide range of marketing activities designed to enhance the value of food products for consumers. Advertising, quality control, packaging, new product development, personal sales, merchandising, trading stamps, coupons, cents-off promotions, trade servicing, public relations, and a host of other activities are all instruments of market development. The goal of market development is to increase consumer satisfaction and, in the process, increase firm or industry profits. Advertising is one of the more visible market development tools. In the modern food industry, millions of dollars may hinge on market development tactics: A slight variation in the taste of a product, a new package design, a penny off, or even a catchy advertising jingle can spell the difference between profit and loss or the success or failure of a business. Many criticize this aspect of food marketing as trivial or wasteful, but such market-making activities seem inevitable in a high-income, free-enterprise economy.

These activities are particularly important in the competitive battles waged between individual food firms and brands. However, market development is also becoming increasingly important in competition between food products (beef versus pork, margarine versus butter), food concepts (fresh versus frozen foods), and food distribution channels (retail food stores versus restaurants). As a result, farmers have also become concerned with market development as a marketing tool.

Market development activities are used either to shift food demand curves rightward or to alter their slope. This is more than just "selling" or "increasing consumption." Consumers can be encouraged to buy more food at lower prices. Market development activities seek to expand consumption without sacrificing prices and profits. This is done by developing and marketing more pleasing products, providing better or more persuasive information for consumers, or building consumer loyalty for the product.

Market development activities are frequently classified by the *4Ps:* product, price, promotion, and place. These are the four variables that the firm can control to accomplish its sales and profit goals. Each firm or industry puts together a unique combination of the 4Ps, termed its *marketing mix.* Supermarkets have a mix different from that of convenience food stores, and the mix of grocery stores differs from that of restaurants, farmer markets, and vending machines. One mix cannot be judged, overall, to be better than another. Each satisfies different consumer tastes and preferences.

We should also distinquish between consumer-directed promotional efforts and trade-oriented promotions. While consumers are more acquainted with such highly visible "pull" promotions as mass media advertising, coupons, packaging, and the like, there is at least as much money spent on trade advertising in the food industry today. These trade promotions are designed to encourage food retailers to give favorable pricing and merchandising attention to a brand or product. They are often called "push" promotional strategies.

As consumer incomes rise and consumers shift their efforts from satisfying survival needs to satisfying sociopsychological needs—such as status, esteem, affiliation, and distinction—market development activities gain in importance. These needs require a wide variety of differentiated products, services, and marketing channels. Few factors have been more important in shaping the modern food industry than the contemporary emphasis on market development.

VARIETIES OF FOOD DEMAND

Although it is convenient to speak of the general demand for a food product, actually there are several states of demand. These are important to the firm's market development strategies.

A *negative demand* represents a situation where most potential consumers dislike the product and would not buy it. This is the condition for many edible materials that are not defined as food: seaweed, dogs, fish worms, insects, and so on. These would be of little interest except that negative demand states can become positive. The tomato was considered poisonous in colonial America, and liver was once fed only to pets. Frequently, new information can convert a product from a negative to a positive state.

A *no-demand state* exists when consumers are unaware of their needs for a product. There was no recognizable consumer demand, for example, for soft margarine, instant coffee, or presweetened cereals prior to their development and market introduction. Most new food products and concepts go through an extensive period of trial, acceptance, or rejection. Advertising, couponing, and samples may help convert a no-demand state into a profitable demand.

Latent demand refers to a general need for which there is no present product. Examples: a calorie-free potato, a low-cholesterol egg, or a safer cigarette. The marketing task here is to develop new or revised products to meet latent needs.

Faltering demand occurs when consumption of a product is declining or failing to grow at an acceptable rate. This may be because of deteriorating quality, the availability of preferred substitutes, high price, or a change in consumer preferences. Lamb appears to be in this category. Faltering products can sometimes be revitalized with new advertising themes or quality improvements. Medical reports on the dubious healthfulness of certain foods sometimes cause faltering demand.

Full demand is the ideal state of demand. Here, the current levels of consumption and prices are satisfactory. The task is to maintain demand at this level in the face of changing consumer preferences and competition from alternatives.

Irregular demand is a fluctuating demand situation, such as for turkey, cider, or cranberries. The task is to level out the peaks for a more stable year-round level of demand. For seasonal food products, this may require considerable cost and effort as well as substantial production adjustments on the part of farmers.

Over-full demand is rare in the food industry but can occur when a product is so popular or in such short supply that it is difficult to maintain adequate stocks without having prices rise to levels that adversely affect future product sales. The solutions lie in better coordination of supplies with demand.

Unwholesome or excessive demand represents a situation where high levels of consumption may prove detrimental to the long-run acceptability of the product. Clearly, there are value judgments involved in labeling the consumption of a product unwholesome. The classic examples are tobacco and alcoholic beverages. Statements declaring that certain foods and consumption patterns are detrimental to the public interest will continue to be made. Countermarketing efforts to cope with these situations include sponsorship of research and development, reformulation of products, consumer education programs, and label indentification.

These states of demand indicate that there is no one universal food demand problem, but from time to time all states are represented in the food industry. Successful marketing development programs require tailoring marketing tactics to specific demand states.

ROLES AND CRITICISMS OF ADVERTISING

Advertising contributes to the market intelligence and information dissemination functions. Traders and consumers must be informed about the availability of products, their attributes and uses, and their prices. Supplying this information can contribute to a smoothly operating market system, increased consumer satisfaction, and lower overall transaction costs of buyers and sellers. As advertising proponents would say, if advertising "sells," it is probably creating utility and can be considered a productive and valuable marketing activity.

In practice, however, advertising is much more controversial than these statements suggest. The supplier and controller of information has power, and information can be manipulated to serve private interests. Advertising is frequently divided into two classes, informative and persuasive (manipulative), but in practice, often it is difficult to distinguish one from another. Good information is frequently persuasive, and persuasive messages usually contain some information. A concern of many is that advertising is simply a huge economic waste, or a marketing function whose cost greatly exceeds its value.

It is easy to criticize advertising as being biased, often tasteless, intrusive, deceptive, or trivial—it is the most visible and vulnerable of all marketing activities. However, as with most institutions that survive and even prosper, advertising does perform some useful functions.

A major debate concerns the effect of advertising on consumer prices. Those who argue that eliminating the cost of advertising would lower consumer prices are focusing only on the cost side of a marketing function, ignoring the benefits advertising can provide in the way of information. The question is whether the benefit or value of advertising to consumers exceeds its cost. Such difficult judgments are probably best left to the market testing of consumers. Also to be considered is the cost of whatever is to replace advertising. For instance, personal selling is usually more expensive than mass-media advertising.

There are also questions concerning the impacts of advertising on competition in the food industry. National advertising has gone hand in hand with the decline in numbers and the increasing size of firms in many industries. This presents an operational-pricing efficiency dilemma. Advertising may permit the growth of firms and markets, thus contributing to economies of scale, but in doing so the number of competing firms may decline, adversely affecting competition and pricing efficiency.

Still another criticism of advertising is that it distorts consumer choices and interferes with consumer sovereignty. Producers and marketing firms are charged with not only creating products but with creating the demand for their products. Proponents of advertising, on the other hand, argue that advertising simply helps consumers to identify their needs and to match them with available products. In this sense, advertising reinforces rather than creates or manipulates consumer preferences. Consumer preferences are so complex that it is doubtful that this question ever will be answered to everyone's satisfaction. It does seem naïve to believe that firms would spend more than $100 billion to sell products if they were not influencing consumers. However, there are sufficient product and advertising failures to suggest that advertisers do not exercise as much control over consumer preferences and choices as is sometimes believed.

Advertising also is frequently credited by its supporters with stimulating new product development and marketing innovations and generally contributing to the American mass-consumption society and to our high standard of living. Advertising probably does foster a favorable innovative climate by encouraging the search for product advantage and by facilitating the exposure of new products to consumers. However, there are other countries with high standards of living that devote considerably fewer resources to advertising than does the United States.

Overall, advertising is neither black nor white but a gray area with many unanswered questions. Even if it could be proved that it is a wasteful marketing activity, it probably would not be eliminated or outlawed. Businessmen presumably have a right to spend their money in wasteful, if not destructive, ways, and advertising is a freedom that is protected by the First Amendment to the Constitution. But this is not to say that there will not be increasing regulation of the content of food advertising.

ADVERTISING IN THE FOOD SYSTEM

Advertising is prevalent at all levels of the food system. The most visible advertising is sponsored by food processors and retailers in magazines, newspapers, and on radio and television. This advertising is intended to influence consumers' food product and brand choices. In addition, there is considerable trade advertising designed to influence the purchase and selling decisions of producers, resellers, processors, and food distributors. The farm input industries, such as chemical, seed, and feed firms, also have large advertising budgets.

Food products have several characteristics that encourage rather large advertising expenditures on the part of food companies. They are relatively low-priced, packaged products consumed quickly and purchased frequently from self-service stores. Although food shoppers exhibit some brand loyalty, they are generally willing to try new food products and brands. Many food products lend themselves to emotional advertising themes and messages. It should not be surprising, then, that advertising is an important food marketing activity.

Food processors are the dominant food advertisers, accounting for about one-half of food industry advertising expenditures and 25 percent of advertising outlays by all manufacturers. Food retailers account for 30 percent of food industry advertising and spend more on advertising than any other group of retailers. The remaining share of the food advertising dollar is sponsored by farmers, commodity and trade groups, and other food firms.

Marketing agencies use different forms of advertising. Food processors tend to use quality and nonprice themes and advertise in national consumer media. Wholesalers and processors lacking strong brand identification are likely to emphasize price and service themes in trade media. Grocery stores spend much of their advertising budget on local, price-oriented advertising—the weekly newspaper ads. Restaurants and fast-food chains also spend large amounts on advertising. Thus, there is a variety of price and nonprice, national and local advertising in the food industry. Overall, in the food industry there is probably as much price-oriented, informative advertising—as well as persuasive/emotional advertising—as in most other consumer product markets.

The 1998 advertising expenditures of several large food manufacturers are given in Table 14-1. The larger advertisers are generally those with highly differentiated and branded products. The firm's advertising expenses expressed as a percentage of sales is an indication of the company's emphasis on advertising as a market development tool. This advertising-to-sales ratio is generally higher in industries where intensive brand competition is found. For example, the ratio is greater for breakfast cereal, soft drink, and candy companies than for meat packing and dairy processing firms.

Food processors' and grocery product manufacturers' advertising receive the most criticism in the food industry. These firms face a highly elastic demand, one that can be shifted rightward with a successful advertising program or shifted leftward by a competitor's advertising. The resulting advertising battle for the consumer's favor produces some valuable information, a great deal of entertainment, some competitive waste, and many headaches for the advertising and marketing managers of these firms. Food processors engage in two broad types of advertising. Product-line or institutional advertising attempts to promote a favorable image for the company or for its general brand. Brand or item advertising emphasizes the merits of a particular product or brand.

Advertising is also important to the development of new food products. Considering that the consumer is already faced with a bewildering array of various brands, any new product must be introduced into the consumer's mind. Advertising and other forms of promotion are effective ways of accomplishing this.

Advertising expenditures are also stimulated by the struggle for relative power between large processors and large retailing units. More and more, retailers wish to produce and sell products under their own labels in order to secure additional customer allegiance. Processors, understandably, want to retain the connection to cus-

TABLE 14-1 Advertising Expenditures of the Leading Food Marketing Companies, 1998

Company	*Advertising Expenditures, $ Million*	*Company*	*Advertising Expenditures, $ Million*
Procter & Gamble Co.	$2,650	General Mills	598
Philip Morris Co.	2,049	Mars. Inc.	548
PepsiCo	2,049	Nestle	534
McDonald's Corp.	1,025	ConAgra	494
Unilever	1,015	Ralston Purina Co.	462
Tricon Global Restaurants	773	Kellog Co.	449
Coca-Cola Co.	770	Kroger Co.	440
Anheuser-Busch Co.	631	Hershey Foods Corp.	443
Sara Lee Corp.	609	U.S. Dairy Producers, Processors	370

SOURCE: *Advertising Age.* Crain Communications, Inc. Chicago, Illinois. http://www.adage.com/dataplace/Ina.index.html. Reprinted with permission.

tomers through their own brands. Therefore, both groups find themselves spending more money to favorably influence the consumer.

Food marketing firms often build their brand marketing programs around two strategies. *Product differentiation* seeks to find a product's unique features that set it apart from its competitors. The object is to build brand loyalty in order to achieve a less elastic demand curve. *Market segmentation* is a complementary strategy concerned with developing unique product variations that will better appeal to different consumer classes or segments of the market. Both these strategies are apparent in food firms' advertising.

FARMERS AND MARKET DEVELOPMENT

Considering advertising's apparent success elsewhere in the economy, it is not surprising that farmers have frequently viewed advertising and promotion as a solution to their price and income problems. Farmers have made numerous efforts to expand the demand for farm products, using advertising, consumer education, foreign market promotions, support for research and development of new foods and uses of agricultural commodities, and encouragement of public food programs, such as school lunches and food stamps.

These promotional programs have been particularly attractive to farmers during periods of surplus food production. However, the results have been mixed. California orange growers seem to have had more success in promoting the Sunkist label than Iowa farmers have had in promoting Iowa beef. Cotton promotion did not initially stop the incursion of synthetic fibers into its market, but more recent attempts at cotton promotion have been more successful. Dairy industry advertising has influenced milk sales and prices in some cases, yet per capita consumption of fluid milk has continued to decline. Promotion obviously played an important role in consumer acceptance of the kiwi fruit.

Can farmers advertise their way out of their problems? Should they? What results can farmers reasonably expect from commodity promotions? How could these efforts be made more successful?

The Rationale for Farmer Promotion

The goal in expanding the demand for farm products is not simply to increase consumption but to improve prices and producers' incomes. Consumption and sales always can be increased by lowering prices and profits. What the farmer means by *expanding demand* is to shift the demand curve rightward so that the same product volume can be sold at higher prices, or so that increased volume can be sold without lowering prices. Also, farmers sometimes attempt to make the demand for their products less elastic by convincing consumers not to make substitutions when prices rise. But simply shifting demand is not sufficient for the success of a farmer advertising program. Like any other marketing activity, advertising has a cost. The farmer wants to influence demand in such a way that profits will be higher even after the advertising costs are paid.

Extent of Farmer Advertising

As sellers of relatively homogeneous products, most individual farmers do not find it feasible to promote their own branded products. Instead, when farmers promote their products, they usually do so through group-sponsored *generic advertising*. This type of advertising attempts to alter the demand curve for an entire class of food, such as beef, milk, or almonds. Examples of generic advertising themes are: "Orange juice. It's not just for breakfast," "Pork: the other white meat," and "The incredible edible egg."

Generic advertising may involve brands, such as the wool industry's Woolmark. However, the emphasis in generic advertising is on expanding total product sales rather than the sales of an individual firm. Generic advertising also sometimes involves regional brands: Maine potatoes, Georgia peaches, and Florida oranges. It may also promote the sale of a local product, such as at statewide roadside markets. The important point about generic advertising is that it is sponsored by a group of producers or firms and all producers and handlers stand to benefit from a successful generic advertising program.

Generic farm product advertising has several purposes. It may attempt to promote the use of one product in place of another. It may try to alter public opinions and attitudes toward a product. Most generic advertising seeks to increase consumer product usage rates or provides information that changes the ways in which the product is used. Generic promotion may also attempt to expand the overseas market for a U.S. food product; for example, the demand for U.S. pork in Japan.

Generic promotion of farm products is usually sponsored by producer groups and commodity organizations. Examples of such organizations include the National Dairy Promotion and Research Board, the Washington State Apple Commission, the Egg Board, the Florida Citrus Commission, and the National Livestock and Meat Board. Dairy products, meats, and fruits have accounted for the largest generic advertising budgets in recent years. Other commodities have smaller promotional budgets.

Effectiveness of Farmer Advertising

The goal of farmer promotional programs is to influence the demand for farm products and improve prices and profits. Some promotional programs have been quite successful, some have produced mixed successes, and others must be judged unsuccessful. It is difficult to generalize and say that all farmers should, or should not, promote their products; each product and farm group must be considered individually.

Several criteria have been suggested for deciding whether farmers can effectively promote their products. Generally, promotional efforts have been more successful for specific, highly differentiated products than for generic classes of homogeneous products. Some brand or regional identification is important in differentiating farm products. Also, the more substitutes there are for a product, the more effective the advertising, even though one goal of advertising is to convince consumers that a product has no good substitutes. For example, advertising "beef" would be more effective than advertising "all meat," and "peaches" are more promotable than "all fruit."

A second criterion is that production and marketing of the product should be in the hands of an organized group of farmers. This is necessary in order to obtain financial support of the promotional program and to prevent supplies from expanding too rapidly in response to higher prices and profits. These events can prove troublesome for a commodity produced by a large number of geographically dispersed producers. For this reason, walnuts, avocados, and almonds, presumably, can be promoted somewhat more easily than apples and peaches. The need for supply control is critical to the success of advertising. Higher prices, sales, and profits will encourage expansion and entry into the market, and in the long run, this supply response could nullify some of the price effects of the advertising. The need for farmer organization is also evident. Unless all farmers in a position to reap the benefits of the promotion share in its costs, free-rider problems will undermine the program.

The advertising program must be closely coordinated with other marketing activities. Advertising is only one component of the marketing mix and is seldom successful without reinforcement from the other components. It must be coordinated with quality control, flow of supplies, prices, product development, and point-of-purchase materials.

It is frequently observed that farm product advertising is most successful for commodities that move to the consumer in fresh form, with little change in identity. These products, of course, are easier to brand than products that lose their identity. They also tend to have a higher farmer's share, so that a shift in retail demand has a considerable impact on farm prices.

A final criterion is that a substantial sum of money be available if mass-media promotions are to have an impact. The effectiveness of advertising is not directly proportional to the dollars spent, and a critical mass of promotional exposure must be achieved in order to penetrate the mass market.

Another concern in farmer promotional efforts is that because most foods are more or less substitutes for one another, an advertising program that shifts demand rightward for one product may shift the demand leftward for another. This *parasitic advertising* may benefit one group of farmers at the expense of another. Farmers need to consider the group impacts of farm product advertising in evaluating the success of their separate programs. The felt need to launch a defensive promotional program to combat parasitic advertising may result in forcing farmers to engage in commodity advertising when it would not otherwise be advisable.

Should There Be Compulsory Farmer Advertising?

The entry of farm groups into advertising ventures and the belief or assumption that such efforts are successful have brought two major reactions. First, there is the attitude that because advertising is successful, it is unfair to support it by voluntary efforts. Under these circumstances, the noncooperators benefit equally from the expenditures of the cooperators. Second, there is the attitude that if promotion is successful for one commodity, it should be expanded to all agricultural commodities.

The first attitude has led to pressures for legislation to make fund collections compulsory for all producers. Various state and federal laws authorize compulsory *check-offs* or assessments of producers to support generic promotion programs. For example, the dairy industry's promotional campaign is funded by a 15 cents per hundredweight assessment on dairy farmers' milk production. Cattlemen contribute $1 per head of livestock to promote beef while pork producers have 25 cents per $100 of hog value subtracted from their receipts to promote pork. These programs and mandatory contributions usually must be approved by a two-thirds referendum vote of all producers. In many programs, producers may apply for a refund of the assessment if they choose to. Nevertheless, these programs are being challenged because many view them as an infringement on the right of farmers to decide for themselves whether to promote their products.

Promotional activities by farm groups, therefore, have entered into the policy picture at the national level. Here is where the second attitude mentioned becomes pertinent. Can the success of an individual commodity campaign be generalized to the total output picture? Can all groups promoting all products persuade consumers to allocate more money for the raw products of agriculture than they would have allocated without the promotion expenditure?

While promotion probably can expand the sales of one food product at the expense of another, its affect on total food consumption is less clear. It is unlikely that promotions can encourage consumers to transfer some funds from nonfood expenditures to purchase food items. At the present time, food in the aggregate does not have close substitutes and there is little potential for this substitution to occur. As a rule, promotion works best when products have more substitutes. Therefore, advertising is most effective for similar, branded products and less effective for generic products.

Most Americans are well fed and are not good candidates for spending more on food. There is a group of consumers who might prefer more or different foods and who are not adequately nourished. However, most of these consumers are not in this condition by choice but because of the lack of income. Persuasion is a poor tool where lack of income is concerned. At present the potentialities of increasing the demand for all agricultural products through a massive, across-the-board advertising campaign appear meager except in areas where close substitutes are available. As synthetic foods become more important, defensive promotional programs can be expected to take on added importance as producer groups struggle to maintain their position.

It appears likely that more and more general commodity advertising will be done by farm groups. Individual groups are concerned only with the sales of their particular commodities. As in individual companies, competitive pressures from other companies that advertise force them to increasingly greater efforts in order to maintain their share of the market. Trade groups threatened by synthetics will develop more aggressive promotional programs. Because most products are sold in a market that is national in scope and will require sizable amounts of money to exploit successfully, more pressures will develop to find ways to collect resources to spend on advertising on a broad, continuing basis.

MINI CASE 14: SHOULD FARMERS PROMOTE THEIR PRODUCTS?

Billy had a tough assignment from his high school agriculture teacher. The debate topic was "Resolved: Farmers Should Promote Their Products." He had to be prepared to argue both the pros and cons of farmer advertising. After dinner, Billy's father, a farmer producing an undifferentiated grain product, helped Billy make a list of arguments against farmer advertising. The next day, Billy asked a dairy farmer why dairy farmers enthusiastically sponsored the largest farmer-sponsored advertising program, now using the slogan *Got Milk?* The dairyman said he thought milk advertising was successful, but he couldn't point to any hard evidence. Finally, Billy talked to a large cattle producer who had developed a successful brand for her beef, *Mary's Lean Beef,* who said that "farmers have to advertise and promote their products, just like any other food company." Billy was confused. Does farmer advertising pay or doesn't it?

Under what conditions can farmers successfully promote their products? What are the purposes of these promotions? How would famers know that the benefits of promotion exceed their costs?

EXPANDING NONFOOD USES FOR AGRICULTURAL PRODUCTS

The domestic demand for food products cannot be expanded rapidly, but what are the possibilities for making nonfood products from agricultural commodities? This idea has always excited the imagination of those who wish to expand the demand for farm products. Farm products have a long history of use in manufacturing. Starches are used for adhesives and textile sizing, vegetable oils can serve as lubricants, and essential oils can be transformed into fragrances and flavors. Industrial uses of corn accounted for 9 percent of the crop in 1996, compared to 2 percent in 1980. Industrial applications of farm products include biodegradable polymers, biodiesel fuel made from vegetable oils, and pharmaceuticals made from milk.

However, the technical practicality of making an industrial product from farm products is not always the same as its economic feasability. For example, a very satisfactory fiberboard for use in the building trades has been developed from cornstalks. However, the cost of collecting the cornstalks from the field makes the product uncompetitive. It has been estimated that the incorporation of 2 to 5 percent starch in pulp and paper products would provide outlets for 40 to 100 million bushels of grain. However, such a procedure is not now considered economical by the paper trade. We can make alcohol for automobile fuel (gasohol) from corn, but corn must be quite cheap to compete in the gasoline market.

Compared with other industries, a relatively small amount of research is being done on the potential industrial utilization of farm products. Agricultural chemists point out that we know much less about the basic chemistry of farm products than we do about coal and petroleum. An improvement in such basic knowledge is a necessary first step to any major breakthrough in increased economically feasible nonfood utilization.

Most analysts in this area conclude that additional expenditures on utilization research and development are necessary not only if we are to expand this demand but also if we are to retain the outlets that now utilize agricultural products. The successful development of new products from agriculture will involve a partnership of private and government research and commercialization. There will be successes and failures. It is like drilling for oil. Many dry holes may have to be explored in order to bring in one producer. The results can prove valuable for farm income and rural diversification.

SYNTHETICS AND AGRICULTURAL SUBSTITUTES

The history of the food industry is one of constant substitution. In the nineteenth century, cane sugar, coffee, and European rapeseed encountered competition from beet sugar, chicory, and whale oil. The twentieth century saw the appearance of cereal sugars, tropical oils, soya, and fish oils. Recent developments include the production of yeasts, fungi, bacteria, amino acids obtained by fermentation, synthetic sweeteners, and petroleum-based fatty acids.

At least part of the expected moderate increase in the demand for agricultural products may be offset by the growth of synthetic foods and food substitutes. *Synthetic products* are manufactured from nonagricultural raw materials, principally petrochemicals. Synthetic sweeteners, shoe "leather," fabrics, and livestock feeds are good examples. *Agricultural substitutes* are manufactured from agricultural raw materials but replace traditional agricultural products. Examples of agricultural substitutes are bacon bits and other meat substitutes made from soybean proteins, artificial milk products, and high fructose corn syrup. In the case of substitutes, the effect is really a transfer of the demand from one agricultural sector to another. The use of margarines, for example, increases the demand for soybeans as the demand for dairy products is reduced. These trends have varying effects on farm producers and agricultural industries and may cause shifts of agricultural resources between them.

To date, losses to synthetics and substitutes have been most extensive among the nonfood agricultural products. Although the trend has now reversed, for many years cotton was losing its market to man-made fibers. Synthetics and substitutes may replace agricultural products for a number of reasons. Manufacturers and consumers can usually be guaranteed a more stable supply of these than agricultural products because they are not subject to the vagaries of weather and disease. It is usually easier to maintain constant quality of man-made goods. Additionally, manufactured goods are often easier to differentiate and cheaper to produce.

Meats provide interesting examples of agricultural substitutes. A meat *analog* is a vegetable soy protein product that is fabricated to resemble meats in texture, color, and flavor. Sometimes even a plastic bone is added to the "soysteak." Imitation dairy products are also now available. A *food extender* is a vegetable protein product that can be added to traditional meat products (hamburgers, meatloaf) to increase their volume. Certain meat analogs and extenders are competitive with red meats in cost per serving (after cooking). This cost factor has encouraged their use in the hotel, restaurant, and institutional markets. The growth of meat analogs and extenders for home food consumption has been slower because of labeling regulations, nutritional requirements, and flavor and texture preferences.

PUBLIC FOOD PROGRAMS

It is also possible to expand the consumption of food by reducing its price or giving it to those with insufficient effective demand. Most societies accept some responsibility to provide food assistance for the disadvantaged. However, there is considerable debate about how to do this and how much assistance is desirable. Publicly subsidized food programs are predicated on humanitarian, moral, and economic grounds.

The federal government is involved in a number of food assistance programs. These efforts began in the 1930s with the emergency distribution of surplus commodities to the needy. Commodities were acquired by the government through efforts to support farm prices and incomes. The programs continue today, although the emphasis has shifted from surplus farm product disposal to improving the diets of poor families and children.

The largest public food programs are shown in Table 14-2. They include the food stamp program, the school lunch programs, the women, infants and children (WIC) program, and commodity distribution programs. These programs expanded rapidly prior to the 1990s but have grown less rapidly in recent years with welfare reform legislation.

Table 14-2 Participation Rates and Costs of Federal Food Programs, 1999

Federal Food Assistance Program	*Number of Individuals Participating (000)*	*Federal Cost, $ Million*
Food Stamp Program	18,183	$17,656
National School Lunch Program	26,946	6,019
Women, Infants, and Children Supplemental Food Program	7,311	3,940
Child/Adult Care Food Program	2,670 per day	1,485
School Breakfast Program	7,371	1,345
Temporary Emergency Food Assistance Program	NA	270
Summer Food Service Program	2,174 per day	268
Elderly Nutrition Program	253	140
Commodity Supplemental Foods	382	98
Food Distribution, Indian Reservations	129	76

SOURCE: Food and Nutrition Service, U.S. Department of Agriculture, http://www.fns.usda.gove/pd/annual.html.

Public food assistance programs affect farmers, food marketing firms, and consumers. To the extent that they involve foods that recipients would not otherwise purchase, these programs can increase the total demand for food. It has been estimated that 2 percent of consumer food expenditures are directly related to food stamps. However, it is recognized that some portion of this food assistance simply displaces what would otherwise have been purchased or frees incomes to purchase nonfood products. In general, purchasing power aid or supplemental income programs, such as food stamps, provide consumers greater latitude in dietary choice but have less impact on total food demand than in-kind programs that distribute commodities directly.

Federal food assistance programs attempt to distribute surplus commodities in a manner that will not interfere with normal marketings or sales. Milk given to children in school probably does not displace home sales of milk. However, a large scale distribution of surplus cheese in 1982 was felt to have disrupted normal cheese sales. While the food stamp program utilizes the existing food distribution system, surplus commodities are often distributed outside these commercial channels, by government or charitable organizations.

SUMMARY

Market development activities—advertising, public relations, product research and development, quality control, and merchandising—are extremely important and expensive elements of food marketing. These demand-influencing activities can alter consumer demand states, affect the intensity of competition in food markets, and influence farm and food prices. Advertising is the most visible and controversial food marketing development activity. Food industry advertising is a blend of persuasive and informative messages conveying both price and nonprice aspects of the marketplace. Food processors dominate the food advertising picture. By comparison, farm groups and related organizations engage in a small amount of generic, and sometimes brand, advertising. The success of farm product advertising has been mixed, depending on the ability to brand the product, the number of substitutes for the product, the organization of farmers and their ability to control supplies, and the amount of money available for farm product advertising. The rate of farm commodity demand expansion is influenced by the development of nonfood uses of farm products, the development of agricultural substitutes and synthetic foods, and the growth of public food programs.

KEY TERMS AND CONCEPTS

agricultural substitutes and synthetics
check-off
demand states
food analog and extender
free rider
generic advertising
market development
market segmentation
marketing mix
parasitic advertising
product differentiation
public food programs
push, pull promotions
the 4Ps

DISCUSSION QUESTIONS

1. Why does market development become a more important aspect of marketing as consumer incomes rise?

2. Identify a food product in each of the demand states discussed.

3. Your friend states, "Advertising is a waste of money." How do you respond?

4. How do you distinguish between informative and persuasive food advertising? Give examples.

5. Why do food processors and grocery manufacturers account for the largest amount of food industry advertising?

6. You are asked by a group of wheat or corn farmers to advise them on whether they should advertise and promote their product. What is your advice?

7. How would you evaluate the effectiveness of a nationwide program to promote increased consumption of milk?

8. How does branding of farm products reduce the free-rider problem in farmer advertising programs?

9. Do you think farmers will do more or less commodity advertising in the future? Why?

10. What are some current nonfood uses of agricultural commodities?

11. Why do farmers support such government programs as food stamps, the school lunch program, and the special milk program? Do consumers?

SELECTED REFERENCES

Armbruster, W. J., & L. H. Myers (Eds.). (1985). *Research on Effectivenesss of Agricultural Commodity Promotion.* Oak Brook, IL: Farm Foundation.

Blisard, N. et. al (March 1999). *Analyses of Generic Dairy Advertising, 1984-97.* Technical Bulletin No 1873, U.S. Department of Agriculture.

Forker, O. D., & R. W. Ward. (1993). *Commodity Advertising: The Economics and Measurement of Generic Programs.* Lexington, MA/Toronto: Lexington Books.

U.S. Department of Agriculture. (February 1995). *Evaluation of Fluid Milk and Cheese Advertising, 1978-93,* Technical Bulletin No. 1839. Washington, DC: U.S. Department of Agriculture.

U.S. Department of Agriculture. (January-February 1996). Industry expands use of agricultural commodities. *Agricultural Outlook*, 22-25.

Ward, R. W., S. R. Thompson, & W. J. Armbruster. (1983). Advertising, promotion, and research. In W. J. Armbruster et al. (Eds.), *Federal Marketing Programs in Agriculture* (pp. 91-120). Danville, IL: The Interstate Press.

CHAPTER 15

MARKET AND BARGAINING POWER

Market power is like the wind. You can feel it but you cannot see it.

OBJECTIVES

After reading this chapter, you will be able to:

1. Define the types of market and bargaining power in the food industry.
2. Identify the sources and impacts of market power.
3. Evaluate the public policy tools by which farmers attempt to increase their market power.

Market power is a perennial area of concern in the food industry. Most firms complain about their lack of power. Food processors maintain they are at the mercy of giant retailers. Food retailers argue that the large food processors have "undue market power." Consumers complain that they are being exploited and are powerless to do anything about it. Farmers, feeling they are the smallest and most unorganized element of the food industry, have historically complained of their lack of bargaining power. Many food industry laws seek to restore bargaining power or to strike a better balance of power among market parties.

What is market and bargaining power? Who has it and who lacks it in the food industry? How is market power achieved and maintained? And, most importantly, what are the consequences of market power in the food industry? These are vital questions in the design of agricultural and food marketing policies.

MARKET POWER IN THE FOOD INDUSTRY

Market power is an intangible characteristic of markets. Unlike the physical marketing activities, it cannot be observed directly, nor can it be measured precisely. It is largely known through its effects on the marketing processes—through prices, contract terms, firm behavior, and other features of markets. None have seen it, but many have felt it.

Market and Bargaining Power

Market power is the ability to advantageously influence markets, market behavior, or market results. While market power is typically associated with influence over prices, it also can take the form of influence over demand, product flows, quality, marketing functions, and other firms' market behavior. Firms seek and use market power in order to achieve their economic goals.

A large retailer or processor with influence over prices and profits through consumer advertising has market power. A dominant food wholesaler who is a price leader for other wholesalers has a degree of market power. A nonfood company that finances discount prices of its food subsidiary is employing market power. A food processor that controls the quality and timing of product flows in a profitable manner through farm contracts is exerting market power. Farmers who succeed in raising prices through supply control programs have gained market power. A consumer boycott of a food product is an attempt to exercise consumer market power. In each of these cases, one marketing agency or group of market participants is influencing, in an advantageous manner, either the market or the market behavior of others. Market power involves shaping and influencing the market process, rather than simply reacting to the market environment.

Bargaining power is a related term and refers to the relative strength of buyers and sellers in influencing the terms of exchange in a transaction. Bargaining power requires market power, but market power is a broader concept, not limited to the buyer-seller situation. Food retailers, for example, may influence farm prices and sales through their merchandising and pricing practices without ever directly negotiating with farmers.

A firm's market and bargaining power can be defined only relative to the lack of market power of others. If all market agencies possess equal market power, none of them has an undue influence on the market; they nullify each other's power. Market power becomes a problem when it is unequally distributed—when one marketing agency can take advantage of its superior influence. Accordingly, there are two approaches to solving market power problems in the food industry: (1) reduce the influence of the more powerful to the level of the weaker; or (2) increase the influence of the weaker to the level of the more powerful. Regulatory and legal measures to maintain competitive conditions at all levels of the food industry illustrate the first approach. Attempts to build "countervailing power," or equal influence, illustrate the second approach.

Market power is frequently associated with the competitive state of markets. Imperfectly competitive firms have market power, but perfectly competitive firms have no power. The association of market power with imperfect competition is correct. However, imperfect competition is not synonymous with market power. Two imperfectly competitive firms may have no market power over one another, if their power is roughly comparable and countervailing; yet both may have superior market power over a third, perfectly competitive firm. Under the conditions of perfect competition there is no market power problem because market influence is equally lacking for all. However, the real world is made up of shades of imperfect competition, and there is opportunity and room for maneuvering and strategies in the attempt of one market party to gain and exercise more market power than another.

It is useful to distinguish between short-term and long-term market power. Marketing agencies may not always take full advantage of their market power. Using short-term market power may conflict with the firm's long-term position, as, for example, when a buyer does not bargain prices to their lowest level for the purpose of building long-term goodwill among suppliers. Similarly, a marketing agency may resist using its full market influence if it feels that to do so would invite more competitors into the industry or might trigger unwanted government intervention.

It is also useful to distinguish between horizontal, vertical, and conglomerate market power. *Horizontal power* refers to the influence that similar marketing agencies have over one another. For example, large retailers and food processors may be able to influence the pricing and output decisions of their smaller competitors. *Vertical power* relates to the influence that vertically related firms in the marketing channel have over one another. *Conglomerate market power* refers to the influence that a firm might have in the food industry by virtue of its ties to nonfood companies.

Types of Bargaining Power

Three types of bargaining power have been identified in the food industry. *Opponent-pain* bargaining power is concerned with the influence of a buyer or seller in a negotiation gained through the ability to threaten or make opponents worse off. Most people associate bargaining power with this form. The influence stems from coercion and the power to inflict damage on others. "Either sell to me or you will have no market," "That's my price, take it or leave it," and "Give us our terms or we will destroy the product" are illustrations of opponent-pain bargaining. In practice, however, the threats are usually more subtle.

Opponent-gain bargaining power represents an alternative to opponent-pain tactics. This power stems from the advantages that one market party can offer to the other in exchange for accepting terms. For example, producer associations may be able to perform certain market activities that increase the efficiency of food processors. Quality control, full-supply contracts, delivery services, and improved scheduling of production are examples of such activities. In this case, neither party to the negotiation needs to be made worse off. Cooperation, rather than threat, dominates the negotiations. How to allocate the benefits of the gain between participants is also a subject of the negotiations.

In practice, both opponent-pain and opponent-gain power stances are evident in most negotiations. A food processor may agree to higher wages if the union will permit the use of new labor-saving technologies (opponent-gain). If the union refuses this offer, the company may threaten to move its plant, and the union may threaten to strike (opponent-pain).

In another type of bargaining power, a buyer and seller may agree to terms that secure a gain from third parties—either consumers, other market agencies, farmers, or the government. In this situation, the two parties to the negotiation cooperate to secure gains (higher prices, government subsidies, protection from imports, and the like) from other sources. For example, corn sweetener processors have an informal alliance with corn producers to maintain high U.S. barriers to sugar imports in order to enhance corn and sweetener prices.

Sources of Market Power

A number of interrelated factors or market conditions have been identified with market power in the food industry:

1. *Size, number, and market concentration of firms.* Ordinarily, large firms are believed to have superior market power in dealing with smaller firms. The key here is the number of alternatives for the weaker market party. Large firms in concentrated markets draw their power from the lack of alternatives that others have in dealing with them. The relationships of firm numbers and sizes to market power are complex in the food industry, where superior vertical power relations might be nullified by countervailing horizontal power relations.

For farmers, organization is a more important influence on market power than simple numbers and sizes of sellers. The greater the organization and cohesion of farm groups, the greater their potential market power.

2. *Supply control.* The most important source of market power stems from the ability to effectively control the amount of product produced and offered to the market. Normally, those who have inferior market power are sellers of products for which it is difficult to regulate production or control supply. These products need to be marketed because of perishability or lack of storage space, and are subject to frequent price fluctuations.

3. *Unequal information.* Information is power that can be used to take advantage of market situations. Typically, firms with the greatest amount of market information have superior market power.

4. *Diversification.* Buyers and sellers who are diversified by products, geography, and marketing functions appear to have more market power than specialized agencies. Diversification contributes to flexibility in market decisions and reduces market risks.

5. *Product differentiation.* Firms with highly differentiated products are in a better position to "manage demand" profitably than firms with homogeneous products.

6. *Control of strategic resources and decisions.* "Gatekeeper" firms control strategic market variables such as brands, consumer loyalty, retail shelf space, or retail prices. Control of these variables appears to give firms some power.

7. *Financial resources.* Firms with large financial resources often can withstand competitive battles and inflict greater opponent-pain power than weaker firms. This is believed to be an important source of market power for the conglomerate food company.

8. *Ratio of fixed-to-variable costs.* Firms with relatively high fixed costs tend to suffer chronic excess capacity and do not respond quickly to changing prices. Because of this inflexibility, these firms are frequently considered to have low market power.

Indicators of Market Power

Market power is difficult to isolate and study in the marketplace. Firms do not proclaim that they have market power! And almost everyone in the food industry complains of a lack of market power, or claims to be the victim of someone else's market power. Unfortunately, there is no litmus test to identify market power. Moreover, there are many misconceptions about market power. "Low," below-cost, or fluctuating prices are not conclusive evidence of superior or inferior market power. In the case of agriculture, these conditions may simply reflect the inelastic demand and supply conditions. Neither is the level or trend in the farmer's share of the consumer's food dollar a useful test of market power. This share can be explained by consumers' low income elasticity of demand for farm products and high elasticity for food marketing services.

Many feel that higher-than-normal profits are an indication of superior market power. In addition to the difficulties of determining "normal" profits and of adjusting these profits according to the levels of associated risks, to use this as an indication of power discounts the role of management skills in profits. Some firms may have higher profits not because they are powerful but because they are better managed. Moreover, firms with market power may show low profits as a result of poor management.

Sometimes market power is attributed to those firms or market levels that quote prices. However, these firms may simply be the point of price discovery as a result of industry tradition or convenience. The firm that quotes prices may be a good price discoverer but may have no power to influence prices or other market results.

Evaluation of Market Power in the Food Industry

Despite the difficulty of identifying and measuring market power in the food industry, some judgments can be made.

There is a general consensus that, compared with food marketing firms, farmers have inferior market and bargaining power. The reasons for this are farmers' numbers, relative sizes, low levels of organization, high fixed costs, geographic and product specialization, and homogeneous products. There is also a concern that decentralization and integration have reinforced this prevailing imbalance of power between farmers and food marketing firms. This imbalance could contribute to relatively low farm prices and incomes, frequent below-cost or fluctuating farm prices, and the shifting of the cost of some marketing functions and risks to farmers. However, other farm market conditions, such as elasticities of supply and demand, also contribute to these conditions and are not necessarily related to the market power of food firms.

Throughout history farmers have used various means to improve their bargaining power—some formal and some informal, some coercive and some voluntary. Some were partially effective and others were completely ineffective. In the nineteenth century, night riders in the tobacco belt burned the barns of farmers who refused to cooperate in withholding the crop from the market. During the 1920s, Aaron Shapiro preached the message of cooperative bargaining power. In the 1930s, the Farmer's Holiday Movement used coercion to gain farm market power. More recently, livestock and dairy producers have attempted to gain market power through supply restriction and dumping. Each period of low farm prices sees a new wave of farmer strikes, witholding threats, and "holding actions" by farmers.

The National Commission on Food Marketing identified two other market power positions in the food industry. Food retailers, it found, possessed market power by virtue of their size, control of private labels and shelf space, direct contact with consumers, and geographic and product diversity. The same study indicated that food manufacturers with strong brands also have superior market power. The commission felt that brand loyalty, large, diversified product lines, a heavy investment in new-product development, and conglomerate affiliations were the sources of this power. Food manufacturers or retailers with strategic market power may assume the role of marketing channel leader and coordinator. There is still much debate about these forms and effects of market power.

Market power is a dynamic feature of food markets. Large, dominant firms are often supplanted by smaller, more vigorous competitors. Food retailers' power may be eroded by vending machines and the fast-food industry. Farmers may gain countervailing power against large buyers through their cooperatives and other marketing efforts. Few power positions have persisted for long in the industry. Even so, transitory power is a public policy concern. New laws and regulations will continue to be developed to regulate market power in the food industry.

In this book we are principally concerned with marketing policies and tools that influence food producers' market power. These range from the government farm price and income programs, discussed in Chapter 21, to the marketing cooperatives discussed in Chapter 13. In this chapter we will explore two other producer market power tools: marketing orders and agreements and cooperative bargaining associations.

MARKETING ORDERS AND AGREEMENTS

Marketing orders and agreements are unique self-help marketing tools intended to improve agricultural producers' prices, incomes, and market power. These marketing tools had their origin in the Agricultural Adjustment Act of 1933 and the 1937 Agricultural Marketing Agreement Act. Through these instruments, farm product producers and handlers are authorized by law to engage in collective marketing activities, the objectives being to enhance the level and stability of producer returns. Marketing orders and agreements strive for more *orderly marketing*—a term that means profitable farmer management of commodities over time, form, and space.

Technically, marketing orders are different from marketing agreements, but both have the same goals and are often used together. The principal difference is that agreements are purely voluntary arrangements between producers, handlers of farm products, and the Secretary of Agriculture on how a particular product will be marketed. In contrast, a marketing order is established by a majority vote of producers—whether or not handlers or buyers approve—and is legally binding on all producers and handlers of the commodity within the marketing order area. Because voluntary agreements often break down, and farm product buyers' interests do not always coincide with producer interests, today virtually all marketing agreements are reinforced by marketing orders.

Marketing orders and agreements may be authorized by either or both state and federal laws. Though not every state has such a law, state orders usually have the same objectives and work in the same fashion as federal orders. Marketing orders are authorized for a limited set of commodities, the most important of which are milk, fresh fruits and vegetables, tobacco, peanuts, naval stores, turkeys, and apples for processing. Marketing orders are prohibited by law for many important commodities, including the major foodgrains and feedgrains, soybeans, livestock, poultry (except turkeys), and eggs. Milk, fruits, vegetables, and nuts account for the majority of marketing orders today.

In 1996 there were 37 federal fruit, vegetable, nut, and specialty crop marketing orders in effect (Table 15-1). These orders covered crops valued at about $4.5 billion in 34 states. These orders were concentrated in the major fruit and vegetable producing states of California, Florida, and Texas. The 11 milk marketing orders covering about 80 percent of all Grade A milk marketed are shown in Figure 24-3.

Marketing order legislation is *enabling,* not *mandatory.* Agricultural producers petition for orders, and orders must be approved by two-thirds of the producers or by producers accounting for two-thirds of production. Separate orders are written for each commodity; and orders usually cover a limited geographic area. Many orders operate in conjunction with marketing cooperatives. The Secretary of Agriculture authorizes and supervises orders and is responsible for ensuring that orders operate in the public interest. Administration of the terms of nonmilk marketing orders is in the hands of a board of directors composed of producers and handlers. Milk marketing orders are supervised by a market administrator appointed by the Secretary of Agriculture.

Marketing Order Provisions

The authority granted under marketing orders is broad and varies to accommodate a wide range of problems and producer objectives. Table 15-1 illustrates the variety of provisions contained in fruit, vegetable, and nut marketing orders. No two orders are alike. The most important provisions and uses of marketing orders are as follows:

1. To classify milk according to its use, set minimum producer prices, and to average (pool) returns to milk producers.

2. To manage the flow of commodities to market—either in total or by grade, size, or timing.

3. To establish producer or handler marketing allotments.

TABLE 15-1 Federal Marketing Orders and Provisions, 1996

	Minimum grade requirements	*Minimum size requirements*	*Pack and container requirments*	*Flow to market*	*Market allocation*	*Reserve pool*	*Producer allotments*	*Research*	*Advertising*
Fruit:									
Florida citrus	X	X	X	X		X			
Texas oranges & grapefruit	X	X	X					X	X
Florida limes	X	X						X	X
Florida avocados	X	X	X					X	X
California nectarines	X	X	X					X	X
California pears, peaches	X	X	X					X	X
Georgia peaches	X	X						X	
California kiwi fruit	X	X	X					X	
Washington apricots	X	X	X					X	
Washington cherries	X	X	X					X	
Wash.-Oreg. fresh prunes	X	X	X					X	
California dessert grapes	X	X	X					X	
Pacific Coast winter pears	X	X						X	X
Hawaii papayas	X	X	X					X	X
Cranberries (10 states)	X	X					X	X	
Wash.-Oreg. Bartlett pears	X	X						X	X
California olives	X	X	X					X	X
Vegetables:									
Idaho-E. Oreg. potatoes	X	X	X						
Washington potatoes	X	X	X						
Oregon-Calif. potatoes	X	X	X					X	
Colorado potatoes	X	X	X					X	
Maine potatoes	X	X	X						
Virginia-N. Carolina potato	X	X							
Georgia vidalia onions								X	X
Walla Walla onions			X					X	X
Idaho-E. Oreg. onions	X	X	X					X	X
South Texas onions	X	X	X					X	
Florida tomatoes	X	X	X					X	X
Florida celery	X	X	X	X			X	X	X
South Texas melons	X	X	X					X	
California almonds	X				X	X		X	X
Oreg.-Wash. filberts	X	X			X			X	X
California walnuts	X	X			X	X		X	X
Far West spearmint oil						X	X	X	
California dates	X	X	X		X			X	X
California raisins	X	X			X	X		X	X
California dried prunes	X	X	X	X	X	X		X	

SOURCE: U.S. Department of Agriculture, Agricultural Marketing Service.

4. To control and equalize the burden of "surplus" production.
5. To regulate the size, capacity, weight, and other dimensions of pack or containers.
6. To set up market information, product inspection and standardization, and market research and development programs.
7. To establish systems for pooling or averaging returns to producers and handlers for different time, form, and spatial markets.
8. To prohibit unfair trading practices.
9. To engage in commodity advertising programs.
10. To regulate the grade, size, quality, and maturity of imported commodities.

Milk marketing orders provide for setting minimum producer prices and even allow a limited form of production controls. However, other marketing orders do not give producers authority to set prices directly or to control agricultural output. Instead, these orders provide producers a vehicle for indirectly influencing farm prices and returns—and, in turn, production—through control of product flows to market. In this sense, marketing orders complement government farm price and income programs, which have directly influenced farm production, prices, and incomes.

Orders as Monopolistic Devices

By providing the machinery for control over marketed supplies, marketing orders confer monopolistic power on agricultural producers that they would not have as individual sellers. Orders transform the farm marketing process from one of independent decisions on the part of many small firms to an imperfectly competitive process. The potential monopolistic gains from this market power result from control over total marketed supplies and control over the flow of supplies to time, form, and space markets.

Marketing order restrictions on quality, size, imports, and other market elements constitute a form of supply control for producers. This power can be used to raise farm prices above what they would be without supply controls. The advertising programs authorized by orders, if successful, have a similar effect on prices.

The control over flow of products to time, form, and place markets presents still another opportunity for producers to enhance farm prices and returns through what is known as *economic price discrimination*. This practice, which is not illegal, rests on the notion that there is not one demand curve of a given elasticity for a commodity but several curves of differing elasticity. There may be a variety of elasticities of demand—according to cities, income levels of consumers, qualities of product, or seasons of the year. By wisely managing the flow of total supplies into these markets, differing prices may be established, resulting in greater total returns than if the same price is charged in all markets. The marketing strategy of price discrimination involves restricting supplies in the less demand-elastic markets and increasing supplies in the more elastic markets. Remember that total revenue rises when prices *increase* in inelastic demand markets and when prices *decrease* in elastic demand markets.

Three conditions are necessary for successful price discrimination. First, there must be more than one identifiable market in time, form, and space, with each market having a different elasticity of demand. This is a common situation for agricultural commodities. Second, a tool, or system, must permit a controlled allocation of the total supplies to each of these markets. The marketing order is tailor-made for this. Third, buyers in each of the markets must be persuaded to pay different prices for the product, and must be prevented from all purchasing in the lowest-priced market. The advertising, grading, sorting, and control of market flows provided by marketing orders can help to accomplish this.

Price discrimination is a valuable marketing tool for farmers. It is also practiced by other food marketing agencies. A food processor who packs both a nationally advertised label and a nonadvertised label is practicing price discrimination. A food retail chain operating full-service supermarkets, "warehouse" economy food stores, and neighborhood convenience stores also has the opportunity to practice economic price discrimination. Marketing orders, then, do not give producers extraordinary monopolistic power, but they do allow producers to do what others are doing in the food industry.

Evaluation of Marketing Orders

Marketing orders and agreements are intended to improve farm prices, returns, and incomes. Their record in doing so is mixed. Many orders have been judged successful on these criteria and have enjoyed a long life. Other orders have been discontinued after a short trial period or after several years. Many proposed orders have failed to obtain the majority producer support necessary to put them into operation.

Each order must be evaluated separately and continuously, as must proposals for new orders on additional commodities. The principal results and achievements of successful market orders might be listed as follows:

1. More orderly marketing of farm products, with somewhat greater stability of farm returns.
2. Modest to inconsequential increases in farm prices in the long run.
3. Some transfer of market power and decision making from marketing firms to producers and greater producer participation in the marketing process. Perhaps a psychological and morale-building gain for producers, who shed their price-taking status.
4. Some increase in farm product differentiation and increased farmer-controlled advertising of commodities.
5. The generating of considerable market information and research, contributing to improved market understanding.

In the 1970s, the Federal Trade Commission and Justice Department expressed concerns about the competitive effects of marketing orders and their impacts on food prices. A number of studies were conducted to assess the effects of orders on efficiency, the distribution of income, and the structure of agriculture. As a result, the Secretary of Agriculture issued a set of marketing order guidelines in January 1982 that discouraged those programs that restricted entry (e.g., allotments), limited supplies, or otherwise represented monopoly controls. The hops marketing order, established in 1966, was terminated by the Secretary of Agriculture in 1986 because of its restrictions on grower sales and entry into production. In 1986 the Secretary of Agriculture also terminated the tart cherry marketing order because of its volume regulations. And in 1996, the Maine potato marketing order was discontinued.

Clearly, marketing orders do not solve all farm market power and price problems, but used with other marketing tools, they can be a valuable marketing device. However, their limitations and problems are evident.

For one thing, orders are not permitted on all products. Even if they were allowed for many of the excluded commodities, they would probably continue to find their widest use among the specialty crops, which are produced in geographically concentrated areas. Marketing orders have been least successful for commodities produced by large numbers of geographically scattered producers.

Nor do orders, except for milk, guarantee producers a price or income. They influence, but are not the only determinant of, prices. Also, orders involve a measure of coercion of the majority over the minority, and this is distasteful to many. Their compulsory nature results from a concern with the free-rider problem in agriculture.

The most important limitation of marketing orders, however, is their lack of control over agricultural production. Marketing orders cannot "cure" overproduction or surplus production. Indeed, to the extent that orders are successful in improving farm prices and incomes, they encourage entry into agriculture and supply expansion. Other farm and marketing programs directed at supply control are thus necessary to ensure the effectiveness of orders and agreements.

Food marketing firms are also influenced by market orders. Farm product buyers lose a degree of market freedom because orders shift some marketing decisions to farmers, and marketing firms are thus obliged to follow the provisions of an order. In addition, food marketing firms' costs, sales, and prices are influenced by supply-restricting order provisions.

There is debate about the extent to which market orders and agreements have affected consumers. Without production controls and price-setting authority, it is unlikely that nonmilk orders could significantly raise consumer prices over the long run. Price and supply stability might be of some benefit to consumers so long as it was not accompanied by higher food prices. The severity of the impact of price discrimination on consumers depends on their willingness and ability to make product substitutions over time, form, and spatial markets. In each of these respects, milk marketing orders appear to have had a greater impact on consumer prices and supplies than nonmilk orders. Milk marketing orders are discussed further in Chapter 24.

Not all producers look favorably on marketing orders, and they seldom have 100 percent grower support. There have been several instances in recent years where producers and market handlers have successfully challenged marketing orders in court. Courts have also struck down the supply control provisions of some marketing orders.

Trends in agricultural production and marketing are challenging the need for marketing orders today. They were developed during a period when small farmers produced undifferentiated products for open markets. Recent changes in specialty product agriculture, including increasing sizes of production units, grower integration into marketing, much larger buyers, production-to-specification, and contract marketing, may one day make orders less necessary.

COOPERATIVE BARGAINING ASSOCIATIONS

Agricultural producers may also attempt to improve their market and bargaining power through the formation of bargaining associations. These cooperatives constitute a horizontal integration of producers and are organized to act as bargaining agents for farmers. Their principal purpose is to influence producer terms of trade through contractual negotiations with the buyers of farm products. Farmer bargaining associations are often compared to collective bargaining arrangements between labor unions and management. However, there are important differences between wage bargaining and farm product price bargaining. The tendency toward decentralization, integration, and increasing size of food marketing firms has resulted in growing interest in both formal and informal farmer bargaining efforts.

Bargaining associations frequently operate in conjunction with farm marketing cooperatives and marketing orders. The major difference between the cooperative bargaining association and other marketing cooperatives is that the associations normally confine their activities to contract negotiations, whereas marketing cooperatives usually perform a wider range of services and functions for their patrons. The difference between cooperative bargaining associations and the market order approach is that the former is a voluntary organization of producers.

There are numerous fruit and vegetable grower bargaining associations. In addition, the National Farmers' Organization and the American Agricultural Marketing Association engage in collective bargaining for their members involving such commodities as fruits, vegetables, popcorn, and livestock.

Producer bargaining associations are most highly developed in milk markets, where bargaining is closely integrated with marketing cooperatives and federal and state marketing orders. Using both opponent-gain and opponent-pain bargaining tactics with processors of milk, producer milk cooperative/bargaining associations often have been successful in securing prices above the minimum levels specified in milk marketing orders.

Producer-Bargaining Legislation

Agricultural producer-bargaining associations are authorized by the 1922 Capper-Volstead Act, which permits farmers to market collectively through a single agent without violating the antitrust laws. This law has an important stipulation that cooperative membership must be voluntary, a requirement that many feel limits the effectiveness of cooperative bargaining associations.

The agricultural marketing order and agreement legislation of the 1930s reinforced the concept of farmer bargaining associations. Public support for farmer bargaining was further demonstrated in the Agricultural Fair Practices Act of 1968. This act prohibits food marketing firms from discriminating against producers who participate in bargaining activities or stopping producers from joining bargaining associations.

Many observers feel that additional laws and public support may be needed in order for agricultural bargaining associations to achieve their full potential. Several such bills were introduced into Congress in recent years, collectively known as the

National Agricultural Bargaining Acts. Their provisions varied but in general they called for a National Agricultural Bargaining Board to oversee, facilitate, and mediate producer–buyer bargaining relations; some form of farm production controls; a broadening of the list of commodities eligible for marketing orders; producer authority to set minimum prices for commodities other than milk; and exclusive farmer bargaining agencies, with complete bargaining rights for all production, as well as power to require all producers to sell through the agency. None of these bills has passed and there is much debate on the need for and the wisdom of these additional producer market power tools.

Problems and Potentials of Agricultural Bargaining

Agricultural cooperative bargaining associations can influence food markets in many ways. Clearly, they have increased the role of farmers in the marketing and pricing processes. Although bargaining associations are not free to set any price they wish, the bargaining association does remove farmers from a pure price-taking status. These associations have also provided members with considerable market intelligence and understanding.

Through opponent–gain tactics, bargaining associations may well have increased operational efficiency in the food industry by encouraging cost-reducing and mutually advantageous marketing arrangements. Through opponent–pain and third-party-gain bargaining tactics, these associations also may have altered the distribution of the consumer's food dollar between producers and food marketing firms and, possibly, increased consumer food prices in the short run.

The market and bargaining power that bargaining associations can provide farmers have several limitations.. These associations do not control agricultural supply. Most associations bargain for terms of trade and leave production decisions to members and to the buyers of farm products. Thus, if the association is successful in improving member prices and returns, it will also encourage increased production—which, over time, will build downward price pressures. Processor integration into agricultural production also complicates the bargaining association's supply-control problem.

The effectiveness of a bargaining association depends on its having control over a substantial volume of the product to be marketed. This is particularly important where witholding threats or strikes are part of the bargaining strategy. The voluntary nature of cooperatives may prevent the co-op from gaining control over a sufficient volume of product to bargain effectively. Some producers will choose not to join the bargaining effort, because they object to giving up pricing decisions to the group, or perhaps because they are motivated by free-rider incentives. These considerations have favored the formation of bargaining associations for commodities that are produced by a relatively small number of geographically concentrated farmers.

Bargaining associations also need alternative marketing outlets to facilitate control of product flows to market. Some associations have vertically integrated into processing activities in order to provide alternative markets if the desired terms of trade are not achieved. Storage facilities serve the same purpose. Marketing order provisions also can assist associations in controlling market flows to achieve the association's goals.

Finally, the self-discipline of members and the skill of the bargaining association leaders are important to the bargaining effort. Bargaining frequently requires some sacrifice on the part of members. Leadership, group cohesion, and the financial ability to make short-term sacrifices for long-term gains are necessary. And the members as well as the leaders should have realistic goals for the bargaining process. Bargaining is a useful marketing tool, not a panacea for all price problems. In particular, bargaining association members and leaders must understand and appreciate both opponent-gain and opponent-pain bargaining power.

MARKETING BOARDS

Marketing boards are powerful marketing agencies with the combined powers of cooperatives, marketing orders, and bargaining associations. They are quite common in Canada and are also operating in Australia, France, England, and Africa.

Marketing boards are essentially producer-controlled organizations with government monopoly power over a broad range of farm production and marketing activities. The chief functions of operating boards are:

1. Collective bargaining and price negotiation, acting as a single agent for all producers of the commodity.
2. Sole marketing agent for the commodity, with broad controls over all aspects of marketing, including the ownership of storage facilities.
3. Sponsorship of market-intelligence activities and market research.
4. Producer pooling arrangements to divide receipts among farmers.
5. Setting production and marketing controls and quotas.

Although there are not yet any operating marketing boards in the United States, and farmers have shown mixed interest in them, in 1966 the National Commission of Food Marketing recommended legislation to authorize food marketing boards as a useful tool for improving farmer bargaining power.

SUMMARY

Market power is the ability to advantageously influence market outcomes in one's own favor. Bargaining power is concerned with influence over the terms of trade in negotiations. The balance of market power in the food industry is a subject of continuing concern. Producers, food marketing firms, and consumers all have a stake in the distribution and impacts of food marketing power. Three sources of gains from bargaining power are the opponent's loss, the opponent's gain, and third parties to the transaction. Several market conditions are associated with market power positions, including sizes and number of firms, supply control, access to information, product differentiation, and financial resources. Agricultural cooperatives, marketing orders and agreements, and bargaining associations are three complementary farmer market power tools. Each can contribute to reducing the farm market power problem through orderly marketing and greater involvement of farmers in the marketing and pricing processes. The success of these marketing tools depends on the organizational strength of farmers, the control over supply that they provide, and the skill with which farmers use these tools.

KEY TERMS AND CONCEPTS

Agricultural Fair Practices Act of 1968
bargaining association
bargaining power
collective bargaining
countervailing power
economic price discrimination
exclusive agency bargaining
flow-to-market market allotment
market order and agreement
market power
marketing board
National Agricultural Bargaining Acts
opponent-pain, opponent-gain power
orderly marketing

DISCUSSION QUESTIONS

1. Explain why control of agricultural output is so important to the long-run success of a bargaining association whereas control of flow to market is so crucial to associations' short-term success.

2. Give an example of opponent-gain and opponent-pain bargaining tactics with which you are familiar.

3. Is the present legislation adequate to achieve equality of bargaining power between producers and buyers of farm products? If not, what additional authority is needed?

4. In what ways is agricultural bargaining different from, and similar to, the collective bargaining of labor unions?

5. Which approach do you favor to solve the farm marketing power problem: (1) reduce the power of farm product buyers; (2) increase the power of farmers? What difference does it make which approach is taken? Which approach is being used today?

6. How would you identify and measure market power in the food industry?

7. How is the consumer interest affected by each of the market power tools discussed in this chapter?

8. Should farmers be required to bargain and market their commodities through a central farmer-controlled agency? What percentage of farmers should agree to this approach in order to make it mandatory for all?

9. Why haven't marketing orders and bargaining associations been more widely used for livestock and grain products?

10. "The price of farm products is determined by supply and demand conditions so farmer bargaining power can have no influence on prices." Comment.

SELECTED REFERENCES

Agricultural Marketing Service website (October 2000). U.S. Department of Agriculture. http://www.ams.usda.gov/.

Brandow, G. (February 1969). Market power and its sources in the food industry. *American Journal of Agricultural Economics*, 1–12.

Powers, N. J. (March 1990). *Federal Marketing Orders for Horticultural Crops,* Agricultural Information Bulletin No. 590. Washington, DC: U.S. Department of Agriculture.

Roy, E. P. (1970). *Collective Bargaining in Agriculture.* Danville, IL: The Interstate Press.

CHAPTER 16

MARKET INFORMATION

The next best thing to knowing something is knowing where to find it.

OBJECTIVES

After reading this chapter, you will be able to:

1. Understand why market information is so important to the functioning of the food system.
2. Distinguish between the different types of food marketing information.
3. Appreciate the problems of collecting and reporting accurate market information.

Market information is a facilitating marketing function, and market intelligence is essential to a smooth, efficiently operating marketing system. Accurate and timely market information facilitates marketing decisions, regulates the competitive market processes, and lubricates the marketing machinery.

Market news, information, and research are the *lifeblood* of markets. Market information agencies take the pulse of the market (are sales active or sluggish?), measure the temperature of markets (are prices rising or falling?), and monitor the market's pressure (are supplies adequate, short, or in glut?). The market's history is recorded in statistical data series, and numerous agencies offer a prognosis or estimate of the market's future health.

Everyone who produces, buys, and sells agricultural products is continuously amassing, revising, and using market information on prices, supplies, demand, and other market conditions. In addition, there are numerous public and private agencies specializing in food marketing information and research.

ROLES OF MARKET INFORMATION

The key purpose of market information is to improve decision making. Farmers use market information when selecting enterprises; changing production plans; making long-term investments; and deciding the when, where, and how of their marketing strategies. Food marketing firms, farmer cooperatives, farm organizations, and legislators also depend on market information for good decision making.

The role of market information is also important in the competitive market processes that regulate product flows and prices in the food industry. Although the perfectly competitive requirement of perfect information is unattainable, in the competitive process more information is better than less. Information is, accordingly, critical to the law of one price and to the price discovery process.

Although it is not widely recognized, market information also contributes to operational efficiency in the food industry. Without the widespread availability of market information, buyers and sellers would need to devote considerably more time and money to market search activities than they currently do. The value of information is evident in markets where firms will pay a high price to specialized agencies for profitable information.

PUBLIC AND PRIVATE FOOD MARKET INFORMATION

There are both publicly and privately sponsored sources of market information in the food industry. Much information is collected by businesses from their own market activities or is purchased from private market news and research companies. Large food companies maintain their own market analysis and research staffs to provide constant market intelligence. This information is usually not published, nor is it available to those outside the firm.

Some private market research companies in the food industry do publish their findings in addition to selling information to clients. The *Leslie Report,* for example, is a monthly crop estimate released just before the monthly U.S. Department of Agriculture crop estimates are made. The "Yellow Sheet," published by the *National Provisioner,* is a daily livestock price reporting service sold by subscription and widely used throughout the meat industry. The *Urner-Barry Report* is a similar daily price report for agricultural products, notably eggs.

Trade associations also compile and frequently publish market information about the food industry. Typically, these associations secure information from individual members on a confidential basis and then release summary averages. Organizations such as the American Meat Institute, the National Canners Association, the National Grocery Manufacturers of America, the Food Marketing Institute, and the American Dairy Association carry on these informational activities.

Other proprietary firms specialize in market research for a fee from clients but do not publish their findings. They may evaluate the market potential for a new product, study how an older product may be revitalized, investigate overseas trade opportunities, design packaging, merchandising, and promotional programs, or perform a number of other services.

Public agencies play a major role in collecting, analyzing, and disseminating U.S. agricultural and food marketing information. The U.S. Department of Agriculture is the most important but not the only one of these agencies. Although there is a trend toward charging for these government publications, a substantial public subsidy is involved in these federal, state, and local market information and news programs. Why should the public support a market information program for farmers and food marketing firms? There are two principal justifications for this public support. First, although farmers and food marketing firms are the direct beneficiaries of the programs, ultimately there are benefits to the consumer as a result of increased market efficiency and enhanced competition. Second, information has been considered a market *equalizer,* which strengthens the farmers' bargaining power when dealing with food marketing firms.

CRITERIA FOR EVALUATING MARKET INFORMATION

To be of maximum benefit, market news and information must meet a number of criteria. First, information must be complete and comprehensive. This is a difficult task in a large country with a geographically dispersed agricultural plant producing more than 200 different farm products and a food marketing system handling more than 15,000 different supermarket items. Food market conditions also change frequently, further adding to the difficulty of providing complete market information.

A reasonably complete description of a food market includes prices, price trends, production, supply movements, stocks, and demand conditions at each level of the market. Food industry decision makers benefit from market outlook assessments and forecasts of future market conditions. Farmers also frequently find it helpful to learn of other farmers' production and marketing intentions. Providing such a mass of information, especially under the constantly changing conditions of food markets, is a formidable and expensive task. As a consequence, there are many gaps in food market news and information.

Accuracy and trustworthiness are also necessary criteria for market information. The credibility of the U.S. Department of Agriculture's market information reports and news services are highly protected assets. By its nature, market information can never be 100 percent accurate, but an honest market appraisal is necessary to earn the trust of information users. Considerable and constant efforts are made to improve the accuracy of market food information and news services.

Information also must be relevant and in usable form. It is not enough to simply collect a mass of numbers and report them. Information must be collected, packaged, and disseminated with the user's interests in mind. Much market information goes unused because it is not in usable, easily accessible form.

Confidentiality is important to the U.S. Department of Agriculture market information and news services. Market prices and supply reports of individual firms are aggregated to provide a general picture of the market without revealing information about any one firm. This protection of sources has proven essential to programs that rely on voluntary reporting of market conditions. Also, care is taken to insure that everyone receives the information simultaneously, so that no one trader can benefit from information that others do not have.

Market information also must be timely, in the sense of being relevant to current decisions, and must be speedily transmitted to users. Much market information is highly perishable. Orbiting earth satellites, computers, and rapid telecommunications contribute to market information timeliness. The immediacy of information needs varies throughout the food system. Futures market traders require minute-to-minute information; daily reports suffice for other traders; and monthly or annual reports are sufficient in other cases.

Finally, it is desirable to have a balance of market information at all levels of the food industry. Each marketing agency should have equal access to all the information relevant to the bargaining and marketing processes.

The importance of these information criteria has become evident in other countries that are developing market economies and are in the early stages of collecting and providing market information.

U.S. DEPARTMENT OF AGRICULTURE INFORMATION PROGRAMS

Although the Departments of Commerce and Labor issue some reports of interest to farmers, food firms, and consumers, in this book we emphasize the U.S. Department of Agriculture's market information services. There are a large number and variety of these programs. The USDA information programs can be divided into several categories: market news reports, market situation reports, outlook and forecasting services, statistical reports, and research reports.

The Federal-State Market News Service

Market news consists of descriptive information on current market conditions, such as prices, supplies, stocks, and demand. This information can be used directly in market decisions. Market news quickly becomes obsolete and requires frequent updating—often hourly.

Prior to the establishment of the USDA market news service, rumor and hearsay were important factors in agricultural markets. Allegedly, farm prices could be "talked down" by an appropriate and well-placed rumor. There was no third-party check against these rumors. The more affluent could afford more information and therefore could take advantage of the less knowledgeable.

The first USDA market news report covered the movement and prices of strawberries at Hammond, Louisiana, in 1915. USDA market news services were begun for meat in 1917, for livestock, dairy products, poultry, hay, grain, seeds, and feeds in 1918, for cotton in 1919, and for tobacco in 1931.

The USDA has separate market news divisions for each of the major products—livestock, dairy products, poultry, grain, fruits and vegetables, cotton, and tobacco. In addition, most states cooperate with the USDA in collecting, disseminating, and sharing the costs of the public market news service.

Market news reporters stationed at the various important markets and production areas of the country collect much of the data for the market news reports. These people are specialists in their individual fields. The market information must be collected from the various individuals and agencies actively marketing the product. The information that agencies give might be incorrect or biased. The news reporter must be thoroughly acquainted with the market and the product so as to appraise the accuracy of the information obtained. For example, a livestock news reporter must be able to grade livestock and be thoroughly acquainted with the methods of the livestock trade. The same is true for reporters of the other commodities.

The federal-state market news service, administered by the Agricultural Marketing Service of the USDA in cooperation with state agencies, employs numerous market reporters who monitor some 450 commodities at more than 1700 markets. The locations of these market news offices are shown in Figure 16-1. These offices are tied together with telecommunications and computer systems by which information can be exchanged almost instantaneously with all other points in the country. Messages placed on the wire at any one point reach all other points on the circuit and, when desired, can be relayed to offices on other circuits. Daily, weekly, and monthly market news reports are available from selected markets. These may be received by mail, telephone, broadcast, Internet, or print media. Direct teletype

FIGURE 16–1 Location of market news offices, 1996. (U.S. Department of Agriculture.)

connection to the market news service, through leased telephone wires, is available for a fee.

Several states have automatic (800 or 900) telephone-answering services that provide instant market news reports around the clock. These are listed in the telephone book under state or federal offices.

Market news reports have a special vocabulary for describing market conditions. *Market tone* refers to the general attitude of buyers and sellers toward prevailing prices; *market supplies* means all stocks currently in trading positions; and *demand* refers to the immediate desire, ability, and willingness of buyers to purchase at current prices. *Bid prices* are buyers' offers, and *offer prices* are sellers' asking prices. *Spot prices* are current cash prices, and *futures prices* are the current prices at which commodities are traded for future delivery.

USDA National Agricultural Statistics Service

The National Agricultural Statistics Service (NASS), formerly the USDA Statistical Reporting Service, is the chief fact-gathering arm of the USDA. Its mission is to report the basic statistical facts of the nation's farm and food industry. It releases some 300 national and 9,000 state reports each year and covers 120 crops and 45 livestock items, as well as farm labor, fertilizers, seeds, prices paid and received by farmers, and other information. In addition to the Washington, DC headquarters, there are 44 state statistical offices.

NASS is perhaps best known for its surveys of farmer production intentions and for periodic crop and livestock estimates during the year. The agency also collects and distributes information on farm prices paid and received, farm costs and returns, farm labor, and agricultural product international trade.

A whole series of reports covers the development of a particular crop. Prior to planting time there are reports of farmers' intentions to plant, which are followed later by reports on the number of acres actually planted. Then, throughout the growing season, there are progress reports and estimates made of the potential size of the crop. In this manner, long before the harvest is ready for market, marketing agencies and producers are well aware of what production volume to expect and can make the necessary plans and arrangements for handling it.

The procedures followed in securing, preparing, and releasing data are elaborate. The questionnaire, which has been designed to secure the needed information, is sent out from the various states to their list of cooperating reporters. When the questionnaires are returned to the state offices, they are checked, edited, and tabulated. The state statistician may even do some spot-checking in the field to verify the information. The final forms are then sent by special delivery to Washington, DC. Here the opening, tabulation, and releasing of the compiled data are done with the greatest of care to prevent information leaks. The following description of the final stages of this procedure will illustrate the great care used:

> On the morning of the release of a report, the chairman of the board, the secretary of the board (who has the key for one lock), one other board member, and a representative of the Secretary of Agriculture (who has the key for the other lock), go to the locked room accompanied by an armed guard. There they unlock the mailbox and take

the state reports to another room. The night before, the venetian blinds on all windows within this corridor have been lowered, closed, and sealed. No one may open or even adjust these blinds while the board is in session. All telephones in the wing have been disconnected, the door at the other end of the corridor is locked, and a guard is on duty outside. These are called the lock-up quarters.

When forecasts or estimates have been adopted for all states for a given crop, they are handed to the computing unit. The tables and stencils are prepared. Mimeograph machines are brought into the quarters the night before so that the report can be processed inside the locked corridor.

After approval by the Secretary of Agriculture, two or three minutes before the release time, the chairman and secretary of the board leave the lock-up quarters and proceed under guard to the release room, looking neither to right nor left and speaking to no one nor acknowledging any greeting, according to regulations. In the release room telephone and telegraph instruments are already connected, and the operators are assembled in a prescribed space, out of reach of the instruments.

The chairman places one report face down beside each telephone and telegraph instrument. At the precise time, a representative of the Secretary of Agriculture says "go" and the various reporters rush to their instruments and begin sending out the reports.

Also at this time the information is sent to the various state offices for their release.

This elaborate procedure is testimony to the value of the information. Estimates of the size of crops and other important information are immediately used in the buying and selling of products. An early information leak might be used by an individual for his personal advantage. To illustrate, suppose the last estimate of the wheat crop had been for a crop of 2 billion bushels. The new estimate to be released, however, will reestimate the production figure to be 1.5 billion bushels. Such a predicted reduction in the crop would no doubt at least temporarily strengthen prices. Someone with advance knowledge could have purchased wheat and profited by the short-run fluctuation.[1]

USDA Economic Research Service

The Economic Research Service (ERS) is the analytical and interpretive arm of the USDA. Its market situation reports on farm costs and efficiency, land and labor markets, food consumption, and foreign trade attempts to interpret and appraise the meaning of market conditions and developments for farmers, food marketing firms, and consumers. ERS outlook reports forecast future prices and market developments. ERS research reports present economic research findings and relationships or trends that will foster a better understanding of food markets.

USDA Agricultural Marketing Service

This agency provides marketing services to producers, manufacturers, distributors, importers, exporters, and consumers of food products. In addition to its market news service, AMS, in cooperation with industry, develops and maintains food quality standards and supervises food quality grading. The agency also manages the

[1]U.S. Department of Agriculture. (December 1949). *The Agricultural Estimating and Reporting Services of the USDA,* Miscellaneous Publication No. 703, p. 13. Washington, DC: U.S. Department of Agriculture.

USDA's commodity purchase program, which provides nutritional assistance for consumers. AMS supervises the Federal promotion and research programs authorized by marketing orders and agreements. The agency also administers the Wholesale Market Development Program, the Perishable Agricultural Commodities Act, the Federal Seed Act and the Plant Variety Protection Act.

USDA World Agricultural Outlook Board

This office has responsibility for monitoring world food supply and stocks. It issues periodic reports on world food conditions and prospects. It provides long-term world commodity projections of wheat, rice, sugar, corn, soybeans, meats, eggs, and cotton.

FOOD MARKETING ON THE INTERNET

The World Wide Web is fast becoming an important farm and food marketing tool. In 1999, 15 percent of the nation's farms bought or sold products on the Internet, and perhaps 2 percent of agricultural product sales were made online in 2000. The Internet is being used to both market farm and food products to consumers (business-to-consumer sales) and in business-to-business marketing. Today's farmer is using the Internet to purchase farm supplies and to find buyers for farm products. Marketing firms, in turn, use the Internet to secure supplies and to sell their products.

The Internet provides several advantages for buyers and sellers and can foster more efficient and competitive markets in several ways. First, it connects numerous buyers and sellers worldwide, greatly expanding the potential numbers of buyers and sellers and making a truly global marketing system possible. Second, it is a very efficient way for buyers and sellers to transact business, considerably lowering search and transaction costs and overall marketing expenses. Third, it reduces barriers to entry of smaller firms and allows them to compete with larger firms on a more level playing field.

Some food marketing websites specialize in providing marketing news and information, while others conduct e-commerce business. Many do both. Some selected sites to visit include:

- AgWeb.com (news, market reports, weather, advice)
- @Agriculture Online (ag news, weather, market information)
- CyberCrop.com (an online cash grain exchange)
- DirectAg.com (sales of seeds, fertilizers, chemicals, animal health products and farm equipment)
- Farmbid.com (online auctions of farm equipment, livestock, crops)
- Foodonline.com (online sales to food consumers)
- Foodtrader.com (worldwide sales of commodities and food products)
- Peapod.com (retail grocery sales online)
- Powerfarm.com (marketing of crop supplies, crop insurace, agricultural equipment)

PROBLEMS OF MARKET NEWS AND INFORMATION

There are several problems regarding the collection, compilation, and dissemination of food market information that should be taken into account by users of the information.

Price Specification

The statement that cattle are selling for $60 per hundredweight is not very useful until other, more specific information is provided, for example: Where? When? What grade? What weight? How? These specifications are necessary to make a price quotation meaningful for decision makers. Price has meaning only with reference to a particular time, form, and geographic market.

The wide range of qualities and uses of agricultural products adds to the food market information problem. The usefulness of price quotations depends on uniform acceptance and application of grades and standards. For example, fresh fruit price quotations cannot be mixed with processed fruit prices. Feeder livestock conditions need to be reported separately from slaughter livestock conditions. Market information programs must be continuously revised to reflect buyer and seller needs as well as changing trade practices.

Net versus Gross Price

Another complicating factor in market information programs is that publicly quoted figures are frequently not the actual price at which commodities are traded. Premium and discount schedules vary from place to place and from buyer to buyer. Some farm prices include allowances for marketing costs such as hauling, packaging, and other marketing activities. Because industry practices vary, making comparable price comparisons is very difficult.

Studies that report significantly higher (or lower) prices in one market as compared with another sometimes neglect the fact that buyer and seller responsibilities differ. Frequently, when all marketing and other costs are considered, net prices among competing markets are more similar than they at first appear to be.

The price comparability problem also arises at the retail level of food markets. Food prices at a cash-and-carry, low-service grocery store should not be compared, superficially, to prices at a full-service supermarket. The product bundles differ at these stores, and a simple gross price comparison fails to take these differences into account.

Information Costs

The costs of gathering and disseminating market information to the public requires that some difficult choices be made. The value of more complete and more accurate information must be weighed against its costs.

Market information is not available for all commodities and is somewhat incomplete for all products. Thus, there are continuing requests for more information. One group would like to see a market information program for riding horses or timber prices. Such requests led to the establishment of a USDA mink market news service in 1970. As new programs are added, old ones are dropped. USDA buckwheat reports were discontinued in 1964 because of the small amount of acreage involved. Should the USDA continue its molasses, hops, and naval stores market reports?

The same amount of information is not available for all levels of the food industry. In general, much more is known about supplies, demand, and marketing at the farm level than at the retail level. Also, more complete information is available for products that are marketed principally through organized, central markets than for those moving through decentralized, integrated marketing channels.

Some sacrifice in accuracy and timeliness probably results from a broadening of the range of commodities covered by USDA market news. For example, as their markets become more geographically dispersed and their commodity responsibilities broadened, market reporters rely more on telephone conversations with buyers and sellers than on personal market observation.

Changing Market Organization

Trends in farming and marketing have also complicated the food market information task. Because of decentralized, direct sales, products now bypass the central terminal markets, where at one time price reporters could fairly easily take the pulse of markets. Market reporters in these days must obtain information from more numerous and more geographically dispersed shipping-point markets, and the cost of obtaining accurate and complete information has risen accordingly.

Changing transportation patterns also alter the market information picture. At one time, railroad freight movement data adequately represented the supply flow of most farm products. Over time, the proportion of products moving by truck, barge, and even air has increased, and market information programs have had to adjust to these trends.

Contractual and ownership integration in farm and food markets present still another problem. Farmers who sign contracts specifying prices and other terms of trade in advance of product delivery need market information different from that needed by farmers who sell on the cash markets at harvest time. In addition, vertical integration of markets tends to "close up" the marketing process and reduce cash price trading. Commodities change hands on the basis of prearranged agreements without being priced in open, organized markets. This adds to the difficulty of assessing market conditions, because much less of the commodity is moving through the channels monitored by the public market information programs.

Voluntary Cooperation

The USDA and the private market news and information programs depend on the voluntary cooperation of buyers and sellers to report prices, supplies, and other market conditions. Because there is no mandatory requirement that they provide this information, many farmers and food marketing firms do not participate in the programs. Statistical techniques can compensate somewhat for the missing information and the resulting statistical bias. However, many observers feel that the voluntary nature of market information programs continues to be one of their principal limitations.

Voluntary reporting presents many problems. For instance, buyers and sellers may quote only those prices and conditions favorable to themselves. Or they may choose to quote "asking" or "bid" prices rather than actual selling prices. That the public market information system is run on a voluntary basis does not mean that it is easily manipulated or that there is fraud, but it does complicate the food marketing information task. Of course, attempts to correct this situation through mandatory market reporting would pose different, but equally difficult, problems—those having to do with buyers' and sellers' privacy rights.

CRITICISMS OF MARKET INFORMATION PROGRAMS

Despite the value of food marketing information programs, these services have been severely criticized from time to time, especially by farmers. Typical criticisms are (1) the forecasts are usually inaccurate; (2) market reports inevitably depress farm prices; (3) market information is of greater value to buyers of farm products than to farmers; and (4) market reports are manipulated.

Many farmers have stated that the USDA crop and livestock intentions reports (intentions to plant crops, to produce a litter of pigs, or feed cattle) are of little value because they frequently prove to be wrong. There have been cases where predicted production varied considerably from actual production, because of inaccurate field reporting, sampling errors, and other factors. Such situations make a persuasive case for improving the methods of collecting market information and estimating future market conditions.

However, even if these sources of error were eliminated, intentions reports would never forecast with perfect accuracy the future of farm markets. In fact, intentions reports are not intended to be forecasts. It is obvious that farmers may change their intentions after learning what other farmers intend to produce. Although this makes the intentions reports "wrong" when measured against actual production, the reports nevertheless contribute to farmers' decisions. Weather and other unexpected events occurring between intentions to plant and harvest time may also make intentions reports appear to be inaccurate.

Farmers also have frequently charged that farm supply reports often depress farm prices. In evaluating this criticism it must be remembered that the purpose of market information is to inform traders and farmers, not to drive prices either up or down. Prices, of course, would be expected to move opposite to changes in reported supplies. Researchers have found that, immediately following USDA market reports, crop and hog prices rise just as frequently as they fall, and that prices *after* the reports are issued often continue in the same direction they had been moving prior to the reports.

Farmers have also expressed the belief that USDA market news and information benefits the buyers of farm products more than it benefits producers. There is some truth to this, because food marketing firms are generally more flexible and can more easily adjust to market conditions than can farmers. However, this does not make a persuasive case against farmers' participation in and support for USDA information services. Farmers *can* use this information in their decisions, and without a public market information program, food marketing firms would have much more market information—and probably even greater bargaining power—than farmers.

Finally, there are persistent charges of bias, misrepresentation, and manipulation of market news and information reports. These allegations are leveled against both private and public market information sources. Again, such charges probably make a stronger case for improving market information services than for discontinuing them.

INFORMATION USERS' RESPONSIBILITIES

The finest information system in the world will contribute little to the efficient working of the marketing system unless it is wisely used by those wishing to buy or sell products. Farmers, for example, must seek out and compare the information available for different outlets if they are to sell to the best advantage. The same is true of homemakers in their buying activities. It is probably an unfortunate truth that our informational system is only haphazardly used. Study after study has indicated that opportunities exist to make money by becoming better informed.

There are many indications that in the future individuals will have to assume more responsibility in collecting and evaluating their own market information. As marketing becomes more decentralized and integrated and nonprice competition increases, centralized market news agencies will furnish less meaningful information. Selling agricultural products becomes more like buying a household appliance. The only way to find out about the market may be to shop around, compare, and then choose!

An educational program for the users of the information is also needed. Only in this way will data be used for the purposes for which they were intended and nothing more. The use of price ranges instead of single specific prices is an example of the misunderstanding of information. In many markets, because of the number and variation of the transactions, the reporting of a specific price is not possible. Only a price range represents the true picture. However, some of the users of the information desire a single quotation and criticize the use of price ranges. This demonstrates the misunderstanding of the mechanics of the marketing and price-making processes. All things considered, securing adequate dissemination and wise use of information present problems no less difficult than those involved in data collection.

SUMMARY

Information and market intelligence are critically important to the efficient functioning of farm and food markets. As a facilitating marketing function, market information improves the decision making of firms and enhances the competitive processes in food markets. Market information must be comprehensive, accurate, trustworthy, confidential, timely, understandable, and fairly distributed throughout the marketplace. The U.S. food industry has an elaborate network of public and private market information services. The U.S. Department of Agriculture provides food market news, market outlook and forecast reports, market situation and analysis reports, and market research and statistical reports. These services have their limitations and are the subject of frequent criticism. Nevertheless, the USDA information programs contribute to market efficiency and a better balance of information between farmers and the buyers of farm products. Market information programs must continually adjust to market trends and user needs.

KEY TERMS AND CONCEPTS

bid and offer prices
demand and supply conditions
market information
market intelligence
market news
market research
market situation reports
market tone
outlook
spot and futures prices

DISCUSSION QUESTIONS

1. Discuss the rationales given for public support of food marketing information programs. Do you see any trends that might alter this public support?

2. Contrast advertising (as a form of market information for consumers) with the market information available to farmers and food marketing firms.

3. It is frequently observed that farm prices change very little after the release of a USDA estimate that greatly differs from previous estimates. How can this be explained by the growth in private market news services?

4. "Farmer planting intentions reports in the spring always turn out to be wrong at harvest time, so they shouldn't be published." Comment.

5. What revisions in the procedures of market news collecting and reporting are made necessary by decentralization and integration of food markets?

6. If you were a farmer, what would determine your willingness to supply accurate information to government reporters?

7. How could a consumer make use of the USDA food market information programs?

SELECTED REFERENCES

Agriculture Network Information Center homepage. (October 2000). http://www.agnic.org/.

Agricultural Marketing Service homepage. (October 2000). U.S. Department of Agriculture. http://www.ams.usda.gov/.

Aldrich, L. (July 1999). *Consumer Use of Information: Implications for Food Policy.* U.S. Department of Agriculture. Agricultural Handbook No. 715.

Economic Research Service homepage. (October 2000). U.S. Department of Agriculture. http://www.ers.usda.gov/.

Information Policy Briefing Room. (January 2001). Economic Research Service, U.S. Department of Agriculture. www.ers.usda.gov/briefing/InformationPolicy/

U.S. Department of Agriculture. (1987). Crop and livestock estimates. In *Major Statistical Series of the U.S. Department of Agriculture,* Vol. 7, Agricultural Handbook No. 671. Washington, DC: U.S. Department of Agriculture.

U.S. Department of Agriculture. (1987). Market news. *Major Statistical Series of the U.S. Department of Agriculture,* Vol. 9, Agricultural Handbook No. 671. Washington, DC: U.S. Department of Agriculture.

World Agricultural Outlook Board homepage. (October 2000). U.S. Department of Agriculture (October 2000).
http://www.usda.gov/oce/waob/wasde/wasde.html.

CHAPTER 17

STANDARDIZATION AND GRADING

OBJECTIVES

After reading this chapter, you will be able to:

1. Explain why we have food grades and standards.
2. Discuss some of the problems associated with developing and implementing a food grading system.
3. Suggest and evaluate recommendations for improving our food grading system.

The sorting of farm and food products into standardized grades is a facilitating marketing function. *Grades and standards* constitute an agreed-upon market language that can greatly simplify the marketing processes and reduce marketing costs. Product grades and standards also furnish an ethical basis for buying and selling. Without some market standards, the rule of *caveat emptor* ("let the buyer beware") prevails, along with confusion and unfairness. Indeed, the standardization of market weights, qualities, and practices is so widespread that it is taken for granted, and the role of standardization in the marketing processes is sometimes unappreciated.

Market standardization has potential benefits for everyone in the food industry. Producers of high-quality farm products can obtain price premiums over lower-quality products when they are differentiated by grades. Food marketing firms can better communicate their needs to farmers, lower their search and transaction costs, and better inform consumers of their market offerings when there are grades and standards. And consumers benefit when grades and standards assist in matching their needs and budgets with the heterogeneous supply of products available in the market place. In a larger sense, grades and standards contribute to the competitive market processes.

Farm commodities are often considered *fungible* (one producer's product is a perfect substitute for another's). As a result, these commodities are standardized and government-graded to a greater extent than any other consumer products. Nevertheless, government grades and standards have not been developed for all food products, and those grades and standards that are available are not universally used. In this chapter we will investigate the nature and prevalence of food grades and standards, their roles in the marketing processes, and the problems of food grading.

STANDARDIZATION IN THE FOOD INDUSTRY

Standards are commonly agreed upon yardsticks of measurement. A number of standards affect the food marketing process.

The U.S. Bureau of Weights and Measures supervises the standardization of weights and other measures. The Public Health Service enforces food sanitation standards. The Food and Drug Administration (FDA) sets food processing standards for "good manufacturing practice," regulates food additive levels, and supervises the nutritional labeling of food. FDA food standards of identity also require that products sold under a common name (for example, mayonnaise) meet certain ingredient specifications. The Environmental Protection Agency enforces pesticide residue standards for food, and the Fair Packaging and Labeling Act sets food package and label standards.

The U.S. Department of Agriculture supervises a broad range of food standardization programs. Its food inspection services certify the safety and wholesomeness of farm products. This department also is authorized by the Standard Container Acts of 1916 and 1928 to establish packaging standards for fruits and vegetables. In this chapter emphasis is given to perhaps the best-known U.S. Department of Agriculture standard program—food quality grades. Federal standards for selected meat products are illustrated in Figure 17–1.

FOOD QUALITY GRADES AND STANDARDS

Quality is a subjective property referring to the usefulness, desirability, and value of a food product. Buyers must constantly judge the qualities of foods in their purchase decisions. Consumers may prefer larger strawberries to smaller ones; steaks with a moderate degree of marbling may be preferred to leaner or fatter steak. A corn processor may prefer one corn variety to another for a particular wet-milling purpose. The fact that buyers are willing to back up their preferences for certain products over others with purchasing power is evidence that product qualities differ.

Because of the biological nature of production, agricultural commodities are produced in a wide range of qualities. There are almost an infinite number of combinations of such food properties as taste, aroma, color, tenderness, texture, size, age, shape, and moisture for each food product. The unique way that these and other food properties are combined into a product determines its quality.

Although we often associate quality with an objective scale ranging from inferior to superior, in reality quality is a highly subjective property of foods that is measured in the minds of buyers. Two products may differ in sensory qualities but be equally attractive to consumers, for example, California and Florida oranges, or

Product Name	Standard
Beef stew	Must contain at least 25% beef.
Chili con carne	At least 40% meat.
Chicken soup	Ready to eat — at least 2% chicken meat. Condensed — at least 4% chicken meat.
Frankfurter, bologna, and similar cooked sausage	May contain only skeletal meat. No more than 30% fat, 10% added water, and 2% corn syrup. No more than 15% poultry meat.
Frankfurter or bologna "with by-products" or "with variety meats"	Same limitations on fat, added water, and corn syrup as products without variety meats. Must contain at least 15% skeletal meat, and the terms "variety meats" or "by-products" must be part of the product name and in the ingredients list.
Ham — water added, cooked, or cooked and smoked	Must be from the hind legs of a hog; picnic hams are from the front legs. Both must contain at least 17% meat protein (fat-free).
Hamburger, ground beef, or chopped beef	No more than 30% fat; no extenders or other added substances except seasonings.
Nuggets	Bite-size, solid pieces of meat and poultry. Usually breaded and deep fat fried.
Nuggets, chopped and formed	Meat or poultry chopped and shaped into nuggets. "Chopped and formed" must be part of the product name.
Pizza with sausage	At least 12% cooked sausage or 10% dry sausage, such as pepperoni.
Turkey pie	At least 14% cooked turkey meat.
Turkey ham	Cured turkey thigh meat.

FIGURE 17–1 USDA standards for popular meat and poultry products. (U.S. Department of Agriculture.)

brown- and white-shelled eggs. These are known as horizontal quality differences. Even when there is general agreement on inferior and superior product forms—that is, when there are vertical quality differences—consumers may not always choose the highest quality. For example, a consumer may purchase a different quality of meat for preparing steak than for preparing stew, but both qualities fulfill a need. The purpose of quality standards is not to assure the marketing of only top-quality products but to facilitate the matching of consumer needs with the different qualities of products available.

Food quality standards are commonly accepted properties that differentiate food products in terms of their value to buyers. Food standards may be physiological—for example, the nutritional value of a commodity—but most food standards are sensory (taste, smell, and so on). The standards vary for individual foods. The properties that differentiate high-valued from low-valued steak are quite different from those differentiating potato quality.

Grading refers to the sorting of unlike lots of products into uniform categories, according to quality standards. The appropriateness and accuracy of the grading process depend on the correspondence between the quality standards and buyer and seller preferences, the range of qualities to be sorted, and the relevance of the sorting to consumer choices. Some selected USDA food grades and marks are illustrated in Table 17-1 and Figure 17-2.

TABLE 17-1 Selected USDA Food Grades

		Nomenclature for			
Food Category	*Food Product*	*Top Grade*	*2nd Grade*	*3rd Grade*	*4th Grade*
Dairy	Butter	U.S. Grade AA	U.S. Grade A	U.S. Grade B	
	Cheddar cheese	U.S. Grade AA	U.S. Grade A	U.S. Grade B	U.S. Grade C
	Instant nonfat dry milk	U.S. Extra Grade	U.S. Standard Grade		
Fruits and vegetables					
Fresh:	Cantalopes	U.S. Fancy	U.S. No. 1	U.S. Commercial	U.S. No. 2
	Cucumbers	U.S. Fancy	U.S. Extra No. 1	U.S. No. 1	U.S. No. 2
	Peas	U.S. Fancy	U.S. No. 1		
	Potatoes	U.S. Extra No. 1	U.S. No. 1	U.S. Commerical	U.S. No. 2
	Watermelons	U.S. Fancy	U.S. No. 1	U.S. No. 2	
Processed:	Fruits	U.S. Grade A or U.S. Fancy	U.S. Grade B or U.S. Choice	U.S. Grade C or Standard	U.S. Grade D or Substandard
	Vegetables	U.S. Grade A or U.S. Fancy	U.S. Grade B or U.S. Extra Standard	U.S. Grade C or U.S. Standard	U.S. Grade D or Substandard
Poultry	Poultry	U.S. Grade A	U.S. Grade B	U.S. Grade C	
	Eggs	U.S. Grade AA	U.S. Grade A	U.S. Grade B	
Livestock[1]	Beef	USDA Prime	USDA Choice	USDA Select	USDA Standard

SOURCE: U.S. Government, *Code of Federal Regulations,* 7CFR 46–58, Washington, DC, 1996; http://www.ams.usda.gov/

[1]The beef grades are adjective grades and should not be categorized as ranks.

It is in the area of quality sorting that some of the greatest food standardization problems arise. What should be the criteria for various grades of quality? How many grades should there be? How uniformly interpreted and widely accepted are the standards for grading from one area to another or from one grader to another? What terminology should be used? Should the standards be compulsory or voluntary?

MARKET IMPACTS OF STANDARDIZATION

If widely used, uniform food quality grades and standards can, potentially, contribute to both operational and pricing efficiency in the food industry.

Grades and standards allow marketing by description, rather than by inspection. By making possible the sale of farm products by sample or description, thus generating more accurate market information, the use of uniform food quality grades lowers buyer and seller search and transactions costs and fosters a more efficient price discovery process. Large quantities of farm commodities, for example, are traded in the futures markets on the basis of standardized grades and contracts without the need for buyer and seller personal meetings or inspection of products. How much more expensive the food marketing processes would be without the trust of traders, which results, in part, from the use of standardized grades.

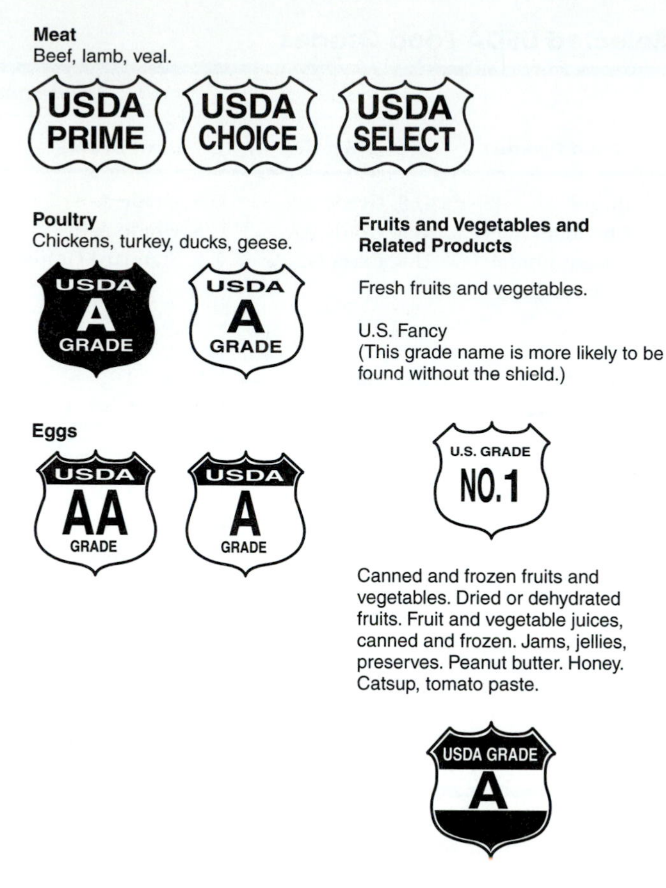

FIGURE 17–2 USDA grade marks found in retail food stores.

Grading also can lower other marketing costs. For example, transportation costs may be reduced by distinguishing between the higher-valued products, to be shipped forward, and the lower-valued products, which can be sold nearer to home. Grading may also reduce market spoilage, separating poorer-quality products from higher-quality ones.

Food grading also can contribute to market competition and pricing efficiency. The product homogeneity resulting from grading can move the market closer to perfect competition, encourage price competition between sellers, and reduce extraordinary profits.

Standardization of quality grades can improve pricing efficiency in two ways. First, the use of grades gives consumers specific information with which to signal their preferences to producers, thus increasing the consumers' sovereignty over the production process. Figure 17-3 shows, for instance, how changes in consumer preferences for a leaner pork product altered the grade distribution of hogs produced and marketed in the 1970s. Second, the use of uniform quality grades provides incentives for producers to adjust to changing consumer preferences. Grade-price differentials reward farmers and marketing firms for shifting their pro-

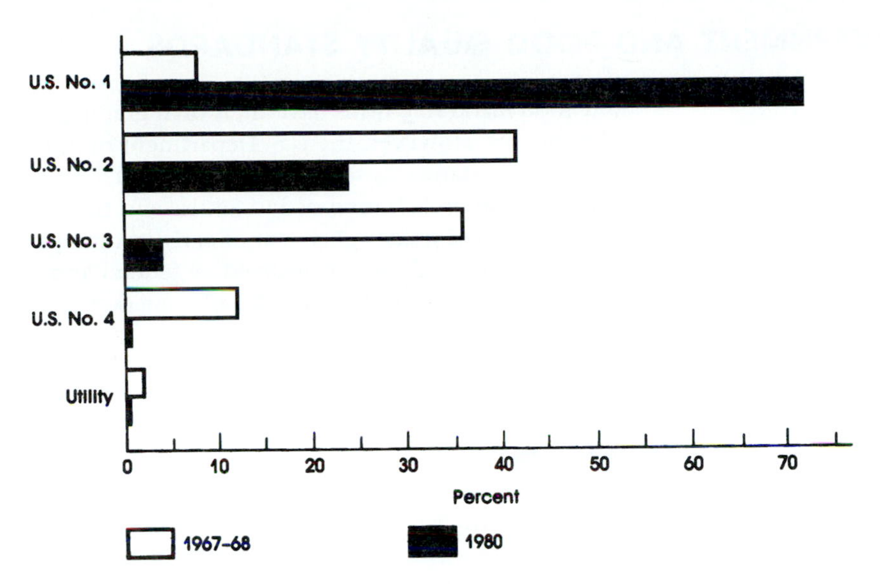

FIGURE 17–3 Grade distribution of hogs marketed, 1967–1968 and 1980. (U.S. Department of Agriculture.)

duction efforts from lower- to higher-valued products. If each producer received the average price of all grades of products, there would be little individual incentive to produce the higher grades. Thus, the use of grades not only provides a standard code that consumers can use to get their message to producers but also stimulates the desired producer response.

Classification by grades also facilitates the cost-reducing concentration processes in food marketing. For instance, without grades, some sort of identifying mark would have to be affixed to a farmer's wheat throughout the market channel in order that the wheat's exact, ultimate value be paid to the farmer. By using the grading system, farm products can be pooled according to grades, making possible lower-cost marketing, without interfering with the communication role of the pricing system.

Uniform grades and standards also can contribute to market development in the form of greater consumer satisfaction and increased producer returns. For example, if by the use of grading there is greater consumer confidence in the product's quality, and a better matching of supplies with consumer wants, the demand curve will shift rightward. At the same time, different prices for each grade of a product can result in a total producer return that would exceed the average price of an ungraded lot of products. This form of price discrimination was discussed in Chapter 15.

Food grades can affect both domestic and foreign sales of farm products. Foreign buyers' concerns with the quality of U.S. grain and alleged fraud and adulteration of grain exported from the United States led to the 1986 Grain Quality Improvement Act. This legislation was designed to improve buyers' confidence in U.S. grain products by providing better information on end-product yield and quality of U.S. grains.

GOVERNMENT AND FOOD QUALITY STANDARDS

Many food marketing firms have their own grading, standardization, and quality control programs. However, the U.S. Department of Agriculture grades are the only universal quality standards for U.S. food products. These grades evolved hand-in-hand with the commercialization of agriculture and the growth of the food marketing system. Some USDA grades grew out of early trading practices; others were developed by trade groups and then formalized by federal and state agencies; and others were developed from research efforts to solve market problems associated with nonstandard grades.

Early History of Standards

In the early days, each grain market had its own grades and grading methods. Different standards for No. 2 corn required that the corn be "dry," "reasonably dry," "have not more than 16 percent moisture," or "have not more than 15.5 percent moisture." Study of the terminology used in grading grain in 1906 disclosed 338 names or grade titles being used. There were 133 designations for wheat alone. There were more than 100 changes in grain grades between 1916 and 1986.

The early cotton trade was also plagued by grade confusion. The term "middling" apparently was adopted from its use in England. Such terms as "good," "fair," and "ordinary" were in general use about 1825. In 1847, efforts were made to adopt a standard classification system, but it failed within a few years.

Livestock was once sold on the basis of the girth of the belly. In the literature of the 1850s hogs were sometimes classified as "fat distillery-fed" hogs and "fat corn-fed" hogs. In most of the early market reports the point of origin was indicated as one method of classifying the animals. Such terms as "prime," "choice," and "good" were not uniformly used either within an individual market or between two different markets.

The lack of accepted standards for fruits and vegetables resulted in especially chaotic trade conditions. These perishable products were shipped long distances, and there was no intelligible basis for price comparison or the settlement of damage claims. Some growers attempted to secure recognition by placing their names on all shipments. In this way they hoped to establish a reputation that would give them market premiums.

The lack of generally accepted standards resulted in many unfair practices and abuses. Not only did producers suffer, but also the middlemen of the trade were often defrauded. In most instances, pressure developed for reform of the grading system from within the trade itself. Trade groups and organizations attempted to systematize nomenclature and grades. Generally, however, substantial progress was not made until the federal government stepped in to coordinate the efforts to improve the grading system. In 1907, Congress appropriated funds to study federal food standardization. The passage of the Cotton Futures Act in 1914 and the Grain Standards Act in 1916 initiated a series of laws that have gradually broadened the area of federal responsibility in promulgating uniform standards. These included the Food Production Act of 1917, the Cotton Standards Act of 1923, the Tobacco Stocks and Standards Act of 1929, and the Tobacco Inspection Act of 1935. The importance of establishing standards was also recognized in the Agricultural Marketing Act of 1946, in which the Secretary of Agriculture was given both broad powers and funds to further the development and administration of standards. In 1981 amendments to this legislation required user fees. Investigations of grain quality abuses in part led to the passage of the U.S. Grain Standards Act of 1976, which required closer supervision of grain exports.

The Present Situation

Food and agricultural commodities are currently graded at the federal level by three agencies. The Agricultural Marketing Service (AMS) of the U.S. Department of Agriculture is the largest, with responsibility for grading meat, poultry, fruits, vegetables, dairy products, cotton, wool, tobacco, and naval stores. The USDA's Federal Grain and Inspection Service (FGIS) administers grading programs for grain, rice, pulses, hay, and straw. The National Marine Fisheries Service grades fish, shellfish, and seafood products.

Federal standards for farm products fall into three classifications—mandatory, permissive, and tentative. *Mandatory standards* are those whose use is compulsory under certain conditions. *Permissive standards* are those officially recommended but whose use is not compulsory. *Tentative standards* are those offered for use but still subject to further study before becoming permissive or mandatory. It is mandatory that grains and cotton that move into interstate commerce, and also those traded on the futures exchanges, be graded according to federal standards. Apples and pears sold in the export trade, tobacco, and naval stores also have mandatory standards.

There are currently about 235 federal grades for food and farm products: 13 dairy product grades, 85 fresh fruit and vegetable grades, more than 225 canned, dried, and frozen product grades, 18 grain and bean grades, 18 livestock product grades, and 155 tobacco product grades. New grades and standards are developed as the need arises, and about 7 percent of these grades are revised each year.

Food grades are widely but not universally used, as shown in Figure 17-4. Grade usage varies by commodity. It is also apparent that the food grades are more widely used at the wholesale level than at the retail level.

Many states cooperate in food grading programs using standards adopted from the federal program. In some cases, however, state grading standards differ from the federal standards, and these grade variations can hinder interstate food shipments, thus increasing marketing costs and food prices. Local grading variations are gradually decreasing through the cooperation of federal and state agencies.

Changes in federal standards, or the development of new ones, come about slowly. Initial suggestions for changes generally come from the trade or from research findings. If a suggestion is considered to be reasonable, the appropriate government agency will confer with interested industry groups to explore the suggestion. Out of the research and suggestions the agency develops tentative grades, which are then tried out. Eventually, if it appears that the proposed standards meet as many of the criteria for good standards as possible in the light of current knowledge, and are usable by the trade, they are issued as the federal standard. This process may extend over a period of several years. The changing of standards, how-

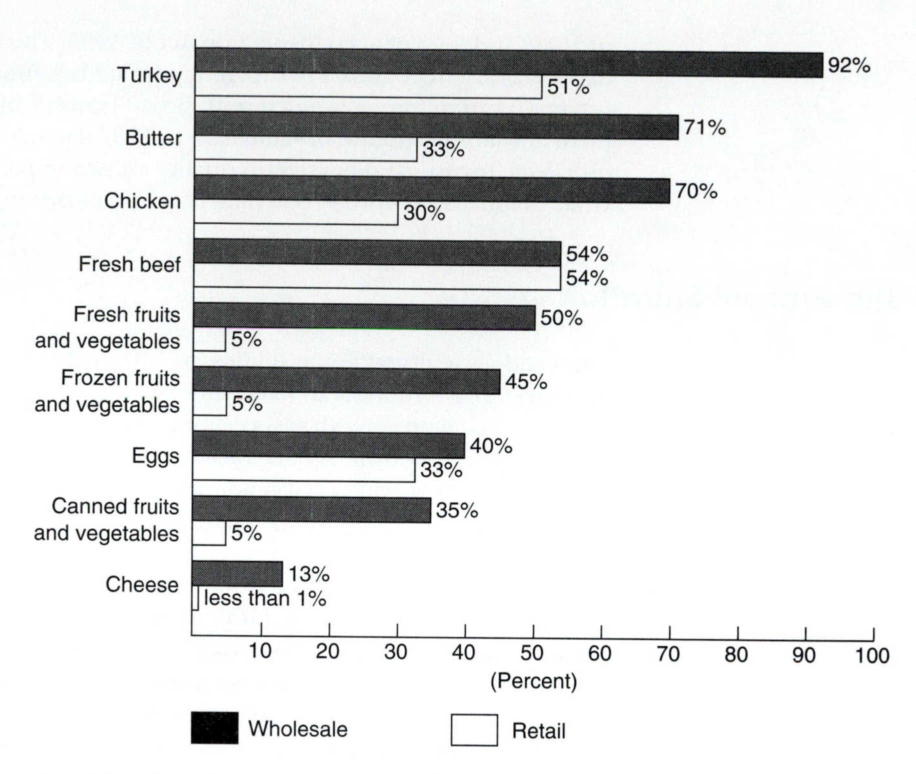

FIGURE 17–4 Proportion of food products graded at wholesale and retail, 1986. More recent data are not available. (U.S. Department of Agriculture.)

ever, is often a difficult and time-consuming process. Many persons have a vested interest in maintaining the existing standard. Others think of the existing standard as being right because it exists. But as long as consumer preferences, manufacturing processes, and production patterns change, so must standards change if they are to achieve maximum usefulness.

The need to change or revise food quality grades and standards can arise for several reasons. For example, the introduction of field-shelling practices for corn altered the quality of grain produced by farmers. In a similar manner, the introduction of high-speed refrigerated transportation changed the quality requirements for fresh produce and meats. Changes in consumer meat preferences and size of carcass requirements by the retail food industry have triggered changes in livestock and meat grades. Cotton standards were changed when the adoption of mechanical cotton strippers caused a change in the amount of leaf and trash particles found in cotton after ginning. The grades for feeder cattle were revised in 1979 in order to better identify the sizes and types of feeders desired.

Mandatory versus Optional Grades

With few exceptions, the use of USDA food quality grades is optional for food marketing firms. This is the reason for the variations in food grading among products shown in Figure 17-4. Some firms and food industries have used the USDA grades more than others. Periodically, there have been proposals to require quality grading of all food products. Proponents argue that mandatory grading is necessary to realize the potential contributions that quality grades can make to consumer choices and marketing efficiency. If grades can improve food market performance, they reason, why should they be optional and only partially used?

The following are reasons given in favor of continuing a voluntary food quality grading program: (1) There are few historical precedents for a compulsory, federal grading system for any product; (2) the cost of food grading might increase significantly if it were mandatory because the present voluntary system is financed primarily by user-firms; (3) USDA quality grades would be redundant with existing brand quality standards; and (4) grades and standards might inhibit product innovation and differentiation of food products.

The latter two arguments raise an important point about the relationship of product standardization and product differentiation. The branding that accompanies food product advertising and differentiation can be viewed either as an alternative form of food quality standardization or as the opposite of standardization, product differentiation.

From the viewpoint of an individual marketing agency, the production process for a differentiated and branded product represents a high degree of standardization. Building consumer loyalty to a branded product requires strict quality control in order that consumers will eventually rely on the brand as a guarantee of quality. However, if every firm develops its own standards and grades, the result will be a lack of standardization among food products, possibly making comparisons more difficult for consumers.

The voluntary food grading system allows marketing agencies to decide between standardized U.S. grades and private brand grades. The effect of this on consumers depends on the nature of consumer preferences for food products. If quality criteria vary widely among consumers, brand differentiation may well be the best method of matching products to consumer wants. But if consumers use quite similar quality criteria in purchasing a product, brand differentiation probably confuses consumer choices, and mandatory grades might be judged desirable.

OBJECTIVES AND PROBLEMS OF FOOD QUALITY GRADING

Food quality grading involves compromises between ideal grades and workable grades; between the needs of consumers for grades and trade needs; and between the costs and benefits of grades.

Criteria for Grades and Standards

The food grading system sets up a channel of communication between food producers and consumers. Ideally, grades will result in a perfect match between the diverse wants of consumers (according to their incomes and preferences) and the heterogeneous qualities of commodities produced and marketed. This sorting and matching process has the potential of increasing both consumer satisfaction and firm profits. Whether these potentials are achieved depends on the choice of quality standards, the design of grades, and the implementation of the food grading system.

The development of a system of perfect and ideal standards is highly unlikely. Each agricultural product presents different problems. Realizing that it is very improbable that any standard will meet them all, the following may be used as criteria upon which to judge the adequacy of standards:

1. Standards should be built on characteristics the users consider important, and these characteristics should be easily recognizable. Grades must be oriented to user opinion of value and not that of a few technical experts.

2. Standards should be built on those factors that can be accurately and uniformly measured and interpreted. If the major part of a standard consists of subjective measurements, uniform application by different graders, or at different points, will be very difficult. Excessive quality variation within a grade reduces the usefulness of the grade itself.

3. Standards should use those factors and that terminology that will make the grades meaningful to as many users of the product as possible. The ideal situation would be one in which the same grade terminology is used at all levels of the marketing channel, from the consumer to the producer. However, this is complicated by the fact that many products have several different uses.

4. Standards should be such that each grade classification includes enough of the average production to be a meaningful category on the market. Though grading standards should be consumer-oriented, they cannot ignore the real facts of production. Consideration must be given to the quality of the product produced. It is of little value to have a standard for the top quality set up in such a fashion that very little of the actual production can meet it.

5. The cost of operating the grading system must be reasonable. Absolute uniformity at any price is not a feasible goal.

Probably the best practical test of the adequacy of standards is their acceptance and use by the various marketing agencies and consumers. If the grading standard is widely used, it is probably true that the standards are fairly adequate and economically meaningful. However, if large segments of the trade or consumers do not use the standards, it usually can be assumed that some of the criteria have not been adequately met.

Problems of Food Grades and Standards

Several practical problems arise in the development of a food grading system and in its implementation in the field. The next sections will discuss them.

Determining Quality Standards

The subjective nature of "quality" makes it difficult to get agreement on universal food quality standards. Which food properties are most important in consumer preferences? Most standards have been determined by food scientists and the trade. As a result, food grades tend to be based on easily measured sensory characteristics: color, size, shape, tenderness, and so forth. No doubt these are important quality criteria. But some critics have suggested that the present quality standards are not complete and accurate representations of all the food properties influencing consumer preferences. It has been suggested, for example, that nutritional value ought to be added to the list of food quality standards. Obviously, the cost and complexity of food grading increases with additional standards.

There is some evidence that the current food grades do not correspond well to consumer preferences. Blind taste studies have shown that some consumers cannot, or do not, discriminate between different grades of foods. Market studies also have shown that consumers are not always willing to pay premium prices for the higher grades of food. Indeed, some consumers seem to prefer the lower grades to the higher grades. Considering the wide range of consumer tastes within the population, perhaps these results are not surprising.

Another criticism of present food grades is that they are convenient for traders but are not consumer-oriented. This is a serious charge, because consumer sovereignty over the food system requires that grades correspond closely to consumer quality preferences. It is true (as shown in Figure 17-4) that food grades are used more extensively at the wholesale than at the retail level, and that brands are often substituted for USDA grades at the retail level. However, it is unlikely that the grades would be used at the wholesale level if they did not correspond in some way to consumer preferences and final product values. Therefore, some might consider it unfortunate that the USDA grades are not more widely used at the retail level; but the motive for this is probably product differentiation, rather than because of defects in the grading system. It may also be true that farmers and food marketing agencies have a vested interest in the current grading standards and are hesitant to change these standards, even when consumer preferences change.

DESIGNING FOOD GRADES

Even if accurate and meaningful quality standards are agreed upon, numerous problems remain in developing a grading system. Measuring product quality against standards, for example, can be troublesome. Physical and chemical properties can be measured fairly accurately and consistently by trained graders. But measurement of sensory qualities is more difficult. These depend on the grader's senses of sight, taste, and smell. The standards for dairy products, for example, depend heavily on the sense of taste, and in such cases variability and error is possible.

Generally speaking, the more mechanical and objective the methods used in grading, the more accepted are the standards by the trade. One of the scientific advancements of recent years has been the replacement of some of the sensory tests with chemical tests and mechanical devices. Photoelectric colorimeters, reflectometers, and other devices have been developed to replace the old color chart comparison method. Tenderometers are now used to measure texture and consistency of peas and lima beans and a few other vegetables. The succulometer is a device used for measuring the juice content of sweet corn. Researchers are even experimenting with a mechanical thumb that will give more uniform results in checking for ripeness in fruit.

When the total supply cannot possibly be graded, or where grading damages the product, sampling variation is a problem. The size and height of egg yolk may be important to consumers, but is difficult to determine by observation of the shell.

Similarly, sweetness is desirable in a cantaloupe, but the grader cannot taste every melon.

There are also problems in determining the number and limits of food grades. Presently, there are eight beef grades and only three chicken grades. How many grades should there be? This is an extremely important question because it can influence the total amount of dollars received from the sale of the total production. Within the limits of the consumer's willingness to pay premiums for certain qualities, the amount that will fall in each grade can be changed. However, agricultural products do not fall into classifications with definite breaks between them. Instead, the quality of agricultural products varies over a wide, continuous range. It has been suggested that most products have a quality distribution very similar to the normal frequency distribution curve, as shown in Figure 17-5.

One of the criteria for good grades is that there be enough of the normal production falling in each grade to make it a meaningful market category. How many grades should there be and where should the boundaries of the grades occur for the commodity illustrated in Figure 17-5? There are some products of very low quality and some of very high quality. Most, however, fall somewhere between these two extremes.

It is evident from this illustration that the grade boundaries will be "zones" rather than clear-cut lines. The more the grade factors are measured subjectively, the wider will be the zone of indecision. This has led to a system of tolerance in standards. For example, grades of fruits and vegetables usually provide for 5 to 10 percent of off-grade specimens.

The quality of the production of a commodity also changes from year to year. The curve in Figure 17-5 might shift either to the right or to the left. One year might find a larger amount of higher-quality products and a smaller amount of lower-quality products. Or the situation might be reversed. It is also possible for producers to change the grade distribution of their production over time, as shown in Figure 17-3. Such conditions make it extremely difficult to maintain consistent standards. Again, this is particularly true when grade factors depend on subjective measurement. For example, if the apple crop is poor in quality, very few apples would meet

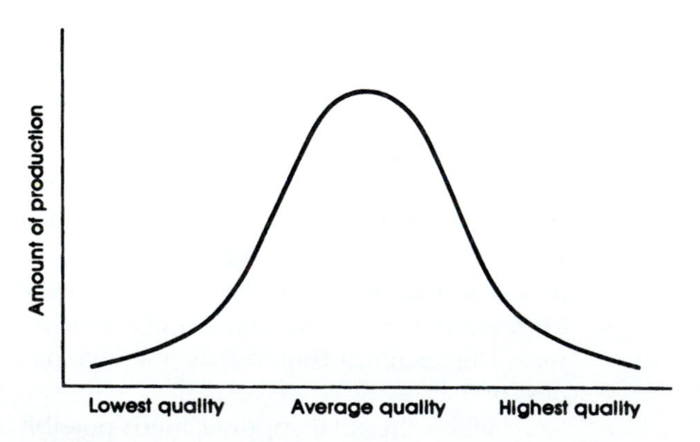

FIGURE 17–5 Average quality distribution of the total production of a hypothetical product.

the top-grade requirements if the standards were rigidly adhered to. Under these circumstances, the pressure is strong to "reach a little farther down" for the top-grade apples.

This tendency to "upgrade" or "downgrade" means that the composition of particular grades will vary from year to year. Under such circumstances, the consumer is faced with a product of a given grade that may differ in quality each year. It has been suggested that preference studies concluding that consumers are not discriminating between grades do not necessarily mean that the standards are incorrectly measuring consumer quality preferences. On the contrary, they may indicate that consumers have found the stated grades an unreliable measure of actual product quality and therefore ignore them.

Grade nomenclature is another problem. In the trade, there is a reluctance to use grade names that suggest inferior quality. Consequently, fruit and vegetable grades may range from U.S. Fancy, to U.S. Extra No. 1, to U.S. No. 1, while the third-lowest meat grade is "Good." These names may prevent consumers from undue discrimination against perfectly good products, but they probably also cause some consumer confusion.

Finally, practical problems arise in designing a single set of grades that are appropriate for trading at all market levels. This is especially true where the farm product undergoes substantial processing. For example, are the carcass grades for beef at the wholesale level also appropriate for grading retail beef cuts? How can consumer preferences for different qualities of bread be communicated to wheat farmers?

IMPLEMENTING FOOD GRADING SYSTEMS

Still other problems arise in implementing the use of food grades in the marketplace. Because grades are initiated at the request of the trade and are voluntary, not all food products have grades, and not all products that could be graded are. There are no quality grades, for example, for bacon.

Another difficulty is that products may change grade in the marketing process. Many agricultural products are perishable. The fact that a commodity was a given quality at one point in the marketing channel does not mean that it will be of the same quality when it reaches consumers.

The problem of quality loss during marketing brings up the question of where in the marketing channel grading should be done. If grades are to fulfill their objective of telling producers what consumers consider desirable, grading must first be done when the farmer sells commodities. Only then will farmers know what the quality and actual worth of their product is. However, if quality deteriorates during the marketing process, this grade will not remain accurate. Therefore, grading must be done as often as needed throughout the marketing process to assure accurate grading when the product reaches the final user.

Sometimes there is also confusion between the requirements for sanitation or edibility and those for quality. For example, meat entering into interstate commerce must be federally inspected to make sure it is fit for human consumption. The packing plants themselves must also meet certain sanitary requirements. But such inspection has nothing to do with the grading of meat for quality.

On the other hand, sanitary requirements are sometimes written into the grading standards. Milk quality standards are a case in point. Though the grade standards consider bacteria count, they also may prescribe the conditions under which the cows must be housed and milked, the way the milk is cooled, and so forth. Regardless of bacteria count, milk cannot meet the grade requirements unless it has been produced and handled under the designated conditions. Such practices confuse the issue. Often, instead of facilitating marketing processes, such standards turn into practical trade barriers and techniques to control production.

FARMERS AND UNIFORM GRADING

Only if the farmer sells on the basis of grades will the fullest benefits of the grading system as a method of consumer-producer communication be realized. The wider the practice of selling on a graded basis, the less is the possibility of fraud and deceit in the selling of goods by farmers. However, it is probably also true that not all farmers stand to gain from selling on a graded basis. The farmers who produce the higher-quality products would gain at the expense of those producing the lower-quality products.

Producing higher-quality products is seldom costless. More careful and often more expensive handling of the product is usually required. In some instances the extra cost would probably outweigh the extra returns. In these situations farm selling on a graded basis has little attraction to producers.

This improvement usually comes about as the producer realizes the things that can be done to produce a crop of higher quality. Many of these things consist of changes in present practices, not the adoption of new or additional ones. For example, picking fruit and vegetables at the proper degree of ripeness can reduce spoilage. Such practices might become more widespread if the producers know what constitutes higher-quality and higher-valued products.

It is in the area of standardization of quality that one of the substantial differences between agricultural and industrial production has existed. Much of industry produces a product *to* a quality specification. If the product does not meet the quality standards, it is either rejected or sent back for correction. In agriculture, however, production quality control is more difficult. The role of standardization, as we have seen, has been to establish grade limits and sort the available production into lots that are as uniform as possible.

It appears that in the future agricultural producers may also be able to produce more closely *to* grade specifications. With improved technology and knowledge in production, it often is possible to follow certain procedures and practices and thereby closely control the quality of the output. After extensive research in plant breeding, one concern developed a new strain of corn that uniformly gave the starch content needed in its manufacturing. Vegetable processors, through the control of the strains of plants and planting and harvesting procedures, can come much closer to securing the quality of product desired. The "fresh-fancy" program in eggs, through controlled production and marketing processes, produces a given quality of eggs. Under these conditions it is possible to certify the flock as a grade A flock and only a sample grading of the eggs is required. Likewise, hog producers, through controlled breeding and husbandry practices, are able to produce a much more uniform-quality product. Biotechnology may also contribute to more uniform agricultural products.

To the extent that quality control in production is feasible, many of the practices and problems associated with heterogeneous and fluctuating qualities will be substantially changed in the future.

MARKETING AGENCIES AND FOOD GRADING

Food grades should fundamentally reflect consumer preferences for different qualities of products. Yet, many food grades seem to be producer-oriented, rather than consumer-driven. It is not a simple task to determine which commodity characteristics, say for livestock, best relate to the preferable consumer characteristics for meat.

It is in the marketing area that many of the conflicting goals of different firms come into focus. A firm may have one view if it is procuring supplies for its own operation and a different one if it is selling its product to buyers. It may be in favor of the simplification of procurement that uniform grading permits, but strongly desire to differentiate its product to its own customers in order to secure some competitive advantage.

The use of federal grades for meat is a good example of these conflicting interests. Large chain retailers like federal grading because it simplifies their problem of obtaining large quantities of uniform products. Smaller packers also like the system of federal grades because it permits them to act as partial suppliers to large chain organizations. Large packers, however, are less enthusiastic in their use of federal grades because the federal grades compete directly with their own grades and brands. Large retail organizations also may not be in favor of uniform grade designations of their products because they would prefer to merchandise under their own brand to their customers.

There is some evidence that food grades and standards have affected market concentration and competition. One study found that federal meat grades contributed to declining meat packer market concentration following World War II, because new firms found it relatively easy to compete with the larger, established packers. It is interesting, too, that eggs and butter are highly graded commodities with relatively low market concentration levels.

No doubt the increasing size and complexity of food processing and merchandising operations will require larger volumes of more uniform lots of commodities. This will probably increase marketing agencies' use of standardized food grades, at least in their procurement activities. However, food marketing firms can achieve a comparable result with integrated purchasing contracts and with greater use of specification buying based on private grades. Therefore, there is no certainty that marketing agencies will increase their voluntary use of USDA food quality grades in the future.

CONSUMERS AND FOOD GRADES

Generally speaking, consumers' goods are not sold on the basis of uniform grading and standards of grade terminology. One of the problems in today's society is that the consumer knows relatively little about the quality of the goods among which he or she makes a choice. In purchasing foods, for example, the consumer is faced with a confusing array of brands and quality terminology. The terms designating top quality vary widely. In canned goods, companies use different brand names to designate different qualities. There is little to guide the shopper who is not familiar with the brands themselves. In a marketing system theoretically based largely on consumer direction, consumers are faced with the problem of how to make their wants known.

Even with the more widely accepted USDA grades that are used on some commodities, many consumers are confused and show a general lack of knowledge. One study revealed that over half of the consumers interviewed could not correctly identify any of the common USDA grades for meat, potatoes, apples, or turkey. Less than 30 percent were aware of what the U.S. Department of Agriculture inspection mark was. Nonetheless, most respondents rated government grades as "very helpful." Many consumers believed that there are bacon grades, but there are none (there are, however, hog carcass grades). Also, many people believed that "Grade A" milk is a quality grade, when in fact it is a local and state sanitation certification.

There have been proposals to adopt a single set of grades for all food products, for example, grade A, B, or C, or grade #1, #2, #3. Such grade-labeling proposals have the benefit of simplicity and might facilitate the consumer's quality evaluation process. However, some people question whether such a system could represent the wide variations in qualities of individual foods. Occasional proposals are made that would require mandatory grade labeling of all foods.

In recent years, attempts have been made to incorporate new criteria into our food grades and standards. The growth of "natural," "organic," and "range-free" farm products has resulted in new certification and labeling standards for crops grown without inorganic chemicals or animals produced in nonconfinement facilities. Because these foods have higher production costs and often command price premiums from some consumers, they must be differentiated in the marketing process. The U.S. Department of Agriculture is developing food standards and label requirements for these products.

Improvement in the area of consumer grading will be slow. Many believe that the average consumer needs help. There is not much agreement as to how to help. Large concerns that have built up preferences for their brands are very reluctant to sanction a uniform labeling system. Much of the emphasis on brands and advertising is based on the fact that the consumer is faced with a bewildering selection of goods and can be persuaded into simple brand loyalty. This does not necessarily mean that inferior products can dominate the market. Even brand shopping cannot prevent consumers from turning away from unsatisfactory products. It does mean, however, that new products and new concerns may have difficulty in securing consumer acceptance even if their product is of good quality.

SUMMARY

Quality standards are measurable properties of foods that differentiate them for consumers. Grading is the sorting of unlike lots of products into heterogeneous categories according to accepted quality standards. Foods are the most extensively government-graded consumer products, but not all are graded. Because government food grades are voluntary and of little value in differentiating the firm's offerings, brands are frequently substituted for USDA food quality grades. Standardized food grades can contribute to operational and pricing efficiency as well as to increased consumer satisfaction and producer returns. There are several problems in translating ideal food quality standards into practical grades for use in food marketing. U.S. food quality grades have been criticized as being inadequate reflectors of consumer preferences, trade rather than consumer-oriented, and confusing for consumers. The food grading system has adjusted and will continue to adjust to changing market needs.

KEY TERMS AND CONCEPTS

food grade
food quality standard
food standard
grading
quality
standards

DISCUSSION QUESTIONS

1. What are the roles of quality grades and standards in the food industry? How well are these capabilities being achieved? Why is food more extensively graded than other consumer products?

2. Do you favor more or less U.S. government grading of foods? Should grading be optional or mandatory?

3. Explain how grading can both increase consumer satisfaction and the total dollars that consumers pay farmers for food.

4. What are the pros and cons of a mandatory, uniform grade-labeling program for foods? How would you react to this as president of a large food processing firm? As director of a consumer's lobby? As a farmer?

5. Should nutrition be added as a food quality standard?

6. How have decentralization and integration affected food marketing agencies' voluntary use of government food grades in their procurement activities? Their selling activities?

7. How would you change food grades?

CHAPTER 18

TRANSPORTATION

OBJECTIVES

After reading this chapter, you will be able to:

1. Explain why transportation is so important and costly in the food industry.
2. Understand why different farm and food products utilize various means of transportation.
3. Make suggestions for improving the efficiency of food product transportation.

Both the American and the world food industry critically depend on physical distribution. U.S. consumers may enjoy a meal of Hawaiian pineapple, California fruits and vegetables, Texas beef, Brazilian coffee, Wisconsin cheese, and Florida oranges. This wide variety of food would not be available without the complex transportation system that serves the food industry. Indeed, adequate and efficient transportation is a cornerstone of the modern food marketing system. America's food travels over 177,000 miles of railroads, 3.2 million miles of intercity highways, and 26,000 miles of improved waterways.

Like all other marketing functions, transportation influences both the numerator and denominator of the marketing efficiency ratio. The movement of farm products from where they are produced to consumption centers creates place utility. The fact that consumers are willing to pay for this utility suggests that it often exceeds the costs of transportation.

Transportation also plays an important role in market development, expansion, and competition. The size of the market area depends on whether products can be moved and the costs of transportation. Improved transportation has expanded the market area for farm products from a local to a national to a worldwide level. In this sense, improved distribution, or lower transport costs, can be a rightward demand shifter for farmers and marketing firms. At the same time, the ability to ship great distances increases competition between firms and areas. Changes in freight rates may alter the competitive advantage of sellers throughout the market.

Transportation also influences other marketing functions and decisions. The availability of transportation facilities affects the storage capacity needed in the food industry. The speed and flexibility of the transportation system can also affect inventory and other storage costs throughout the food system. Transportation costs affect the location of food processing plants and food distribution warehouses. Finally, transportation expenses contribute to the size of the food marketing margin and thus influence farm and consumer food prices.

TRANSPORTATION FOR THE FOOD INDUSTRY

The estimated cost of long-distance transportation of farm and food products was nearly $24 billion in 1997—$92 for each person in the United States. Intercity truck and rail transportation expenses are the third-largest component of the food marketing bill, after labor and packaging costs. Transport costs account for about 6 percent of all food marketing costs. The addition of air, water, and local trucking costs would probably raise the food transportation bill to about 11 percent of total marketing costs.

Farm products depend on transportation for the creation and preservation of their value. Food transportation costs—a measure of place utility—usually constitute a higher share of the retail price than is the case for nonfoods. This varies, however, even for foods. As shown in Table 18-1, the contribution of transportation to a food's final monetary value varies by the distance it is shipped, its perishability, and its bulkiness.

The food industry presents a number of special transportation needs. Food production is geographically dispersed, and the task of linking the several food production centers with the consumption centers is no simple one. A flexible transportation network is needed to compensate for the inflexibility of geographically

TABLE 18-1 Contributions of Intercity Transportation Costs to Retail Food Prices, 1997

Product	*Intercity Transportation Costs as a Percent of Retail Price*
California oranges	10.5
California lettuce	9.5
California tomatoes	9.5
Northeastern potatoes	6.0
Frozen orange juice	3.0
Broilers	2.4
Pork	1.6
Eggs	2.3
Beef	1.4

SOURCE: U.S. Department of Agriculture.

specialized agriculture. The production and transportation patterns for wheat are shown in Figure 18–1.

The variability of agricultural production also complicates the food transportation picture. Food transport needs vary widely—annually and by seasons—and are as unpredictable as farm supplies. A large crop puts severe strains on the capacity of the transportation system, just as a small crop results in excess capacity of transportation equipment.

The biological and bulky nature of farm products makes special demands on the transportation system. Food and feedgrains require high-capacity equipment, live animals need special equipment, and perishable products require rapid transportation. Few other industries have such diverse transportation needs.

Because of these conditions, farmers have always been concerned with the adequacy and costs of the farm product transportation system. Pioneer agriculture developed hand in hand with the national transportation network. Early plank roads were constructed to facilitate the movement of farm products to market. Rural roads were later paved for the same reason. The first laws regulating transportation rates in the nineteenth century were the result of farmers' complaints that they were the victims of monopolistic railroads. In 1935, farmers sought and gained special regulatory treatment of trucks hauling agricultural commodities. More recently, dock strikes, freight rate increases, railroad abandonment, and shortages of railroad cars have again focused the attention of farmers on the transportation system. The expansion of international trade in agricultural products in recent years has also increased the importance of transportation to the food industry.

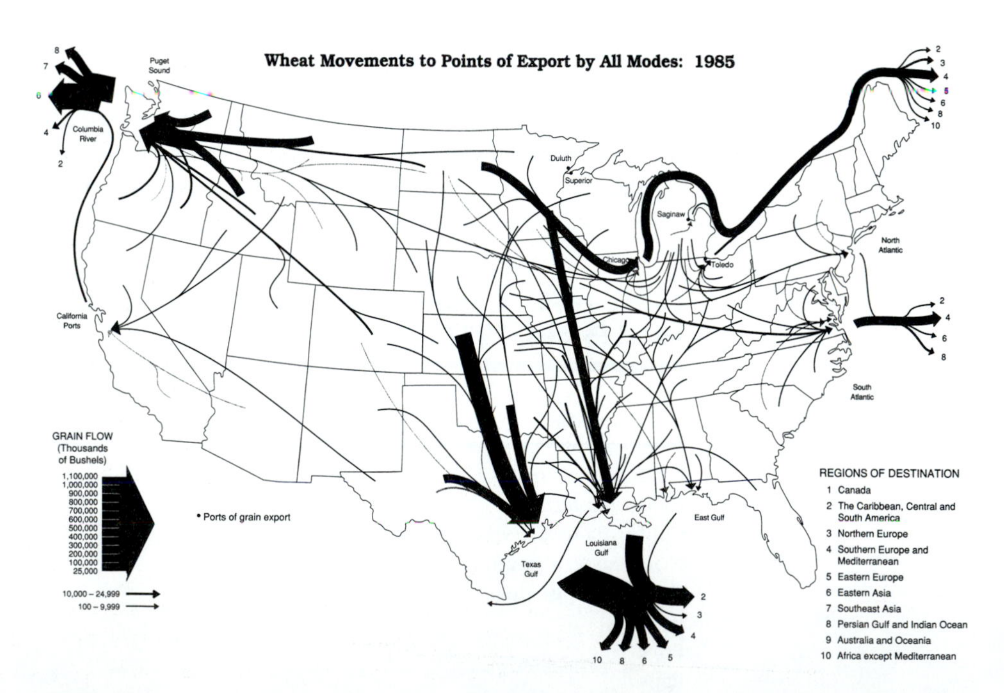

FIGURE 18–1 U.S. Wheat Movements to Points of Export, by Mode of Transportation, 1985. (Source: Hill, L. D. and D. Timmerman. Grain Movements to Points of Export in 1985. Agricultural Experiment Station, College of Agriculture, University of Illinois at Urbana-Champaign.)

ALTERNATIVE MODES OF TRANSPORTATION

Agricultural and food products are transported by virtually every sort of carrier except pipelines, as shown in Figure 18–2. However, rail and truck transportation are the dominant modes of transportation for farm and food products, with water and air carriers playing a small but important role for certain commodities. Farm and food commodities account for perhaps 15 percent of all railroad freight, 10 percent of all truck freight, and 5 percent of all barge freight.

Rail and Truck Transportation

A major trend in transportation has been the diversion of freight from railroads to trucks. Between 1950 and 1985, the railroads' share of *all* freight ton-miles fell from 57 percent to 37 percent, whereas trucks' share rose from 16 percent to 25 percent. The transportation of farm and food products has followed a similar pattern. In 1929, trucks accounted for 24 percent of the food transport bill, but this had grown to 70 percent by the 1980s.

Railroads, nevertheless, continue to be important shippers of farm and food products. They still haul about 50 percent of the grain, but they now transport less than 10 percent of the fruits and vegetables and less than 1 percent of the livestock

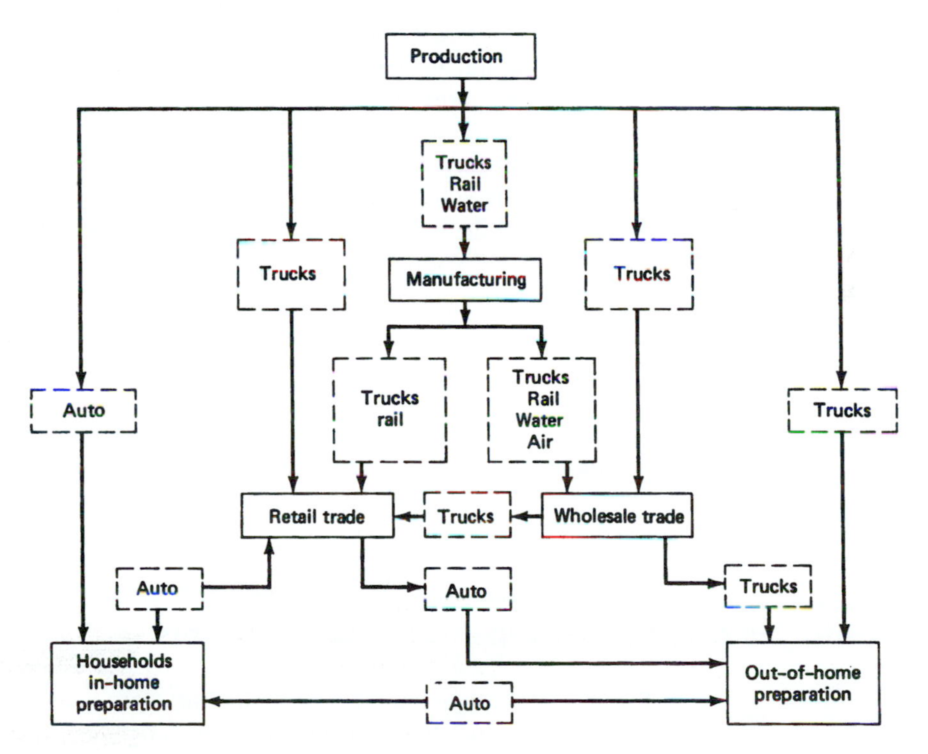

FIGURE 18–2 The farm and food transportation system. (*Energy Use in the Food System,* U.S. Federal Energy Administration.)

and poultry products. Trucks now dominate in the short-haul movement of food products and are also important in long-haul shipments. This trucking trend is related to a number of developments, the most important of which have been lower trucking freight rates compared with rail rates, and the greater speed and flexibility of trucks as a result of the development of the interstate highway system. About one out of every three trucks hauls farm-related items. Many farmer-owned trucks are used for short hauls. It has been estimated that farmers own almost 30 percent of all trucks in the United States.

Water Transportation

When compared to that hauled by truck and rail, the volume of agricultural commodities carried on the country's inland and coastal waterways is small. However, where speed is not important and the commodity has great bulk and weight, water transportation is often the cheapest method. A 1500-ton barge can carry the equivalent of 15 hopper rail cars or 63 semitrailer truckloads. Tows of 30–40 barges are common on the major waterways. Barge capacity expanded rapidly in the 1978–1982 period.

The Great Lakes and the Mississippi River system are the most important water commerce arteries in the United States. The Mississippi River system accounts for more than half of the soybeans, wheat, and corn moved by barge. The Snake and Columbia Rivers, flowing through the Northwest, have also become important waterways for grain. Puget Sound and California ports are important in exporting a wide range of food and agricultural products to the Pacific rim and Asian nations.

Domestic farm product exports move primarily by ocean vessels. Houston, San Francisco, Baltimore, and New Orleans are important grain ports. The 1959 opening of the St. Lawrence Seaway provided an alternative route for Midwest grains flowing to Europe. The volume of trade moving through the Great Lakes ports expanded in the 1970s but declined in the 1980s as a result of high freight rates, restrictions on vessel sizes, and winter weather conditions.

Air Transportation

Though agricultural air freight has increased rapidly, the quantities of agricultural commodities moved by air freight are very small indeed. Cut flowers, nursery products, and fresh fruits and vegetables are the biggest single agricultural users of air freight.

Advantages of Different Types of Transportation

Farmers and other handlers of agricultural products have a choice among several types of transportation. A major decision is between the use of the railroad or truck. Among the types of truck transportation, common carriers, exempt carriers, or self-owned trucks are available.

TABLE 18–2 Advantages and Disadvantages of Alternative Food Transportation Modes

Transport Mode	*Major Advantages*	*Major Disadvantages*
Railroad	1. Lowest cost for long haul 2. Can handle large volumes 3. Transit privileges available	1. Inflexible route 2. High cost for short hauls 3. Service problems 4. May be car shortages when needed
Regulated motor carrier	1. Flexibility of roads 2. Government regulated	1. Rates lower than rail but higher than exempt carriers 2. Prefer truckload lots
Exempt motor carriers (nonregulated)	1. Low rates 2. Highly flexible routes 3. Will handle small lots	1. No government supervision of financial responsibility or reliability
Motor carriers owned by shipper	1. Flexible routes 2. Good control by shipper	1. Large investment 2. Difficult to attain full utilization
Air freight	1. High speed	1. High cost 2. Inflexible routes 3. Airport waiting time
Water	1. Low cost 2. Can handle large volumes	1. Inflexible, limited routes

However, some movement of agricultural products is provided by all kinds of transport, which indicates that each has its advantages and disadvantages in doing a particular job. In Table 18–2 these advantages and disadvantages are summarized. A close study will show that costs, flexibility of service, and dependability of equipment are major areas of difference. Selection is based on which of these factors are of prime importance. A study of the frozen orange juice concentrate industry concluded, for example, that trucks were widely used in the expansion period, when markets were being developed and speed and flexibility of service were important. However, the use of rail transportation increased later, when large-volume shipments were essential and speed was not so important.

TRANSPORTATION REGULATION AND FREIGHT RATES

The Interstate Commerce Commission, created in 1887, supervises the transportation industry. This agency has broad authority to regulate freight rates, competition, and transportation routes. Historically, this regulatory authority was based on the notion that transportation is vital to the national interest, that there are natural monopolistic tendencies in transportation, and that there would be destructive competition among competing carriers in the absence of government regulation. There is considerable debate today over these assumptions, and whether they continue to justify governmental regulation of transportation rates and routes. There are also differences in regulations for the various transportation modes.

Railroad Freight Rates

The railroad freight rate structure of the United States is very complex. Freight rates are not always directly proportional to distances traveled. It is sometimes cheaper to ship a carload of grain a long distance over one route than a shorter distance over another route. Also, separate rail rates have been established for different commodities. For example, grain rates per ton-mile differ substantially from canned goods rates per ton-mile.

In general there are two basic types of rates, the class rates and the commodity rates. *Class rates* are those established for a limited number of broad categories into which thousands of different products can be placed. This eliminates the necessity of establishing individual rates for each different product. *Commodity rates* are those established specifically for an individual commodity, considering the needs and problems of shippers. Such rates are established for large-volume, low-valued items, such as coal, ore, and grain. Most agricultural products move under commodity rates. Over four-fifths of the total carloads of all products is moved under commodity rates. Commodity rates are generally lower than class rates, an example of price discrimination in freight rates.

There are also rate differences depending on whether the shipment is in carlot (c.l.) or less-than-carlot (l.c.l.) amounts. The rate is generally lower on the carlot than the less-than-carlot shipment. This preferential rate for carlots was one of the reasons for the development of early cooperative livestock shipping associations. As the railroad was the only method of moving livestock to major markets, the association assembled small lots from individual farmers so they could ship to the terminal markets at carlot rates. Lower rates for *unit trains* (whole trains of one commodity) have also favored farm cooperatives.

Rates are also differentiated on the basis of whether they are local or through rates. Because of the terminal expense involved, longer hauls usually are moved at a lower rate per mile. The through rate is less than the sum of several local rates covering the same distance.

There also are several special services available to rail shippers. Two such services of importance to agriculture are the transit privilege and the diversion and reconsignment privilege. The *transit privilege* permits a shipper to stop a shipment en route to permit some processing or operation to be performed and then to reship it to the original destination, still at the original through rate. This privilege is widely used by the grain trade. Wheat en route from a western point to the East can be stopped for cleaning, grading, and milling. The flour can then continue onward at the original through rate from the initiating point to its eastern destination.

The *diversion and reconsignment privilege* allows a shipper to change the destination of the product either while en route or after arrival at the originally desired destination. Within limits, reconsignment and diversion can take place and the through rates apply to the new destination. This service is one of the major reasons why shippers choose rail over truck transport. Often produce is sold after it is en route so that the exact destination cannot be known at the time it is shipped. Then, too, the original market for which the shipment was intended may become glutted and prices break so that it would be more profitable to reship to another market. The use of diversion and reconsignment aids better allocation of supplies of perishables and reduces the spoilage and waste that accompany market gluts.

Historically, rail freight rates were regulated by the U.S. Interstate Commerce Commission. Since 1980, these rates have been deregulated and are now set by competitive market forces, rather than by government.

The trend in rail freights for agricultural commodities over the 1950–1992 period is shown in Figure 18–3. The 1950–1957 rise in rates reflected the railroads' and the ICC's resistance to competitive pricing. As a result of these rises, much of the agricultural commodity freight was shifted to trucks and barges. The Transportation Act of 1958 gave railroads wide authority to reduce rates to levels competitive with alternative transportation modes. Consequently, rail freight rates fell substantially from 1957 to 1966, as shown in Figure 18–3. However, rising wages, fuel prices, and other costs have resulted in increased agricultural rail rates from 1966 to the present. In recent years, railroads have also increased the costs of transit and diversion privileges as well as *demurrage,* the charge for extra loading or unloading time. Overall, these developments continued to encourage the diversion of agricultural commodities from rail to truck transport. However, the deregulatory Railroad Revitalization and Regulatory Reform Act of 1977, along with 1980–1981 legislation that further relaxed the regulatory climate for railroads, has permitted the railroads to regain some lost traffic in recent years. U.S. freight costs fell 25 percent between 1986 and 1996. This resulted not only from deregulation of the transport industry but also

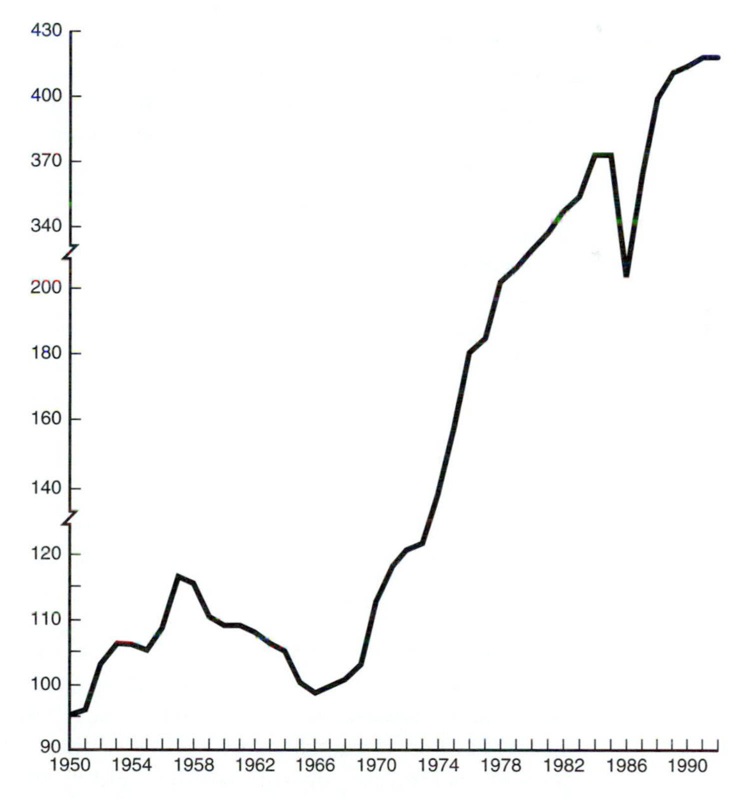

FIGURE 18–3 Railroad freight rate index for agricultural commodities, 1950–1992. (U.S. Department of Agriculture.)

from containerization and improved refrigeration technologies, which permitted a shipment to jump from truck to ship to train without being emptied.

Trucking Freight Rates

Three varieties of trucks haul agricultural commodities. Trucks operating on regulated schedules and routes, and that are available for hire, are classified as *common carriers.* Another classification of motor carriers consists of trucks privately owned by manufacturing or marketing firms for their own use. Finally, many farm product producers own and operate their own trucks.

There are other types of transportation middlemen. The *truck broker* is a special type of agent middleman whose function is to help match available trucks with available cargoes. Truck brokers also help establish rates in the negotiation between the exempt carriers and shippers for a percentage fee of the gross freight costs. The *freight forwarder* is another specialized middleman who aggregates small lots of commodities into larger shipments for more favorable transport rates.

FOOD PRICES AND TRANSPORTATION COSTS

As a marketing cost, transportation expenses influence farm and retail food prices. Increases in the cost of transportation fall primarily on farmers in the short run. An increase in freight rates either reduces the price that food marketing firms can offer farmers for their products, or it increases the farmers' cost of getting products to market. However, in the longer run, as farmers adjust production to prices and profits, the increased transport costs are partially passed on to consumers. The sharing of a freight rate increase between farmers and consumers depends on the elasticity of the supply and demand curves.

Freight rates react to changing farm prices in much the same way as other marketing costs. These rates are relatively stable during the initial stages of rising agricultural prices. However, freight rates eventually accelerate with farm prices, and may even continue to rise after farm prices begin to fall. The failure of freight rates to adjust simultaneously to changing farm prices contributes to agricultural price variability.

This freight-price relationship becomes particularly critical for those shippers located long distances from markets. The inflexibility of the rates often explains why a sharp market price decline for perishables, for example, will result in California growers either destroying their crops or leaving them unharvested. After freight and harvesting costs are met, the result may be an actual net dollar loss to the grower. So the crop is not harvested.

The severity of these problems for farmers depends on their access to alternative transportation carriers. Although it is true that today's farmer has more transportation alternatives than in the past, many farmers have limited transportation alternatives. For example, many of the country elevators to which farmers deliver grain are served by only one railroad, and many of these elevators are located on a recently abandoned rail line.

REDUCING THE FOOD TRANSPORTATION BILL

The American transportation system is good by most standards, but it is by no means perfectly efficient. Railroad cars stand idle 90 percent of the time; perhaps 40 percent of all trucks are running empty at any given time; railroad cars and trucks are lost, delayed, or arrive with damaged cargo; and trucks and rail cars stand idle for hours waiting to be unloaded. It has been said that it now takes longer to move certain food commodities by rail from the West Coast to New York City than it did in 1953.

As in all efforts to reduce marketing costs, attention must be paid to both the numerator and the denominator of the marketing efficiency ratio. It would be an easy task to reduce food transportation costs if consumers were willing to drive 100 miles to pick up their groceries from central food depots, or if consumers would forego a wide variety of year-round fresh food products. But these economies would also reduce consumer satisfaction and, probably, overall marketing efficiency. Nevertheless, a number of areas show promise as possible ways to improve the operational efficiency of the food transportation system, and reduce transportation costs, without sacrificing consumer satisfaction.

Technological and Other Improvements

There have been several innovations in farm and food product transportation. The increased availability and reliability of mechanically refrigerated rail cars and trucks, the increased number of covered hopper cars, the double-bottom (two-unit) truck, and larger-sized rail cars are examples of cost-reducing transportation technology. The interstate highway system, the St. Lawrence Seaway, and inland waterway improvements also have contributed to transportation efficiency. At the consumer level, freeways and the increasing number of two-car families have increased the mobility of food shoppers.

Railroads have innovated in the development of combination truck-rail transportation systems. The trailer-on-flatcar (TOFC) consists of truck trailers that can be "piggybacked" on special rail flatcars. There are also railroad cars that have both rubber and train wheels and can be converted from road to rail in less than five minutes. A new railroad cattle car with feed and water facilities and room for the cattle to lie down has been developed for long-distance shipment of cattle from the Southeast to western feedlots.

Another improvement is palletized shipping. Many products are placed on pallet platforms at the manufacturing point and moved all the way to the retail outlet on the same pallet, thus reducing labor requirements and reducing the chance for damage of products. Containerization is a further refinement of palletization. This consists of packing many smaller packages into one large container. Sometimes special equipment is installed to preserve a particular environment, particularly for meats and fresh fruit or vegetables.

The unit train concept, that of an entire train carrying a single product, has reduced transportation costs for grain products. Unit trains can be handled directly and efficiently but require exceptionally large shipments. A single, 125-car unit train can carry the equivalent of 3,700 acres of corn. Computers are now being used extensively to route and schedule complex shipments.

The trucking, railroad, and barge industries have all become more energy-efficient in recent years. Lighter-weight materials, improved equipment designs, and more efficient engines have improved fuel efficiency. Tow boats are using less expensive heavy petroleum fuels. Studies have shown significant savings from replacing mechanical refrigeration with carbon dioxide snow.

Regulation and Competition

Competition between alternative forms of transportation has played a role in holding down rising food industry freight costs and in improving services. The rivalry between the rails and trucks, for example, resulted in downward pressures on freight rates as well as the development of new services. Railroads have attempted to improve their speed and services to regain lost traffic. The "piggyback" combination rail-truck-and-water service is an example of this. There are also "fishyback" and "birdyback" services that combine alternative transportation modes.

Despite this competition, the transportation industry was one of the most highly regulated of all industries. In the 1970s, the purposes and effects of transportation regulations were reconsidered. Those favoring deregulation argued for allowing free market forces to determine transportation rates and service levels. Those arguing for continued regulation emphasized the need to prevent excessive competition among carriers and to preserve alternative transportation modes for shippers.

Two important laws were enacted in 1980 that deregulated the transportation system to a significant extent. The Staggers Rail Act of 1980 provided railroads with much greater rate flexibility in competing for traffic. It permitted railroads to offer short-term discounts to shippers and to sign contracts offering large-volume shippers special freight rates. The act also permited railroads to purchase and operate motor and water carriers and removed barriers to railroad mergers. The Staggers Act also made it somewhat easier for railroads to abandon unprofitable branch lines. The Motor Carrier Act of 1980 similarly deregulated trucking, facilitating the entry of new carriers and reducing other federal regulations influencing trucking rates and routes. This new regulatory climate has substantially affected farmers and food marketing firms, sometimes adversely, as when the only rail line serving a local grain elevator is abandoned. Increased transportation rate and route flexibility may raise rates for some agricultural shippers. However, other shippers and receivers may receive better, lower-cost services.

Increased Capacity Use

There are several ways to improve the capacity utilization of existing transportation equipment, including the elimination of excessive duplication of some transportation facilities and the better arrangement of routes so as to assemble a full load more efficiently. The pickup of two or more commodities might be combined to secure a larger load with less travel. Careful planning in many instances might eliminate an empty return trip, which often occurs when commodities are hauled to market.

The local assembly of products from farm to initial market offers another area in which more effective use of transportation equipment could be obtained. This has been illustrated by several studies of milk assembly. These studies found that trucks were traveling long distances to pick up small amounts of milk. Many times two or more truck routes would be traveling down the same road, each carrying less than capacity. Consolidation and better planning were offered as ways to reduce hauling costs substantially. However, the technique of bulk handling of milk has tended to solve this problem by permitting farmers to store up larger quantities of milk, which then can be transferred to tank trucks on an every-other-day basis. Hauling costs to farmers are then often reduced. This is a good example of a new technological development opening up new solutions to an old problem. The pressures to economize have exposed many other areas in the marketing system where the same type of savings could be made.

Reduced Spoilage and Damage

Reducing spoilage and damage can considerably reduce transportation costs. Many of the remedies to reduce spoilage and breakage are remarkably simple. Three factors account for much of the wide variation in damage claims presented to railroads: differences in the value of the carload because of the density and bulk of the product, types and suitability of the containers, and the degree and efficiency of loading and bracing methods used to prepare the car for shipment.

Many of the damage claims for fruits and vegetables occur because of unsuitable and faulty containers and poor loading practices. These factors are subject to control and improvement. For example, it was found that cantaloupe crates loaded on end had only about one-third the breakage of crates loaded on their sides and lengthwise of the car. The tying of a single wire around lettuce crates was found to reduce crates damaged in shipment. Experimentation is now going on with new containers and new packing methods for many commodities. Many of the methods and containers used have not changed from the early days of railroad transportation. Imagination coupled with a lack of reverence for the status quo can produce cost-saving results.

Changing the Product

In this area are probably some of the greatest potentialities for attacking the transportation cost problem. High damage claims to a substantial extent hinge on the nature of the product itself. The product should not be accepted as a given, unchangeable fact. It, too, can be changed.

High perishability is one of the basic reasons for expensive transportation. But poor-quality products are more perishable than top-quality products. An expanded program of farm selling on grades might result in products being more closely graded before shipment. Then, only those products best able to stand up during movement would move long distances. Those that might deteriorate more quickly would be sold in the nearby markets.

Bulky, low-valued shipments can also be changed. The shipment of frozen concentrated orange juice in place of the whole fruit is an example of one kind of possible change. The production area slaughter of livestock and the shipment of carcasses or prime retail cuts rather than live animals is another example of a possible change that can be made. In general, production area processing will result in less bulky, higher-value, and often less perishable products for shipment to consumption areas.

Continued High Food Transportation Bill

Though a fresh and imaginative approach can perhaps lower costs, the food transportation bill will remain high. The inherent nature of agricultural production and products makes for an expensive transportation situation. The great distances between production and consumption areas will still exist in a large country. The job of assembling the products from scattered small production units will remain an expensive operation. Many commodities will still be perishable, resulting in high spoilage and extensive use of refrigeration. The seasonal nature of agricultural production will continue to create peak transportation demands in some seasons. Costs of the transportation agency will not suddenly become flexible so that rates can respond quickly to price level changes. Finally, the rising costs of labor, energy, and other inputs into the transportation function will keep upward pressures on freight rates.

SUMMARY

Transportation is the key link in the food systems' marketing chain, connecting geographically specialized farmers and an urbanized consumer population. This marketing function constitutes 6 percent of the food marketing bill and contributes significantly to the creation and preservation of place utility for consumers. Owing to special product and production characteristics, agriculture and the food industry have unique transportation needs. Most farm and food products move by rail and truck, and there has been a shift in tonnage from rails to trucks. However, the deregulation of the railroads in the 1980s has improved the competitive position of the rails. Barges are important in hauling grain to the major ports. Air transportation serves a limited and special role for a few select high-valued farm products. Transportation costs influence both farm and consumer food prices and quality as well as the costs of storage and inventory. Although there have been some improvements in transportation efficiency, they have not been sufficient to prevent rising freight rates. Consequently, the food industry transport bill will continue to rise.

KEY TERMS AND CONCEPTS

agricultural exemption
alternative transportation modes
class, commodity rates
diversion privilege
freight forwarder
Interstate Commerce Commission

common carrier
deregulation
demurrage
transit privilege
unit train

DISCUSSION QUESTIONS

1. What are the principal features of the alternative transportation modes that would affect a food marketing firm's choice of transportation? A farmer's choice?

2. Railroads have sometimes cited public subsidies for water, air, and motor carriers as the reasons for their competitive disadvantage in attracting freight. What are these subsidies? Are the railroads correct?

3. What factors determine a farmer's choice of transportation?

4. Although pipelines are very efficient at moving some products (oil, gas) long distances, they are not used to transport farm products. Do you see any potential for pipeline transportation of farm commodities? How about blimps?

5. Improvements in transportation efficiency are sometimes said to lower firm's inventory costs. Explain how this occurs.

6. Debate the proposition: "Rate regulations should be removed from all transportation carriers so that the free market sets rates."

7. How could a change in freight rates change the comparative advantage of a country? Of a U.S. region?

CHAPTER 19

STORAGE

OBJECTIVES

After reading this chapter, you will be able to:

1. Understand why storage and inventory operations are so important in the food industry.
2. Explain why and when farmers, marketing firms, and consumers should perform the storage marketing functions.
3. Make recommendations for improving the efficiency and lowering the costs of food storage.

Storage operations occur at every level of the food industry. Because sales and purchases rarely happen at the same time, every food marketing firm performs some storage and warehousing. Farmers are assuming increased responsibility for commodity storage. Consumers also store considerable quantities of food in refrigerators, freezers, and pantries. The storage marketing function is associated with the creation of time utility. This is an important source of value in the food industry, where supply and demand are seldom in immediate balance. Storage operations are necessary to bridge the time gap between periodic harvests and marketings and relatively stable usage of food on a year-round basis.

Storage is interrelated with other marketing functions, such as transportation, processing, financing, and risk bearing. In a sense, farm products are being "stored" at the time they are in transit or are in the processing operation. The relationship of storage and transportation is particularly critical at harvest time. A shortage of transportation facilities during a harvest glut backs up grain at the farm and at the local elevator, resulting in falling cash prices. Processing fresh products by canning or freezing them is another form of storage. And because storage operations delay sales and subject the firm to inventory risks, financing and risk bearing are considered part of the storage function.

A number of issues concern the food storage function. How large should U.S. food stocks be? World food stocks? Who should own these stocks? How should food stocks be managed and financed? These are important public policy questions. For the firm, there are other problems. How can storage and inventory costs be reduced? What level of stocks is necessary for efficient plant operation? Should storage capacity be increased?

FOOD STOCKS, CARRYOVER, AND RESERVES

There are several kinds of food storage, serving various purposes. A certain level of supply, or *working inventory,* is necessary for an efficient marketing process. These stocks keep the marketing pipeline full, contributing to full-capacity operations and preventing supply disruptions. Both consumers and food marketing firms maintain these working inventories for convenience and efficiency.

Seasonal food stocks are a related form of food storage. Over the marketing year, these are held to balance out supplies with demand. Seasonal stocks are necessary for products that are harvested in a short time but are consumed throughout the year. Both farmers and food marketing firms hold seasonal food stocks. Consumers may also build these stocks by increasing purchases of in-season commodities for later consumption.

Carryover stocks refer to the amount of commodity left over from one marketing year to the next. Annual production and consumption seldom balance precisely, and there may be carryovers ("old crops") or shortfalls, going into the next harvest period. These carryovers then become an addition to the supply available for consumption in the following year. The 1991–1999 production, carryover stocks, and usage of corn are shown in Figure 19–1. For a single-harvest period crop like grain that is consumed year-round, the seasonal stock pattern is illustrated in Figure 19–2. A commodity with a year-round production and consumption pattern, such as meat, will have a very different annual storage cycle.

In recent years, concern has increased regarding *reserve* or *buffer food stocks.* This form of commodity storage is intended to balance food supplies with demand over the long run and between food surplus and deficit-producing countries. Such food reserves are seen as a way to support and stabilize farm prices and protect against severe food shortages worldwide.

A fourth form of food storage might be termed *speculative stocks.* Farmers, food marketing firms, and consumers may at times hold larger than normal food stocks when they expect prices to rise. These speculative stocks would then increase in value and result in an inventory profit.

A distinction should be made between *voluntary* and *involuntary storage* of food stocks. A farmer or marketing firm may voluntarily increase stock levels for efficiency or speculative reasons. However, at other times stocks may rise as a consequence of an unanticipated harvest glut or a slow demand. These stock level changes are involuntary. During the 1950s, for example, the United States unintentionally built up a grain reserve as a result of farm price and income policies. The difference between voluntary and involuntary stocks is that the former is a purposeful, intentional change in food stocks, and the latter is accidental and unanticipated.

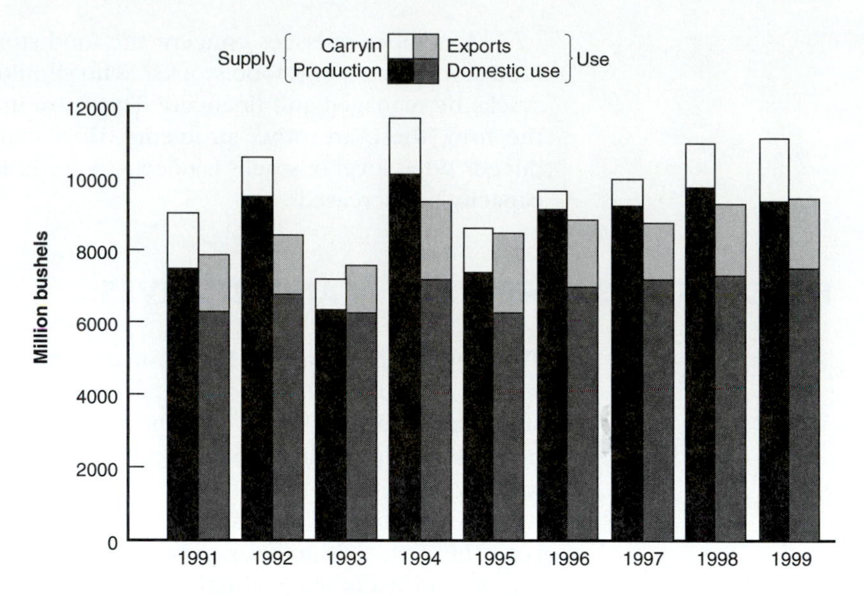

FIGURE 19–1 U.S. corn supply, usage, and carryover stocks, 1991-1999. (U.S. Department of Agriculture.)

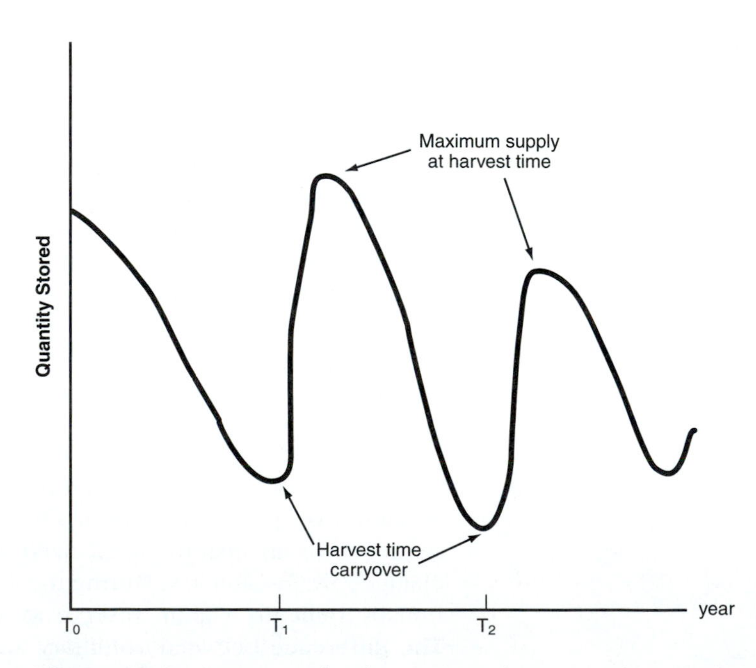

Figure 19–2 Storage and sales pattern for a storable, single-harvest commodity.

STORAGE OPERATIONS

Farm and food commodities can be stored at several places in the food system. Over time, changes in food production and in storage technology have altered food storage patterns.

Storage Locations and Capacities

It is extremely difficult to estimate the amount of food in storage at any time, or even to measure the total food storage capacity available. A considerable proportion of the total food storage takes place within the channels of trade in the form of inventories. Civil defense studies have estimated that on a minimum 2,000-calorie basis there would be 15.1 days' supply of foodstuffs on hand in retail food stores, 1.9 days' supply in restaurants and other eating places, and 16.1 days' supply in public warehouses, private processors, and wholesalers. If we adjust these figures to the approximately 3,000-calorie level of consumption that exists in the United States, they indicate about a three weeks' supply of food in the wholesale and retail marketing channels. As would be expected, more than 80 percent of this food is in cans, bottles, and packages; the remainder is in fresh and frozen products. With notable exceptions, consumers do not store large quantities of food at home.

Processing plants are also an important part of the storage structure, either holding substantial amounts of the raw material before it is processed or the finished product after processing. For example, a cereal manufacturer stores large amounts of grain but generally does not keep a large inventory of finished cereals on hand. On the other hand, a large tomato canner must can the entire year's supply at tomato harvest time and carry holdings in the form of the canned product.

Perishable commodities, of course, require more specialized and expensive storage facilities than nonperishables. This includes public warehouses, private warehouses, and freezer-locker plants that rent space to consumers. Private refrigerated space has been increasing more rapidly than public space, and each now accounts for about 50 percent of the total space. Frozen food storage capacity has also increased rapidly in retail food stores. In addition, almost all homes have mechanical refrigeration, and many families have a home freezer.

The storage capacity of farms is particularly difficult to estimate because general-purpose farm structures are frequently used to store grain. Farm storage capacity has undoubtedly increased over the years as government grain storage operations have declined and more on-farm storage facilities were built. There is probably more grain storage capacity on U.S. farms than off-farm in the marketing channels. Farmers in a few grain surplus countries, including the United States, appear to own the bulk of world food reserves. Most of the worldwide grain stocks held by food marketing firms and other countries are working, carryover stocks.

Changing Seasonal Storage Patterns

Changes in production and utilization patterns may also change storage practices. For example, seasonal variation in egg production has decreased, with the consequence that the storage holdings of eggs as a percent of production have been getting smaller and smaller. On the other hand, though livestock production also has less seasonal variation with passing time, seasonal variability in meat storage is increasing. The growth of processed meats has tended to increase the usefulness of frozen holdings.

Technological developments may also change the places and methods of storage. Prior to the perfection of quick-freezing techniques, for example, the fruit preserve industry could operate only at the time of harvest. Now, however, it can operate on a year-round basis by securing most of its supplies in the frozen form. Before refrigeration equipment was developed, the basic unit of storage was the canned finished product. Now food is stored in the frozen raw form. The ability to partially process and bulk store a commodity like tomatoes allows processors to stretch out the canning season over a longer time period.

Public Warehouse Supervision

Early agitation for some type of public supervision of commodity storage arose when farmers were unable to secure adequate loans on crops they wished to store. Such a situation often forced the farmer to sell a crop immediately upon harvest in order to secure needed money. The 1916 United States Warehouse Act gives the Secretary of Agriculture authority to supervise warehouses operating under its provisions.

Whether a warehouse wishes to operate under the provisions of the Warehouse Act is optional. However, if it desires to so operate, then the business must meet certain requirements. The warehouse must pass an inspection of its condition and facilities, and the operator must post a bond and furnish a guarantee of financial responsibility.

All commodities stored in a supervised warehouse are inspected for quantity, condition, quality, and insurance coverage. The owner is then given a warehouse receipt showing the exact nature of the commodity in storage. Because the commodity has been verified by a third party, these receipts are usually accepted by banks as collateral for loans. In this way the person who does not wish to sell the product immediately has a source of reasonable credit by which to secure needed cash. Here, then, is another illustration of the close interrelationship of various marketing functions. The public warehouse system is a very important factor in financing the marketing processes. The negotiable nature of the warehouse receipts also greatly facilitates the transfer of title and reduces the amount of physical handling.

WHO SHOULD STORE?

Like the other functions in the marketing system, storage is a necessary function. But it is not always clear who should do it or where it should be done. Obviously, the storage within the trade channels, in the form of inventories, is the responsibility of the various firms in the system. But the storage throughout the season may be provided by several agencies—the farmer, the commercial storage operator, the food processor, the speculator, and others.

Costs of storage are not necessarily the same for all participants in the marketing channel. Though the theoretically perfect seasonal price fluctuation covers the cost of storage, it does not necessarily cover the operations of highest cost. Even though storage is necessary, it is not equally profitable for all individuals at all levels of the marketing channel to store.

Each level of the marketing system tries to push much of the storage function to another middleman in the system. Retailers struggle to hold inventories at the lowest level consistent with serving their customers. Processors attempt to schedule production in order to avoid tying up their capital in costly storage stocks.

Should the farmer store his own products? No generalized rule can be made, and the case of each commodity must be decided on its cost-return conditions. In some circumstances it may pay farmers to build storage facilities and provide their own storage. In other cases, it may be best for producers to rent commercial storage. In still other circumstances, it may be best for producers to sell at harvest and let some other agency of the marketing system perform the storage operation. Nor can the profitability of farm storage be generalized for all years. The outlook for general business conditions or farm prices may make storage desirable for one year and not for another.

The Commodity Credit Corporation (CCC), an agency of the U.S. Department of Agriculture, is the government agency that buys, stores, and sells farm commodities in connection with federal price and income support programs. Its activities are carried out by the Agricultural Stabilization and Conservation Service (ASCS), another agency of the U.S. Department of Agriculture. The CCC has a line of credit with the federal treasury that enables it to support farm prices through purchase and storage programs. Its purchase and storage activities have lessened in recent years with more market-oriented farm programs designed to provide direct payments to farmers rather than to raise prices through market interventions.

The federal government continues to provide U.S. farmers financial assistance in storing grains. A farmer may take out a non-recourse or marketing assistance loan from the government rather than sell at harvest. The grain is stored in either farm or commerical facilities. This allows the farmer to avoid selling at low harvest prices and possibly to gain storage profits. Farmers may also store grains under the longer-term Farmer-Owned Reserve Program and receive storage payments from the government. A loan deficiency payment (LDP) is a government payment to grain farmers who agree to sell their crop at harvest rather than obtain a government storage loan.

IMPROVING FOOD STORAGE

The storage function will always be a complex and expensive one in the food industry. Like other marketing functions, storage cannot be eliminated, but the costs of storage can be reduced by a decrease in storage activities and increased operational efficiency. In determining the total costs of holding commodities, five possible categories of costs must be considered:

1. *The costs necessary to provide and maintain the physical facilities for storage.* These costs would include such items as repairs, depreciation, and insurance against loss.

2. *The interest on the financial investment in the product while it is in storage.* Whether the money is actually borrowed or not, this is a cost that should be assessed at the rate of interest that would have to be paid if money were borrowed during the storage period.

3. *The cost of quality deterioration and shrinkage during storage.* Many commodities either deteriorate in quality or shrink in volume—or both—while in storage. In a few cases, some stored commodities, such as corn, may increase in quality while shrinking in volume. In such cases, storage may result in a net gain instead of a net loss for this particular factor.

4. *The loss that may result from poor consumer acceptance of the stored versus the fresh product.* Consumers now view canning, freezing, and drying as accepted value-adding processing and storage operations. Frozen meat may only be accepted by consumers at a price discount, even though its quality as measured by the grading system is not lessened. There is also consumer resistance to storage eggs as opposed to fresh eggs, though the quality as measured by the grading system may be the same. This is not a problem in all commodities. Nor do such consumer preference patterns remain unchanged. For example, the widespread presence of home freezers is overcoming the resistance to frozen meats.

5. *The risk that the price of the product might unexpectedly decline.* Under these circumstances the product might have to be sold at less than its value at the time it was placed in storage. The possibility of a favorable movement in prices, on the other hand, is a major factor in encouraging speculative storage.

Increasing the Productivity of Storage Facilities

The productivity of grocery wholesalers has not risen as fast as that of farmers and food processors. The probable reasons for this were an increase in the number and variety of food products stored and the shifting of some marketing functions from retailers to wholesalers. Nevertheless, there are ways to improve food storage efficiency in the future.

The most promising are increasing labor efficiency by reorganization of handling methods and by additional mechanization and computerization of inventory control. Often, a by-product of increased efficiency is an increase in the storage capacity of a given area.

The concept that storage is not static but is an integral part of the movement of goods has directed increasing attention toward problems of handling. Much of the work (and costs) of warehousing occurs during the unloading and loading operations. Attention to this area has led to the construction of one-story warehouses with ample loading facilities in contrast to the old multiple-story buildings. The use of pallet storage, along with the fork-lift truck, conveyor systems, automatic dumping devices, and other mechanical devices is giving some storage operations a long-needed face-lifting. One multiple-story warehouse originally handled a maximum of five trucks a day and 40,000 cases a month, with a work force of fourteen. After redesigning and installing new equipment, it could handle ten trucks a day and 50,000 cases a month, with a workforce of seven.

For some commodities, such as eggs and hogs, production has gradually shifted to a more uniform year-round pattern. Such changes reduce the amount of storage needed. In other cases, product developments have changed the nature of the storage operation. The rapid development of frozen orange juice concentrate has shifted the emphasis from that of maintaining fresh fruit to that of properly maintaining the frozen product. The trend toward more processed, table-ready meats has changed some of the meat storage requirements. The development of frozen and dried eggs in addition to shell eggs has increased the flexibility of egg storage operations. Some varieties of crops have been found to be better for storage than others.

Basic seasonal price patterns can be affected by changes in storage costs. Any changes that will reduce the costs of holding a commodity will tend to reduce the amount of seasonal price variation. This helps explain why a relatively new crop often has a greater amount of seasonal variation, which declines as the years pass. In a new situation, storage facilities may not be well organized. The costs of storage are therefore quite high. Then, as the situation improves, storage costs may decline. The seasonal price pattern then tends to become less pronounced. Such developments and their effects on the amount of seasonal price variation are, of course, not limited to new commodities. The price pattern of any commodity may be altered by changes in the storage situation.

Improved business management techniques have also contributed to improved inventory and production control methods. This means that fewer goods per dollar of sale need to be kept on hand at retail and wholesale levels. Processors, through improved scheduling of activities, can produce more to meet orders. Just-in-time inventory control can reduce storage costs. These developments speed up the flow of goods and thereby reduce the amount of needed storage. For example, inventory on hand in a supermarket has always been an expensive proposition. Although the cost of "stocking out" of products is high, so is the cost of the money tied up in inventory. Modern management techniques, sometimes referred to as "efficient consumer response," now allow computerized inventory and ordering systems that provide daily ordering and store deliveries. Bar codes and optical scanning devices at the checkout counter "sense" what products have been purchased and automatically reorder appropriate products from the warehouse to restock the shelves. These techniques reduce the need for large investments in inventory and facilities and may help reduce the cost of marketing.

Reducing Product Deterioration

Wine and cheese may improve with storage and age, but most food products deteriorate in quality and decline in volume while in storage. Arresting these quality and quantity changes also has potential for reducing storage costs. Great advances have been made in discovering the best storage conditions for individual products. Different commodities have different temperature and humidity requirements for optimum maintenance of quality. Temperatures of 30° to 32°F are usually recommended for apple storage, but it has been found that Grimes Golden apples hold up best at 34° to 36°F. For most fresh products the recommended relative humidity is 80 to 90 percent, but 70 percent is better for onions and nuts. Temperature and humidity conditions are now known to be important even for storage of canned goods, to control deterioration and rusting.

Food science is making increasing contributions to quality control. One example is the potato sprout inhibitor. Potatoes sprayed with this chemical when placed in storage will not sprout readily. Another example is the use of a "modified atmosphere" where proportions of various gases are controlled in gas-tight rooms to retard deterioration.

Practices followed in preparing products for storage have an important effect on their storability. The deterioration rate of many products can be reduced if the temperature is reduced as soon after harvest as possible. The discovery that cuts and bruises received during harvesting reduce storage life has led to redesigning of machinery to reduce sharp edges and dropping distances. The use of polyethylene liners in boxes has been found to cut quality losses of such products as pears, sweet cherries, and apples.

The proper farm use of insecticides and fungicides has reduced storage rotting caused by insect injury. Experimentation with the use of antibiotic chemicals and irradiation has uncovered some potentialities in retarding the spoilage of various commodities. Some antibiotics have been approved for use in the preservation of poultry. And new ways may be found to retard the spoilage of red meats. The degree of ripeness at harvest has also been related to length of storage life. In grains this knowledge has led to various methods of quick-drying immediately after harvest.

Storage and Risk Bearing

The futures market can assist in managing the financing and risk-bearing operations associated with food storage. Whether or not money is borrowed on a stored product, the interest lost on the value of the stored commodity is a cost to the farmer or the food marketing firm. Frequently, it is easier to borrow working capital on a hedged commodity. Moreover, the storage hedge can be used to transfer the price risk from the hedger to the speculator (see Chapter 20). This, too, can result in decreased food industry storage costs.

SUMMARY

Because of imperfect coordination between supplies and demand, the need to maintain pipeline-level stocks, and the perishable nature of the product, food storage and warehousing are important marketing functions. Food stocks can be classified into working inventory stocks, seasonal stocks, carryover stocks, and buffer stocks. Stocks are held at all levels of the food industry and also by households and by government. Recent changes in farm policy have encouraged farmers to hold and control more farm product stocks. Although changes in seasonal food production have in some cases reduced food storage needs, food storage capacity is increasing overall, both on-farm and off-farm. Storage operations play an important role in stabilizing prices. Costs of physical facilities, interest costs, quality deterioration, and risk costs all affect food storage expenses. These costs may be reduced through increased warehousing productivity and improved storage management.

KEY TERMS AND CONCEPTS

- carryover
- Commodity Credit Corporation
- efficient consumer response
- farmer-owned reserve
- food reserve
- free stocks
- loan deficiency payment
- U.S. Warehouse Act
- voluntary stocks
- working inventory stocks

DISCUSSION QUESTIONS

1. Explain how food storage operations contribute to time utility, and show how the law of one price operates over the storage season.

2. It is sometimes said that food processing is an alternative to storage. Explain why this is so.

3. How could improved coordination of food supplies and demand reduce the costs of food storage?

4. What are the principal differences between inventory stocks, seasonal stocks, carryover stocks, and buffer stocks?

5. Specify the alternatives farmers have to selling or storing their crops at harvest. What factors influence the storage decision?

6. Examine the most recent farm bill for provisions relating to on-farm storage and food reserves.

7. "The U.S. government should not be involved in food storage operations." Comment.

CHAPTER 20

RISK MANAGEMENT AND THE FUTURES MARKET

I invest. He speculates. They gamble.

OBJECTIVES

After reading this chapter, you will be able to:

1. Understand how the futures market can assist firms in protecting against price risk.
2. Explain how and when farmers and marketing firms use hedges and options to protect themselves.
3. Appreciate the differences between hedges, options, and forward contracts.
4. Evaluate criticisms of the futures markets.

Storage, financing, risk bearing, and market intelligence are important marketing functions. Commodity futures markets are marketing tools for farmers and food marketing firms that can aid in the performance of these functions.

TYPES OF MARKET RISK

Risk is inherent in the ownership of goods. And as with the other functions of marketing, risks must be borne by someone. They cannot be eliminated. Kinds of risk may be classified as follows:

1. *Product destruction* from natural hazards, such as fire, wind, pests, spoilage, and so on.

2. *Product deterioration in value* resulting from (a) quality deterioration or (b) price variations due to changing consumer preferences, changes in the supply situation, or changes in general business conditions.

Product Destruction

Those engaged in food marketing must face the possibility that fire, pests, or other forces may suddenly damage or destroy the products they have on hand or in storage. A marketing firm, especially if it is a large one, may build up its own funds to cover such a possibility. However, firms may also transfer this risk to an insurance company for a fee.

Insurance companies are specialized risk bearers that spread the risk over a wide area and groups of people or businesses. Most marketing firms find buying insurance more economical than attempting to provide for their own protection. In the marketing of agricultural products, insurance companies of varying types are of considerable importance. They range from those insuring products in transit from farm to market to those insuring the inventory of the retail store.

Product Deterioration in Value

Every marketer runs the risk of physical deterioration of a product while it is owned or stored. Aside from using the best technical equipment and knowledge available, there can be little transference of this risk from the owner of the products. A rapid change in temperature or a breakdown in equipment are possibilities. Such factors increase the marketing costs of perishables.

Value deterioration that stems from price changes may originate from many sources. Risks from changes in consumers' preference or acceptance is greatest in fashion lines. For example, clothes that may be high-priced one year will bring much less the next because they are out of style. Of course, most food products do not face such rapid preference shifts. Food preferences usually change much more slowly over a longer period of time. However, price changes that occur because of large annual shifts in production and in the general price situation are common for agricultural products. A handler who purchased large amounts of eggs for storage in the spring at what was considered "safe prices" may face a highly unfavorable fall egg market because of an unforeseen deterioration in general business conditions. An untimely frost may greatly increase the value of current apple storage holdings by sharply reducing the crop. A series of rains may make or break the wheat crop, and thereby greatly affect the fortunes of those who are holding large inventories of wheat in storage.

Wide and unpredictable fluctuations in the volume of available products cause food processors, handlers, and retailers additional and costly uncertainties. Successful advertising and merchandising require a reliable flow of uniform products. Equipment and personnel must be geared to an optimum volume of flow. Because the volume of products available to agricultural marketing agencies is predetermined by the level of farm production, uncertain production levels mean uncertain cost levels to many marketing firms.

It is in this area of price and cost change that agricultural marketing agencies probably face their greatest risk. Many devices are used either to minimize this risk or to shift it from one person or firm to another.

Because much of the risk may stem from lack of knowledge or inaccurate knowledge, efforts that improve the gathering and dissemination of market news and statistics and the standardization of products may reduce risk. Much of the government price-supporting activity is a mechanism for transferring price risks from producers and handlers to the tax roles of society. Government storage activities, which often accompany such supports, help reduce the risks inherent in wide variations of available market volume.

The efforts toward vertical integration of the marketing channel also attempt to reduce or transfer risk. Many of the integration contracts with farmers arrange for a fixed return, and thereby transfer the risks from changing prices to the integrator. The integrator in many instances attempts to obtain some effective control over the kind and amount of production available and thereby help reduce the risks involved in market activities.

Also, products are often sold "in advance"; that is, the price is fixed in the present for delivery at a specified future date. Elevators may sell grain, and once the grain is loaded in cars on the railroad siding, their responsibility ceases. Vegetable canners often sell their packs immediately to other middlemen who then must assume the price risk on inventories. Fresh produce dealers may obtain their supplies from growers with whom they share returns and thereby share market risks.

Some of these devices help reduce the risk that accrues and that people must bear in total. Others are devices for shifting the risk from one agency to another within the market channel. There are also ways by which the risk of price change may be shifted to those outside the market channel proper. The principal mechanism for contacting outside risk bearers and financing is the futures market.

THE FUTURES MARKET

The cash price for agricultural commodities—often called the spot price—is today's price for products delivered today. The *futures price* is today's price for products to be delivered in the future. Pricing and product delivery usually occur simultaneously in the cash markets, but these two events take place separately in the futures market. Futures contracts are not limited to agricultural commodities. Magazine subscriptions, pick-of-the-litter stud service, and even college tuition are forms of futures contracts in which prepayment is made for goods or a service to be delivered in the future at a guaranteed price.

The futures market is a mechanism for trading promises of future commodity deliveries among traders. As such, it is a unique risk-management and profit tool for farmers and food marketing firms.

Futures Market Exchanges

Commodity exchanges are marketplaces designed to facilitate trading in futures contracts. There are twelve organized commodity exchanges in operation in the United

States and several more operating in other countries. The Chicago Board of Trade and the Chicago Mercantile Exchange are the largest, accounting for perhaps 75 percent of all commodity futures trading.

The commodity exchanges are somewhat like the stock market. They are composed of member-traders who are authorized to buy and sell futures contracts for the public. There is a trading floor, where buyers and sellers meet; a governing board that sets and enforces the rules for orderly trading; and a clearinghouse that facilitates trading and delivery of commodities. Most commodity exchanges are found in cities that are major transportation centers and through which a substantial portion of the product moves (Chicago, New York, Kansas City, Minneapolis, and Winnipeg, Canada).

The most visible aspects of the exchanges are the exchange floor and the trading "pits," where the actual trades are made. Less visible but equally important is the communication network that links brokers throughout the world to the traders in these pits. The commodity trading floor is a physical place, but it represents a worldwide market. Some fifty different commodities are traded on futures markets, and more are being added each year. Some of these are shown in Table 20-1. Futures market daily price quotations can be found in the business sections of most newspapers, and reports of exchange prices are available in the offices of most commodity brokers as well as on the Internet.

Commodity exchanges are quite close to the perfectly competitive market. At any moment there are thousands of buyers and sellers of futures contracts participating in the market, and there are an even greater number of potential participants. Prices are established through open trading on the floor of the exchange, where all buyers and sellers are represented either personally or via electronic communication through their brokers. Most information on developments affecting futures prices is public, and the prices themselves are communicated worldwide.

TABLE 20-1 Selected Agricultural Commodities Traded on Futures Markets

Commodity	*Contract Size*	*Primary Exchange*
Crops		
Cotton	50,000 lb.	New York Cotton Exchange
Corn	5,000 bu.	Chicago Board of Trade
Soybeans	5,000 bu.	Chicago Board of Trade
Wheat	5,000 bu.	Chicago Board of Trade
Livestock		
Cattle	40,000 lb.	Chicago Mercantile Exchange
Feeder cattle	44,000 lb.	Chicago Mercantile Exchange
Hogs	40,000 lb.	Chicago Mercantile Exchange
Semiprocessed Products		
Soybean oil	60,000 lb.	Chicago Board of Trade
Soybean meal	100 tons	Chicago Board of Trade
Pork bellies	38,000 lb.	Chicago Mercantile Exchange

SOURCE: Information obtained from various exchanges.

The futures market has grown rapidly in the past 25 years. This market is expected to continue to grow as more traders are attracted to it and as more farmers and food marketing firms become better acquainted with the market and its role as a marketing tool.

The Futures Contract

The futures contracts that are traded on commodity exchanges are promises to deliver or accept delivery of a specific commodity at a specified time in the future. No physical commodity changes hands when the contract is traded and priced. Deliveries are made against the contract when it matures (or becomes due) in the month for which it is named. Payment for delivery of the commodity is at the price determined when the original trade was made, often several months prior to delivery. Thus, a seller of a futures contract guarantees delivery of the commodity at the price agreed upon when the contract is traded, and the buyer is assured of receiving the commodity in the specified month at this price. In this way, futures contracts allow *forward pricing* of commodity deliveries.

These contracts represent standardized quantities and qualities of commodities, as indicated in Table 20-1. For example, a Chicago May corn futures contract may trade at $2.50 on January 5. This means that a seller has contracted to deliver 5,000 bushels of No. 2 yellow corn to a specific warehouse in May. Upon delivery of the warehouse receipt for the product, a buyer who bought a contract in January will pay the seller $2.50/bushel. The exchange clearinghouse guarantees the delivery of the product against the contract.

Each commodity has different delivery (maturity) months. For instance, wheat can be contracted for delivery in July, September, December, March, or May. Other commodities have other delivery months.

Where do futures market contracts come from? Here the futures market differs from the stock market. Anyone may legally "create" a futures contract to buy or sell commodities at any time by simply calling a brokerage house and doing so. That farmers and merchants can make a promise to deliver 5,000 bushels of corn in the future seems reasonable. However, it is surprising to many that doctors, lawyers, and even students may also legally create and trade these contracts.

Trading and Pricing of Futures Contracts

The futures market has been described as a place where sellers make promises to deliver something they do not have to buyers who promise to accept delivery of something they do not want—and both legally break their promises. A strange market indeed!

There are two kinds of futures market traders: *speculators* and *hedgers.* Speculators are traders who attempt to anticipate and profit from futures price movements. Speculators generally have neither the capability nor interest in fulfilling their futures contracts by taking or making delivery at contract maturity. Hedgers also attempt to profit from anticipated price changes, but they usually can take or make delivery of the commodity at contract maturity. However, like speculators, hedgers seldom allow futures contracts to mature. There are generally many more speculators than hedgers in the futures market.

Futures contract promises can be fulfilled in either of two ways. The commodity can be delivered or accepted at contract maturity, or the promise can be nullified by an offsetting futures market transaction prior to contract maturity. Sellers can buy back their promise, or buyers can sell out their promise. In doing so, it is not necessary to locate the original buyer or seller of the contract. To nullify a contract, it is sufficient to simply make an equal and opposite transaction from the original. Most futures contracts are nullified prior to contract maturity, and relatively few contracts result in product delivery.

It is generally more profitable to nullify a contract than to deliver or take delivery. A trader who sells a May futures contract for $2.50 on January 5 would make a ten-cent profit (ignoring brokerage fees) per bushel if the contract could be bought back later at $2.40. Similarly, a trader who bought a November futures contract in June at $4.00 and later sold it at $4.20 would also make a profit. Of course, the traders on the other sides of these transactions would have lost equivalent sums of money.

The attraction of futures market speculative trading lies in the potential for frequent, large, and somewhat unpredictable, commodity price swings. The uncertainty of futures commodity prices assures that there will be a large number of both buyers and sellers of futures commodities at all times. Buyers will always be able to find sellers, and vice versa.

Some specialized terms and names have evolved in futures trading. Sellers of a contract are said to be in a *short* position—they owe the commodity. A buyer of a contract is said to be *long*—and is committed to take delivery of the product. *Bulls* are traders who feel prices will rise, so they "go long" (buy futures). *Bears* feel prices will fall, so they "go short" (sell futures). It takes a bull and a bear to make a transaction.

In the futures market, speculative profits can be made from either rising or falling prices. A bull makes money by buying today, watching the price rise, and then selling out at the higher price. The price rise is the profit. A bear does the opposite, selling today and buying back the contract later at a lower price. Speculative profits, then, depend on the traders' correct anticipation of price movements. Clearly, for every futures contract traded, there must be a bull and a bear, a long and a short trader, and a winner and a loser (unless the price stays constant).

What determines the price of a futures contract? Like any other commodity that can be traded, a futures contract price is determined by how much buyers are willing to buy it for, and how much sellers are willing to sell it for. In turn, these attitudes toward buying and selling futures contracts reflect the expected value of the actual product represented in the contract at maturity. Why would anyone pay $2.50 for a contract calling for November delivery of soybeans if they thought the cash price of soybeans would be less than $2.50 in November? Conversely, why would anyone sell a promise to deliver soybeans next November if they thought soybeans might sell for more than $2.50 next November? Because there is the possibility of delivery, or accepting delivery, of a futures promise, futures and cash prices of a commodity are closely related.

Hence, a futures price is somewhat more complex than a cash price. A cash price reflects "what is" today. A futures price reflects the market as traders think it will be.

Because of the time lag between pricing and delivery of a futures commodity, and the uncertainties of agricultural prices, there is always a wide difference of opinion as to the "correct" futures price. As commodity conditions and attitudes toward buying and selling futures contracts change, the futures price will also change. These differences of opinion and unpredictable price changes ensure that there will always be plenty of bulls and bears in the market at the same time. Prices will adjust to make this so. More buyers than sellers will raise prices, transforming some buyers into sellers, and more sellers than buyers will lower prices, changing some bears into bulls.

Futures trading involves two costs. There is a brokerage fee for executing orders. There is also a *margin* requirement, which is a form of earnest money. For example, a trader may post a $500 margin when making an initial buy or sell of a corn futures contract. This can be used by the broker to cover losses and brokerage fees, or it is returned to the trader in the event of profitable trades. This margin "down payment" amounts to only a small portion of the value of the corn represented by the contract, say 5 percent if corn is selling for $2.00/bushel. This ability to trade a large value of product (say, $10,000 of corn) with limited capital is known as *financial leverage.* Futures markets are highly leveraged, meaning that there are opportunities for large profits on investments—or large losses.

RELATIONSHIPS BETWEEN CASH AND FUTURES PRICES

The difference between a cash and a futures price is called the *basis.* The basis is critical to an understanding of the futures market. In fact, agricultural commodity prices are often described by their basis. For example, a local grain buyer may quote a spot grain price of "30 cents under," meaning that today's cash price is 30 cents under the nearby futures price. Commodity traders often are more interested in the level or change in basis than in actual cash prices.

There are three normal relationships between cash and futures prices for a storable commodity, like grain. First, because actual commodities can be delivered against futures contracts at contract maturity, the cash price of a commodity at the delivery point will be almost the same as the futures price at that time. That is, the basis will approach zero at contract maturity, when "the future becomes the present."

Second, prior to contract maturity, futures prices are normally above cash prices by the cost of holding the commodity until contract maturity. This storage cost increases as contract maturity approaches. Therefore, the basis is said to "narrow" (or "strengthen") as the contract matures. This also follows from the fact that cash and futures prices must approach each other at maturity.

Third, cash and futures prices tend to move up and down together because both are affected in the same way by changes in supply and demand. A change in market conditions that increases the cash price will also tend to raise the futures price. The two markets do not move together perfectly because of errors in price discovery, expectations, and other random variations. Nevertheless, the tendency toward similar cash and futures price movements means that there are periods when the basis will be constant despite changing cash and futures prices.

These relationships are illustrated in Figure 20-1, showing the cash and futures prices for corn. As shown, the cash and futures prices do come together at contract maturity; the futures price is generally above the cash price; the basis narrows as

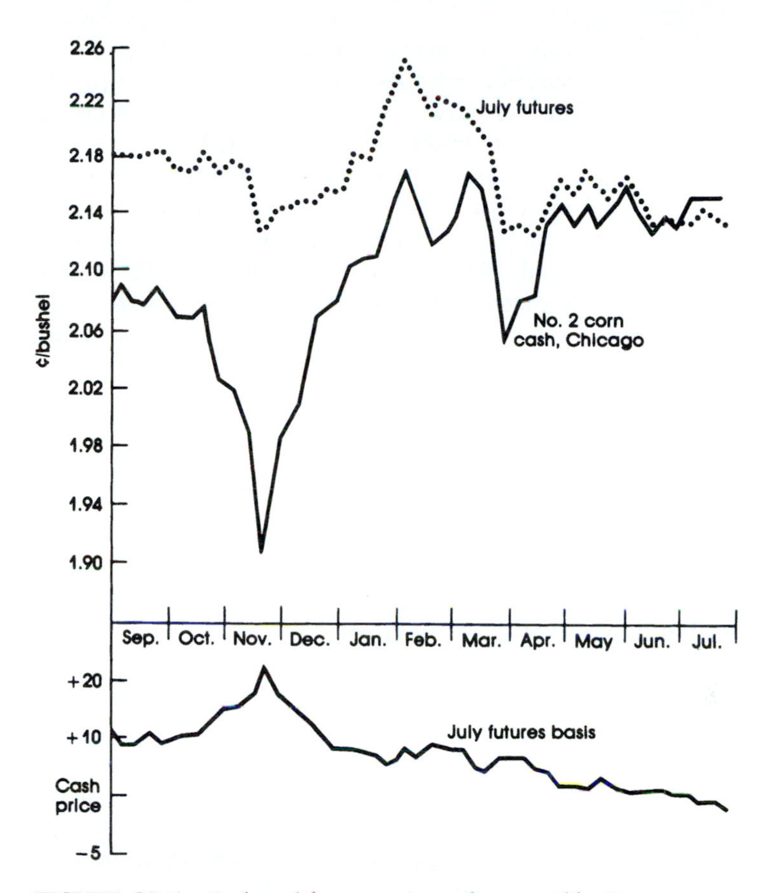

FIGURE 20–1 Cash and futures prices of corn and basis.

contract maturity approaches; and there are short-term periods when the basis is relatively constant. Traders often calculate the basis by subtracting cash prices from futures prices, making the basis a negative value.

One other important point is that the basis level and the basis changes tend to follow predictable patterns from year to year. In fact, basis is more predictable than cash prices. For example, the July corn basis might average 10 cents in November at harvest, then narrow to 2 cents in June. Or, a farmer might know from experience that, at harvest time, local elevators normally offer a cash price for soybeans of "35 cents under the November futures." Knowledge of these historical basis patterns is vital to the use of the futures market in risk management.

HEDGING AND RISK MANAGEMENT

Price risk is inherent in the ownership and handling of agricultural commodities. By the nature of their activities, farmers and food marketing firms are exposed to unpredictable price swings. A food processing company that holds commodities for a few days, weeks, or months during the manufacturing and marketing processes is subject to an inventory price risk. An elevator owner buying cash grain for later sale assumes a price risk between the time of purchase and sale. A farmer who plants in the spring with uncertain knowledge of fall harvest prices is in a sense speculating on cash prices and is facing a risk.

Farmers and food marketing firms can make profits by speculating in cash prices or by performing utility-adding marketing functions. However, speculating in cash prices involves the risk that prices might fall rather than rise. Because agricultural prices are highly volatile, many firms cannot afford to speculate on cash price changes—or choose not to do so—preferring instead to profit only from their conventional marketing functions. For these firms, the futures market provides an opportunity to limit their exposure to price risks.

A hedge implies a protective mechanism. A futures market hedge is such a risk-management device. It involves the temporary substitution of a futures market transaction for a cash transaction. The mechanics of a hedge consist of making equal and opposite transactions on the cash and futures markets in order to protect the firm against adverse cash price movements.

The Hypothetical Perfect Hedge

The process of hedging can be explained by the operations of the owner of a grain elevator. The owner buys cash grain from farmers and ships it to a terminal market for cash sale one week later. Owners of grain elevators normally operate on a small profit margin per bushel, and they attempt to make their profits from handling charges rather than from speculative market positions.

The elevator owner is long in the cash market when it originally purchases the grain from farmers. Until this grain is sold, the owner is exposed to a cash price risk because the price of grain could fall by the time it is sold one week later. To protect against this possibility, the owner could hedge as follows:

Date	*Cash Market*	*Futures Market*	*Basis*
March 19	Buy 100,000 bu. @ $2.00	Sell 100,000 bu. @ $2.50	$ .50
March 25	Sell 100,000 bu. @ $1.90	Buy 100,000 bu. @ $2.40	$ .50
Gain or Loss	$–.10	$+.10	

March 25 cash price	=	$1.90
+ Futures gain	=	.10
Total return	=	2.00
– Original cash cost	=	2.00
Net profit or loss	=	- 0 -

When cash grain was purchased, futures were sold to "set" the hedge. When the cash grain was sold, the futures contracts were purchased back. The cash market declined, as the owner of the elevator feared it would, and there was a 10 cents per bushel loss on the cash transaction. However, the futures market also declined, and the opposite position there resulted in a 10 cents per bushel gain. On balance, the futures gain offsets the cash loss, and the owner of the elevator receives $2.00/bushel from the combined transactions.

This illustrates the *perfect hedge,* where the gain in one market exactly offsets the loss in the other. Thus, the futures hedge "protected" the owner from the 10-cent decline in the cash market. Notice that the owner did not deliver against the contract, even though delivery could have been made. The manager had no intention of making this delivery and knew the grain would be sold in one week. Most hedgers do not deliver against their contracts but turn their hedges in the same manner as speculators, buying back a previously sold position. The futures position is simply a temporary substitute for a planned cash sale.

There are two reasons why this hedge was effective. First, the cash and futures prices both declined. Therefore, the opposite positions provided compensating gains and losses. The second reason for the success of this perfect hedge was that the basis remained constant at 50 cents on both days. A constant basis, as indicated previously, is common for short-run periods such as this, but is by no means guaranteed. If the basis changes during the hedge, it will not be a perfect hedge, but it can still provide some protection against falling cash prices. For example, if the basis were to rise from 50 to 55 cents, so that on March 25 the futures contracts were purchased at $2.45, the futures transaction would offset 5 cents of the 10-cent cash market loss—a useful, but not perfect, hedge.

Basis changes, then, are an important influence on the success of hedges. Hedgers are more concerned with basis than with actual cash prices. Cash prices could have fallen from $20.00/bushel to $2.00/bushel and the hedge would still have been perfect if futures prices fell from $20.50 to $2.50, keeping the basis constant at 50 cents. Because basis is not perfectly predictable, it is often said that hedging involves speculation in basis rather than in cash prices. However, because basis is somewhat more predictable than cash prices, this can represent a risk reduction for the hedger.

What if cash and futures prices had risen instead of falling? In that case the cash market gain would have been exactly offset by the futures market loss, if the basis had remained constant at fifty cents. And the elevator owner still would have received $2.00/bushel on the two transactions. Of course, the elevator would have been better off without the hedge when prices rise, but this is hindsight. Without the hedge, the owner would run the risk of the price falling. This illustrates an important point about hedging: A hedge not only protects the hedger against falling prices, but also prevents the hedger from benefiting from rising prices. The hedger's cost of protection against falling prices is that speculative profits must be sacrificed if the cash markets should rise.

Where does the hedger's price protection come from? The answer is that another hedger—or, more probably, a speculator—bought from the elevator owner the 100,000 bushels of corn contracts on March 19 for $2.50/bushel, and still another futures trader sold corn contracts back to the elevator at $2.40 on March 25. The futures market provides the hedger with price protection when prices fall. In turn, the hedger pays the cash price rise to the futures market when prices increase.

The example just mentioned illustrates a *short hedge,* when a sale of futures is made as a temporary substitute for the cash sale of a commodity. A *long hedge* is used when the firm makes a promise today to deliver commodities, not yet owned, to the cash market at a specified future time. For example, an exporter might agree today to deliver 100,000 bushels of grain to a foreign port next month at a price of $4.50/bushel. For this hedge, the firm is initially short in the cash market, and a long position is taken in futures to set the hedge. The cash transaction is completed one month later when the firm purchases the 100,000 bushels at the going price and delivers them to the port. The hedge is "turned" by selling the original long position in the futures market.

Both of these hedges highlight the importance of equal and opposite positions on the two markets. They also illustrate the "no-profit" nature of the perfect hedge. The purpose of these hedges is to gain risk protection, not to increase profits. Unhedged cash market positions offer a greater opportunity for profit (and for loss) than hedges. For people who like price risks and can afford the accompanying potential losses, the futures market also offers the *Texas hedge*—one that doubles the firm's risk position by taking similar positions (long or short) in both cash and futures markets. This is not really a hedge.

The Storage Hedge

Although the perfect hedge provides valuable insights into the hedging process, it is not an accurate picture of how hedging is practiced by firms. The perfect hedge is a no-profit situation and would be unattractive to most firms. Nor could farmers use these short-term perfect hedges. In the real business world, hedging is both a risk-management and a profit-making tool. Businesses normally use the storage hedge when commodities are to be held for a period of time during which the basis is expected to narrow. This hedge has two purposes: (1) to protect the firm against adverse cash price movements, and (2) to assist the firm in earning *carrying charges* (storage costs, interest, and insurance) during the storage period. The storage hedge can be used by farmers and food marketing firms. Storage facilities are expensive to maintain, and to get the most from these investments the facilities should be used each year. However, there is no guarantee that the seasonal price rise over the storage period will be sufficient to cover these carrying charges. The storage hedge provides the firm with an opportunity not only to protect itself from a cash price decline but also from the possibility that cash prices at the end of the storage period may not be high enough to compensate the firm for its storage expenses.

The storage hedge is based on the expectation that the basis will narrow for a storable commodity as the futures contract matures. For example, suppose our elevator owner has storage costs of 35 cents per bushel from November to June, and fears that the seasonable cash price rise may be less than this. The elevator owner could refuse to store corn this year, but most would prefer to use their storage facilities to earn a profit, if possible. To do this, a storage hedge would work as follows:

Date	*Cash Market*	*Futures Market*	*Basis*
November 1	Buy corn @ $2.00/bu. and store for spring sale	Sell July corn contract @$2.50	$ .50
June 1	Sell corn @ $2.30/bu.	Buy July corn contract @ $2.40	$ .10
Gain/loss	+.30	+.10	$+.40

June 1 cash price	=	$2.30
+ Futures gain	=	.10
Total return	=	$2.40
– Original cash cost	=	2.00
= Return to storage	=	$ .40

Without this storage hedge, the elevator would receive $2.30/bushel for grain in June—a 5-cent loss on storage operations. The hedge allows the elevator owner to realize a 40-cent gain to storage. This price gain is composed of a 30-cent cash price gain and a 10-cent futures gain. Where did the 40-cent storage return come from? It was paid to the elevator owner by futures market traders who bought the contracts in November and sold them in June. An important point about the storage hedge is that it extracts storage charges from those holding long positions in the futures market. The change in basis (40 cents, in this case) will always be equal to the carrying charge return from the storage hedge.

In actual practice, a firm would decide whether, and when, to set a storage hedge based on its knowledge of the historical spring basis. Because the firm's return to storage is always equal to the basis change, a firm that expected basis to narrow to 10 cents at its lowest point next spring would set the hedge in the fall at the point of widest basis. The elevator owner would not set the hedge until the November basis was at least 45 cents, ensuring a 35-cent basis change to just cover storage costs. The firm can "lock in" any profit above a November basis of 45 cents by using the storage hedge. Thus, storage hedges are motivated by profit seeking as well as risk protection.

To illustrate, suppose that the elevator owner observes a 48-cent basis for July futures when purchasing and storing corn in November. The 10-cent expected June basis is then promising the elevator owner a 3-cent profit above storage costs ($.48 – .10 – .35 = .03). The owner can lock in this profit with a storage hedge, or wait to see if the November basis will widen more and provide a greater storage profit. If three days later, the November basis has widened to 50 cents, the futures market is promising the firm a 5-cent profit for storage ($.50 – .10 –.35 = .05). If the elevator owner feels this is as high as the November basis will go, he can set the storage hedge and lock in the 5-cent profit. Thus, hedging is a form of speculating on basis; it does not eliminate the need for intelligent market decisions. Even after the storage hedge is set, there is no guarantee that the basis will follow historical patterns precisely.

The Preharvest Hedge

This hedge is appropriate for farmers during the period between planting and harvesting a crop. It allows farmers to lock in a profitable selling price before or after the crop is planted and prior to harvest.

The preharvest hedge requires that the farmer be knowledgeable regarding the *local harvest basis*—the difference between local cash prices and the nearby futures price at harvest time. This basis differs widely and therefore must be studied for each area of the country. The local basis is more predictable for some areas than others. Basis charts are available from local elevators and brokers.

Suppose that on March 1 the farmer has to decide how much corn to plant. March cash prices are of little help in estimating the following November cash price. But the December futures market is an estimate of what traders believe will be the cash price of corn in December. However, the futures price is a Chicago price, and the farmer will be selling to local elevators. The farmer can subtract the historical harvest-time basis from the December futures price to estimate the harvest cash price that the local elevator will offer. This is the "target price" to which the farmer will target production. In order to lock in this estimated harvest cash price, the farmer can use a preharvest hedge as follows:

Date	*Cash Market*	*Futures Market*	*Nov. Basis*
March 1	Plant at estimated Nov. cash price of $2.60 ($3.00 – .40)	Sell Dec. futures @ $3.00/bu	$ .40 (expected)
November 1	Harvest and sell locally @ $2.40	Buy Dec. futures @ $2.80	$.40 (actual)
Loss/Gain Results:		+.20	

Nov. 1 cash price = $2.40
+ Futures gain = .20
Total return = $2.60
Estimated return = $2.60

The 40-cent historical basis provides the farmer with a spring estimate harvest cash price of $2.60. If the farmer fears that cash prices might be lower than this at harvest, he could sell December futures in March and buy them back in November when the crop is sold locally. The futures market gain can be added to the $2.40 cash price in November to bring the two transactions back to $2.60—the price the market was promising and the farmer locked in, back in March.

The success of the preharvest hedge hinges on an accurate prediction of the harvest basis. To the extent that this basis follows its historical pattern, the hedge allows farmers to lock in harvest prices in advance. It does *not* provide the farmer with any influence on cash prices. Nor can it be helpful to the farmer if the difference between the December futures and the historical harvest basis is an unsatisfactory cash price. The farmer is still a price-taker.

FORWARD CONTRACTS AND FUTURES CONTRACTS

Hedges permit farmers to forward price their commodities prior to harvest or sales from storage. However, many producers prefer to forward price their purchases or sales using contracts rather than hedges. In the spring, farmers can sign these contracts with local elevator owners for the delivery of grain at harvest time, at a prearranged price. This is a forward contract, and works somewhat like the preharvest hedge. The farmer can lock in a favorable price for a crop in the spring when planting decisions are being made. This provides a forward price with no basis risk and no futures market transactions.

The forward contract has two disadvantages when compared with the preharvest hedge. The reason why the buyer can offer a guaranteed price to the farmer in the spring is because once the spring contract is signed, the elevator owner sets a preharvest hedge. So the first disadvantage of the forward contract is that the farmer is, in effect, asking the elevator owner to do the hedging. And the charge for this service is reflected in the contract price. The second disadvantage is that farmers have less flexibility with forward contracts than with hedges. A farmer can remove his hedge if it becomes certain that cash prices at harvest will be higher than the estimated hedge price. On the other hand, farmers must deliver against the forward contracts, regardless of the cash price.

AGRICULTURAL OPTIONS

Options are another marketing tool that can provide price insurance and marketing flexibility for farmers. Agricultural options have been traded since 1984, following a 48-year Congressional ban on them. Like futures contracts, options are traded through brokers on organized exchanges. In fact, the "option" refers to the right to do something with a futures contract.

An option gives the owner the right, but not the obligation, to buy or sell a futures contract at a certain price for a specified period of time. The guaranteed price at which the contract can be bought or sold is called the *strike price.* A *put* option conveys the right to sell a designated futures contract at a specified strike price. A *call* option conveys the right to purchase a futures contract at a designated strike price. The price that must be paid for these option rights is termed the *premium* and is determined in the option markets by what buyers and sellers of options feel they are worth. An option buyer would be willing to pay a premium of at least 30 cents for a put option with a strike price of $7.00/bu. if he believed the price of the futures contract would fall below $6.70 sometime during the life of the option. Buying and exercising this put option when and if the futures price is below $6.70 would permit the owner to make a profit after paying the 30-cent premium. The seller (writer) of the option, on the other hand, is expecting the futures price to remain above $6.70, making it unattractive for the buyer to exercise the option. The buyer of the option pays the premium to the seller whether or not the option is exercised.

Examination of the commodities quotations in a financial newspaper will show the premiums of puts and calls for numerous futures contracts with differing strike prices. These premiums change hourly and daily with trading.

The buyer of an option has three alternatives: (1) exercise the option, taking a buy or sell position in the futures market; (2) allow the option to expire without exercising it; or (3) offset the original purchase with an option sale. Unlike a futures contract trader, the buyer of an option is not obligated to take delivery of the commodity or to nullify the option with an offsetting trade. In most cases farmers will resell the put option, rather than exercise it, in order to avoid the costs associated with taking a futures market position.

Farmers might purchase a put option (the right to sell a futures contract at a strike price) as protection against falling market prices. For example, suppose that the farmer stores corn in November and observes the following prices.

Date	Cash Market	Futures Market	Option Market
November 1	$2.00/bu.	$2.50/bu.	Buy put option with $2.50 strike price, 20 cent premium
January 10	Sell @ $1.80	$2.30	Sell put option with $2.50 strike price, 40 cent premium

January 10 cash price = $1.80
+ Premium gain = .20
Total return = $2.00

In this case, the option premium increased as the price of the futures contract fell below the option strike price. This gain in premium provided the farmer with downward price protection and returned $2.00 to the farmer rather than $1.80. The farmer could also have exercised the option on January 10 by taking a sell position in futures at the $2.50 strike price while also buying a futures contract at $2.30. This also would have provided 20-cent downward price protection, ignoring margin costs and brokerage fees.

When prices fall, as in the above example, options provide price protection roughly comparable to that of forward contracts and hedges. However, options are often superior to contracts and hedges when prices rise, as shown below:

Date	Cash Market	Futures Market	Option Market
November 1	$2.00/bu.	$2.50/bu.	Buy put option with $2.50 strike price, 20 cent premium
January 10	Sell @ $2.40	$2.90	Let option expire

January 10 cash price = $2.40
− premium = .20
Total return = $2.20

Here the option owner would not exercise the put option but would participate in the rising price trend, after paying the 20-cent premium. In contrast, a farmer who forward priced in this rising market with a contract or a hedge would not participate in the rising price trend.

Table 20-2 Comparisons of Different Producer Marketing Alternatives

Comparison Criteria	Cash Sale	Forward Contract	Futures Market Hedge	Agricultural Options
Downward Price Protection	Low	High	High	High
Profit Opportunity	High (if prices rise)	Low	Low	High
Marketing Flexibility	High	Low	Medium	High
Degree of Difficulty	Easy	Easy	Medium	Complicated

MINI CASE 20: WHAT IS THE BEST WAY TO MANAGE AGRICULTURAL RISK?

Bill Brown, Jack Williams, and Clay Thomas meet each week to discuss common farming problems. And maybe complain a little about the weather. Lately, all they seemed to talk about was low prices. "There's nothing we can do about these terrible prices," said Jack. "Farming is a risky business and you take your chances." Bill agreed, "It's always been that way, and I don't see how to change it. We don't control the weather, the insects, the supply of our products, or anything else that determines our prices." Clay agreed, "You take what you get in this business, and you don't always get what you want or need." Bob Howard, a local grain marketing advisor, overheard the group and asked, "What would you fellows give me if I told you some ways to set your prices in advance and be sure that you covered your costs and made a profit?" "We'd buy you another cup of coffee and maybe a doughnut," said Bill, Jack, and Clay. "You got a deal," said Bob.

How can farmers forward price their products? What marketing tools can help them do this? How should they choose between alternative risk management tools, such as forward contracts, futures contracts, and options?

Farmers can also use call options in their marketing strategies. These options provide producers with protection against rising input prices or allow them to benefit from rising seasonal prices without storing or owning the commodity.

Thus, we have reviewed four ways that producers can sell a storable commodity like grain. These alternatives are compared in Table 20–2, using some realistic but simplifying assumptions. The producers' choice of these marketing alternatives will depend upon knowledge, experience, attitude toward risk, and other factors.

FUTURES MARKET PARTICIPANTS

There are several different kinds of traders in futures markets, and these markets perform a number of roles. There are speculators in futures prices and speculators in basis (hedgers). There are traders who wish to assume risks, and others who wish to avoid risks. There are hedgers who fulfill their contracts by delivery of the commodity, and hedgers who nullify their contracts positions in the same manner as speculators. There is no typical hedger. One firm may use hedges to protect the value of inventories, another to reduce the risk of an increase in raw material costs, and still another to establish the price of its products in advance.

The futures market serves a variety of purposes. It is a speculative medium, a price-discovery mechanism, and a risk-transfer device. It can also serve to lock in profitable prices and to facilitate the financing of stored commodities. Despite this versatility, or perhaps because of it, the futures market is not universally understood. Many firms do not use the futures market. Most hedging is done by grain dealers, livestock buyers, and others to whom farmers sell, but not all buyers of farm products hedge.

Only a small portion of farmers trade futures contacts, and many of these do so for speculative reasons rather than for hedging. Among the reasons farmers give for not using the futures market are (1) lack of familiarity with trading, (2) farm too small, (3) trading too risky, and (4) lack of capital for trading. Lack of understanding, and perhaps distrust, probably also limit farmers' use of the futures market.

THE FUTURES MARKET CONTROVERSIES

Controversy has surrounded organized futures trading since it began in America at the Chicago Board of Trade in 1865. The speculative activity of these markets, the close relationship of cash and future prices, the fact that more commodities are traded than are produced, and the occasional stories of futures market manipulation have resulted in continuing concern with these markets on the part of the trade, the farmer, and the public.

Criticism of the futures market has often led to regulation and sometimes prohibition of these markets. In the 1800s, many states passed laws limiting futures trading. In 1958, after considerable debate, Congress passed a law against onion futures trading on the grounds that excessive speculation resulted in extreme and unwarranted price fluctuations. Over the years, the critics of futures markets have been successful in securing increased federal regulation of the commodity futures markets.

Attitudes toward these markets range from violent opposition to what their detractors call "organized gambling" to staunch defenders of the futures markets as the best example of a perfectly competitive market and the last outpost of a free economy. Many of the criticisms are unfounded, but there are serious questions about these markets. Are they necessary? Do they perform a useful function? How do they affect cash prices, marketing costs, and farm prices? What would be the result of laws prohibiting futures trading?

Most serious observers agree that the futures markets play a valuable role in risk transfer, equity financing, price discovery, and forward pricing. But there are other ways to organize these functions and activities in the food industry. Moreover, many commodities appear to be marketed efficiently without the benefit of futures trading. The legitimacy of these markets must rest on their positive contributions outweighing whatever drawbacks they have. We shall examine some of the alleged criticisms and contributions of these markets.

Are Speculators Necessary?

Some critics concede the importance of the hedging and risk-transfer roles of these markets but advocate eliminating speculation in futures trading. Speculators, they argue, know nothing of agriculture or farm prices, so their judgments should not be permitted to enter into the futures pricing machinery. They might somehow "misprice" commodities, or the psychology of the speculators might result in exaggerated price swings detrimental to the efficient marketing of commodities.

This line of reasoning suggests that hedgers and speculators are somehow different, and that one is more legitimate than the other. We have seen that both are speculators—one in prices, one in basis. The criticism also neglects the symbiotic relationship of hedgers and speculators. Speculators make active markets for hedgers, and assume the hedgers' risk. Hedging may be the rationale for futures markets, but to be effective for hedging, a futures market needs large numbers of speculators.

Do speculators trade in futures commodities without regard to cash prices, and could they somehow "misprice" commodities? We have seen that cash and futures prices move together in somewhat predictable fashion, and become one at contract maturity. Speculators cannot ignore these price relationships. Speculators do make wrong decisions about prices. However, for every "wrong" decision that loses money there is a corresponding "right" decision on the part of other speculators who make money.

A related concern is that more contracts are traded than there is product to deliver. However, we have seen that hedgers as well as speculators nullify contracts before they mature. This is necessary if hedging is to be useful to traders who buy and sell other than at contract maturity. The futures market exists to transfer risks, not necessarily products. The many broken promises in the futures market make it an unusual—but not crooked—market.

Finally, critics sometimes argue that speculators follow mob psychology—everyone buying and selling at once—which exaggerates price swings. Again, every "buy" position is matched with a "sell" position; for every bull there must be a bear; and for every price-increasing action (buy) there must be a price-decreasing action (sell). The market is symmetrical. Also, the threat of contract delivery maintains discipline among traders.

Do Futures Prices Determine Cash Prices?

The contention that futures prices fluctuate wildly, without regard to fundamental market conditions, is usually linked with a statement that futures prices "determine" cash prices. Undoubtedly, the practice of basing cash prices on futures prices is widespread. Most traders reinforce the suspicion that futures prices determine cash prices when they quote a spot price of "10 cents under the futures price." Nevertheless, the statement is not true. Because cash and futures prices are systematically related by the basis, to say that cash determines futures makes as much sense as to say that futures determine cash. Both prices are jointly determined by the same market forces. Another reason for the illusion that futures prices determine cash prices is that the futures price often *anticipates* events that later affect cash prices. Consequently, events may affect first the futures price and then the cash price.

This is not to say that futures trading has no influence on cash prices. One of the most important contributions of futures exchanges is that of establishing a sensitive price-registration machinery. All kinds of news and statistics are focused on the trading floor and are then incorporated into prices. Futures markets are close to perfect competition and thereby contribute to an accurate price-discovery process. These markets have been compared to the historical terminal markets as reference points for pricing throughout the food marketing system. By comparison, many imperfectly competitive cash markets are relatively sluggish in responding to market developments.

Studies of the effect of speculators on short-run futures price variability provide mixed results. Some research shows that speculators increase price variations, whereas others suggest that speculators moderate price variability. However, there is general agreement that futures trading diminishes seasonal price fluctuations of storable commodities and reduces the magnitude of price changes from one season to another. On the other hand, futures markets may exaggerate short-run price variations.

Effects on Marketing Costs

Because risk bearing has a cost, the shifting of risk through futures trading can potentially lower hedgers' marketing costs. In turn, lower marketing costs could result in either higher farm prices or lower retail prices. Whether the shifting of risk from food industry hedgers to speculators outside the industry actually lowers hedgers' marketing costs depends on what speculators can extract from hedgers in the way of risk-handling charges. Most observers have concluded that speculators willingly assume and finance hedgers' price risks at *no cost* whatsoever to the food industry. Therefore, marketing costs are reduced by the costs of risk for a net increase in operational efficiency for the food industry. In effect, the speculators are subsidizing the hedgers' price risks.

IS SPECULATION A FAIR GAME?

Above all, the futures markets are speculative vehicles. Most speculators are not even aware of their role in the hedging and risk-transfer operations. Their only interest is in profits and losses.

As an almost perfectly competitive market, the futures markets are "fair." No one is prohibited from trading, and all who trade do so willingly. Each trader recognizes—or should be aware of—the opportunities for gains and the possibility of losses.

As a group, speculators do not break even. Total gains are equal to total losses—a break-even proposition—but brokers are paid regardless of profits or losses. Despite stories of enormous speculative gains, the brokers are probably the only participants in the futures market who consistently make a profit. Among speculators, the consensus is that the great majority of speculative trades result in losses. The speculative gains are concentrated among a few, large traders. Even for professional speculators, trading success requires a few highly successful trades to offset the numerous losing trades.

Does Hedging Work?

The perfect hedge is rare. And most hedges are motivated by a desire to earn profits rather than to avoid losses. But, clearly, hedging can transfer risks and thus be a useful marketing tool. Basis *is* more predictable than cash prices. And hedging is widely—if not universally—practiced by the trade. The reason that more firms do not hedge is probably the result of a lack of understanding of the process rather than because of any deficiencies in the mechanism.

The role of hedges in financial management is also important. Firms can usually borrow working capital on the value of hedged inventories. Farmers also can secure equity financing by fixing the selling price of commodities in advance through storage or preharvest hedges.

Hedging does not guarantee profits and is not a mechanical process. Firms must know whether, when, and how to hedge. It is a marketing tool that must be used selectively. Routine hedging of a crop every year will be no more profitable for a farmer than automatically selling the crop at harvest every year.

Can Futures Prices Be Manipulated?

The history of futures trading is full of reports of price fixing, rigging, corners, power plays, and market manipulation. These allegations continue today. The markets attract big spenders, and the profits to be won by the successful manipulation of "wheeler-dealers" are enormous. The market is fueled by greed, which is only kept in check by competitive forces and by regulated trading rules.

There were several famous market "corners" in the late-nineteenth century. A *corner* results when a single individual holds a large portion of the outstanding long positions and also controls most of the commodity that could be delivered by the short sellers at contract maturity. As maturity approaches, the shorts can either buy back their speculative positions or purchase grain and deliver it against the contracts. However, they face the same seller in both cases, and must pay that person's price—which is steep!

Today, the manipulation of futures markets is difficult and illegal. Recent regulations on the quantity of contracts that can be held by one individual, increases in the number of delivery points, and more careful scrutiny of these markets for anti-competitive behavior limit the opportunities for manipulation of markets. Overall, the futures markets are probably no more subject to manipulation than any other market.

PUBLIC REGULATION OF FUTURES TRADING

Public concern with futures markets—and especially alleged "excessive" speculation—has resulted in state and federal regulation of these markets. The regulations complement the very extensive self-regulating efforts of the commodity exchanges.

The Grain Futures Act of 1922 was the initial federal regulation of futures markets. In 1936, this law was amended to the Commodity Exchange Act. These laws granted authority to the federal government to prevent market abuses, price manipulation, cheating, and fraud in futures markets.

From 1936 until 1975, futures trading was regulated by the Commodity Exchange Authority (CEA), a division of the U.S. Department of Agriculture. Regulatory authority over these markets was shifted to the Commodity Futures Trading Commission (CFTC) by the Commodity Futures Act of 1974. In 1982, the CFTC was given the authority to initiate and regulate option trading. Every year new laws and regulations are passed to insure that the futures markets operate fairly and honestly.

SUMMARY

There is inherent price risk in producing, handling, and processing agricultural commodities. The futures market exchanges are highly developed markets, resembling the perfectly competitive markets, where speculators and hedgers transfer price risks and compete for profits. A hedge is a risk-management and profit-seeking marketing tool that involves the temporary substitution of a futures market transaction for a cash market transaction. The predictability of the difference between cash and futures prices—termed the *basis*—is essential to effective hedging. Both farmers and food marketing firms may use storage hedges and options. Farmers, in addition, may use preharvest hedges or contract sales to forward price their commodities during the production process. Futures markets provide important pricing information in today's decentralized food markets. The commodity futures markets are much criticized, but they play necessary and important roles in risk management, financing, market intelligence, and the price discovery process.

KEY TERMS AND CONCEPTS

basis
bears, bulls
call option
carrying charge
Commodity Futures Trading Commission
corner
delivery month, contract maturity
economic (price) risk
forward pricing
futures contract
futures market exchange
hedging, hedgers
leverage
long
margin
perfect hedge
preharvest hedge
put option
round turn
short
speculators
spot price
storage hedge
strike price

DISCUSSION QUESTIONS

1. Discuss the following markets as futures markets: (a) a college education; (b) a layaway purchase from a department store; (c) a newspaper subscription. What are the price risks in each case?

2. What are some possible reasons why there are no organized futures markets for automobiles, homes, and other consumer items?

3. What does it mean to purchase an option to buy real estate? How is this similar to an agricultural option?

4. Show that a rising basis for a storage hedge provides less than full price risk protection when prices fall.

5. On October 5, 1947, the President of the United States criticized the commodity futures markets by saying, "The cost of living in this country must not be a football to be kicked about by gamblers in grain." Comment on this statement.

6. It is frequently felt that amateur commodity speculators lose so often because they "cut their profits short and let their losses run," whereas the professionals "cut their losses short and let their profits run." How do you account for these differences in trading behavior?

7. On a chart, plot the basis of a grain commodity and a livestock commodity for three months. What are the differences in basis patterns for these?

SELECTED REFERENCES

Commodity Marketing: Options on Agricultural Futures. (1984). Chicago: Chicago Board of Trade.

Farm Risk Management Briefing Room. (January 2001). Economic Research Service, U.S. Department of Agriculture. www.ers.usda.gov/Briefing/RiskManagement/

Harwood, J. et al. (March 1999). *Farmers Sharpen Tools to Confront Business Risks.* Agricultural Outlook. Washington, D.C.: U.S. Department of Agriculture, pp. 12–16.

Harwood, J. et al. (1999). *Managing Risk in Farming: Concepts, Research, and Analysis.* Agricultural Economics Report 774. Washington, D.C.: U.S. Department of Agriculture.

Hieronymus, T. A. (1977). *Economics of Futures Trading* (2nd ed.). New York: Commodity Research Bureau.

Introduction to Agricultural Hedging. (Home Study Course). (1988). Chicago: Chicago Board of Trade.

Kenyon, D. E. (April 1984). *Farmers' Guide to Trading Agricultural Commodity Options,* AIB-463. Washington, DC: U.S. Department of Agriculture.

Paul, A. B., R. G. Heifner, & J. D. Gordon. (May 1985). *Farmers' Use of Cash Forward Contracts, Futures Contracts, and Commodity Options,* Agricultural Economic Report No. 533. Washington, DC: U.S. Department of Agriculture.

Risk Management Agency. (November 2000). U.S. Department of Agriculture. http://www.rma.usda.gov/.

Schnept, R., R. Heifner, and R. Dismukes (April 1999). *Insurance and Hedging: Two Ingredients for a Risk Management Recipe.* Acricultural Outlook. Washington, D.C.: U.S. Department of Agriculture, pp. 27–33.

PART V

THE GOVERNMENT AND FOOD MARKETING

CHAPTER 21

GOVERNMENT PRICE, INCOME, AND MARKETING PROGRAMS

Beginning with the 1985 Farm Act and continuing with farm legislation in 1990 and 1996, a series of fundamental changes in agricultural policy has altered the nature of government intervention in agriculture and moved the sector toward greater market orientation.

Economic Research Service, U.S. Department of Agriculture

OBJECTIVES

After reading this chapter, you will be able to:

1. Define the *farm economic problem* and explain why we have a national farm policy.
2. Appreciate the various ways by which governments intervene in farm prices, incomes, and markets.
3. Understand how contemporary farm policies, especially the Freedom to Farm Act of 1996 and the Farm Bill of 2001, differ from earlier farm programs.
4. Explain how government farm policies impact farmers, food marketing firms, and consumers.

Farm policies are attempts to achieve broadly held social goals through the instruments and powers of government. Because the rural and food economies are considered so important, some form of government involvement in farm and food markets is found in most societies. In the United States, government farm policies have been shaped by the American "farm problem"—often defined as unstable farm prices and incomes, periodically declining farm prices, and low farm incomes relative to nonfarm incomes.

Government intervention in prices, incomes, and markets is always controversial. There are philosophic conflicts, as well as differences in opinion, as to what the problems are, how government could contribute to the solutions, and the precise form the policy should take. Two polar views are (1) that the government should play *no direct role* in influencing farm markets or farm incomes (the laissez-faire, free-market approach), and (2) that the food industry should be *regulated like a public utility,* with strong government controls. Actual farm policies are somewhere between these two views. Government involvement in farm prices and incomes in the United States appears to alternate between increasing and decreasing intervention. These cycles are triggered by changes in the level of farm prices, shifts in the balance of political power, and changing attitudes toward government intervention.

Although farm policies usually are directly addressed to farmers' problems and production agriculture, they invariably influence the marketing of food in important ways. Price-support programs, land-retirement or set-aside policies, and government surplus-purchase programs not only affect farm production and prices, but also influence all the other actors in the food industry. For example, programs that restrict farm output and raise farm prices influence sales of farm supply firms, affect the capacity utilization of food processing facilities, and result in changed consumer food prices. At the same time, many farm policies require direct intervention in food marketing activities: trade policy, demand-expansion programs, processing taxes, and the like. Thus, it is practically impossible to isolate "farm policy" from "food and marketing policy" today.

RATIONALE FOR GOVERNMENT MARKET INTERVENTION

Throughout this book, a number of the dimensions of the "farm problem" and the "farmer's marketing problem" have been discussed. Because these are frequently cited as justifications for government involvement in agriculture, they are reviewed here.

Many farm price and income problems stem from the nearly perfect competitive nature of agriculture, the inelastic demand for most farm products, and the low income elasticity of demand for food. As price-takers, farmers attempt to improve profits by expanding production or lowering costs. This shifts the agricultural supply curve rightward along an inelastic demand. Farmers cannot easily adjust output to falling prices, and they face a cost-price squeeze because the prices of purchased inputs do not adjust downward as much as farm prices. Although the inelastic demand gives farmers an incentive to restrict supply and raise prices, their large numbers, geographic dispersion, and widely varying economic situations make organization of farmers difficult. It is also difficult for farmers to influence the demand for food as other businesses do with advertising, promotion, and other market development tools.

In contrast to these conditions, the farmer sells products to an imperfectly competitive food marketing system that adds more than 50 percent of the total value of food products, readily adjusts quantity and prices to changes in consumer demand, engages in extensive market development activities, and has varying degrees of control over the procurement and final selling prices of food products.

In addition to these farmer marketing problems, the uniqueness of food as a product has fostered government involvement in its production, marketing, and consumption. The availability of an adequate and reasonably priced food supply influences social stability and international power relations. As a strategic commodity, nations may elect to trade off some economic efficiency for self-sufficiency in food production. Society may also elect to interfere with "free markets" where it is judged that these will lead to unacceptable income or product distribution. Should the rich person's cat get the milk needed by the poor child? Should food prices be permitted to rise when the effects will be more damaging to the poor than to the rich? Should consumers be permitted to choose poor diets if this influences national health, worker productivity, and unemployment? Do farmers have a right to a guaranteed price above costs of production? Obviously, value judgments are involved here, and these will color society's attitudes toward government food market policies.

Farm policies also sometimes appear to be in conflict, for example, when farm price supports encourage farmers to expand output at the same time production controls are attempting to reduce agricultural supplies. These apparent inconsistencies may result from multiple policy goals, unanticipated policy results, the attempts of policy makers to satisfy more than one constituency, or the need to consider both long-run and short-run goals. Agricultural policy is never simple and seldom straightforward in design or execution.

Government intervention in the marketplace has usually been justified by a concern that markets and prices were not allocating resources in a way that satisfies the preferences of individuals and society. While it is difficult to envision a marketplace and a market exchange process that is completely free of any government rules, rights, and obligations, we must recognize that government cannot solve all the problems of markets. Markets may sometimes fail in allocating resources accurately and satisfying everyone's preferences, but governments are also imperfect. Often, the choice is between imperfect markets and imperfect governments.

THE EVOLUTION OF U.S. FARM PRICE AND INCOME POLICY

Although farm policies change, the basic instruments of farm policy—price fixing, income supplements, supply control, government purchases and storage, and demand expansion programs—have been used throughout history. They were part of the farm programs of the early Egyptians, Romans, and Greeks. And they are found in the most recent farm bill. Twentieth-century U.S. farm policy can be best understood when viewed in the context of the trends in farm prices and income as shown in Figure 21-1. These trends have both shaped and have been shaped by government farm policies.

The 1920s

Attempts to influence farm prices in America can be traced to 1621, when each Virginia planter was restricted to 1,000 tobacco plants of 9 leaves each. However, modern U.S. farm policy began in the 1920s against the setting of the great farmer movements of the late 1800s, the Populist political movement of the early 1900s, and the post–World War I depression.

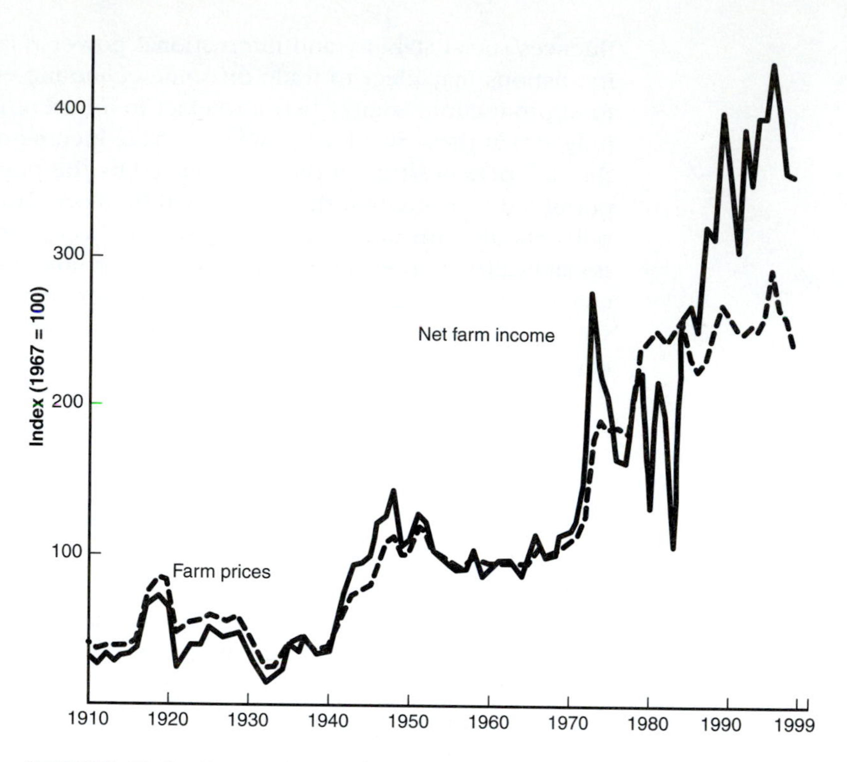

FIGURE 21–1 Farm prices and net farm income, 1910–1999. (U.S. Department of Agriculture.)

Within two and a half years after World War I, prices farmers received had fallen 52 percent from their postwar peaks. Prices paid by farmers had declined only 18 percent. Net farm income in 1921 was only 39 percent of its 1919 peak. Though prices received recovered somewhat, the relationship between farm prices received and prices paid remained about 15 percent below the prewar relationship throughout the 1920s.

The cry for agricultural relief arose out of this difficult price and income situation. Several bills aimed at relieving the situation were poured into the legislative hoppers by farm-sensitive Congressmen. However, the one major plan that was debated throughout the decade of the 1920s was based on separating the domestic and foreign markets. Domestic prices were to be pegged at a "fair" level. All that could not be sold domestically at that level was to be purchased by the government. This excess amount would then be sold on the world market for what it would bring. High tariffs were to protect the domestic markets from foreign imports. The government loss between the domestic support prices and the world prices was to be shared equally, through various methods, among agricultural producers. The net price to farmers would then be the domestic support prices minus the losses suffered in the world market. This basic idea was introduced into Congress as the McNary-Haugen Bill. The bill was approved twice by Congress, in 1927 and in 1928, but was vetoed both times by President Coolidge.

Though the McNary-Haugen plans never became law, they merit attention because they represented the opening guns of the battle to guarantee a "fair" price for agriculture. The assumption was that domestic food demand was inelastic, whereas the world demand was elastic, or at least less inelastic than domestic demand. By restricting amounts and raising prices, the total returns from the domestic market could be expanded. If the world market demand for U.S. farm products was more elastic, selling more in the world market would increase the total returns. Because the domestic price level would be substantially above the world price level, the total dollars for a crop would be increased. In addition, such a system still permitted products to sell at the world price levels in the export trade even though domestic prices were protected at much higher levels.

In 1929, with the passage of the Agricultural Marketing Act, emphasis changed from the two-price approach of the McNary-Haugen plans to one of "orderly marketing." The basic philosophy of this law was that agricultural problems resulted from disorganized distribution methods. If products could be stored by farmers and released in an orderly fashion throughout the year, higher price and income levels could be secured. To attain this goal, farm cooperatives were to be encouraged vigorously.

The Federal Farm Board, established to put the program into effect, was given an initial appropriation of $500 million to undertake loan-storage programs to stabilize farm prices. However, the newly established board was just taking its first tottering steps when the depression struck. By 1933, it had spent all its funds, acquired large stocks of commodities, and prices had still fallen drastically. The Board died when Congress did not grant it additional funds.

The 1930s

Starting in 1930, prices began to slide again. By early 1933, prices received by farmers were 62 percent below 1929 levels. Prices paid, again lagging behind, were only 32 percent lower. Net farm income in 1932 was 70 percent below 1929. For those whose memories are only of the recent price levels, a look at the depression prices of several commodities will help in gauging the seriousness of the situation.

U.S. average farm prices for several commodities in 1932 were as follows:

Per bushel		*Per cwt.*	
Corn	$0.32	Milk	$1.28
Oats	0.16	Hogs	3.34
Wheat	0.38	Cattle	4.25
Cotton	0.07		

Though we are concerned here with agricultural developments, we must not overlook that the whole country was in an economic depression. Businesses failed; banks shut their doors; one out of every four workers was without a job. The agricultural policy that developed was only a part of a broad program that included social security and unemployment insurance for workers, special legislation for banks, and the National Industrial Recovery Act for other businesses.

The legislative developments of the 1930s signaled a change in the basic approach to the problems of a depressed agriculture. During the 1920s, the two-price system was offered as a device by which our agricultural production could have a protected domestic market and yet secure what it could get from the world market. The Federal Farm Board proposed to solve the problem by helping farmers to level out the peaks and valleys of their marketings and thereby stabilize the prices of agricultural products. The basic approach now to be followed was to reduce production. In this manner, demand and supply were to be brought into balance at prices that were "fair" to farmers. Though many camouflaging terms were used, the basic concept of curtailing production prevailed.

The Agricultural Adjustment Acts of 1933 and 1938, along with the 1936 Soil Conservation and Domestic Allotment Act and the 1937 Agricultural Marketing Agreement Act, were the forerunners of modern American farm programs. The Federal Agricultural Reform Act of 1996 was an amendment to the 1938 Agricultural Adjustment Act. These depression-born farm policies attempted to help farmers through supply control and "orderly marketing" measures. The Commodity Credit Corporation (CCC) was founded to make storage loans to farmers and to support farm prices at above equilibrium levels through government purchase and storage activities. Acreage controls, import and marketing quotas, and soil conservation programs were authorized to limit food supplies, and income supplements were granted to farmers for compliance with these programs. Marketing orders and agreements permitted farmers to control the quantity and quality of farm products marketed and, in the case of milk, set minimum prices. Surplus food distribution, the school lunch program, and food stamps were instituted in the 1930s to attack the demand side of the farm problem.

1940–1952

U.S. farm prices and incomes rose dramatically from 1940 to 1952. Food production and supplies were curtailed by the diversion of resources and output to defense needs. Food demand was fueled by World War II, the inflation following this war, the Marshall Plan, and the Korean conflict. These were prosperous years for agriculture, and only minor adjustments were made in the 1933–1938 farm programs. Consumers experienced wartime food rationing and food price controls.

The 1950s and 1960s

During this period, the scientific and technological revolution in agriculture resulted in greatly expanded food production capacity. These increased supplies pressed against an inelastic and more slowly growing demand for food. Combined with persistently rising farm input costs, the result was a downward pressure on farm prices that intensified the cost-price squeeze in agriculture. Farmers reacted predictably to this situation by readily adopting the newest output-increasing and cost-reducing technologies and by further expanding output. This was called the *agricultural treadmill* because farmers had to run fast just to stay in place. By and large, consumers received the benefits of this situation through declining relative prices of food. Farm prices fell drastically in 1953 and 1954 and then stabilized from 1954 to 1957, propped up by acreage controls, subsidized exports, price supports, and government purchase programs.

A number of farm policies were developed in the 1954-1967 period to improve farm prices and incomes. The Agricultural Trade Development and Assistance Act of 1954 (P.L. 480), also known as the "food for peace program," attempted to bolster farm prices by exporting surplus commodities to needy countries. The 1956 soil bank program authorized the government to rent cropland from farmers, thus withdrawing this land from production.

By the end of the 1950s, the efficient agricultural machine had reduced farm prices 21 percent from 1951 levels; annual net farm income had fallen from $15 billion in 1950-1953 to $12 billion in 1957-1959; P.L 480 food exports (some would call them gifts) were running at more than $1 billion per year; and the federal treasury costs of the farm surplus program were estimated at $9 billion in 1960. The value of government-owned food stocks rose from $1.3 billion in 1952 to $7.7 billion in 1959, and by 1961 the cost of storing a bushel of grain purchased by the government exceeded the original farm value of the grain.

The 1960s saw a movement toward stronger farm supply control programs and the payment of direct income supplements to farmers. Some supply control programs were mandatory, and others were voluntary. Under voluntary programs farmers were offered incentive payments to comply with supply controls to avoid surplus production at the support price levels. The acreage diversions and set-asides—particularly for feedgrains, wheat, and cotton—resulted in a slow rise in farm income and the gradual working down of stocks in the 1960s. These programs withdrew large amounts of land from farming. By 1972, 62 million acres—20 percent of the nation's cropland—were idled.

The 1970s

The 1970-1980 decade was a turbulent period for U.S. agriculture and farm policy. The worldwide demand for U.S. farm products expanded rapidly in the early 1970s. This resulted from increased global populations and incomes, a realignment of foreign exchange rates making U.S. farm products less expensive to foreign buyers, worldwide crop failures, and purchases of U.S. farm products by the U.S.S.R. and the Peoples Republic of China. At the same time U.S. crop yield increases were slowing from the impressive rates of the 1960s.

As shown in Figure 21-1, these conditions pushed farm prices and incomes rapidly upward in the 1972-1980 period. Moreover, since farm production costs lagged behind these price rises, real net farm income in 1973 reached its highest level since World War II (Figure 21-2). Not all farmers benefited, however. The higher grain prices meant losses for livestock farmers and triggered a massive liquidation of the cattle herd in the second half of the 1970s. Consumer food expenditures increased by $47 billion between 1972 and 1975, largely the result of higher farm prices.

These relatively high farm prices and incomes generated expectations of continued long-term agricultural prosperity. The supply response was predictable: an increase of 47 million acres farmed between 1969 and 1978 as all the diverted acres were brought back into production and large amounts of pasture were converted into cropland. By the end of the decade U.S. farm supply was again outrunning demand, stocks were building, acreage was being idled, and farm costs had caught up with, and exceeded, prices.

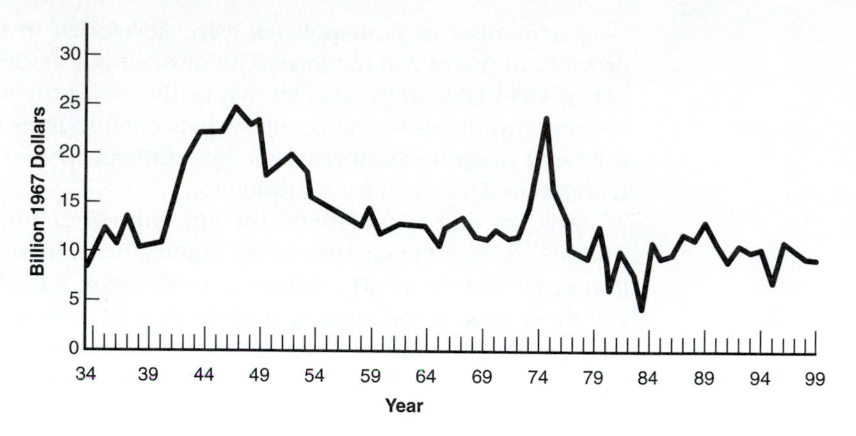

FIGURE 21–2 Real (after-inflation) net income from farming, 1934–1999. (U.S. Department of Agriculture)

The 1980s

U.S. farm policies of the 1980–1990 period were addressed to the price and income problems of farmers. It was assumed that too many resources were devoted to farming, resulting in excess capacity and surplus food production. Government intervention was deemed necessary to bring about a better balance of supply and demand and improve the welfare of farmers relative to the nonfarm population. Two alternative views of this historical situation shaped the farm policies of the 1980s.

One view was that there was no longer excess capacity in the U.S. farm sector. Proponents of this view pointed to the 1972–1975 period as an example of what an unregulated, "full production" farm economy could do. This view argued for a continued movement of agriculture to a nonregulated, market economy with much less government involvement in farm markets in the future.

The contrasting view was that the early 1970s farm price and income experience was unique and nonsustainable, masking continued excess capacity and surplus production in U.S. agriculture. Moreover, proponents suggested that the instability introduced into U.S. farm product markets by increased agricultural exports warranted continuing government involvement in agriculture. Support for these views is shown in Figure 21–2, indicating that real net incomes in farming declined in the early 1980s. Figures 21–1 and 21–2 also suggest that farm income instability was a growing problem in the 1980s.

The Food and Agriculture Acts of 1981 and 1985, setting farm policy for the 1982–1989 period, represented a compromise of these two views. The goal was to bring about a better balance between farm supplies and demand by acreage controls, selective reductions in price support levels, and increased agricultural exports. There was continued movement toward a more market-oriented agriculture with flexible price supports, less government purchases and storage, and a greater reliance on direct subsidy payments to farmers.

There were several new policy instruments included in the 1985 farm bill. The *Conservation Reserve Program* (CRP) idled 10 percent (40 million acres) of the nation's crop land by paying farmers to put land into a ten-year reserve. *Generic certificates,* entitling farmers to claim government stocks or to pay off government loans, were issued to farmers as payment for cooperating with government acreage set-asides and as price supports. These certificates provided farmers greater flexibility in their marketing decisions as well as another source of income. The 1985 policy required *marketing loans* for rice and cotton and authorized these loans for wheat, feed grains, and soybeans. With these, farmers were encouraged to sell their crops at market prices rather than to deliver these to the government as payment for loans. The *Export Enhancement Program* (EEP) was designed to expand sales of U.S. farm products to other countries through export assistance and subsidy measures.

The 1990s

The 1996 farm bill (Federal Agricultural Improvement and Reform Act) continued the reforms of agricultural policy begun in the 1990 bill. This legislation removed the historical linkage between farm income support payments and farm prices. Farmers of the major grain crops entered into "production flexibility contracts" whereby the federal government agreed to make declining annual payments to participating farmers over the 1996–2002 period. Farmers were also given more flexibility in crop decisions but had to comply with soil conservation plans and wetland protection provisions. Nonrecourse commodity loans were replaced by marketing assistance loans sometimes called *loan deficiency payments.* Dairy price supports were reduced. Increased emphasis was given to export market development, especially for higher value-added foods. The 1996 farm bill also stressed conservation by continuing the Conservation Reserve Program and adding an Environmental Quality Incentives Program. The food stamp program was continued and reforms were mandated.

2001-

The current farm bill was written by the Congress in 2002. As always, this policy attempted to balance government intervention with free-market principles. It was influenced by the feeling of many that the 1996 act was not successful and farmers needed to return to some form of price and income support. It was shaped by continued trends in farm and food markets. These included the further industrialization of agriculture with continued productivity gains and surplus production, the growing consolidation and integration of agriculture and food marketing firms, the public perception that modern agriculture is becoming more and more like other forms of business, the prospect for international trade in agricultural products, and increasing interaction between environmental and agricultural policy.

PARITY

Throughout the 1920s, farmers sought an economic *parity,* or equality, with other sectors of the economy. The Agricultural Adjustment Act of 1933 established the criterion for a fair price for agriculture. Parity, or fair price, was the price that would give agricultural commodities a purchasing power over supplies that farmers buy that was equivalent to the purchasing power these commodities had in a base year. The base period for the great majority of products was established as the five-year period from August 1909 to July 1914. This period was chosen because it was believed that farm and nonfarm prices were in reasonable and fair balance during those years.

As time passed, it became obvious that the parity price formula had several shortcomings. There was no provision for taking into account any changes that had taken place since 1910–1914 in either the demand or the supply of the commodity. For example, the present parity or fair price of horses under the parity formula would still be the price comparable to the purchasing power horse prices represented in 1910–1914. But with modern mechanized farming and the reduced demand for horses this was a ridiculous goal for a fair price. The example is an extreme one, but it shows the problem of handling change when only a fixed period in the past is used.

Use of the parity concept has diminished over time. Since 1977, cost of production has been more important than parity in the setting of farm support prices. The 1985 farm bill was the first not to mention the parity concept. Dairy product prices are still linked to parity, but the Secretary of Agriculture has considerable flexibility in the parity level that is set for milk prices.

ANALYSIS OF FARM PRICE AND INCOME SUPPORT PROGRAMS

Historically, farm programs employed three techniques to directly influence farm prices and incomes: (1) storage loans, (2) government purchase and disposal programs, and (3) direct farm income supplements. These alternative approaches to the farm price and income problems vary in method of operation, government costs, and other details. They are illustrated in Figure 21-3.

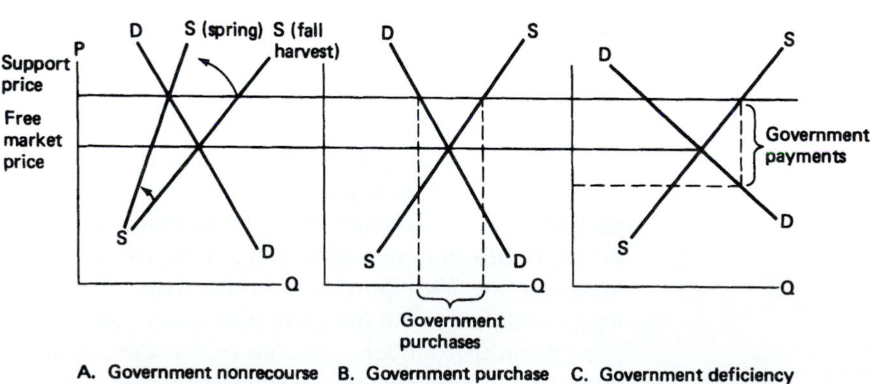

FIGURE 21–3 Analysis of government farm price and income programs.

Nonrecourse Storage Loans

Under the terms of a *nonrecourse loan,* a farmer obtains a government loan on the crop at harvest, when the crop was stored. The *loan price,* or rate, is the price level at which the government supports the commodity. By taking the loan, the farmer is assured of this minimum price for the crop.

The nonrecourse loan assists the farmer in paying for storing the crop and prevents the farmer from having to sell for cash at the harvesttime low prices. If, later in the season, the cash market price rises above the loan rate, the farmer may reclaim the crop, sell it at the going market price, repay the loan and interest, and thus take advantage of seasonal price rises (Figure 21-3A). If the seasonal market price never reaches the loan rate, the farmer delivers the crop to the government's Commodity Credit Corporation as payment for the loan.

The cost to the government of storage loan programs depends on the loan rate, the interest rate charged on the loans, the seasonal price rise, and the amount of the crop that is delivered to the CCC against the loans. These programs have been important forms of farm price supports for most grains, tobacco, and cotton. They were continued in the 1996-2002 farm bill in the form of marketing assistance loans, or loan deficiency payments, which minimize loan forfeitures when prices are low relative to loan rates.

As shown in Figure 21-4, the CCC loan rate has historically set a price floor, but not a ceiling, for grains. However, the issuance of generic certificates in 1985 broke this pattern, and the loan rate no longer sets the minimum price of grain.

Government Purchase and Storage Programs

The government can also support farm prices through direct purchase, storage, and disposal programs. For these programs, the government purchases that quantity of the commodity necessary to make the price rise to the desired support level. This is shown in Figure 21-3B. These government stocks can be sold later (in periods of

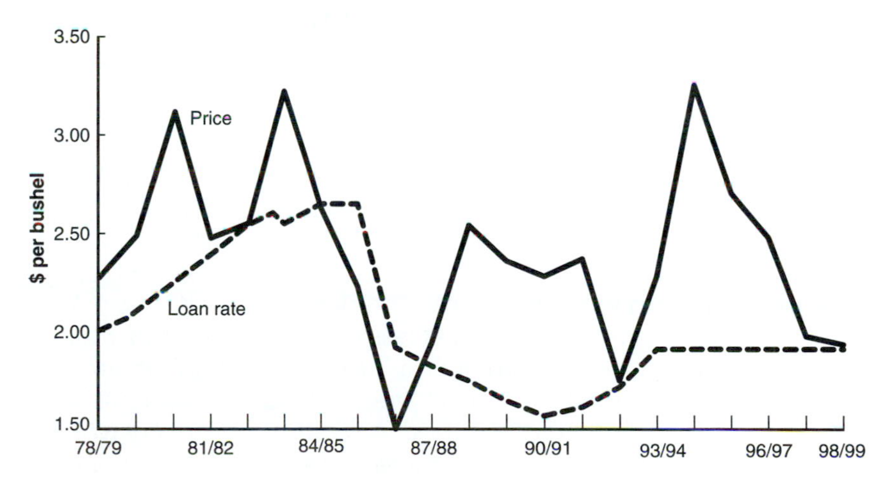

FIGURE 21-4 Average farm price and loan rate for corn, 1978-1999. (U.S. Department of Agriculture.)

higher prices), disposed of through noncompeting outlets, or, as a last resort, destroyed.

This type of farm program led to the high stock levels of the 1950s and the early 1980s. As a result, government purchase programs were tied to supply control provisions, such as acreage allotments and marketing quotas. In the 1960s there was also an expansion of government food programs—such as school lunches—to which these stocks could be channeled. In the 1980s the high stock levels were reduced by the use of generic certificates and CCC auctions. Government purchase and storage programs were not included in the 1990 and 1996 farm bills.

The costs of these programs obviously depend on the elasticity of supply and demand for farm products. As supply and demand become less elastic, the government cost of supporting prices at any level declines. Government purchase, storage, and disposal programs are the chief way of supporting U.S. milk prices.

Direct Farm Payments

The post-1973 farm bills represented a movement away from stock-building government purchase programs and toward direct government price supplements. *Deficiency payments* are made to farmers when prices are below the *target price* or support level set by the Congress. This approach is illustrated in Figure 21-3C. Farm supplies are permitted to clear markets at equilibrium levels, and the deficiency payment goes to the farmer directly from the government, rather than through the medium of higher market prices. The farmer's deficiency payment varies inversely with the market price. Eligibility for these deficiency payments may or may not require participation in supply control programs. In the 1996 farm bill, direct farm income supports were called *production flexibility contract payments.*

The government cost of direct farm payments depends on the level of the support price relative to the free market clearing price and on the elasticity of the sup-

MINI CASE 21: WRITING THE NEW FARM BILL

Senator Colt, Chairman of the Senate Agricultural Committee, and Congressman Comworst, Chairman of the House Agricultural Committee, were discussing the testimony at the recent hearings on the new Agriculture Bill. "Half of the farmers and farm groups testifying wanted less government interference in farm markets and less government influence on farm prices and incomes," said Colt. "That's true," said Comworst, "but the other half said that farmers can't survive without more government price and income assistance." "Do farmers want free markets and less government assistance or more government involvement in their markets?" asked Colt. "It's difficult to see how they can have both," puzzled Comworst.

Why do farmers and their representatives differ in their attitudes toward free markets and government farm programs? How can these differences be addressed in crafting a farm bill which serves all farmers?

ply and demand curves. In the 1970s and 1980s, direct payment and loan programs were the principal forms of government farm price supports for grains, tobacco, cotton, and dairy products. These programs have never been used to support livestock or specialty crop prices. The 1996 farm bill removed the linkage between government payments and farm prices.

IMPACTS OF FARM PROGRAMS ON MARKETING

Each of these agricultural price and income programs has affected food marketing firms, the costs of performing food marketing functions, and the retail cost of food to consumers.

The programs are intended to raise farm prices and income—or at least prevent them from falling. To the extent that this goal is accomplished, food marketing firms and consumers pay either higher food prices or higher taxes in order to support farm programs. In this sense, the programs represent a transfer of income from the nonfarm to the farm sector. What do consumers, marketing firms, and society get for this? Here there is considerable difference of opinion. Although many of the farm programs have price and income enhancement and supply control as objectives, agricultural output has continued to grow. Some might say that the programs are a reasonable price to pay for an abundant, relatively inexpensive American food supply and an agricultural sector where incomes are brought closer to nonfarm incomes. Others might point out that the almost perfectly competitive nature of agriculture would guarantee an abundant and even cheaper food supply even if there were no such programs. The debate quickly comes down to, What is a "reasonable" price for consumers to pay for food? And what is a "fair" return for the farmers' productive effort?

The precise levels of farm prices and incomes are probably not of great concern to food marketing firms. These firms can adjust their selling prices to farm prices within a broad range. However, food marketing firms' costs are influenced by the quantity of farm products produced. Thus, these firms would have reason to support output-increasing farm policies.

Food marketing firms also benefit from the price- and quantity-stabilizing effects of farm price supports and stock programs. Stability in farm markets assists these firms in accurately forecasting product costs and in making future investments in plants and equipment. Government subsidy and storage programs also probably reduced the costs of storage and risk-bearing functions for food marketing firms. The sales of these firms benefit, too, when the government supports prices by purchasing food products, such as milk or other foods, or attempts to expand the demand for food through food stamps. Some food processors also benefit from school lunch and milk programs.

Do consumers benefit from these government farm programs? The answer appears to be *yes*, in the case of programs that increase and stabilize food supplies and reduce consumers' real food costs. Most consumers do not carry large food stocks and are somewhat flexible in their purchases. Fluctuating food prices pose no great burden for consumers as long as some prices are falling while others are rising, because acceptable substitutions can be made. Farm price and income programs do transfer purchasing power from the nonfarm sector to farmers. In recent years the food stamp budget alone has exceeded the costs of all farm income supplements by a wide margin.

Supply Control Programs

The supply control programs of recent years have been voluntary. Farmers have been given incentives to set aside or retire acreage, but they were not required to do so. The inelastic demand for farm products and the relatively perfectly competitive nature of many farmers have made mandatory production controls appealing to many as a way to raise farm prices and incomes, reduce agricultural surpluses, and reduce the costs of government farm policies. Mandatory production controls make it illegal for farmers to produce or sell more than their allotted amounts of commodities.

Mandatory production controls in agriculture have been debated for years, and with a few exceptions (cotton, 1939–1943; tobacco and peanuts, 1945 to present; wheat and rice, periodically in the 1950s), they have not been adopted. Mandatory production controls are widely used in Canada, Australia, and the European Union. It is generally agreed that these programs could raise farm prices and incomes in the short run. However, aside from their restrictions on farmers' freedom, they would also increase consumers' food prices, reduce economic activity in the farm supply and food marketing activities, and encourage food imports while discouraging U.S. farm exports.

SUMMARY

Throughout history and in most societies, governments have intervened in farm production, prices, and incomes. The United States has been no exception. Contemporary farm policies date from the 1920s, and have been shaped over the years by farm price and income trends, changing attitudes toward government's role in farm and food policy and the changing political constituency. Government loan, purchase, and direct payment programs are the chief techniques by which government has influenced farm prices and incomes in the post-World War II period. The trend has been toward a greater market orientation for producers with direct government or deficiency payments to farmers when prices fall below support levels. The 1996–2002 farm bill separated government farm payments from farm prices. These farm price and income programs can benefit food marketing firms and consumers through increasing output and stabilization of food prices.

KEY TERMS AND CONCEPTS

agricultural treadmill
Conservation Reserve Program
deficiency payment
export enhancement
farmer-owned reserve
generic certificates
loan deficiency payment
loan rate
marketing loan
nonrecourse loan
parity
production flexibility contracts
set-aside policies
support price
target price

DISCUSSION QUESTIONS

1. Discuss the pros and cons of government farm price and income policies.
2. In Figure 21-1, trace the impacts of farm price and income levels, wars, and the urbanization of the American people on farm policy.
3. How would knowledge of the costs of storing grain from harvest to the following spring be helpful in setting the nonrecourse loan rate?
4. Demonstrate that government purchase programs are less costly to the government than deficiency payments when the demand and supply curves are highly inelastic.
5. Farmers frequently complain that government storage programs depress farm prices. Do you agree or disagree?
6. How do the buyers of farm products benefit from government programs that increase the supply of agricultural products and stabilize farm prices?
7. Why have farmers rejected mandatory supply controls when these have been subjected to a referendum vote?
8. Why is there declining political support for agricultural government price and income programs?

SELECTED REFERENCES

Farm and Commodity Policy Briefing Room. (January 2001). Economic Research Service, U.S. Department of Agriculture. www.ers.usda.gov/briefing.FarmPolicy/

Effland, Anne B. W. (March 2000). U.S. farm policy: the first 200 years. *Agricultural Outlook.* U.S. Department of Agriculture, pp. 21-25.

Farm Service Agency website. (December 2000). http://www.fsa.usda.gov/ This agency administers such programs as production flexibility contracts, commodity loan programs, the Commodity Credit Corporation (CCC) farm product purchase and storage programs, and the Conservation Reserve Program (CRP). There is an FSC office in every country.

Glickman, D. (April 1996). Agricultural policy for a new century. *Agricultural Outlook,* U.S. Department of Agriculture, 2-4.

Moorehart, M., B. Kuhn, and S. Offutt. (May 2000). A fair income for farmers? *Agricultural Outllook,* U.S. Department of Agriculture, pp. 22-26.

Tweeten, L. (1995). The twelve best reasons for commodity programs: Why none stands scrutiny. *Choices*, American Agricultural Economics Association, 4-7.

CHAPTER 22

FOOD MARKETING REGULATIONS

Which do you prefer: imperfect markets or imperfect governments?

OBJECTIVES

After reading this chapter, you will be able to:

1. Discuss why there are so many food marketing regulations.
2. Differentiate between restrictive, enabling, and facilitating regulations.
3. Classify existing and new regulations into six categories.
4. Discuss the impacts of food marketing regulations on farmers, marketing firms, and consumers.

Determining the appropriate role of government regulation in the food industry is a highly controversial topic. Why is public regulation of food markets necessary at all? Is more or less regulation needed in the food industry? Do the benefits of market regulations in the food industry justify their costs? Is regulation improving the food industry, or is the industry overregulated? These are important food marketing regulation issues.

The scope of government market regulations has increased over the years. Many of these newer regulations apply to the food industry. There are an estimated seventy-seven different federal agencies regulating some aspect of marketing. The newest regulatory missions of the federal government include occupational safety, environmental protection, and energy conservation and development. Each of these regulatory areas has increased the role of government in the market economy. Few food production or marketing decisions today are wholly free from some government regulation.

The food industry is among the most highly regulated of all consumer product industries because of its importance to consumers, farmers, and society in general. Many food industry regulations stem from consumers' concerns with the wholesomeness, quality, and cost of the food supply and from the efforts of farmers and consumers to maintain a workably competitive food marketing system.

Here, food marketing laws and regulations are grouped into six classes:

1. Regulations to maintain competition and prevent monopoly
2. Regulations to control monopoly conditions
3. Regulations to facilitate trade and provide public services
4. Regulations to protect consumers
5. Regulations to directly affect food prices
6. Regulations to foster economic and social progress

ISSUES IN MARKET REGULATION

The statement, "There are too many (or too few) food marketing regulations" is guaranteed to start an argument. Everyone has a unique opinion on the need or justification for market regulations. Those opposing regulations point to their inhibiting effects on innovation, efficiency, and competition, as well as to the costs of regulations, which are borne either by food firms (in the form of lower profits), by consumers (in the form of higher prices), or by government. Those who are pro-regulations emphasize the need to "harness" private profit-seeking efforts to the public interest and to protect firms and consumers from anticompetitive market practices.

Few people are opposed in principle to *all* government intervention in food markets. Most would agree that an orderly and dynamic economy requires a set of rules that are devised and enforced by society through the instrument of government. But many people would also agree that too much regulation can stifle the profit seeking and independence of decision making that is so vital to the decentralized free enterprise system. Marketing policy by the public sector seeks to achieve a level of regulation that fosters an innovative, efficient, and progressive food marketing system which serves the public interest. These are not necessarily conflicting objectives, but in practice, the balance between them has been difficult to maintain.

Regulation is perhaps a misnomer for government involvement in food markets. This involvement may be restrictive, permissive, or facilitating. There is a tendency to associate public laws and regulations with restrictions: Do not monopolize, do not advertise deceptively, do not fix prices, do not discriminate among buyers, and the like. These kinds of regulations do restrict market freedoms.

Government laws and policies can also be permissive, however. For example, this is true when government permits farmers to sell their farm products and bargain collectively for price—acts that might otherwise be considered restraints of trade. Permissive or enabling legislation may also encourage farmers to restrict farm product supplies in order to raise farm prices. Such restriction would not be permitted for nonfarm food marketing firms.

Facilitative market regulations, such as grades and standards, regulation of public food markets, and the market news service, increase the efficiency of the food marketing machinery. These services are made equally available to all, and they have none of the compulsory features of restrictive regulations. Perhaps the largest food regulatory program of all—the meat inspection program, which ensures the wholesomeness of red meat and poultry products—should be classified as a facilitating market regulation.

Decisions regarding regulations are usually not as simple as to regulate or not to regulate, or, to have free markets or controlled markets. Often, the freedom of one group must be limited in order to increase the freedom of others. Frequently, regulations also must subordinate the private interest to the public interest. There are difficult political choices to be made when regulation benefits one group at the expense of another. As a result, market regulatory policy in the food industry represents a blend of objectives and interests, and usually is never wholly satisfactory to either industry, consumers, or society.

Differences between the intent and the actual effects of food marketing regulations are another problem. Some food marketing regulations appear to have the opposite effect from that intended. For example, regulations designed to protect the owners of small grocery stores from the impact of chain stores in the 1930s appear to have hastened the demise of the smaller stores. Similarly, restrictions on horizontal integration of food marketing firms in the 1950s and 1960s seem to have led to vertical integration and to conglomerates. There is by no means any assurance that a food marketing regulation will have only the intended effects. There is also some concern that the agencies authorized to enforce food marketing regulations may, by altering the implementation of a law, affect its regulatory function.

Many food marketing regulations attempt to mold the industry along the lines of perfect competition. Valuable as the perfectly competitive model is, it is increasingly considered an inadequate criterion for judging food industry performance. No market can attain perfectly competitive conditions, and it is doubtful that consumers would prefer these markets to the imperfectly competitive markets of real life.

REGULATIONS TO MAINTAIN AND POLICE COMPETITION

The rationale for maintaining workably competitive food markets is straightforward. The essence of competitive markets is the free interaction of buyers and sellers in arriving at mutually advantageous trades. The prices determined by these trades then allocate resources and products to their highest-valued uses and reward producers in proportion to their productive contribution. Business practices that interfere with the freedom to buy and sell, that artificially raise prices by restricting output, or that insulate a company from the discipline of the competitive marketplace violate these objectives and reduce market efficiency. Although these business practices may be profitable for one firm or for whole industries, they are not conducive to the welfare of society as a whole.

The fundamental economic proposition of the United States from its beginning has been that the marketplace should be operated primarily by individual, private, competitive enterprises. Out of this organization would come the maximum economic good for all. Throughout the first hundred years or so of our history, it was assumed that such business organization was normal and automatic if government would just let things alone. However, with the price decline and economic troubles that followed the Civil War, public pressure forced investigation into charges of monopoly and collusion. A Congressional investigation into charges of collusion among the large meat packers was in part responsible for the Sherman Anti-Trust Act. This act, which was passed in 1890, became the cornerstone of federal antimonopoly policy. It stated that "every contract, combination in the form of trust or otherwise, or conspiracy in restraint of trade or commerce among the several states, or with foreign nations is hereby declared to be illegal."

It was not enough, however, simply to declare actions illegal. The subsequent history of legislation in this area has been largely to determine what are "undesirable" or "unfair" methods of competition. Both the Clayton Act (1914) and the Federal Trade Commission Act (1914) spell out how firms may and may not compete.

Food marketing firms have frequently been the target of government antitrust actions. In 1892 the government attempted to dissolve the American Sugar Refining Company, which owned 90 percent of the industry's producing capacity. A 1911 antitrust suit divided the American Tobacco Company into four separate firms. A 1920 antitrust suit stripped meatpackers of their ownership and control of stockyard facilities and prevented packers from engaging in other sectors of food processing and distribution. More recent antitrust activity in the food industry has been directed toward mergers that might lessen market competition.

In 1936 the Robinson-Patman Act attempted to establish rules against price discrimination. Equal treatment must be given to all buyers unless the discounts can be shown to be derived from economies in manufacture, sale, or delivery. The intent of this law was to prevent large concerns from using their market power to secure unfair price advantage over their competitors. The food industry has played a major role in the establishment and interpretation of the Robinson-Patman Act. A 1930s investigation that found food processors making discriminatory price concessions to food chains was a major reason for the 1936 law. Between 1950 and 1965 there were more than 200 cases charging violation of the Robinson-Patman Act by food industry firms. Fluid milk processors, feed and grain manufacturers, fruit and vegetable processors, bakeries, and grocery retailers are frequently indicted—and sometimes convicted—of violating this act.

With the development of large-scale retailing, the cry arose to protect the small retailer. Many states passed so-called fair trade acts that permitted manufacturers to control the retail selling prices of their products. In this way large retailers were not permitted to undersell the smaller retailer if they wished to continue handling the manufacturer's products. These state regulations were made legal in interstate commerce by the Tydings-Miller Act (1937). This urge to protect the small operator has also resulted in the regulation of selling margins for many products. With the growing practice of the food retailer to use pricing of various items and product lines as a competitive weapon, the question of below-cost selling arose. Now, however, it is not so much the question of protecting small retailers but rather of protecting specialized processors against the power of the possible discriminatory actions of the large retailing firms. In 1951 a court held that resale price maintenance (allowing food manufacturers to set retail prices for their products) as then practiced was illegal. Movement was quickly underway in Congress to repair the damage. The McGuire Act (1952) restored the legality of resale price maintenance. However, most states have repealed their resale price maintenance laws and manufacturers find it difficult to enforce these prices even where it is allowed. Today, with the exception of a few states where milk cannot be legally sold below cost, food manufacturers are not directly setting the retail prices of food products.

Mergers among lines of business have occurred increasingly in recent years. A business merger may be a method by which one firm absorbs its competition and, therefore, can secure more market power. On the other hand, a business merger may be a way in which two relatively weak firms can join together to become more effectively competitive with other more powerful firms. A blanket prohibition of mergers under these conditions would prevent the desirable along with the undesirable developments. A 1950 amendment to the 1914 Clayton Act declared illegal any merger—whether horizontal, vertical, or conglomerate—that tends to substantially lessen competition or to create a monopoly. This law has been frequently used in challenging food industry mergers, especially in dairy processing and food retailing.

The food industry has been the subject of numerous unfair trade practices regulations. Court records show charges of price-fixing, boycotting, exclusive dealing arrangements, and predatory pricing practices by food firms. A large number of price-fixing cases, for example, have been brought against dairy and bakery firms.

When one studies the trends in interpretation of the regulations in this area of maintaining and policing competition, it becomes obvious that there is considerable difference of opinion as to what is desired. Monopoly is illegal. But a business does not run around with a sign on it saying, "I am a monopoly!" Both the Robinson-Patman Act and the resale price maintenance acts have protected the "little fellow" as a way of maintaining competition. The resale price maintenance laws seem to take the position that *numbers* of business establishments are the key to desired competition.

The legal activities in this general area sharply point up our uncertainty as to the answer to some critical questions. Effective competition requires that several firms offer competing alternatives to consumers. But how many firms are enough? Optimum technological or operational efficiency may require firms of great size. How much cost in operational efficiency must we pay to have desirable economic efficiency in the food system?

REGULATIONS TO CONTROL OR OFFSET MONOPOLY CONDITIONS

Throughout history many have felt that vigorous competition is not always desirable. This has been generally true for public utilities, such as light, power, and telephone, and in other industries such as public transportation. The large, fixed investments in these industries mean that competitive duplication would be very costly. For these industries, the attempt has been made to simulate the results of competition through public regulation. It was toward this end that the Interstate Commerce and Federal Communications Commissions were established. Such commissions have been given regulatory powers, the most important of which is the supervisory power over rates charged.

In still other areas of the economy, Congress has recognized that weak and disorganized groups must do business with strong and concentrated industries. Or, in some cases, small and disorganized businesses are at the mercy of marketing forces, with wasteful or disruptive consequences to the public. In these instances Congress has attempted to create "power balance" by granting organizing privileges that would otherwise be in violation of antitrust laws. The Webb-Pomerene Act (1918) legalized monopolies in the export trade. The Fisherman's Collective Marketing Act (1934) permits associations of fishermen and planters of aquatic products to bargain collectively for their products. The Capper-Volstead Act (1922) recognized the legal rights of farmers to act together cooperatively. The Agricultural Marketing Agreement Act (1937) provided that producers and handlers could band together to promote orderly marketing of their products. Under the provisions of this latter law have developed the federally supervised milk markets and the various fruit and vegetable marketing orders that regulate the quantity and quality of farm products marketed.

Legislation such as the Agricultural Fair Practice Act of 1967 and the Agricultural Marketing and Bargaining Act of 1969 ensure the right of producers to join together in a cooperative bargaining association. Handlers must negotiate with these groups. Producers are protected against discrimination from buyers.

The balance between preventing unfair competition and monopoly and controlling or creating a counterbalancing monopoly is a delicate one indeed. The problems of adequate and intelligent enforcement of regulations in the former area are great. The danger in the latter area is that of creating monopolistic power that may become more harmful than the situation it was intended to balance.

Over the years, many have advocated that cooperatives take a more active part in the marketing of food. However, court decisions in the early 1960s found some cooperative merger and acquisition activities in violation of the antitrust statutes. Associations that are attempting to bargain collectively for farmers outside of the marketing order structure also may lack the necessary legal authority to do so.

REGULATIONS TO FACILITATE TRADE AND PROVIDE SERVICES

The facilitating marketing functions are standarization, financing, risk bearing, and market information. Because these influence the overall efficiency of the food marketing process, they are frequently provided by government agencies at public expense.

The Packers and Stockyards Act (1921) sets up the supervisory machinery to control marketing practices and charges on terminal livestock markets. The United States Warehouse Act (1916) provides for the licensing and supervision of warehouses and their operations. The Produce Agency Act (1927) attacks fraudulent practices of commission agents and brokers. The Perishable Agricultural Commodities Act (1930) provides for the licensing of commission dealers and brokers handling fresh fruits and vegetables. All of these laws have attempted to establish standards of practice and operation of markets and firms. More recently, these laws have provided recourse for farmers when marketing firms are slow to pay farmers or when the firms go into bankruptcy owing farmers money.

Many laws have been passed affecting product handling. Laws regulating weights and measures of containers have aimed for uniformity in this important area. There also have been a series of acts that established the authority of the United States Department of Agriculture to study and issue standards and grades for agricultural commodities. The Federal Seed Act (1970) prohibits the sale of seed contaminated with noxious weeds and requires that all agricultural seeds shipped interstate be truthfully labeled and advertised.

The Federal Grain Inspection Service was established by the U.S. Grain Standards Act (1976) in order to assure integrity in the inspection, weighing, and handling of U.S. grain products and to promote the orderly marketing of that grain at interior and export locations. The Farmer-to-Consumer Direct Marketing Act (1976) encourages the development of local farmers' markets and direct farmer sales to consumers. The Beef Promotion and Research Act (1985) enables cattle producers to establish, finance, and conduct a program of research, information, and promotion for the purpose of developing markets for beef. The Pork Promotion, Research, and Consumer Information Act (1985), the Watermelon Research and Promotion Act (1985), the Wheat Research and Promotion Act (1970), and the Cotton Research and Promotion Act (1966) perform similar functions for those commodities.

Legislation has established the market news service and the various grading and inspection services. The Poultry Product Inspection Act (1957) and the Wholesome Meat Act (1967) provide for inspection of all interstate sales of meat and poultry for wholesomeness and truthful labeling. Another series of laws has sanctioned the collection and dissemination of statistics of various commodities and the agricultural census.

Most federal legislation can affect only those commodities that enter into interstate commerce. Therefore, in most of these fields, states have developed counterpart legislation to cover trading within the states.

REGULATIONS TO PROTECT AND PROMOTE CONSUMER HEALTH AND SAFETY

There is recognition today that in a modern, complex society the consumer cannot possibly be expected to have complete knowledge about all choices and purchases. Regulations in this area generally strive to assure adequate and accurate information so that consumers can make informed decisions and protect themselves from products and practices that might be harmful. The Federal agencies with responsibility for food safety and quality are shown in Figure 22-1.

The principal regulations in this area were initiated by the Federal Food and Drug Act (1906, expanded 1938) along with the related laws of the states. Generally, the purpose of these laws is to prevent shipments of adulterated or misbranded foods, drugs, and cosmetics. The administrator of the act also has the power to establish minimum quality and fill of container for most packaged goods. In addition, the law provides for labeling of contents.

The increased use of chemicals in food productions and processing has led to several amendments to the original Food and Drug Act. The Pesticide Chemical Amendment (1954) sets tolerance levels for pesticide residue on food after harvest. The Delaney clause to the Food and Drug Act (1958) prohibited adding to food any substance that has been shown to cause cancer in man or animal. The Food Quality Protection Act (1996) replaced the Delaney Clause and provides for testing of farm chemicals. The FDA Color Additives Act (1960) controls coloring agents used in food processing.

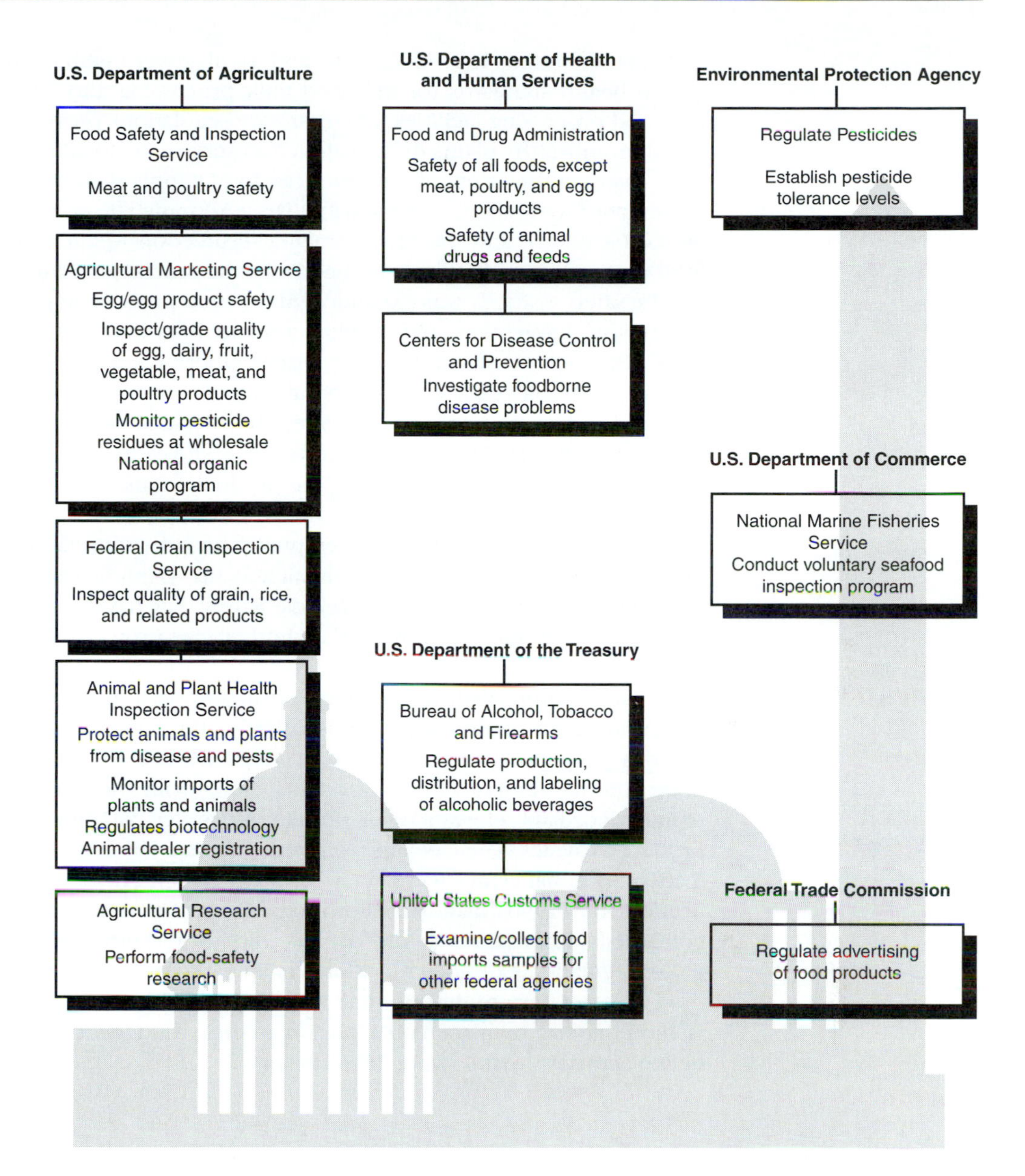

FIGURE 22–1 Federal agencies involved in food safety and quality, 2000.
SOURCE: U.S. Department of Agriculture.

The Wheeler-Lea Act (1935) set up a code of ethics for advertising. It makes false and deceptive advertising an unfair method of competition and thereby illegal. The Meat Inspection Act (1907) authorizes inspection of animals, meats, and packing establishments to assure that products will be fit for human food. More recent legislation extended inspection to poultry products and processing plants. The Egg Products Inspection Act (1970) assures that eggs and egg products are wholesome and unadulterated. Individual state and local governmental units have set up required health standards for milk and milk processing and for various slaughtering and food processing facilities. Phytosanitary regulations prevent the shipment of diseased or unhealthy plants or animals across state or national boundaries.

Insuring the safety of the American food supply is the responsibility of the Department of Agriculture, the Food and Drug Administration, the Department of Commerce (seafood), and the Environmental Protection Agency. As a result of several food contamination incidents in the 1990s, the federal government's role in monitoring the safety of foods was expanded. Many food processing plants are now required to develop, implement, and monitor a *hazardous and critical control point plan* (HACCP). Each plant must identify places where disease-causing microrganisms, as well as chemical residue and physical hazards such as metal fragments, enter the food chain. The managers must then establish critical or maximum levels to prevent these problems. HAACP was applied to the seafood-processing industry in 1995, the meat and poultry industry in 1996, the produce industry in 1997, and the fruit juice industry in 1998.

In the 1970s, there was an emphasis on providing factual information about products for consumers. Legislation such as the Truth in Packaging Act, the Wholesome Meat Act, and the Motor Vehicle Safety Act have been designed to provide consumers with more information and otherwise protect them from deception and danger.

Nutritional labeling provides consumers with information on the nutritional values and costs of alternative foods. In 1973, the Food and Drug Administration initiated a program whereby food processors declare the calorie, protein, carbohydrate, and fat content of their products, along with the percentage of the U.S. recommended daily allowances of protein and seven vitamins and minerals that the product provides. The labeling requirements were voluntary, but became mandatory if any nutritional claims were made for the product. The Nutritional Labeling and Education Act (1990) mandated nutritional labels for almost all processed foods and contained strict definitions of "low-fat," "light," and other nutritional claims. A 1993 USDA regulation extended nutritional labelling to meats and poultry.

It is not easy to describe the nutritional/health attributes of a food in a simple fashion that can be placed on a label in an easily understood form. The 1990 food labeling act uses the following definitions:

Calories:

Calorie free	Fewer than 5 calories per serving
Low calorie	40 or fewer calories per serving
Reduced or fewer calories	At least 25 percent fewer calories per serving than a reference food

Fat:

Fat free	Less than 0.5 grams of fat per serving
Saturated fat free	Less than 0.5 grams per serving saturated fat and less than 0.5 grams per serving of trans fatty acids
Low fat	3 grams or less of fat per serving

Low saturated fat	1 gram or less per serving; not more than 15 percent of calories from saturated fatty acids
Reduced or less fat	At least 25 percent less per serving than reference food
Reduced or less saturated fat	At least 25 percent less per serving than reference food

Organic foods, while still occupying a small market niche, have grown rapidly in recent years, and both the producers and consumers have called for regulations and labels certifying these products. The Organic Food Products Act (1990) called for national standards to define "organic foods" and assure consumers that food marketed as organic meets certain standards. In 2001, new U.S. Department of Agriculture standards for labeling organic foods became effective. These ban synthetic pesticides and fertilizers in the growing of organic foods and antibiotics in meat labeled organic. The regulations also require labels to indicate the percentage of the product which is organic.

With the increasing product differentiation and branding in the food industry, there again have arisen proposals to further regulate advertising and other merchandising practices. There is some opinion that advertising, though not overtly dishonest, often goes beyond acceptability and good taste. The proliferation of different package sizes—often making it very difficult for the average consumer to make reasoned choices—has led to proposals for standarization and limitation of this merchandising practice.

The dividing line between protector and dictator is a fine one. The intent of these laws has been to protect consumers from fraud and danger that they could not reasonably detect for themselves and to give them knowledge upon which to make reasoned choices. There is always the danger that such regulations may be administered in such a way as to actually make the choice for consumers. Such interpretation may stifle change and initiative if industry must serve the regulatory administrator instead of the consumer.

Are consumers really being protected, or is the progress of technological change being unnecessarily handicapped? Should it be the role of these regulations to decide what is best? Or should regulations provide the basis for informed consumer action and then leave the decision up to the consumer? The controversy over what to do about cigarettes illustrates this issue. With the establishment of the connection between cigarette smoking and health problems, many wished to prohibit the manufacture of cigarettes. The "warning and information" approach, however, was taken. Cigarette manufacture was not prohibited, but appropriate warning labels must appear on the product. Beginning in 1971, advertising of cigarettes on television was banned because of their potential health hazard. The USDA dietary guidelines, as represented by the familiar food pyramid, is another example of government taking an informational rather than a restricting approach to food choices.

Consumerism—or the consumer movement—is a wide range of government, business, and consumer activities designed to foster and protect the consumer's rights to safety, to information, to choose freely, and to be heard. Legal protection of these rights for the food consumer dates back to two 1906 laws, the Food and Drug Act and the Meat Inspection Act. Since then, the food industry has served as a proving ground for the consumer movement in the areas of advertising, labeling, packaging, product safety, and other consumer concerns.

Today's consumer expects the food industry to function efficiently and honestly. The citizen-consumer also expects to have a voice in public policy decisions affecting food production, food prices, and food distribution. As a result, the "farm policy question" has been broadened to the "food policy question." Food consumers also have rising expectations for the food industry. These may or may not always be realistic, feasible, or economical, but they cannot be ignored.

The level of consumer satisfaction with the food industry in general and with specific food products can affect farmers and food marketing firms. Dissatisfaction may trigger boycotts, substitutions of one food for another, and, most importantly, increased consumer support for government intervention in food markets. Intervention may take the form of price controls, restrictions on industry marketing programs, advertising and labeling controls, import-export control policies, and even profit controls.

REGULATIONS THAT DIRECTLY AFFECT FOOD PRICES

Government can influence farm and food prices in many ways. Government regulation of public utilities has led to direct control of rates and prices. Railroad rates, public utility charges, livestock yardage, and numerous other prices and fees have been regulated by government. The rationale for intervention in the pricing system is that monopolies can be permitted only if there is assurance that they operate in the public interest. But there has also been growing concern that government price regulations may in some cases protect inefficient firms and result in higher prices than would be the case without government regulations.

Many macroeconomic and most farm price and income policies also directly or indirectly influence food prices. Food price supports, supply control, import-export, and domestic demand-expansion programs, such as the Food Stamp Program, fall into this category. These programs, of course, may have other objectives and effects outside of and in addition to their impacts on food prices.

There have been periodic attempts of government to directly control food prices through price control or freeze programs. The United States experienced food and other product price controls in World War I (1917–1918), World War II (1941–1946), the Korean conflict (1950–1953), and during the 1971–1973 period. These were periods of rapid price inflation when the government felt an obligation to intervene directly in prices. The 1941–1946 price controls were also accompanied by food rationing. During these periods, retail food prices and marketing margins were "frozen" or controlled whereas farm prices were free to vary. Of course, a retail price freeze influences the derived farm price.

The government's responsibility and ability to directly regulate food prices are highly debatable points. Control of food prices is an awesome economic power. Because prices allocate resources and influence all food industry decisions, there are significant market consequences of price regulation. Many developing countries have employed food price controls as a way of maintaining low food prices for urbanized populations. Such policies may also produce low farm prices and shortages.

REGULATIONS TO FOSTER ECONOMIC AND SOCIAL PROGRESS

Each interest group that pressures for preferential legislation generally justifies its actions as being necessary for economic and social progress. Usually such laws fit better into one of our categories. However, there have been legislative developments that have addressed themselves specifically to the general purpose of "holding the carrot in front of the horse" so that our economy will move forward.

Patent and copyright laws give innovators a monopoly for a period of time so that they may reap personal gain from their creative efforts. Such an incentive system, which has a long history in English law, was recognized by the U.S. Constitution and set forth in the original patent act of 1790. The intent is to give to the innovator the right of monopoly exploitation for seventeen years, after which the innovation becomes available to the public for general use. Such incentive for profit has no doubt fostered research and development. With increasing food processing and more synthetic products, patent and copyright provisions are of growing importance in the marketing of farm products. The Plant Variety Protection Act (1970) grants patents to developers of new plants and varieties that reproduce by seeds. The challenge is how to furnish incentives for the continuing parade of new and better things and methods without, at the same time, preventing the active search for improvement by anyone in any area of endeavor.

Another very important group of laws is that providing for public support of education and research. In 1862 Congress established the Department of Agriculture. The Morrill Land-Grant College Act (1862) provided the basis for the extensive U.S. system of government-supported higher education. This was followed by the Hatch Act (1887), which provided for the establishment of agricultural research experiment stations in conjunction with the land-grant colleges.

In 1914 the Smith-Lever Act broadened the educational system in agriculture by establishing the agricultural extension system. Through a widespread system of county agents and various types of trained specialists, the findings of the experiment stations are quickly brought to the farmers' attention. This act was followed in a few years by the Smith-Hughes Act, which provided federal support for teaching vocational agriculture in the public schools. The resulting agricultural research-educational team today is widely admired and imitated around the world.

This educational and research structure has been augmented by a number of laws making funds available for research in specific areas. Early work was aimed at farm production problems. However, the Research and Marketing Act of 1946 laid the groundwork for a broad research attack on marketing problems. One very pertinent contribution of this legislation has been to expand the horizons of agricultural marketing research and extension beyond the point of first sale. The complete producer-to-consumer area is now recognized as the field in which to work. Official recognition has been given to the fact that the actions of food processors, wholesalers, and retailers are the business of those who are concerned with agricultural marketing. To a large degree the improvements in marketing will depend on research progress in this enlarged area of study.

Governmental agricultural research is carried on at about 400 field locations, including federal field stations and laboratories and cooperative work at the individual state experiment stations. The states maintain about 325 research centers, including experiment stations and substations. Together the federal and state governments employ more than 10,000 agricultural research scientists. The general framework for financing agricultural research consists of the states matching the federal appropriations.

THE REGULATORY AGENCIES

The U.S. Department of Agriculture has been assigned dual regulatory and service roles in the food industry and serves farmers, food marketing firms, and consumers. The United States Department of Agriculture administers the food grading and inspection programs as well as such laws as the Packers and Stockyards Act, the Warehouse Act, the Produce Agency Act, and the Perishable Agricultural Commodities Act. This department is also involved in farm price support and income programs, and has the responsibility for the food stamp, the school lunch, and milk programs. The laws authorizing agricultural production and marketing research, and the adult education/extension programs, are also the province of the Department of Agriculture.

USDA's Food Safety and Inspection Service inspects all meat and poultry sold in interstate and foreign commerce. Its Plant Protection and Quarantine Service prevents the importation of food-destroying pests and weeds into the country. And USDA's Veterinary Services Division has the responsibility of keeping foreign diseases out of the country, eradicating outbreaks of animal diseases, and ensuring the humane care of livestock.

The Justice Department and the Federal Trade Commission have responsibility for the laws regulating competitive practices in the food industry. Anticompetitive mergers, unfair trade practices, price fixing, and deceptive advertising are the concerns of these agencies. The Food and Drug Administration is responsible for ensuring the safety and wholesomeness of the food supply. The Commodity Futures Trading Commission has responsibility for ensuring fair trading and orderly growth of the commodity futures market.

Numerous other federal agencies affect food marketing firms and activities in a less direct manner. The Interstate Commerce Commission, the Internal Revenue Service, the Federal Energy Agency, the Environmental Protection Agency, the Federal Communications Commission, the Federal Power Commission, the Occupational Safety and Health Administration, the Securities and Exchange Commission, and the National Highway Traffic Safety Administration all play a role in today's food industry. Consider the effects of a 55 mph speed limit, energy rationing and higher fuel costs, waste-water effluent limitations, or the changing tax structure on food marketing.

At the state and local levels, public health agencies, state departments of agriculture, and other agencies oversee a vast number of food marketing programs ranging from grocery store sanitation to commodity promotion programs. The growing number of state and city food regulations is adding to the complexity of the food marketing process. Most state and local regulations follow federal laws, but some local regulations may erect trade barriers and artificial restrictions on interstate commerce in food products.

THE FUTURE OF FOOD MARKETING REGULATIONS

We can now return to the initial questions of the where, the what, and the how much of regulation. Careful thought brings the conclusion that there is a place for government regulations and laws. Often it is not the intent of regulations, but their enforcement and administration, that cause problems. The role of government is not passive or static. Nor does the fact that certain laws are on the books make them currently good or adequate laws. A progressive economy is one of change. Legislation once considered vital may no longer be useful. New developments may bring about the need for new regulations even in fields where such actions were once considered improper.

Increasingly, those interested in the performance of agricultural markets must give attention to the proper role of government. At one time agricultural marketing was considered to operate under nearly perfect competitive conditions. However, this is no longer a realistic assumption. We have seen the large-scale enterprises that exist in food marketing. We have noted the rapid development of large and powerful food marketing organizations. We have discussed the implications of more direct marketing channels and integration for the bargaining power of farmers and other participants in the marketing system.

How many firms are enough? What trade practices should be allowed? Encouraged? How much protection for individuals and firms is needed? Questions like these are ones that leaders in agricultural marketing must address. The real challenge is to find the "proper" solution among the many alternatives that are always available.

SUMMARY

Food marketing is a highly regulated industry. Food marketing regulations attempt to regulate competition and monopolistic conditions, facilitate trade, protect consumers, directly influence food prices, and foster economic and social progress. These regulations may be restrictive, permissive, or facilitating. The food marketing regulatory environment undergoes continual change in the search for the appropriate blend of regulations that will achieve public and private objectives. Although many food marketing regulations restrict freedoms, others provide valuable services for farmers, food marketing firms, and consumers. An overall evaluation of food marketing regulations requires consideration of both their costs and benefits.

KEY LAWS AND REGULATIONS

Agricultural Fair Practice Act
Agricultural Marketing and Bargaining Act
Capper-Volstead Act
Clayton Act
Egg Products Inspection Act
Organic Food Production Act
Packers and Stockyards Act
Patent Act
Perishable Agricultural Commodities Act
Phytosanitary regulations
Plant Variety Protection Act

Farmer-To-Consumer Direct Marketing Act
Federal Trade Commission Act
Federal Seed Act
Food and Drug Act
Grain Standards Act
HACCP
Hatch Act
McGuire Act
Morrill Land-Grant College Act
Nutritional Labeling and Education Act
Produce Agency Act
Research and Marketing Act
Robinson-Patman Act
Sherman Anti-Trust Act
Smith-Hughes Act
Smith-Lever Act
Truth-in-Packaging Act
Tydings-Miller Act
United States Warehouse Act
Webb-Pomerene Act
Wheeler-Lea Act
Wholesome Meat Act

DISCUSSION QUESTIONS

1. Why is it necessary for government to establish basic "rules" for orderly economic and marketing activity?

2. Businessmen frequently complain that there are too many government regulations to comply with and that these add unnecessarily to the price of food. Evaluate this charge.

3. Give an example of a restrictive, a permissive, and a facilitating food marketing law.

4. Comment on the observation that no government regulations would be necessary if businessmen were honest and concerned with the public welfare.

5. Give an example of a food marketing law that has unintended effects in addition to its stated goals.

6. Comment on the feasibility and desirability of regulating all food marketing firms so that they are perfectly competitive.

7. Why has the food marketing system so often been charged by government agencies with monopolistic and anticompetitive practices?

8. Economists generally agree that direct price controls on food products will not be successful. Why then have they been used during periods of wartime?

9. Do you think there will be more or fewer food marketing regulations in the future?

SELECTED REFERENCES

Crutchfield, S. R. (May–Aug. 1999). *New Federal Policies and Programs for Food Safety.* Food Review, Promoting Food Safety. U.S. Department of Agriculture, pp. 2–5.

Food Safety. (1998). *Agricultural Fact Book, 1998.* U.S. Department of Agriculture.

Food Safety Briefing Room. (January 2001). Economic Research Service, U.S. Department of Agriculture. www.ers.usda.gov/briefing/foodsafetypolicy/.

PART VI

COMMODITY MARKETING

CHAPTER 23

LIVESTOCK AND MEAT MARKETING

The U.S. meatpacking industry consolidated rapidly in the last two decades, as today's leading firms built very large plants and many independent packers disappeared.

U.S. Department of Agriculture,
Agricultural Economic Report No. 785, 2000.

OBJECTIVES

After reading this chapter, you will be able to:

1. Appreciate how the production and product characteristics of livestock and meat influence the marketing of these products.
2. Describe the changing market patterns and distribution channels of the meat and livestock sector.
3. Evaluate how the changing livestock and meat industries are affecting producers and consumers.
4. Explain how changing consumer preferences are influencing the meat industry and how the industry is responding to these changes and attempting to influence demand for meat products.

Livestock production is an important segment of U.S. agriculture. As shown in Figure 23-1, livestock products account for almost one-half of farm cash receipts. The cash receipts from the sale of cattle and calves is the largest for any single farm enterprise. Meat and livestock products account for about one-third of the consumer's food bill.

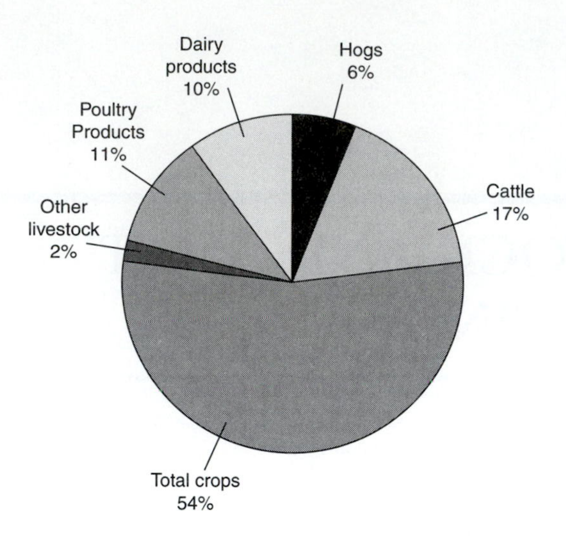

FIGURE 23–1 Cash receipts from farm marketings by commodity, 1997. (U.S. Department of Agriculture.)

Here we are concerned with the red meat animals—cattle, calves, hogs, sheep, and lambs. This is one of the most complex sectors of the food industry. Ranchers and farmers, feeders, meat packers, processors, wholesalers, and retailers are interrelated through a complex chain of markets involving a large number of marketing activities, functions, and institutions.

It is very difficult to draw a line between farming and marketing in the combined feed-livestock sector of the food industry. Livestock are protein converters, transforming vegetable protein into animal protein. These "protein factories" are a form of food processing. Moreover, livestock feeding can be viewed as an alternative way for the grain farmer to market grain. In this sense, the grain-livestock farm, so common in the Cornbelt states, is in reality a very sophisticated, vertically integrated farming and marketing protein-conversion operation.

The red meat livestock segment of the food industry has undergone substantial change over the years and continues to change. Decentralization, integration, specialized cattle feeding, the growth of supermarkets and chain stores, and improved transportation, grading, market information, and product quality have been important trends in this industry. More recently, increased world demand for meats and accelerated U.S. meat exports and the growth of boxed or fabricated beef have been important factors affecting the meat industry.

LIVESTOCK PRODUCTION

The nature and structure of livestock production has had a significant impact on marketing patterns in this industry. Livestock production consists of a series of operations involving breeding, raising young animals, and feeding these animals to market weights. These operations can all be performed on one farm, as illustrated by the traditional grain-livestock farm. Or, they can be performed by specialized farms and firms, as has occurred in the beef industry where cow-calf operations, feeder cattle raising, and cattle feeding are performed by different firms often located in different parts of the country. This specialization greatly complicates livestock marketing, contributes to a longer marketing channel, and has given rise to completely new markets, such as the feeder cattle auction markets. As an illustration, Figure 23-2 shows the various stages in pork production and their interrelationships within the marketing channel.

The geographic location of livestock production is determined by the regional availability of key production resources, such as pasture, rangeland, feed grains, and farm labor. Livestock are produced in every state, but there are concentrations of production (Table 23-1). Because feedgrains are a major animal production cost, livestock production is closely linked to grain production.

Another important characteristic of livestock production is the diversity of producing units. There are a large number of very small producers and a few very large ranches and livestock feeding operations. Enterprise specialization also varies widely. Many producers specialize in either beef, pork, or sheep production. Some combine these enterprises with feedgrain production, and others do not. There are producers who view the livestock enterprise as an income supplement, and use it to occupy excess labor, buildings, or other resources. This diversity of producing units has contributed to the difficulty of organizing livestock producers for collective market actions.

As a result of the long production periods, and because of the tendency to adjust future production to current prices, livestock production is subject to output and price cycles. Farmers and ranchers periodically produce too much livestock to get what they consider to be reasonable prices, and livestock prices are frequently below the cost of production.

Because of the biological lags involved, producer adjustments in livestock production often result in unexpected market effects. In order to expand future meat

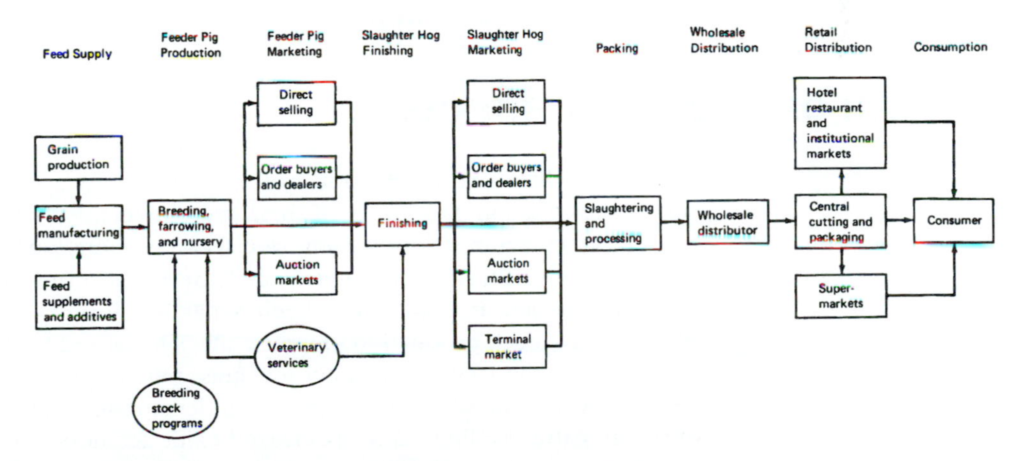

FIGURE 23–2 Elements of pork production. (R. N. Van Arsdall and H. C. Gilliam, "Pork," in L. P. Schertz, et al., editors, *Another Revolution in U.S. Farming.* U.S. Department of Agriculture. Agricultural Economics Report No. 441, December 1979, p. 193.)

TABLE 23-1 Top-Ten Leading States Marketing Livestock, 1997

Percent of U.S. Livestock Sold					
Cattle and Calves		*Hogs*		*Sheep and Lambs*	
Texas	18%	Iowa	22%	Colorado	20%
Kansas	13	N. Carolina	16	Texas	14
Nebraska	12	Minnesota	10	Wyoming	10
Colorado	6	Illinois	8	California	7
Oklahoma	6	Missouri	6	S. Dakota	5
Iowa	5	Indiana	6	Montana	5
California	3	Nebraska	6	Iowa	4
S. Dakota	3	Ohio	3	Utah	4
Missouri	3	Oklahoma	2	Idaho	4
Idaho	2	Kansas	2	Oregon	4
Top-Ten States	71%	Total	81%	Total	77%

SOURCE: U.S. Department of Agriculture.

supplies in response to anticipated profits, livestock producers must hold back animals from the market in the near term to build up the breeding herd. This supply expansion thus *increases* meat prices in the short run. Conversely, when low prices and profits signal livestock producers to reduce production, the herd selloff *increases* meat supplies and further reduces prices in the short run.

Other characteristics of livestock influence their prices and marketing patterns. Livestock and meat products are homogeneous and therefore difficult to brand. Farm animals are also bulky and expensive to transport. Finally, livestock is a moderately perishable commodity. However, meat can be "stored" in cold storage or on the hoof, and livestock sales can be deferred to some extent. By varying the finish and weights of livestock, the producer has alternatives to immediate sale. However, there are limits to these options because when deferred sales eventually come to market, there may be weight and age price discounts.

GROWTH OF SPECIALIZED FEEDLOTS

Historically, livestock production has taken place on a large number of relatively small farms and ranches. Many of these were diversified crop-livestock farms, combining the animal breeding, raising, and feeding operations. These integrated cattle and hog farms continue today, particularly in the eastern Cornbelt. But increasingly, the grain feeding operation has been separated from the breeding and finishing phases of livestock production in large-scale, specialized feedlots.

Prior to the 1950s, most beef was grass-fed or grain-fed in farmer-owned, small feedlots. As a result of the rapid growth in demand for grain-fed beef and economies of size in cattle feeding, large specialized cattle feedlots developed in the West and the Plains states after 1950. Owners of these feeding operations purchase grain and feeder animals for finishing to market weights. Feedlots are highly specialized protein conversion factories, with only vestigial ties to traditional farming.

Cattle feedlots are a highly concentrated segment of the livestock industry. Prior to 1962, almost 64 percent of the fat cattle marketed were fed in farmer-owned feedlots with an annual capacity of less than 1000 head. Today, less than one-fourth of all cattle are fed in these smaller feedlots. Cattle feedlots with capacity exceeding 1000 head now account for over 90 percent of fed-cattle sales. Some 400 of the largest feedlots, some with over 100,000 head capacity, market 50 percent of the nation's fed cattle today.

Pork production has not yet reached this level of specialization and market concentration. But a similar shift of hog and pig production and feeding from numerous, small, diversified farms to fewer, large-scale, year-round, intensive breeding and feeding operations is occurring. The declining numbers and increasing sizes of pork producers are shown in Figure 23-3.

Most sheep are still raised in small farm flocks on mountain pastures and open ranges. However, a few producers are experimenting with large-scale confinement sheep feedlots.

New markets and marketing patterns have developed in the livestock industry with the growth of these intensive livestock feeding operations. Feeder animal markets facilitate the transfer of livestock from specialized cow-calf or hog farrowing (birthing) operations to the finishing feedlots. This involves private sales, terminal markets, or auction markets. Many feeder cattle and pigs are sold through auction markets. There are specialized middlemen buying and selling these feeder animals, referred to as order buyers, cattle dealers, commission men, and salaried buyers. Other intermediaries, known as *stockers* or *backgrounders*, coordinate the flow of animals from breeding operations to the feedlot.

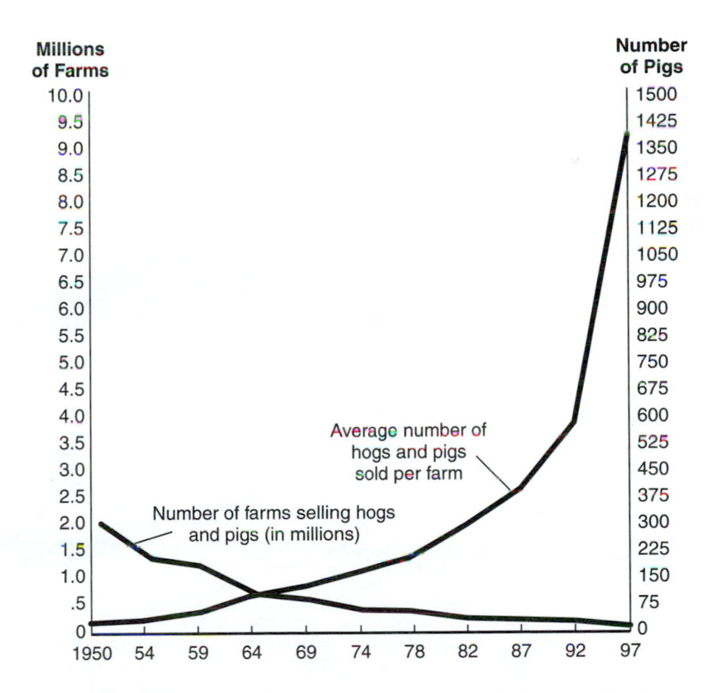

FIGURE 23–3 Number and average size of farms selling hogs and pigs, 1950–1997. (U.S. Census of Agriculture.)

The geographic specialization of the stages in livestock production has resulted in large-scale movements of animals and feed around the country. The clockwise movement of feeder cattle is illustrated in Figure 23-4. Northwestern states provide feeders for the western Cornbelt feedlots, and the southern states are important sources of feeder cattle for the western feedlots. A "tourist" calf born in Georgia may spend some time foraging in the Texas high plains and then visit a Cornbelt feedlot before becoming a steak in Chicago.

Feeder animals have presented the livestock industry with an entirely new set of marketing problems and opportunities. There is concern for the operational efficiency of the numerous and small auction markets, the bargaining power of the small feeder animal seller against the feedlot buyer, and the pricing efficiency of these local markets. Electronic marketing systems have been developed for feeder cattle and feeder pigs. While there are USDA grades for feeder animals, they are not widely used, and the purchase and sale of feeder animals remains very much an art.

Intensive livestock feeding operations have also altered the marketing system for slaughter cattle and hogs. It is generally agreed there are significant economies of scale in these operations. These efficiencies have undoubtedly influenced the costs and prices of meat. The largest cattle feedlots are located in the West, lengthening the distance to the populous eastern markets. Most large cattle feedlots are corporations, and their need for capital has attracted many nonfarm and nonfood industry investors. The expansion of intensive, confinement hog feeding operations also increased pork producers' reliance on debt financing and stimulated producer contracting by feed companies and meat packers. Seasonal variation of cattle and hog production, and thus meat production, has declined with the growth of large feedlots.

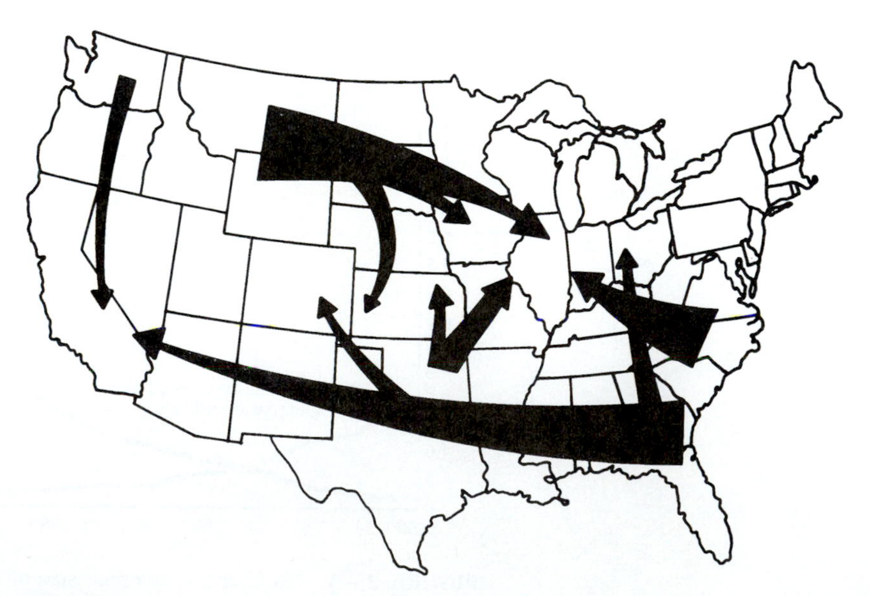

FIGURE 23–4 Major directional movements of U.S. feeder cattle.

LIVESTOCK PRODUCTS AND MEAT CONSUMPTION

Livestock and meat are complex products. An animal carcass is in reality a bundle of products, each with different markets, demands, and values. A steer is not all steak, as shown in Figure 23–5. A 1,000-pound steer typically dresses out to a 620-pound carcass. This carcass yields about 540 pounds of retail meat products. About 40 percent goes into the making of hamburger and 60 percent into higher-value cuts. The carcass yields other by-products, such as hides, pelts, lard, and offal, which have value in the making of clothing, foodstuffs, fertilizers, and industrial products.

Meats are versatile foods available in several forms. Meats can be purchased ready to eat, ready to cook, or in forms requiring substantial preparation. Much meat is sold fresh, but such processed items as canned meat, sausage, frankfurters, and lunch meats provide the consumer with a wide choice. Food processors, called *fur-*

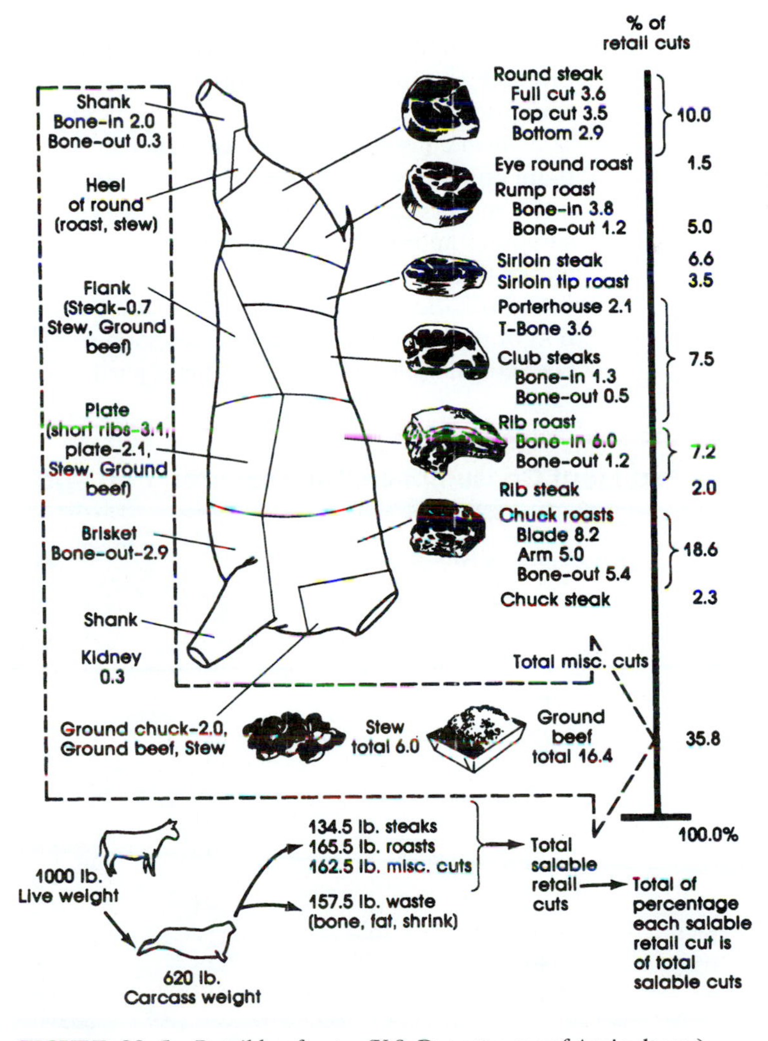

FIGURE 23–5 Retail beef cuts. (U.S. Department of Agriculture.)

ther processors, combine meats with other foods to give the consumer an even wider variety of meat dishes.

Many factors affect consumer demand for red meats, including prices, income, population, and tastes and preferences. Per capita consumption of these products generally increased from 1950 to 1970, as shown in Table 23-2. However, most of this gain was the result of growth in beef consumption as pork consumption and consumption of lamb fell. Red meat expenditures as a percentage of consumers' disposable income continues to fall.

The upward trend in U.S. beef demand from 1950-1976 was one of the most dramatic changes in food demand ever witnessed. Per capita beef consumption increased from 50 pounds in the early 1950s to a peak of 94 pounds in 1976. This demand surge caused a virtual revolution in beef production and marketing. The income elasticity of demand for beef is strongly positive, and this was a period of rapidly increasing consumer incomes. Other factors contributing to increased beef demand included the growth of the fast-food industry, grocers' frequent use of beef as a price special, the versatility and convenience of beef, the changing age-distribution of the population (young consumers eat the most beef), and perhaps the promotional programs sponsored by the beef industry. Table 23-2 also shows that beef prices were relatively stable in the 1950s and 1960s with real prices falling.

Beef consumption generally fell from 1976 to 1993. This decline was related to more rapidly rising beef prices, slower rises in consumer incomes, competition from other meats and foods, and perhaps general dietary changes. Per capita beef consumption appears to have stabilized in the 1990-1998 years and was rising at the end of the 1990s.

Pork did not experience a demand surge like beef. While per capita consumption of pork exceeded beef consumption in the early 1950s, it has been lower than beef consumption since then. This is partly explained by the declining carcass yield

TABLE 23-2 Red Meat Consumption, Expenditures, and Prices, 1950-1998

	Per Capita Consumption (Pounds)				*Retail Price ($/lb.)*		*Consumers' Expenditures for Red Meat as a % of Income*
Year	*Beef, Veal*	*Pork*	*Lamb, Mutton*	*Total, Red Meat*	*Beef*	*Pork*	
1950	57	64	4	135	.75	.54	5.5
1955	72	62	4	148	.67	.54	4.7
1960	69	60	4	133	.82	.55	4.6
1965	78	55	3	136	.80	.65	4.2
1970	86	55	3	151	1.02	.77	4.1
1975	90	43	2	143	1.55	1.35	4.1
1980	78	57	1	147	2.34	1.39	3.7
1985	81	51	1	144	2.29	1.62	2.6
1990	65	46	1	112	2.81	2.13	2.7
1995	65	49	1	115	2.59	1.95	NA
1998	65	49	1	116	2.53	2.43	NA

SOURCE: U.S. Department of Agriculture.

and use of lard as the industry switched to producing a leaner hog. Pork consumption fluctuated around 50 pounds per capita over the 1975–1998 period. There are several explanations for the lack of growth in pork consumption. Pork has a lower income elasticity and a more inelastic demand curve than beef. Consumers are also more inclined to substitute beef and chicken for pork when relative pork prices rise than to substitute pork for these meats when relative pork prices fall. Also, pork has not received as much promotion from grocers or producers as has beef. Pork producers began aggressively promoting pork products in 1986.

Lamb and mutton appear to have a faltering demand state. While very popular among certain segments of consumers, American consumers eat little lamb. One consequence is that some grocers do not stock lamb, display it prominently, or feature it. Thus, many consumers are not made familiar with this meat, further contributing to its declining demand.

LIVESTOCK ASSEMBLY OPERATIONS

Livestock are usually purchased for one of three reasons: (1) for breeding herds; (2) for placement in feedlots; or (3) for immediate slaughter. The term *livestock assembly* refers to bringing together animals for any of these purposes. There are several livestock assembling agencies, which are described in the following paragraphs.

Local cooperative associations originally functioned largely as shipping agencies that collected the small lots from farmers and shipped them forward to terminal markets in carload lots. Many have taken on a broader service and often merchandise their livestock directly to packers and other buyers. Some producer groups engage in bargaining activities.

Country dealers are independent operators who buy and sell livestock for a profit. They usually buy livestock from the farmer and then resell to packers or some other market agency. Some maintain a small yard. Others merely operate a truck and pick up the livestock from the farmer.

Livestock are also offered for public sale on an *auction* basis. Some auctions operate as primary outlets and sources for feeder livestock and breeding animals. Others are used as outlets for slaughter livestock and are patronized by packers, dealers, and other types of buyers.

Concentration yards are often referred to as "local markets" and may be operated by independent or cooperative owners or by packing companies to buy livestock. Very often it is difficult to clearly distinguish between such outlets and the country dealer who maintains yard facilities and operates on a large scale. *Order buyers* purchase feeder and fed livestock for other buyers for a fee, from terminal markets, auctions, and at country points.

Terminal public markets are large central markets where livestock are received and cared for and where the privileges of buying and selling are available to all who may wish to use them. The yard facilities are owned by a stockyard company. This organization is often referred to as operating a "livestock hotel." It does not enter into the buying and selling of the livestock, but merely furnishes the physical facilities for which it receives payment in the form of *yardage fees.*

The farmer normally consigns his livestock to commission merchants who then act as the farmer's selling agent. These agents receive commission charges as payment for their work. Commission organizations may be either privately or cooperatively owned. Sometimes the private commission companies are organized in what are called livestock exchanges. These organizations attempt to self-police the practices used and carry on other activities of mutual concern to their members.

Some livestock *packing and meat processing plants* are located near the large terminal yards and secure most of their livestock through those facilities. Others, either in the production areas or at the central markets, may maintain their own buying yards at the plant itself or within the production area. Still others purchase their livestock from dealers, concentration yards, auctions, and direct from producers. In some instances *retailers* or frozen food locker plants buy and slaughter their own livestock.

The sale of livestock from one farmer to another is an important livestock outlet, especially for feeding and breeding stock.

Most livestock buyers and sellers still have access to a variety of these markets and agencies. However, in recent years producers have been concerned about a reduction in the number of local buyers and sales alternatives. This lessening of competition could affect livestock prices. The *number* of buyers is not the only determinant of competition in these markets, of course, and two firms can compete just as vigorously as ten. Also, long-distance phone calls, fax, email, World Wide Web, and shipments can quickly expand the sellers' marketing options and maintain local competitive discipline.

DECENTRALIZATION IN THE LIVESTOCK AND MEAT PACKING INDUSTRIES

One of the most important trends shaping livestock marketing channels in the past 100 years has been decentralization. This increased direct sales from producers to meat packers and moved packing plants from the terminal markets to livestock producing areas. This re-shaped the livestock industry and is still influencing the sector.

Growth and Decline of Terminal Markets

During the early development of the livestock industry, railroads dominated the transportation picture. This encouraged the development of large meat packing centers in Cincinnati, Chicago, St. Louis, and Omaha. The cooperative shipping associations and the large livestock yards grew out of this situation. Because freight rates were higher for less-than-carload lots, the shipping association furnished a means by which farmers, each with a few head of livestock, could assemble a carload to be shipped at one time. At the "end of the line," large terminal markets grew to serve the packers.

Soon after World War I these yards came under government supervision with the passage of the Packers and Stockyards Act. The object of this act was to assure the maintenance of competition and fair dealings. Among other things, the act provided for general supervision of the buying and selling practices of firms operating on the market. In 1958 the Packers and Stockyards Act was amended to extend its coverage to any agency handling interstate livestock. This extension of supervisory coverage was necessary to keep pace with the changing patterns of marketing.

The terminal markets reached their peak in 1922, when there were 78 major livestock terminals. Many meat packers built their plants adjacent to these markets. There were perhaps 20 livestock terminal markets operating in 2000.

Direct Marketing and Decentralization

Since 1923, livestock terminal markets have declined in importance. Livestock sales made directly between the producer and packer, without using the services of terminal market facilities, are called *direct sales.* Because this shifts the point of livestock pricing from centralized terminal markets to numerous country points, direct sales are termed *decentralized marketing.*

The livestock decentralization trend is shown in Figure 23-6. Between 1975 and 1996, meat packers' cattle purchases from terminal markets declined from 33 percent to 14 percent of all purchases. Similar trends are evident for hogs and sheep. Decentralized market channels are represented by packer purchases from producing area auction markets and from direct or country dealers.

Direct sales and some technological improvements in meat packing and processing plants encouraged meat packers to move their plants from terminal markets to livestock-producing areas. The newer meat packing plants have been built in the western Cornbelt states. This decentralization of meat packing facilities was largely completed by the 1970s.

Reasons for Change in Livestock Assembly

The basic reasons for the general change to more direct marketing channels were treated in Chapter 12. The particular reasons for livestock marketing decentralization are summarized as follows:

1. The expansion of corn and livestock production into the West placed the production areas still farther from the consuming areas, lengthening the transportation distances for bulky livestock.

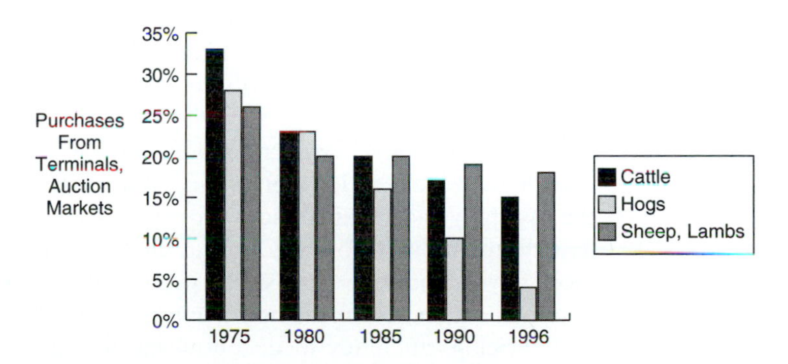

FIGURE 23–6 Decentralization of livestock markets. (U.S. Department of Agriculture, Grain Inspection, Packers, and Stockyards Administration.)

2. The freight rate structure and improvements in refrigeration favored the shipment of meat over the shipment of live animals. This was particularly true when the saving in shipping weight of carcasses compared with live animals was considered.

3. Originally, lower wage rates encouraged the growth of small local packers in the production area. Later, the national packers also built new plants there, both to take advantage of these economies and to meet competition.

4. The development of trucks, along with improved roads, greatly encouraged local marketing, reducing the dependence on the railroads.

5. The farmers' longstanding distrust of the central market and its consignment method of selling encouraged them to favor markets closer to home.

6. The recent development of much larger farm production units has increased the feasibility for packers to send their buyers directly to the farmer to acquire livestock.

Centralized versus Decentralized Marketing

It is not possible to say that centralized marketing is either superior or inferior to decentralized marketing. Both marketing arrangements have advantages and disadvantages, and each appeals to various buyers and sellers of livestock for different reasons. The trend toward decentralization suggests that there may be some advantages and/or preferences for this marketing channel. However, the recent slowing of the trend toward decentralization may indicate that both types of markets now have equal, but different, advantages for producers and meat packers.

Each producer and packer chooses the most advantageous market channel, based on needs, preferences, and economic returns. Their choices will reflect prices, marketing costs, and net returns of the two alternatives, and neither high prices nor low marketing costs alone determine what that choice will be. The important thing is that all buyers and sellers are free to choose among real alternatives.

Nevertheless, there are several reasons why many farmers have preferred direct sales to terminal market sales. Direct sales usually require fewer marketing services and therefore fewer out-of-pocket marketing expenses for farmers. Also, farmers sometimes feel that in direct sales there is less shrinkage (loss of product and value in marketing), and that direct sales are more convenient. In decentralized marketing farmers maintain physical control over animals until they are priced and sold, and this, too, probably contributes to the preference of many for this marketing channel.

Even so, some farmers continue to use terminal markets. In some cases, this may be because they have no choice. But small producers frequently need and value the assembly and other services provided by terminal markets. Some farmers contend that competition is more vigorous in the terminal markets. It is true that there may be more buyers and sellers physically present at the terminal markets; however, modern communication networks and the internet link all local and terminal markets today.

Pricing efficiency in decentralized livestock markets is a legitimate concern. Decentralization has intensified the market information problem for livestock producers. Collecting and disseminating useful market news for price discovery at the numerous decentralized markets is a problem. It is difficult for the USDA Market News Service to physically cover these widely scattered markets and to gather accurate, voluntary information from traders. There are continuing experiments with electronic, computerized trading in the livestock industry. These attempt to preserve the pricing efficiency of centralized trading within decentralized markets.

MEAT PACKING AND PROCESSING

The meat packing industry is a disassembling industry. Whereas other manufacturers combine simple raw materials into a complex, composite product, meat packers break down a complex raw product—livestock—into its constituent parts. This reverse manufacturing process nevertheless adds form utility and value to livestock products.

The meat packing industry is composed of slaughterhouses, where livestock is slaughtered and red meat is further processed, and specialized meat processors that do not slaughter but manufacture sausage, luncheon meats, and other prepared products. The combination slaughter-processors tend to be larger than the specialized meat processors and have experienced more rapid sales growth as a result of the increased demand for fresh beef.

All meat sold across state lines must be slaughtered under the supervision of federal inspectors. This is not a form of quality grading; rather, it ensures that animals are disease-free, fit for human consumption, and that the plant meets sanitary requirements. The 1967 Wholesome Meat Act required all intrastate meat inspection programs to equal or exceed the federal standards. Today, more than 95 percent of all cattle, hogs, and sheep and lambs are slaughtered under federal inspection.

New agribusiness firms have entered the fresh meat industry in recent years (for example, IBP, Cargill, ConAgra), in most cases through mergers or acquisitions of older meat packing companies. The largest firms are now diversified "protein suppliers" marketing beef, pork, and often poultry in contrast to the specialized firms of the past. Many of the meat processing firms have discontinued their slaughtering operations, finding it more efficient to purchase their raw products from the fresh meat packing firms. Increasingly, fresh meat packers are attempting to brand their products.

The livestock slaughter and processing industries are comprised of a few dominant, large companies that market a large share of the industry's output and numerous small and medium-sized companies. As shown in Figure 23-7, the four-firm market concentration ratios in these industries rose dramatically from 1970 to 1999. This resulted from mergers, technological change, and plant closings, and there was a change in the identity of the top firms during this time. Concentration in many local markets is likely higher than these national figures.

While meat packers have not integrated forward into retailing operations, they have moved forward and become further-processors. Also, there has been backward integration of meat packers into livestock feeding. This takes the form of packer ownership of feedlots, or custom feeding by feedlots under contract to meat packers. The U.S. Department of Agriculture estimated that 20 to 25 percent of cattle were vertically integrated from producer to meat packer in 1997.

Cattle feeding by packers is more prevalent in the West than elsewhere. Meat packers integrate into feeding operations in order to control product quantity and quality and to reduce procurement costs. Farmers and independent feedlot operators are concerned that this integration and *captive supply* provides packers with

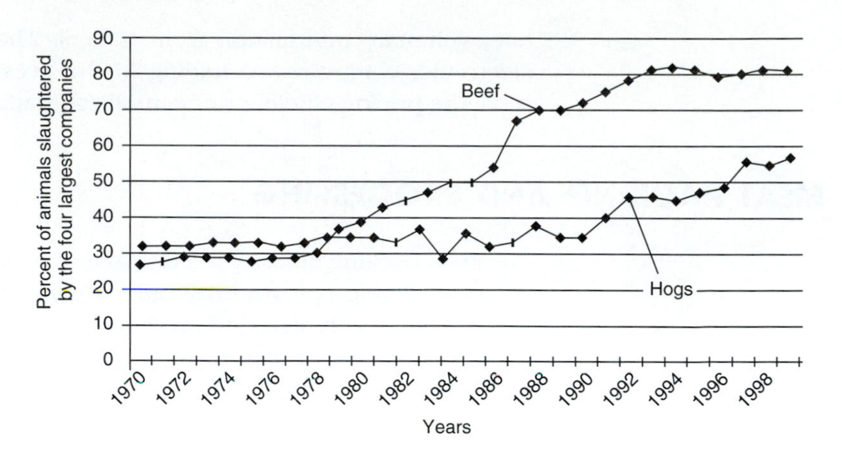

FIGURE 23–7 Concentration of U.S. beef and hog slaughtering 1970 to 1999. (Packers and Stockyards Administration, U.S.D.A.)

market power over livestock prices, but many producers prefer the market risk reductions from contracting. Some food retail chains engage in cattle feeding and slaughtering. Probably less than 5 percent of lambs and 20 to 25 percent of hogs are vertically integrated with packers.

Historically the meat packing industry has experienced lower earnings and profits than other food processing sectors and other manufacturers. Some factors contributing to this include a tendency toward excess capacity, the entry of several new firms into the industry in recent years, susceptibility to wide swings in livestock prices, the lack of differentiated, brandable products, and a rather high proportion of costs accounted for by wages. In recent years, meat packing profits have shown some increase as more efficient plants and merchandising methods were developed. However, profit levels in the industry remain below those of several food industries.

MEAT WHOLESALING AND RETAILING

The meat wholesaling sector consists of numerous merchant wholesalers, brokers, jobbers, and purveyors as well as the branch sales offices of the meat packers and the distribution warehouses owned by food retail organizations. The tasks of this sector are to move enormous quantities of product over long distances and to maintain a high degree of freshness for perishable products.

As a result of the chain store and supermarket movements, integrated retail chain store organizations are the dominant meat merchandisers today. Most fresh and processed meats move directly from packers and meat processors to chain store warehouses and then to retail stores. The role of the independent meat wholesaler has accordingly declined, although independent meat wholesalers and jobbers still serve smaller retailers and the away-from-home food market.

There is considerable variation in the ways that beef is being wholesaled by meat packers to food retail operations. Traditionally, packers delivered fresh, refrigerated carcasses direct to retail stores, where the final retail cuts and products were prepared. However, many chain stores have set up a central meat cutting, processing, and fabrication facility for serving several area stores. In this case, the packer ships directly to the central meat cutting facility, and the chain store then distributes the meat products to individual grocery stores. Some retailers have shifted all butcher operations to these central processing plants; others maintain store butchers. In still another variation, referred to as *boxed beef* or *tray-ready beef,* packers break down the carcass into primal cuts (e.g., rounds, loins), subprimal cuts (e.g., top rounds, bottom rounds), or even final retail cuts, then ship these, in vacuum packs, to either chain store warehouses or directly to retail stores.

Central meat processing by chain stores can result in substantial operational economies. Although this centralization offers both labor and material savings, there are problems nevertheless. Labor unions have attempted to prevent central meat processing, and some consumers have complained about the loss of personal butcher service in the retail store. There are also problems in maintaining meat condition ("bloom") as the distance between central cutting and the retail store increases. Therefore, the operational economies of boxed beef must be balanced against its impact on consumer acceptance and satisfaction.

Grocery stores, restaurants, and institutional feeding operations account for 90 percent of all retail beef sales. Meats represent about one-third of grocery store sales. Beef alone contributes 10 percent of foodstore sales and almost 50 percent of grocery store meat sales. Chain stores frequently feature meats in their promotions, and price specials are common to attract consumers to a store. Foodstore and restaurant meat prices are much more stable than farm or wholesale prices.

STANDARDIZATION AND GRADING OF LIVESTOCK AND MEAT

Meat and Livestock Grades

Livestock and meat products are relatively heterogeneous products varying widely in quality and value. Animal production is also geographically dispersed, and the specialization within the industry requires several market transactions as animals and meats move through a complex marketing channel to final markets. For these reasons, it is important to have a grading system that accurately describes products in a uniform and meaningful manner. Livestock and meat grades and standards contribute to operational and pricing efficiency by facilitating the buying and selling process, transmitting valuable information to market participants, and providing price incentives for producers to tailor their products to consumer demand. Nevertheless, livestock and meat grades—and their usage by the trade—are often cited as a major problem for this industry.

Feeder Livestock Grades

Federal grades and standards exist for feeder cattle, feeder hogs, and feeder lambs, but they are not widely used. These animals are usually purchased by experienced buyers on the basis of personal inspection at a livestock auction or commission house. An entire lot is typically priced on a head or weight basis. There is little agreement among buyers on the standards of quality for feeder animals. Some buyers purchase by breed, and others by frame size.

The USDA grades for feeder cattle are based primarily on animal conformation. These grades were revised in 1964 and 1979 and the grade names were changed to conform to the grades for slaughter animals: prime, choice, select, standard, and so on. Feeder pig grades were last revised in 1969 and also conform to slaughter hog grades (U.S. No. 1, No. 2, etc.). Because these standardized grades are little used, it is sometimes difficult for feeder producers and feedlot buyers to communicate with each other, and the cost of buying and selling animals is higher. In addition, there are no consistent and meaningful price differentials between different lots of feeder animals, which dulls the producer's incentive to improve feeder animal quality.

Slaughter Animal and Carcass Grades

USDA grades and standards also exist for slaughter animals and meat carcasses. Unlike the USDA mandatory meat wholesomeness inspection program, these quality grades are voluntary. Meat packers may use them in purchasing livestock or selling carcasses, but many meat packers have developed their own, unique grading systems.

Two factors are considered in grading beef, pork, and lamb: (1) the "quality" or "palatability" of the lean meat, and (2) the "cutability" or percentage of lean meat yielded by the carcass. Beef quality grades include prime, choice, select, standard, commercial, utility, cutter, and canner. About two-thirds of all beef carcasses are graded. Pork is graded number 1, 2, 3, 4, and utility.

There are indications of improvements in the quality of livestock marketed in recent years. The percentages of beef and pork meeting the highest grades has steadily increased. Part of this is explained by changes in the grading standards over these years, but most of it was due to quality improvements. There are rather consistent price differences for these grades, and it appears the grades have provided incentives to improve beef and pork quality.

Meat packers purchase perhaps 50 percent of their livestock on a live-weight basis. This requires buyers to estimate the carcass yield and quality of live animals in order to arrive at an offer price. Buying is an art, rather than a science, and there are significant variations in buyer and seller estimates of live animal carcass values. This variation is a frequent source of conflict in the livestock marketing channel.

Because of this problem, many meat packers provide livestock producers an alternative way of selling their animals. In *carcass basis purchases,* buyers and sellers agree to a schedule of prices for various grades and weight ranges, but the final value of each animal is determined after slaughter and grading. This may involve USDA or individual packer grades. Some producers and meat packers prefer carcass basis sales because it more nearly matches the price of each animal with its value to the meat packer and to the final consumer. It also increases the amount and quality of market information available to buyers and sellers.

The selling of livestock on a carcass basis has increased but is still not widely practiced. As shown in Table 23-3, 47 percent of cattle, 52 percent of hogs, and

TABLE 23-3 Meat Packer Purchases of Livestock On a Carcass Basis, 1965–1996

	Percent of Purchases on a Carcass Yield and Grade Basis		
Year	*Cattle*	*Hogs*	*Sheep, Lamb*
1965	11	3	5
1970	19	5	10
1975	24	9	11
1980	28	11	28
1984	31	14	21
1990	38	12	30
1996	47	52	52

source: Packers and Stockyard Statistical Report, U.S. Department of Agriculture, Packers and Stockyard Administration, annual.

52 percent of sheep and lambs were sold by producers on a carcass basis in 1996. Larger packers are more likely to use carcass basis pricing, and Cornbelt packers use this purchasing system more frequently than other packers.

Despite producers' distrust of the meat packers' grading methods and the length of time it takes to price livestock and pay producers, carcass basis sales will probably continue to grow. An alternative has been to improve live-weight pricing procedures. For example, it has been established that the weight, dressing percentage, and fat-back thickness account for a large percentage of the variation in hog carcass value. There is reason to believe that livestock salesmen and buyers, who are trained to give attention to these factors instead of the more conventional standards, can improve their estimate of live carcass value. New federal standards for grading live animals have attempted to reflect these factors.

The problem of federal meat grading lies in the voluntary nature of the grades and the battle for market power between national packers and retail chain store organizations. Large chains prefer federal grades because they help them in securing their needs for large volumes of uniform products. Smaller packers also usually favor uniform federal grading because it permits them to sell to these chain buyers on an equal footing with larger packers. Large packers, however, see such grades as a direct competition with their own efforts to grade and differentiate their products—by which they attempt to secure merchandising acceptance from consumers and thereby offset the market power of the chains.

LIVESTOCK FUTURES AND OPTIONS

With larger, more capital-intensive operations and increased price variability in recent years, livestock producers have become more interested in risk management. A number of new futures contracts, as well as options, have been developed for this purpose. Futures and option markets now exist for live cattle and hogs, feeder cattle, and frozen pork bellies.

Speculative trading and hedging in these livestock and meat contracts is similar to that described in Chapter 20. Producers, livestock buyers and sellers, meat packers and processors, and meat wholesalers and retailers can use these markets to hedge their price risks and forward-price their products. A livestock producer may hedge anticipated grain purchases by buying grain futures and forward-price livestock sales by selling livestock futures.

The livestock basis (difference between cash and futures price) is more variable and less predictable than the basis for a storable commodity like grain. Furthermore, this basis is as likely to be positive as it is negative. These basic patterns make livestock hedging somewhat more complex than grain hedging, but effective livestock hedging is possible. Livestock hedgers must know whether to hedge, when to place the hedge, and when to lift the hedge. Livestock options, another method for forward-pricing, have been available since 1984.

MEAT MARKETING COSTS

As a perishable product sold largely in fresh form, beef and pork have relatively high farmer shares. The farmer's share of the consumer's beef dollar averages 50 percent while the pork farmer's share averages about 35 percent. These shares tend to rise and fall with farm prices. The farmer's share for beef is higher than for pork because a larger portion of beef is sold in fresh form.

Figure 23-8 shows the breakdown of the retail beef dollar into two marketing spreads and the farmer's share. The farm-to-wholesale spread includes all the costs

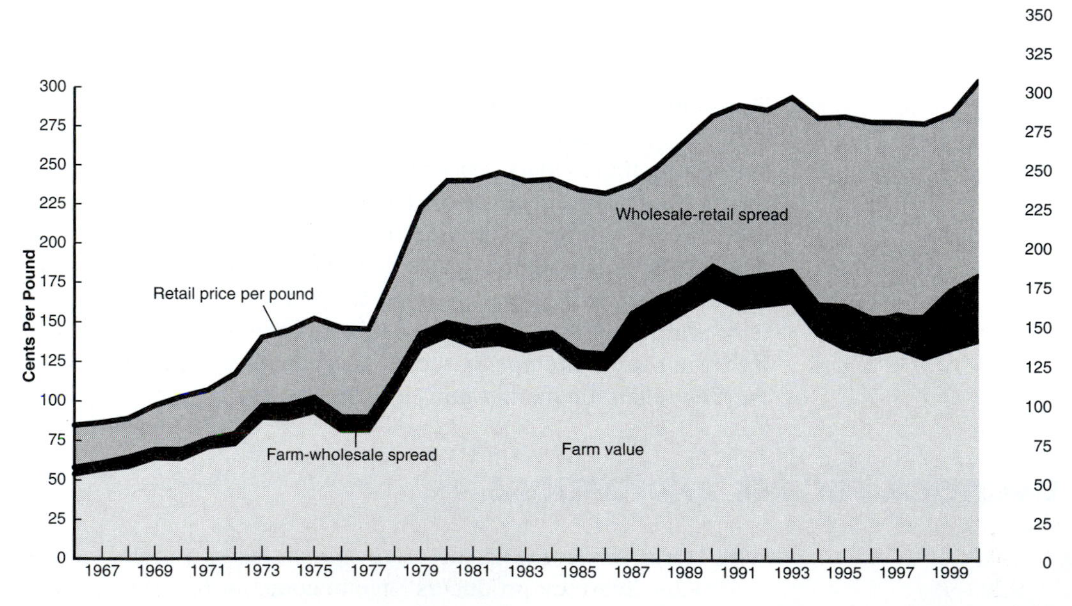

FIGURE 23–8 Beef prices, marketing margins, and farm value, 1966–2000. (U.S. Department of Agriculture.)

involved with purchasing and slaughtering the animal. The wholesale-to-retail spread includes the cutting, processing, and packaging operations, transportation, and retail merchandising functions. The wholesale-retail spread for beef (and also for pork) is the larger of the two and has grown the most rapidly in recent years. Improved efficiencies in the slaughter operation have moderated the size of the farm-carcass spread.

Retail meat prices probably fluctuate more than the prices of other foods, and meat retail prices do not always closely follow farm prices. Producer and wholesale meat prices follow each other quite closely and change minute by minute with changes in market conditions. Retail meat prices, on the other hand, are often set a week or so in advance and do not change from day to day.

MEAT MARKET DEVELOPMENT

With declining or slow growth in meat demand, all segments of the meat industry have recognized that more attention must be given to market development, including promotion and export market enhancement. The industry has developed innovative and successful advertising programs, such as the producer-sponsored *Pork, the Other White Meat* and the *Beef, It's What's for Dinner* campaigns. The meat and livestock industry has also successfully expanded exports of U.S. meat products. Exports of both beef and pork began expanding in the 1980s and have continued growing. This was in part because of the North American Free Trade Agreement (NAFTA) and the reduction of world trade barriers by the General Agreement on Tariffs and Trade (GATT) and World Trade Organization (WTO). The livestock and meat industry has also achieved product improvements through research and development, quality control, better genetics and management practices, advanced food processing technologies, and improved supply chain management. Biotechnology promises to affect future livestock and meat production costs and quality.

SUMMARY

The livestock sector of the food industry is a complex set of markets, activities, products, and marketing functions. It is an integrated production and marketing system, ranging from almost perfectly competitive feeder animal production to highly concentrated feedlots and to imperfectly competitive livestock slaughter and meat processing firms. The chief trends in U.S. livestock production and marketing have been changing geographic location, the development of large-scale feeding operations, decentralization of the meat packing industry, rapid growth and concentration of meat processing firms, and central meat processing by retail organizations. Each of these has left its imprint on livestock marketing channels. The major problems in livestock marketing are pricing efficiency in integrated, decentralized markets, changing red meat consumption trends, and methods of livestock grading and pricing.

KEY TERMS AND CONCEPTS

backgrounder
boxed beef
carcass basis purchases
cattle feeder
central meat processing
cow-calf producer
direct sales
feeder animals
feedlots
further processors
meat packer
stocker
yardage

DISCUSSION QUESTIONS

1. How does livestock feeding differ from other farming activities?
2. Why are so many different types of marketing agencies involved in livestock assembly?
3. What is the significance of the statements that livestock are protein-conversion factories and alternative grain markets?
4. Some groups have called for a reduction in the feeding of grain to livestock so that this food would be available for the hungry world. Examine this suggestion.
5. What do you think will happen to meat consumption in the next ten years?
6. Frequently it is suggested that decentralization was responsible for the decline in the market concentration ratio of meat packers over the 1950–1977 period. Explain this.
7. Why do you think farmers and packers have not accepted carcass grade and yield selling more rapidly?
8. How do you think livestock will be produced, and meat marketed, in the future?

SELECTED REFERENCES

American Meat Institute. *Financial Facts about the Meat Packing Industry.* Washington, D.C.: American Meat Institute, annual publications.

Brewster, G. W. et. al., (Fourth Quarter, 1997). "Challenges to the Beef Industry," *Choices,* pp. 20–25.

Competition In the (Meat and Livestock) Marketplace. (December 2000). Grain Inspection, Packers and Stockyards Administration, U.S. Department of Agriculture. http://www.usda.gov/gipsa/programspsp/competition.html.

Livestock, Dairy and Poultry Situation and Outlook. Economic Research Service, U.S. Department of Agriculture. Published monthy and available at http://usda.mannlib.cornell.edu/reports/erssor/livestock/ldp-mbb/2000/.

Livestock Market Structure Briefing Room. (January 2001). Economic Research Service, U.S. Department of Agriculture. www.ers.usda.gov/Topics/view.asp?T=103230.

MacDonald, J. M. et. al. (February 2000). *Consolidation In U.S. Meatpacking,* Agricultural Economics Report No. 785. Washington, D.C.: U.S. Department of Agriculture. Summarized in "Consolidation in Meatpacking: Causes & Concerns," Agricultural Outlook, June–July 2000. U.S. Department of Agriculture, pp. 23–26. Consolidation In U.S. Meat Packing website at: http://www.ers.usda.gov/whatsnew/issues/meatpacking/index.html.

Martinez, S. W. (April 1999). *Vertical Coordination in the Pork and Broiler Industries: Implications for Pork and Chicken Products.* Agricultural Economic Report No. 777. Washington, D.C.: U.S. Department of Agriculture.

McCoy, J. H., & M. E. Sarhan. (1988). *Livestock and Meat Marketing* (3rd ed). Westport, CT: AVI Publishing.

McDonald, J. M. (May 1999). "Concentration and Competition in the U.S. Food and Agricultural Industries," *Agricultural Outlook.* U.S. Department of Agriculture.

Meat Prices and Price Spreads Briefing Room, pp. 26–29. (December 2000). Economic Research Service, U.S. Department of Agriculture. http://www.ers.usda.gov/briefing/meatbrif/.

Packers and Stockyards Statistical Report (annual). Grain Inspection, Packers and Stockyards Administration, U.S. Department of Agriculture. http://www.usda.gov/gipsa/pubs/stat98/stat98.htm

Red Meat Yearbook. (January 2001). U.S. Department of Agriculture. http://www.ers.usda.gov/data/sdp/view.asp?f=livestock/94006/.

Stasko, G. F. (1997). *Marketing Grain and Livestock.* Ames, Iowa: Iowa State University Press.

U.S. Department of Agriculture. (April 1996). Perspectives on global meat trade. *Agricultural Outlook,* 18–20.

U.S. Department of Agriculture. (1973). *The Lamb Industry: An Economic Study of Marketing Structure, Practices, and Problems,* Packers and Stockyards Administration No. 2. Washington, D.C.: U.S. Department of Agriculture.

CHAPTER 24

MILK AND DAIRY PRODUCT MARKETING

There are only one or two people in the country who really understand how milk is priced.

OBJECTIVES

After reading this chapter, you will be able to:

1. Discuss the major trends affecting milk production and marketing.
2. Better understand and appreciate our complex milk pricing policies.
3. Distinguish between milk marketing orders and dairy product price supports.
4. Explain the major trends in the structure, conduct, and performance of the dairy product marketing channels.

The American dairy industry illustrates a number of food marketing trends, issues, and principles discussed in earlier chapters of this book. Dairy farmers have experienced technological change and milk output per cow has expanded rapidly in recent years. At the same time, consumers have shifted their preferences for dairy products. Dairy farmers and companies have spent considerable sums of money developing new products and promoting dairy products. Cooperatives continue to play an important role in dairy marketing. Integration and mergers are continuing to shape the modern dairy industry. Food distributors have become a dominant processor and marketer of dairy products. Historically one of the most regulated sectors of the food industry with much government influence on milk prices, the dairy industry is currently deregulating and reducing the role of government in dairy pricing. The dairy industry of the 21st century promises to be very different from that of the 20th century.

MILK PRODUCTION AND USE

U.S. dairy farming accounts for about 10 percent of cash farm sales. As in other segments of agriculture, increased specialization and commercialization of dairy farms have altered traditional production and marketing patterns. Total milk production generally increased from 1924 to 1964, as shown in Figure 24-1, declined over the 1964-1975 period, and then increased from 1975 to 1999. In recent years the number of milk cows has stabilized while the output of milk per cow has continued to increase.

The sideline dairy enterprises on the general farm and the small dairy farm are disappearing. They are being replaced by larger, more specialized dairy operations. In the late 1950s, there were nearly 2 million dairy farms with an average herd size of less than 20 cows. In 1997 there were 117,000 dairy farms with herds averaging 78 cows. Large dairy farms (500 or more cows) accounted for 27% of all dairy cows

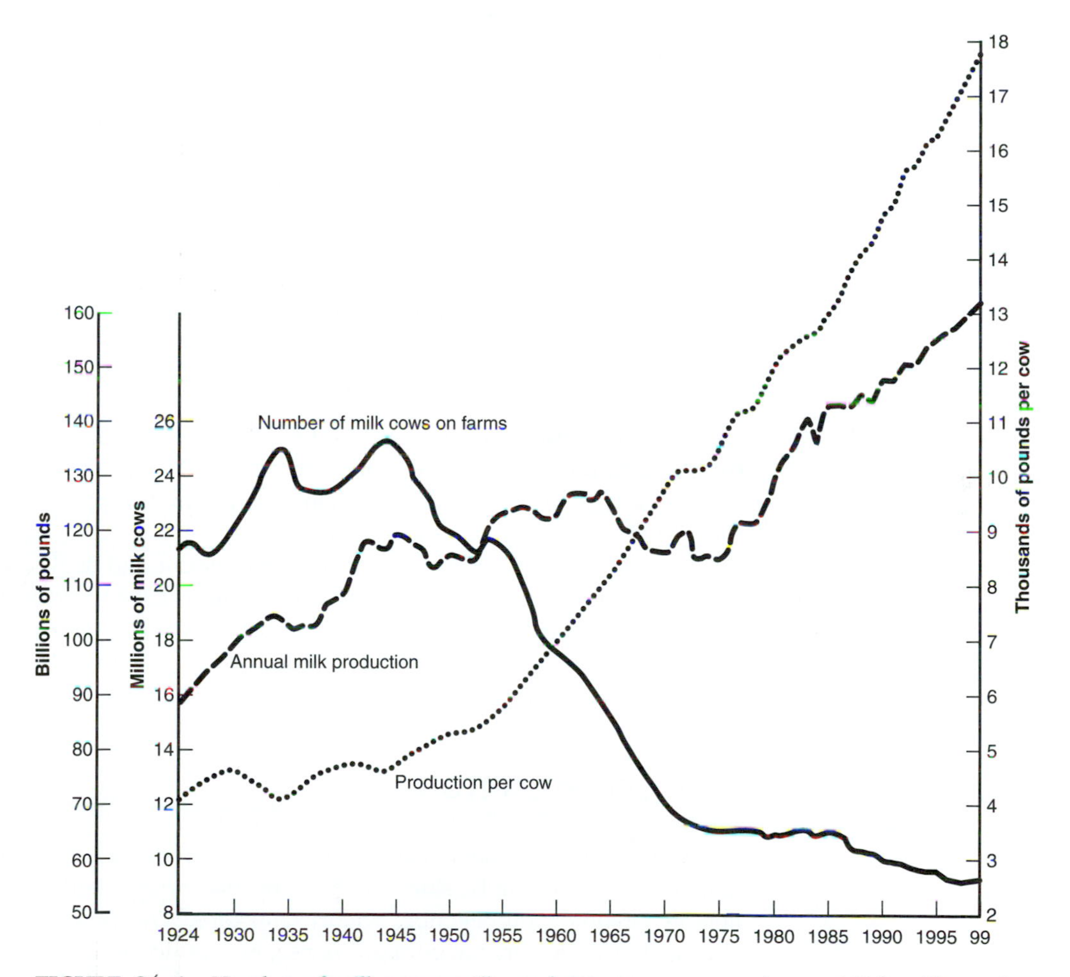

FIGURE 24–1 Number of milk cows, milk production per cow, and annual U.S. milk production, 1924-1999. (U.S. Department of Agriculture.)

and 31 percent of total milk production. Herds of 500 or more cows are common in the South and the West.

The geographic pattern of milk production has also changed. The Pacific and Mountain region shares are rising while the share of production by states east of the Mississipi river is declining, though still larger than the Western share. California became the leading dairy state in 1993. Wisconsin, New York, Minnesota, and Pennsylvania account for more than 50 percent of all milk produced.

Farmers have several options in the use of their milk: (1) feed it to calves; (2) consume it in the farm household; (3) separate it into skim milk and cream and sell the cream only; (4) retail the milk directly to consumers; and (5) sell the whole milk to dairy processors. Over time, important changes have occurred in farmers' use of these marketing alternatives. Today, less than 1 percent of milk is retailed direct from the farm to consumers. Milk used on the farms where it is produced declined from 25 percent of total milk production in 1929 to 1 percent in the 1980s. In the 1920s about one-third of all milk was separated on-farm and only the cream sold, but today there is little on-farm separation. As a result of these trends, 98 percent of milk produced on farms today enters the commercial marketing channels as whole milk.

The milk marketing system has been influenced by the characteristics of the production process and the products. The resources used in dairy farming are highly specialized with few alternative uses. There are also rather high capital costs and long biological lags involved in expanding or contracting the dairy cow herd. As a result, farm resources are somewhat slow in adjusting into and out of milk production. In the short run, the marketing system and consumers must adjust to the available milk supply. This supply consists of a veritable river of milk flowing from two-a-day milk "harvests" into perishable and nonperishable product forms. There is a premium on rapid and efficient marketing of dairy products.

Although the seasonality of milk production is less than it used to be, milk output still varies over the year. Production peaks in the late spring, when cows are on pasture, and reaches a low in the fall. Milk consumption, on the other hand, is highest in the fall and winter when children are in school. A similar problem occurs in the short run. Cows produce milk every day, whereas consumers purchase most of their milk at the end of the week. Coordinating milk supplies with demand is therefore a difficult marketing task.

In the last few decades, improved equipment and facilities, artificial insemination, and improvements in breeding and nutrition have increased dairy productivity. New technologies such as bovine somatotropin (bST), computerized dairy feeding systems, total mixed rations, bypass proteins, and on-farm ultrafiltration and reverse osmosis have also increased milk output and productivity.

MILK PRODUCTS AND CONSUMPTION

The principal milk products are fluid milk and processed dairy products. Fluid products include whole milk, skim and low-fat milk, cream, half-and-half, and such products as flavored milk, eggnog, yogurt, and dips. The processed dairy products are cheese, butter, ice cream, cottage cheese, nonfat dry milk, whey and evaporated and condensed milk. On the average, about 37 percent of all farm milk is sold to consumers in fluid form and 63 percent in processed forms.

Per capita consumption of total dairy products declined from a peak of 564 retail pounds (milk equivalent) in 1970 to 540 pounds per capita in 1977–1981. Per capita consumption of dairy products then rose to a record high 601 pounds in 1987 and returned to the 575 pound level in the 1990–1995 period.

Various dairy products have had different consumption trends in recent years, as shown in Figure 24–2. Yogurt, low-fat fluid milk, sour cream and dips, ice cream, and cheese were the principal growth products over the 1980–1998 period. Consumption of fluid whole milk, nonfat dry milk, and cottage cheese declined over this period. These consumption trends reflected the availability and relative prices of dairy product substitutes and changes in consumers' food preferences. As an example of the beverage substitutions consumers have made over time, consumption of milk and soft drinks were about equal in 1967 at 25 gallons per capita. By 1994 per capita consumption of fluid milk was still at 25 gallons while consumers drank 52 gallons of soft drinks.

The dairy industry is continuing to experience product changes. It is technologically possible to market reconstituted milk, a product made by combining nonfat dry milk, milk fat, and water at a local bottling plant. This product provides substantial economies of transportation and storage. Ultra-high temperature (UHT) processing of milk results in a fluid product that does not need refrigeration and has a six-month shelf life. "Filled" dairy products and synthetic cheese substitute nondairy ingredients for dairy products. There are experiments underway with frozen milk concentrate.

These demand trends vividly illustrate the dynamic nature of consumer food preferences. Many in the dairy industry assumed there were no good substitutes for dairy products. Although this may be true from a nutritional standpoint, consumers have made economic substitutions: coffee, tea, and soft drinks for milk; margarine for butter; and coffee whitener for cream. All foods are vulnerable to this kind of competitive market erosion.

Dairy farmers have long been the most important generic farm product advertisers. The American Dairy Association is an organization of dairy farmers, processors, and equipment manufacturers that has been promoting milk consumption since 1940. The Dairy Stabilization Act of 1983 created two dairy product promotion pro-

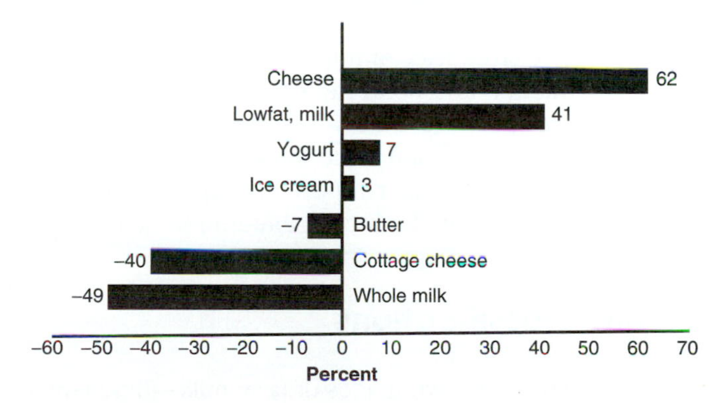

FIGURE 24–2 Changes in per capita dairy product sales, 1980–1998. (U.S. Department of Agriculture.)

grams. The National Dairy Promotion and Research Board oversees the milk and dairy product promotion, nutrition education, and research programs. This is funded by a 15 cents per hundredweight fee paid by dairy farmers and managed by Dairy Management, Inc. The Fluid Milk Promotion Act of 1990 established a second promotional program for milk processors with a 20 cents per hundredweight assessment on milk processors. This program developed the popular milk mustache and Got Milk campaigns. Studies have shown a 4 to 7 dollar return to each dollar invested in milk production.

COUNTRY ASSEMBLY OF MILK

At one time, the country assembly of milk was performed by trucks picking up cans of milk from individual farms for delivery to local milk plants. This was a costly means of milk marketing, because each truck picked up only a small amount of milk for each mile traveled. This farm milk assembly operation was revolutionized by the development of bulk milk handling systems. Milk producers installed large cooling tanks on the farms for receiving the milk from the milking machines. The milk is then pumped directly into tank trucks. The bulk milk delivery system eliminated the laborious loading and unloading of milk cans at the farm and at the dairy plant. It also permitted the accumulation of larger amounts of milk at the farm, which could be picked up every other day instead of daily. Because of the large investment in bulk milk handling equipment, an added push was given to the trend toward larger dairy farms and processors.

Improvements in transportation facilities and milk handling methods also have enlarged the procurement area of dairy processing plants. Once, milk markets 30 to 40 miles apart were separate markets. Today, bulk milk moves as far as 2000 miles, and packaged milk may move over 200 miles to market. This development has brought once geographically isolated dairy farmers and processors into closer competitive contact. Milk prices and policies in one market are now influenced by potential competition from other markets.

MILK PRICING

The farm price of milk is influenced by several factors, including: (1) the cost of production and domestic supply of milk; (2) the consumer demand for milk in its various product forms; (3) the federal government dairy price support programs; (4) federal and state milk marketing orders; (5) dairy farmer cooperatives; and (6) dairy product import and export policies. No one of these dominates in the pricing of farm milk. Each contributes in some measure to the complex milk pricing process.

Milk Grades and Classifications

There are two grades of farm milk—fluid Grade A and manufacturing Grade B. Grade A milk meets strict sanitary standards and is eligible for sale to the consumer as beverage milk. Grade B milk meets somewhat lower standards, which are acceptable because it undergoes processing at higher temperatures than pasteurized fluid milk. Although Grade B milk can only be used in making processed dairy products, such as cheese and butter, Grade A milk can be, and is, used to produce either fluid products or processed dairy products. This routing of Grade A milk into processed dairy products provides a "market bridge" between the prices of milk used for fluid and processed dairy products.

Milk has different values in its various uses. Milk that is used for fluid products (Class I) has a higher value than the milk used for processed dairy products (Class II for soft products like ice cream and cottage cheese and Class III for butter, cheese, and nonfat dry milk.) There are two reasons for this. First, the retail price of fluid products is higher because it costs more to market highly perishable and bulky fluid milk than it does to market the processed forms of milk, such as cheese and butter. Second, the elasticity of demand is lower for fluid products than for processed dairy products. This provides an incentive for the industry to allocate some of the Grade A milk to processing uses. This legal form of price discrimination results in a higher price for fluid products and a lower price for processed products than would be the case if milk supplies were divided equally between the two uses. Because of these milk-value differentials at the retail level, milk dealers can pay higher prices to farmers for milk used in fluid products than for milk processed into cheese, butter, and other products.

A classified pricing scheme is used to ensure that these milk-value differentials are transmitted to farmers. Grade A milk meets the sanitary standards for use in fluid products and can also be used in the production of dairy products such as cheese and butter. Grade B milk meets somewhat lower sanitary standards and can be used only in making manufactured soft or hard dairy products. The percentage of milk marketed as Grade A has expanded over the years, amounting to 95 percent of all milk in 1995. This means that a growing volume of Grade A milk is being used to manufacture dairy products.

The Dairy Price Support Program

Historically, the federal dairy price support program provided the foundation price for all milk used in making processed dairy products. The Department of Agriculture's Commodity Credit Corporation offered to purchase butter, American cheese, and nonfat dry milk at prices that would return the designated support price to farmers for Grade B milk. Because of the market bridge formed by the use of Class II fluid milk in processed products, this dairy support price program also influenced the price of Grade A milk. The products purchased by the federal government to support the farm price of milk were either stored or diverted to noncompeting markets, such as export markets, the school lunch program, or other surplus disposal markets.

CCC stocks and government purchases of milk and dairy products for donations (school lunch, exports, etc.) rose rapidly in the 1976–1983 period. These government purchases were absorbing the increased milk production stimulated by high price supports during these years. In the early 1980s the large government support expenditures and huge storage stocks stimulated changes in government programs, and government dairy product stocks fell. Price support levels were reduced and commercial dairy product usage began to increase. The Dairy Termination Program also reduced government dairy product stocks in the mid-1980s.

The Dairy Termination Program (DTP), which began in 1986, was a unique measure designed to reduce dairy product surpluses. Dairy farmers participating in this program received government payments in return for their agreeing to end all milk production, slaughter their dairy cows, and idle their facilities for a five-year period. This program stabilized the number of milk cows in 1984–1986. However, milk production resumed its growth rate quickly in the late 1980s.

A 50-year-old dairy price support program was eliminated in 2000. It was replaced by a non-recourse loan program (see Chapter 21) for dairy processing companies. Among other things, these reforms increased price risks for dairy farmers, took the government out of the dairy product purchase and storage business, and restructured the milk production and processing sectors. The effects of these dairy policy changes will be felt for many years.

The Federal Milk Marketing Order Program

The farm price of Grade A milk has been influenced by both the dairy price support program and federal milk marketing orders (Chapter 15). The most important provisions of these orders are those that regulate the minimum prices that dealers must pay farmers for Class I and Class II Grade A milk.

The federal milk marketing order program is authorized by the Agricultural Marketing Agreement Act of 1937. Under this legislation, producers in a designated area of milk production and consumption—known as a *milkshed*—may petition the Secretary of Agriculture to issue a market order governing the pricing of milk in the area. In most cases, two-thirds of the producers in the area must approve the terms of the order in a referendum before it is implemented. Each order is administered by a market administrator appointed by the Secretary of Agriculture. The 33 federal milk marketing order areas were consolidated into 11 order areas in 1999 (Figure 24-3).

Federal milk marketing provisions affect two important pricing decisions: What are the appropriate Class I and Class II prices and how will the average price paid to producers be determined? The Class II market order price is always at or near the Minnesota–Wisconsin base price for manufacturing grade milk upgraded by a *basic formula price* (BFP). Manufacturers located elsewhere cannot pay more than this price for Class II milk and remain competitive with processed products shipped from Wisconsin and Minnesota. Thus, the area of greatest surplus and lowest price tends to set the national farm price of Class II milk.

The Class I market order price east of the Rocky Mountains is determined by adding a Class I differential to the Class II milk price in each market order area. This differential represents the cost of transporting bulk milk from the surplus area to each market order area. This procedure for calculating the Class I differential serves two purposes. First, it provides for automatic adjustments in Class I and Class II prices as supply and demand conditions for milk change. If prices were set solely on the basis of administrator decisions or hearings, problems could arise in obtaining timely adjustments to changing market conditions. Second, the Class I differential results in a national milk price structure conforming fairly well to what would be expected in a perfectly competitive market. Prices are lowest in the region of greatest surplus, and prices are higher in more distant markets, reflecting transportation

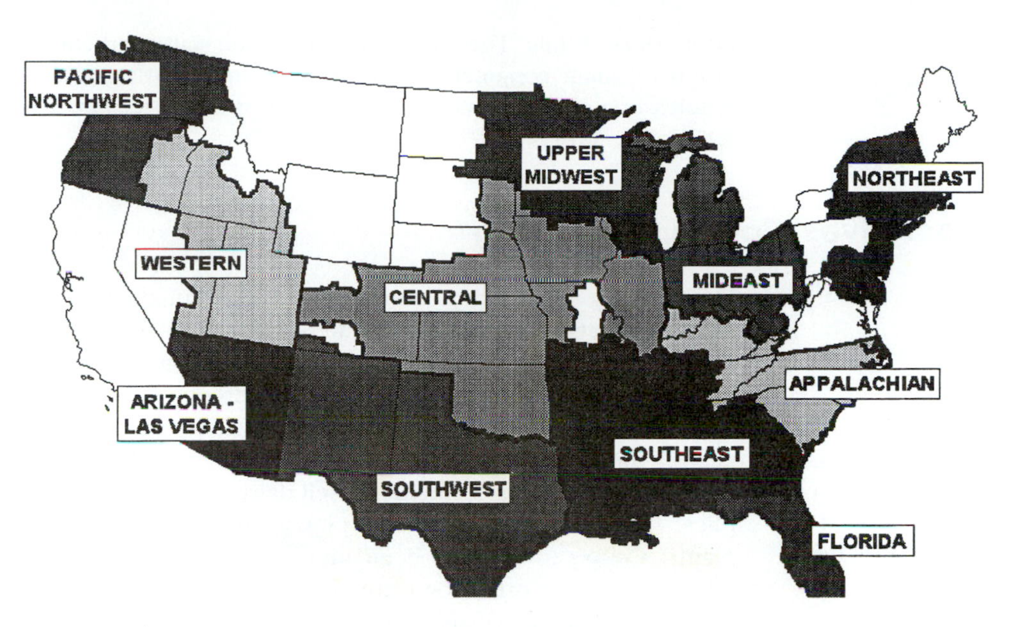

FIGURE 24–3 Federal milk marketing order areas, 2000. (U.S. Department of Agriculture.)

costs and local supply and demand conditions. California and the western states have a somewhat different pricing system.

There is still the problem of determining the farmer's price of milk in each market, where some milk is priced at the Class I level and other milk at the Class II level. Under the terms of federal milk marketing orders, the farmer is paid a *blend price* for milk, regardless of how dealers use that milk. The blend price is a weighted average of the Class I and Class II price of milk in each market area.

In this way, no individual producers will gain or lose because of the way their milk is used. The actual price which the dairy farmer receives for milk is called the *mailbox price.* This differs from the milk marketing order blend price as a result of premiums and discounts which individual producers receive for producing milk with higher qualities. For example, dairy farmers receive premiums or discounts for the components of their milk, such as butterfat, protein, or other solid nonfats. This is called *component pricing* and is an increasing trend in the dairy industry. In the mid 1990s trading in milk futures contracts and options began, providing producers with a marketing tool to manage their milk price risks.

One of the major pricing problems is to arrive at an effective incentive plan that will encourage a more uniform seasonal pattern of milk production. Markets that have a great amount of seasonality are continually faced with "surplus" milk that must be sold through lower-valued outlets. Marketing orders and dairy processors often provide for some system of payment that will encourage farmers to produce milk on a more uniform year-round basis.

Federal milk marketing orders do not directly provide for controls on milk production or the uses of milk. However, Class I and II prices and the dairy price support levels do influence the prices of milk and ultimately affect producers' supply decisions and milk handlers' use decisions. Neither do federal orders set wholesale and retail prices of milk and dairy products, although these are clearly influenced by the farm prices of milk. There are, however, some state marketing orders operating in addition to, or in conjunction with, federal milk marketing orders, that can regulate wholesale and retail fluid milk prices. Most states terminated these controls in the 1970s.

Cooperatives' Role in Milk Pricing

Farmer dairy cooperatives dominate in the production and assembly of farm milk. Dairy products account for about 30 percent of total agricultural cooperatives' sales. More than 80 percent of all dairy farmers belong to dairy marketing cooperatives and dairy cooperatives market almost 90 percent of all milk produced. These cooperatives perform numerous marketing activities, including serving as the sole bargaining agent for members in negotiating prices with dairy processors. They also perform quality control functions, sell dairy supplies, and manage surplus milk supplies. Many cooperatives operate pickup routes and some operate dairy processing plants. Cooperatives have been more important in manufactured dairy products while proprietary firms dominate in milk and ice cream products.

The number of dairy cooperatives has declined significantly in recent years, and the volume of business per cooperative increased substantially. This resulted from the consolidation of many small, local dairy cooperatives into large, regional cooperatives. Moreover, many of the dairy co-ops have pooled their marketing efforts and have become even larger federations of regional cooperatives that coordinate milk marketing efforts over larger areas. In 1998, Dairy Farmers of America (DFA) was the result of a merger of four regional cooperatives. Land O' Lakes, California Milk Producers, and California Dairies, Inc. are large milk coops.

Dairy farmer cooperatives and cooperative federations have become more active bargaining agents for their members. They have increasingly assumed the tasks of coordinating milk supplies and the procurement activities of milk handlers. In doing so, they have increased their bargaining power and frequently negotiated prices above the minimums specified in the federal market orders. Some dairy cooperatives only negotiate prices for their members' milk. Others are engaged in milk processing activities and bargaining.

Dairy Product Trade

Because of perishability and bulk, there is no foreign trade in fluid milk. However, there is an international market for dairy products such as butter, cheeses, dried milk, and casein. The United States, the European Economic Community, Australia, New Zealand, and Argentina are the world's major dairy product exporters. During the 1965–1985 period, high price supports in both the United States and the EEC caused world milk production to grow much more rapidly than demand. As a result, the EEC and the United States have used export subsidies and foreign aid donations in an effort to stimulate the export of their surplus dairy products.

Since 1991 the Dairy Export Incentive Program has increased American dairy product exports. This program subsidizes exporters who purchase U.S. dairy products and ship them to eligible countries. This reduces dairy product surpluses and may raise producer prices.

The world dairy industry has historically had very high trade barriers. Some processed dairy products are imported into the United States. Normally, imports amount to only about 1 to 2 percent of total domestic production (milk-equivalent). U.S. dairy product imports are limited through the use of quotas and tariffs.

During the 1970s, U.S. imports of casein, a milk protein, increased rapidly because there were no import restrictions on this product, which has not been produced in the United States since 1968. Casein, however, has become an important ingredient in making imitation cheese products, coffee whiteners, bakery products, and pet foods. New Zealand, Australia, and Europe are the principal U.S. suppliers of casein.

Rationale for the Milk Pricing System

The reader may wonder why such an elaborate and complex system of milk pricing is needed. Is it necessary and desirable for the government to regulate milk prices this closely? Why is milk given such special attention?

Several explanations are usually given. First, milk has always been considered a special and necessary food. Second, there has been public concern for the competitive position of the dairy farmer in areas where the perishability of milk gives buyers a strong negotiating position. Finally, it has been felt that special pricing arrangements are necessary to coordinate milk supply with demand in both the short and the long run. Trends in the dairy industry have raised serious questions about many of these assumptions and about the need for continuing the special treatment of milk prices. As noted, the industry was deregulated in the late 1990s.

TABLE 24-1 Retail Prices of Dairy Products and Margarine, 1950–1995

	All Consumer Prices	*All Foods*	*Dairy*	*Fluid Milk Products*	*Butter*	*Cheese, American Process*	*Ice Cream*	*Margarine*
				1967 = 100				
1950	72	74	73	72	83	65	103	100
1955	80	82	80	81	81	72	98	94
1960	89	88	88	91	87	76	101	89
1965	94	94	90	90	89	85	95	97
1970	116	115	112	112	104	116	105	106
1975	161	175	157	153	124	176	152	225
1980	247	255	227	208	228	230	218	250
1985	322	310	258	228	263	267	259	301
1990	390	386	318	287	247	305	293	316
1995	454	433	333	289	200	345	298	312
Percent Change, 1980–95	84	70	47	39	−12	50	37	25

SOURCE: U.S. Bureau of Labor Statistics.

It is probably true that the price support program, milk marketing orders, and dairy cooperatives have raised the farm and retail prices of milk above the levels that would exist in their absence. However, these programs have also contributed to the stability of the milk market and to orderly marketing. Moreover, consumer prices of dairy products do not seem to be out of line with other prices. Table 24-1 shows that dairy product prices in general, and fluid milk prices in particular, increased less rapidly than "all retail food prices" and "all consumer prices" over the 1950-1995 years. By this comparison, the dairy industry has performed reasonably well for consumers.

FLUID MILK CHANNELS

Fluid milk marketing methods have changed considerably over the years. In the 1930s, over 75 percent of all milk was home delivered, seven days per week, in quart-sized glass containers. During World War II, as an economy measure, delivery was cut to every other day. In the 1950s, only 50 percent of milk was delivered to the home, three times per week, and disposable containers were introduced. By the 1970s, 90 percent of all fluid milk was sold in grocery stores, most milk containers were disposable, and a wide variety of container sizes was available. These trends illustrate how more efficient marketing techniques tend to displace less efficient ones. Home delivery of milk is more costly than sales through stores, and there are economies to marketing milk in larger packages.

Fluid milk processing is characterized as an oligopolistic industry. Most of the major cities are served by a few national and regional dairy firms. Market concentration of dairy companies is not high on a national basis, although concentration often is high in local markets.

The number of fluid milk companies has continued to decline. There are substantial economies of size in fluid milk processing and distribution, and the decrease in the number of plants was the result of dairy company mergers and plant consolidations. Typically, the pattern has been for larger dairy companies to buy smaller ones, or for the smaller dairies to discontinue operations altogether.

Merger activity in the dairy industry has been a public policy issue. Between 1951 and 1955, the largest eight dairy processors absorbed an average of 71 dairy companies each year. This acquisition rate slowed to 27 mergers per year between 1956 and 1961, and to only four mergers per year in the 1962-1974 period, the result of Federal Trade Commission challenges to dairy industry mergers on the grounds that they lessened competition in the industry. Blocked from growing via the horizontal integration route, the dairy companies entered into conglomerate mergers in the 1960s and 1970s. By the early 1980s, many major national dairy companies received less than 50 percent of their sales from dairy products. The industry continues to experience declining numbers of firms and increasing firm sizes.

Most dairy companies perform their own distribution functions. These include milk delivery to grocery stores, restaurants, and institutions. There are also independent distributors who pick up milk at the dairy plant dock for delivery to retailers, schools, or homes. Competition is keen for these accounts. Packaged milk is largely an undifferentiated product; therefore, dealers compete on the basis of price and services. Generally, retail chain stores prefer to do business with the larger dairies that can service an entire chainstore division with central delivery and billing. Small dairies service independent retailers, the institutional market, and home delivery routes.

Food retailers have integrated vertically into milk processing and branding, and are now important milk packagers. A related development has been the trend toward packaging milk under the retail chainstore's label. Many dairies resisted this, fearing it would undermine their own branding efforts. But the threat that retailers might operate their own fluid milk plants encouraged many dairies to engage in private labeling of milk—an example of market power and channel coordination on the part of the chainstores. In the 1980s the nation's three largest food chains disposed of their milk plants. However, they continue to market milk under their own brands. Private labels accounted for two-thirds of retail milk sales in 1995.

These trends in fluid milk distribution have greatly altered the competitive environment of the dairy industry. No longer are milk markets and companies geographically isolated. Where once a small community may have had only one source of milk, it may now find two or three national giants competing against one another. The small, family dairy plant has been displaced by larger, more efficient operations. And even the national dairies find themselves in competition with the chainstore organizations.

PROCESSED DAIRY PRODUCT CHANNELS

Processed dairy products, such as cheese and butter, are less bulky and perishable than fluid milk. This means they can be produced near areas of concentrated milk production and shipped to distant markets. Minnesota, Wisconsin, and California account for a large share of U.S. butter and cheese production and nonfat dry milk production. Processed dairy products can be stored for considerable periods of time, and this also influences their marketing and pricing patterns.

Dairy processing plants may be independent proprietorships, cooperatives, or corporately owned by multiplant national and regional dairy companies. The large dairy companies produce a broad range of products, and the small companies usually manufacture a limited product line. Some dairy plants produce more than one product, but many butter and cheese plants and condensories tend to specialize in one product. A number of foreign firms have recently bought into the U.S. dairy industry.

Dairy cooperatives are important dairy product processors. In recent years, cooperatives have accounted for about two thirds of butter production, 43 percent of natural cheese, and more than 80 percent of dry milk products. These processing operations provide cooperatives with greater flexibility in the marketing of members' milk.

Because of plant closings, mergers, and consolidations, there has been a marked decline in the number of plants manufacturing dairy products, as shown in Table 24-2. The technological developments in dairy processing—which sharply reduced costs along with increasing volume—are propelling the trend toward fewer and larger dairy processing plants.

TABLE 24-2 Number of Plants Manufacturing Selected Dairy Products, 1940–1992

	Number of Plants			
Type of Plant	*1977*	*1982*	*1987*	*1992*
Butter	136	74	49	32
Cheese	791	704	644	576
Ice cream, froz. desserts	612	552	541	456
Fluid Milk	1,924	1,190	916	746
Dry Condensed, Evaporated milk	266	204	186	214

SOURCE: U.S. Bureau of the Census.

Processed dairy product marketing channels have undergone decentralization, like so many other food industries. Earlier, cheese and butter were shipped by processors to central terminal markets where wholesalers and jobbers sold the products to retailers. The large meat packers were once major cheese and butter wholesalers. Ice cream was made locally and distributed through ice cream parlors in earlier years. The chainstore and supermarket revolutions, along with the growth of large, national dairy companies, altered these market channels and encouraged more direct sales. Today, supermarkets are the chief outlet for processed dairy products, and dairy processors ship directly to chainstore warehouses or to individual retail stores.

Despite this decentralization, central markets still serve as important price-discovery points for cheese and butter. Direct sales of butter are frequently negotiated based on price quotations of the Chicago and New York Merchantile Exchanges. The National Cheese Exchange, Inc., in Green Bay, Wisconsin, serves as the wholesale price-discovery point for American cheese. Concern with the price-discovery process has grown as direct sales have increased and as less of the supply passes through these central markets.

Marketing contracts are widely used throughout the dairy product distribution channel. In addition to the contracts that dairy farmers sign with cooperatives, and the full-supply contracts in force between cooperatives and processors, butter and cheese are normally sold under contracts that specify what the basis of pricing is and who pays the shipping costs. Processors also compete aggressively for dairy product sales contracts with schools and government installations. There are, in addition, cooperative sales agencies, such as Land O'Lakes, which has marketing contracts with some 100 member cooperatives for distributing their butter and cheese under a common label. These contracts facilitate the orderly marketing of dairy products, but they also displace open market trading arrangements.

SUMMARY

The dairy industry is undergoing substantial changes in market organization. Declining numbers and increasing sizes of dairy farms, changing consumer preferences for various dairy products, a tendency toward fewer but larger dairy product handlers and processors, chainstore integration into dairy processing, decentralized marketing, and increased cooperative activity in assembly and processing markets have all altered the marketing patterns and competitive environment of the dairy industry. Product quality and price regulations make the dairy industry one of the most highly regulated food sectors. Farm milk prices—and indirectly, consumer prices—are influenced by a complex system of competitive and regulatory arrangements, including federal price supports, import regulations, and classified market order prices. Dairy farmer cooperatives have assumed most of the assembly market functions and also play an important role in negotiating farm milk prices.

KEY TERMS AND CONCEPTS

blend price
bulk handling
classified pricing
component pricing
mailbox price
milkshed
Minnesota-Wisconsin milk price

DISCUSSION QUESTIONS

1. What effect has the declining number of dairy farms and dairy cows and the increasing milk output per cow had on the milk marketing system?

2. Which developments in consumer markets explain the trends in dairy product consumption?

3. Investigate the market development efforts of the American Dairy Association. What effect do you think these programs have had on dairy product consumption?

4. What are some of the reasons why dairy cooperatives integrated into milk assembly markets? Did dairy processors resist this integration? Why or why not?

5. Explain how the dairy industry can improve profits by adopting a two-price system for milk at the retail level. Why is a classified pricing system needed at the farm level to transmit these price differentials to farmers? Why is a blend price used to ensure that all farmers receive the same price within a market area?

6. Use a supply and demand graph to show the effect of the dairy price support program on farm prices of milk when (a) competitive forces would result in a farm price below the support level; and (b) milk processors bid the price of milk above the support level.

7. Under what conditions will the classified blend price promote a higher farm price of milk than would occur if processors paid farmers a simple average of the Class I and Class II milk prices?

8. What factors govern the decision of how much milk will be sold as fluid products and how much will be processed into dairy products?

SELECTED REFERENCES

Bailey, K. W. (1997). *Marketing and Pricing of Milk and Dairy Products in the United States.* Ames, Iowa: Iowa State University Press.

Blayney, D. P. and A. C. Manchester. (May–August 2000). "Large Companies Active in Changing Dairy Industry," *Food Review,* Volume 23, Issue 2. Washington, D.C.: U.S. Department of Agriculture.

Blayney, D. P. and A. C. Manchester. (February 1998). "U.S. Dairy Product Markets Restructuring," *Agricultural Outlook.* Washington, D.C.: U.S. Department of Agriculture.

Blizard, N. (January–February 1997). *Generic Dairy Advertising: How Effective? Agricultural Outlook.* Washington, D.C.: U.S. Department of Agriculture.

Blizard, N. et. al. (March 1999). *Analyses of Generic Dairy Advertising, 1984–97.* Technical Bulletin No. 1873. Washington, D.C.: U.S. Department of Agriculture. http://www.ers.usda.gov/epubs/pdf/tb1873/index.htm

Dairy Briefing Room. (January 2001). Economic Research Service, U.S. Department of Agriculture. http://www.ers.usda.gov/Topics/view.asp?T-103206.

Federal Milk Marketing Reform website. (December 2000). Economic Research Service, U.S. Department of Agriculture.
http://www.ers.usda.gov/whatsnew/issues/milk/index.html.

Livestock, Dairy and Poultry Situation and Outlook. Monthly. Economic Research Service, U.S. Department of Agriculture.
http://usda.mannlib.cornell.edu/reports/erssor/livestock/ldp-mbb/2000/ldp-m78.inf.

Manchester, A. C. and D. P. Blayney. (September 1997). *The Structure of Dairy Markets: Past, Present, Future.* Agricultural Economics Report No. 757. Washington, D.C.: U.S. Department of Agriculture.
http://www.ers.usda.gov/epubs/pdf/aer757/index.htm

Short, S. D. (September 2000). *Structure, Management, and Performance Characteristics of Specialized Dairy Farm Businesses in the United States.* Agricultural Handbook No. 720. Washington, D.C.: U.S. Department of Agriculture.

U.S. Department of Agriculture. (November 1993). *An Evaluation of Fluid Milk and Cheese Advertising,* Technical Bulletin 1828. Washington, D.C.: U.S. Department of Agriculture.

CHAPTER 25

POULTRY AND EGG MARKETING

Able to sell chicken at a much lower price per pound than beef, chicken companies further boosted chicken consumption with adept marketing programs.
U.S. Department of Agriculture report, *Food Review*, May–August 2000.

OBJECTIVES

After reading this chapter, you will be able to:

1. Explain how the poultry industry has been able to provide increased supplies at declining real prices for food consumers.
2. Appreciate how promotion, technology, and new product development have increased the demand for poultry products.
3. Understand why the poultry and egg industry is considered the most industrialized sector of the food industry.
4. Better understand some of the marketing problems of the modern poultry industry.

The poultry industry consists of three products—eggs, chickens, and turkeys—and also ducks, geese, emus, ostriches, and game birds. Although these are distinct commodities, they have had similar production and marketing trends in recent years. Poultry products account for about 5 percent of consumer food expenditures and 10 percent of total farm sales. Chickens represent 62 percent of farm poultry sales, eggs 21 percent and turkeys 14 percent. Ducks and geese are a small but growing segment of the poultry market.

The popular images of a few chickens scratching in the barnyard and the farmer's wife earning "egg money" are obsolete today. The poultry industry is considered one of the most industrialized sectors of agriculture. Modern poultry production units and methods resemble an assembly-line, factory-type operation rather than traditional, land-based agriculture. The industry is one of large, specialized, and highly integrated firms. Production and processing are automated and mechanized for maximum operational efficiency. The boundary line between farming and marketing activities has largely been obscured by integration and decentralization. More than any other development, industrialization has shaped the contemporary marketing channels and processes of the poultry industry.

PRODUCTION PATTERNS

Growth and specialization have been the hallmarks of the poultry industry's development. The 1940–1999 expansion in poultry and egg production is traced in Table 25–1. The production of broilers and turkeys increased significantly between 1960 and 1995. Egg production also advanced over this period. *Broiler* is the trade term for the young chickens, grown solely for meat, that are usually sold in retail stores today. They are also known as "fryers." As is evident in Table 25–1, the growth in chicken production has been in broilers. The output of farm chickens, a by-product of the egg-laying flock, has declined. The development of separate meat-type and egg-laying chickens is an example of specialization in the poultry industry.

The growth in poultry output, as shown in Table 25–1, was caused by both an increase in demand and improved efficiency in poultry production. The supply curve shifted rightward with cost-reducing and yield-increasing improvements in poultry production technology following World War II. As shown in Table 25–2, the amount of feed needed to produce 1 pound of broiler meat fell from 2.9 pounds in 1955 to 2.1 pounds in 1980. The amount of feed required to produce one dozen eggs fell from 5.8 pounds in 1955 to 4.2 pounds in 1980. Efficiency gains in poultry and egg production have continued in recent years at a less dramatic pace. Shorter growing periods and lower mortality rates have also increased the productive efficiency and output of poultry and eggs.

These production efficiencies generally favored larger operations and encouraged the growth of more specialized poultry producing units. As a result, poultry and egg production has become concentrated in the hands of fewer and larger producers. The number of farms producing eggs has also declined and the average size

TABLE 25–1 Trends in Poultry Production, 1940–1999

	Million Pounds			
	Chickens	*Broilers*	*Turkeys*	*Eggs (Million)*
1940	2,158	413	502	39,707
1950	2,310	1,945	817	58,954
1960	1,346	6,017	2,202	61,602
1970	1,190	10,819	2,198	68,212
1980	1,205	15,541	3,071	69,683
1985	1,025	18,810	3,702	68,645
1990	985	25,631	6,043	68,134
1995	992	34,222	6,540	74,772
1999	NA	29,741	5,297	69,996

SOURCE: U.S. Department of Agriculture, *Agricultural Statistics,* annual.

TABLE 25-2 Production Efficiency in the Poultry Industry, 1955–1980

Item	*1955*	*1965*	*1975*	*1980*
Eggs per hen/year	192	218	232	242
Pounds of feed per:				
dozen eggs	5.83	4.95	4.25	4.20
100 pounds broiler	285	236	210	208
100 pounds turkey	470	476	333	305
Hours of labor per:				
100 hens	175	97	—	53
100 eggs	0.9	0.4	—	0.2
100 broilers	4.0	2.0	—	0.5
100 pounds turkey	4.4	1.3	—	0.4

SOURCE: F. Lasley, *The U.S. Poultry Industry,* U.S. Department of Agriculture, Agricultural Economic Report No. 502, July 1983, p. 15. More recent data not available.

of egg producers has increased. These trends have increased the concentration of poultry production and processing.

There also has been a shift in the geographic location of poultry production. Broiler and turkey production have shifted from the northern to the southern states. Lower labor and energy costs and milder weather were factors in this shift. As shown in Table 25-3, seven states account for two thirds of all broiler production, and three of these—Arkansas, Georgia, and Alabama—produce more than 43 percent of all broilers. The DelMarVa (Delaware-Maryland-Virginia) area is also an area of concentrated broiler production. North Carolina, Minnesota, Virginia, Arkansas, California, and Missouri produce over 64 percent of U.S. turkeys. Egg production is more widely distributed but is concentrated in the North and South Central states. Like cattle and hog feeding, poultry production is closely interrelated with the feed-grain economy. Feed accounts for more than 60 percent of egg, broiler, and turkey production costs.

The seasonality of poultry supplies has been reduced as confinement production has become concentrated in fewer hands and as the industry has shifted to

TABLE 25-3 Leading States in Poultry Production, 1997

Broiler Chickens		*Turkeys Sold*		*Egg Production*	
	Percent		*Percent*		*Percent*
Georgia	15%	N. Carolina	18%	California	9%
Arkansas	15	Minnesota	15	Ohio	9
Alabama	13	Virginia	8	Pennsylvania	8
N. Carolina	9	Arkansas	8	Iowa	7
Mississippi	8	California	8	Indiana	7
Texas	6	Missouri	7	Georgia	6
Top 6 States	66%		64%		46%

SOURCE: U.S. Census of Agriculture, 1997.

warmer states. Egg and broiler output still peak in the spring, and turkey production is predictably highest in the fall. These seasonal patterns are caused by the biological characteristics of poultry and the seasonality of demand. However, large-scale broiler production in the South has reduced the seasonal variability of broiler production and prices. Seasonal egg production and price variations also are less pronounced than formerly. This has affected the need for the storage marketing function and the capacity utilization of poultry processing plants. The development of new turkey products has also reduced the seasonality of turkey production and consumption.

Poultry and egg producers can increase output rapidly. There is typically excess capacity in this industry that can be utilized to expand production when prices and profit expectations warrant. A new batch of broilers can be produced about every seven weeks, and two or more flocks of turkeys may be produced each year. On the other hand, this excess capacity and the specialized facilities used in poultry and egg production often hamper industry supply adjustments to low prices and profits. Consequently, poultry and egg producers frequently find their prices below costs of production.

POULTRY PRODUCTS, CONSUMPTION, AND PRICES

Poultry is marketed in a variety of product forms and convenience states. Prior to World War II, most poultry was sold to consumers in the live or the New York dressed state (only blood and feathers removed). By 1975, most broilers and turkeys were marketed as eviscerated ice-packed or frozen ready-to-cook (RTC) poultry. This product is available with varying degrees of built-in convenience services: whole birds, cut-up birds, poultry parts, and self-basting poultry. This product diversity contributes to consumer satisfaction, but it also increases the complexity and cost of the marketing process.

Some poultry is marketed in the fresh unprocessed state, but many eggs, chickens, and turkeys are further-processed. Eggs are processed—by firms called "breakers"—into dried, frozen, and liquid egg products. These are used as ingredients by food processors in baby foods, noodles, bakery mixes, scrambled egg mixes, mayonnaise, ice cream, and numerous other products. The share of eggs marketed as either liquid, frozen, or dried egg products increased from 16 percent in 1985 to 27 percent in 1996. This resulted from foodservice firms' preferences for pasteurized egg products rather than fresh eggs. The pasteurized eggs reduce kitchen labor and the dangers of pathogens in egg dishes. Further-processed turkey and chicken products include cut-up parts, breaded and precooked parts, frozen dinners, poultry rolls and roasts, soups, hot dogs, pot pies, and deboned products. These are growth markets for the poultry industry.

Product Characteristics

The perishability and homogeneity of chicken and eggs are key factors in their marketing. Most of these products have a relatively short shelf-life. This requires special transportation and merchandising facilities for marketing poultry. Perhaps 2 to 3 percent of poultry and 7 percent of eggs are lost in the marketing channel because of spoilage and breakage. The perishability of poultry products also results in an urgency to market them that limits producers' and handlers' flexibility. Birds must be sold when they reach the proper market weight and maturity, and eggs when they are fresh, regardless of market conditions and prices. Errors in the supply-demand balance are expensive.

Traditionally, poultry products were considered relatively homogeneous and undifferentiated products at both the farm and retail levels. However, branding and brand advertising are now common for eggs and further-processed poultry products. Through brand advertising, large poultry firms and retailers are attempting to build consumer loyalty in order to remove their products from the generic category. The larger firms have led in these efforts to differentiate poultry products.

Market Development

The poultry industry engages in extensive market development and demand-expansion programs. The growth of the processed egg product industry and the development of further-processed poultry meat products are examples. The development and promotion of year-round turkey products is another example. The rapid growth of the fast-food chicken industry has also influenced poultry markets. The industry's recognition that consumers desire a younger, smaller, and more tender chicken than was produced as a by-product of the egg-laying flock, and the resulting expansion of the broiler industry, is a classic example of the marketing concept and industry market development. Most of this market development was sponsored by processors and the fast food industry. Under the provisions of the 1974 Egg Research and Consumer Information Act, the industry established the American Egg Board to carry out an industry-financed, coordinated program of research, consumer education, and promotion to improve and develop markets for eggs and egg products. The Egg Nutrition Center provides science-based information on the role of eggs in human nutrition and health.

Global trade in poultry products has expanded significantly in recent years. U.S. exporters are extending the successful domestic marketing strategy of selling poultry parts and further-processed poultry to international markets. This expanded U.S. poultry trade has been stimulated by world trade liberalization, the export enchancement program, and low U.S. production costs.

International trade has become increasingly important for the American poultry industry. In 2000, exports accounted for one-sixth of U.S. poultry production. Russia has become the top market for U.S. exports of chickens and turkeys. Japan and Mexico are also large importers of these products. Foreign consumers prefer dark poultry meat rather than the white meat which is now preferred in America.

Chicken plays an important role in grocery store and restaurant marketing strategies. Chickens and chicken parts are often featured as price specials by grocers in order to differentiate the retail store and attract customers. Frequently they are sold below cost. Restaurant managers have also found that adding chicken to menus can improve sales and profits. This aggressive price competition and marketing have no doubt contributed to increased chicken consumption.

Prices and Consumption

The poultry industry provides a dramatic illustration of how food production efficiency gains are passed on to consumers under highly competitive market conditions. As shown in Table 25-4, farm and retail prices of poultry and eggs fell from 1946 to the mid-1960s. These declining prices resulted from the rightward-shifting farm supply curve and from a competitive market environment that transmitted those lower farm costs and prices to consumers in the form of increased supplies and lower retail prices. Even though increased feed costs and inflation halted the decline in broiler, egg, and turkey prices in the late 1960s, the 1981-1987 retail prices of these products were not far above their 1946-1950 levels. Few products can match this price record. In real, after-inflation dollars, the price of eggs fell from 49 cents per dozen in 1967 to 27 cents in 1986. The real price of broilers fell 9 cents over this period.

Table 25-4 also shows the trends in consumption of poultry and eggs over the 1946-1995 period. Per capita consumption of chicken and broilers more than tripled and turkey consumption expanded six-fold during these years. This was a result of declining real prices, increases in income, more eating out, the development of new products, and the declining price of chicken and turkey relative to other protein sources. Per capita poultry consumption exceeded beef consumption for the first time in 1993. The price-quantity relationship for broilers, shown in Figure 25-1, illustrates the law of demand for broilers.

Per capita egg consumption has not responded to falling real prices and rising income in the same fashion. Instead, egg consumption declined steadily from the 1951-1952 peak of 47 pounds per capita to 30 pounds per capita in the 1991-1995 period. This trend has been attributed to changed attitudes toward breakfast, diet and health considerations, and a basic change in the food preferences of an urbanized population. The recent stabilization in egg consumption is because of increased use of eggs as food ingredients and the development of fast-food egg-based products.

TABLE 25-4 Poultry Consumption and Prices, 1946-1995

	Chicken			*Turkey*			*Eggs*		
Years	*Farm Price Cts./lb*	*Retail Price Cts./lb*	*Per capita Consumption (pounds)*	*Farm Price Cts./lb*	*Retail Price Cts./lb*	*Per capita Consumption (pounds)*	*Farm Price Cts./doz.*	*Retail Price Cts./doz.*	*Per capita Consumption (number)*
1946-50	.32	.60	19	.36	NA	3	.40	.70	–
1951-55	.27	.57	21	.33	NA	5	.41	.65	–
1956-60	.18	.44	26	.25	NA	6	.37	.56	–
1961-65	.14	.38	30	.21	NA	7	.25	.55	317
1966-70	.14	.40	37	.21	.51	8	.43	.55	314
1971-75	.20	.52	41	.30	.66	9	.44	.68	292
1976-80	.25	.66	47	.39	.84	10	.56	.82	271
1981-85	.30	.74	54	.42	.98	11	.63	.90	260
1986-90	.33	.86	64	.39	1.01	16	.62	.89	245
1991-95	.33	.94	79	.39	1.15	18	.63	.95	236

SOURCE: U.S. Department of Agriculture.

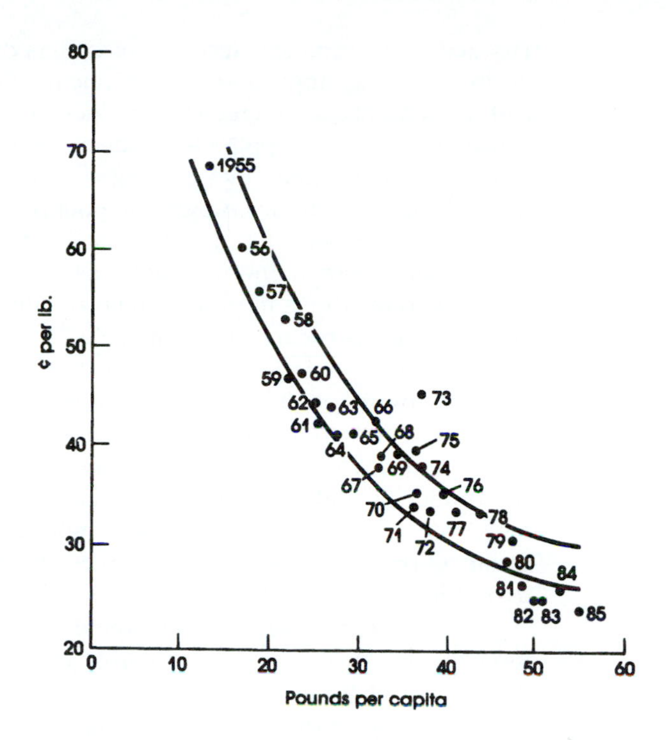

FIGURE 25–1 Per capita broiler consumption and real retail prices. (From Lasley, 1983)

INTEGRATION IN THE POULTRY INDUSTRY

The poultry industry provides dramatic examples of the integration of farm production and marketing activities. Vertical coordination (Chapter 12) of the sequential poultry markets—from input suppliers to distributors—is highly developed in this industry. Integration has progressed to the point where large, multifunction producer-processor-distributor firms are the dominant force in poultry and egg production and marketing. These firms may be independents, cooperatives, subsidiaries of feed manufacturers, meat packers, retailers, or affiliates of conglomerate corporations.

The poultry industry has passed through three stages of integration. Prior to the 1930s, poultry production was a highly integrated operation, combining a breeding flock, a hatchery, grain farming and feed mixing, growing, and delivery services. In the 1930s and 1940s, specialized breeding, hatchery, feed mixing, and packing operations evolved. In the 1950s and 1960s, these separate operations were recombined to form the modern, highly integrated poultry production and marketing firm.

Several forces fueled the integration of poultry and egg production with input suppliers and marketing firms. First, new cost-reducing production technologies encouraged larger poultry producing units, the expansion of industry capacity, and the use of supervisory management for standardized production tasks. Because increased poultry production means an expanding market for feed, feed manufacturers financed producers by supplying capital for building houses, buying equipment, chickens, and feed. Eventually, feed manufacturers developed contracts that assisted growers in expanding output and in marketing the finished products. Independent packers and wholesalers also entered into production arrangements in order to control quality and assure adequate supplies for efficient plant operation. Large producers also found it necessary to integrate into packing and distributing in order to assure markets for their products. Continued progress in poultry genetics, including biotechnology, poultry nutrition, disease control and resistance, and poultry processing technologies will fuel further growth in firm sizes and consolidations in the future.

The three forms of production-marketing integration prevalent in the chicken and turkey industrys are shown in Table 25-5. Vertical integration of production and marketing has progressed further for broilers than for turkeys and eggs, but the trend is upward for all three commodities. In the case of owner integration, the producing and marketing firms are one and the same. This form of integration has been more important for eggs than for broilers and turkeys. In grower-contract marketing, the producing and marketing firms retain their separate identities, but producers enter into contracts with distributors who specify outlets, timing of delivery, and quality of products. Cooperatives typically use marketing contracts to integrate poultry production and marketing.

Grower-contract production, the dominant form of integration for broilers and turkeys, calls for the producer to grow the birds for the distributor—under closely supervised conditions and for a guaranteed return. For most poultry production contracts, the integrator supplies some inputs (chicks, feed, medication, field supervision) and the grower provides other inputs (housing, water, fuel, labor, and so on). The grower's return is not tied directly to the market price; it is a specified return (such as 5 to 7 cents per bird) plus incentive premiums for superior feed-conversion efficiency. Thus, growers bear less price risk under contract production arrangements than under nonintegrated or contract marketing arrangements. But in return for this, the grower loses some freedom of production and marketing decisions.

TABLE 25-5 Form of Vertical Integration in the Poultry Industry, 1955-1994

	Coordinated chicken				*Coordinated turkey*			
Year	*Integrator owned*	*Grower-contract production*	*Grower-contract marketing*	*Total*	*Integrator owned*	*Grower-contract production*	*Grower-contract marketing*	*Total*
				Percent				
1955	2.0	87.0	1.0	90.0	4.0	21.0	11.0	36.0
1960	5.0	90.0	1.0	96.0	4.0	30.0	16.0	50.0
1965	5.5	90.0	1.5	97.0	8.0	35.0	13.0	56.0
1970	7.0	90.0	2.0	99.0	13.0	42.0	18.0	73.0
1975	8.0	90.0	1.0	99.0	20.0	47.0	14.0	81.0
1977	10.0	88.0	1.0	99.0	28.0	52.0	10.0	90.0
1980	n.a.	n.a.	n.a.	n.a.	28.0	52.0	10.0	90.0
1982	12.0	87.0	0.0	99.0	28.0	54.0	8.0	90.0
1990	n.a.	n.a.	n.a.	n.a.	28.0	55.0	5.0	88.0
1994	14.0	85.0	0.0	99.0	32.0	56.0	5.0	93.0

SOURCE: U.S. Department of Agriculture estimates.

POULTRY AND EGG MARKETING CHANNELS

Decentralization and direct sales have proceeded hand-in-hand with the integration of the poultry market channels. Integration and decentralization have closed up the poultry and egg marketing channels. Products may move the entire length of the channel without changing hands or entering into open market trading arrangements. Most of the assembly and wholesale distribution marketing functions have been assumed by the integrated poultry production and marketing firms. This has reduced the role and numbers of country buyers, independent merchants, wholesalers, brokers, and jobbers. Economies of size, increased density of geographic production, and continuing integration are further reducing the number of poultry marketing firms and increasing their sizes. There is widespread concern for the competitive effects of these trends on the poultry industry.

Egg Channels and Agencies

The market flows of eggs are shown in Figure 25–2. Most eggs are sold in fresh form to retail stores or to institutional buyers. A smaller but rising portion go to breakers, are used in hatcheries, or are consumed on-farm or exported.

The key role of the assembler-packer is evident in Figure 25–2. In the late 1950s, independent wholesaler-distributors made 67 percent of the sales to retail buyers. By the 1980s, integrated assembler-packer-distributors had accounted for almost all of the sales to retail buyers. The integrated egg packing firm also shifted the location of the grading and cartoning operations from the central terminal markets, where they were performed by wholesalers, to country packing plants. There are

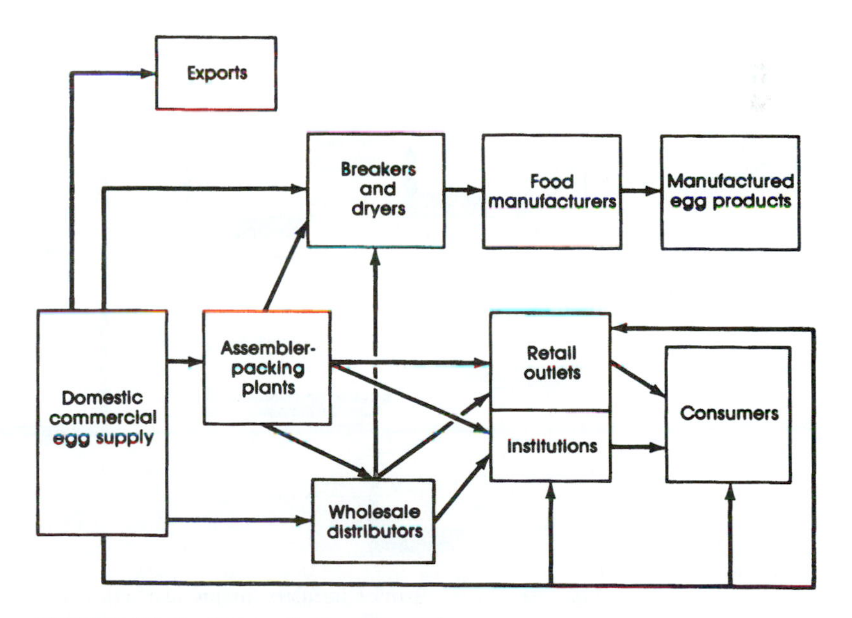

FIGURE 25–2 Egg marketing channels.

significant economies of size in egg packing and distributing, and the number of plants is declining while plant size is growing. In 1996 there were 61 firms producing three-fourths of total egg output with the largest 10 firms accounting for 36 percent of eggs sold.

Poultry Meat Channels and Agencies

Poultry processing firms also have become larger and more specialized, and the slaughter-packing operations have moved from consumption centers to production areas. The broiler marketing channel is illustrated in Figure 25-3. Much of this channel is integrated by the slaughter-packing-distributing firms. The number of chicken and turkey processing plants did not change from 1967 to 1992. However the size of these plants increased dramatically. Further, the share of chicken-broilers sales accounted for by the largest four firms increased from 15 to 20 percent in the 1954 to 1977 years to 45 percent in 1992. By 1996 the largest chicken processing firm controlled 22 percent of total slaughter and the largest 10 firms accounted for 70 percent of industry sales. By 1996 the 10 largest turkey processors controlled 74 percent of their market and 27 firms produced almost all of the turkey output.

MARKETING COSTS

As products marketed largely in the fresh form, poultry products have relatively high farmer shares. In the 1997-1999 period, egg producers received 50 percent of the consumer's dollar while broiler producers received 51 percent of the retail dollar. The distribution of the poultry consumers' dollar among the various marketing agencies and functions is shown in Table 25-6.

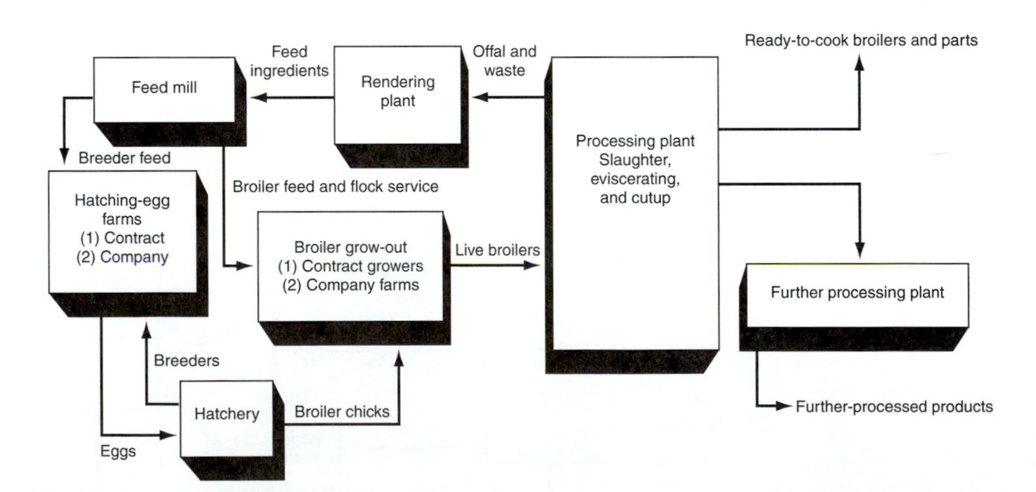

FIGURE 25–3 Broiler industry major marketing channels. (U.S. Department of Agriculture.)

TABLE 25-6 Farm Value, Marketing Costs, and Retail Prices of Poultry Products, 1955-1994

Product	*Years*	*Farm Value*	*Assembly*	*Processing*	*Transport*	*Retailing*	*Retail Price*
Eggs	1955-1959	65%	2%	13%	7%	13%	100%
	1975-1979	65	1	12	6	14	100
	1985-1989	62	1	13	6	18	100
	1992-1994	57	1	14	7	21	100
Broilers	1955-1959	58	2	11	7	22	100
(RTC)	1975-1979	54	2	13	8	23	100
	1985-1989	55	2	11	8	24	100
	1992-1994	53	2	13	8	24	100

SOURCE: U.S. Department of Agriculture. (More recent data not available.)

INDUSTRY PROBLEMS

Despite its market successes, the poultry industry is experiencing many marketing problems, including quality control problems, market stability problems, and price discovery problems.

Quality Problems

Because of the highly perishable nature of the product, poultry quality and grading have been a concern for the industry. Poor handling by producers and marketing agencies can result in loss of freshness and quality as products move through the marketing channels. Better handling techniques and refrigeration have improved egg quality and food safety. Mass merchandisers require consistent quality, and many egg production contracts contain quality control provisions. There is evidence of a trend toward production of a higher quality egg product today. Controlling salmonella organisms is a continuing problem in poultry and egg processing. Quality preservation and the control of microorganisms may take several forms for poultry meat: ice packing, chilling, CO_2 dry packing, irradiation, and freezing.

There are federal grades and standards for poultry products, but they are not universally used. The difference between Grade A and Grade B poultry depends on appearance, finish, and meatiness. The age of the bird also is a factor in tenderness. Although age is not a grading factor, this information is legally required on all poultry labels. Eggs are graded for quality and size, on the basis of albumen thickness (as measured by the height of the yolk and white) and shell condition. Egg grades include U.S. Grade AA or Extra Fancy, Grade A, and Grade B. Sizes are U.S. Jumbo, Extra Large, Large, Medium, Small, and Peewee. "Undergrades" are used in egg breaking operations. Mass scanning and sampling techniques have replaced the costly method of hand candling eggs to evaluate their interior quality. There is controversy over whether the egg quality standards truly reflect consumer preferences. Some studies have shown that consumers cannot discriminate between egg grades in blind tests or are unwilling to pay premium prices for the higher quality grades. There also is concern that grading terminology may be confusing to consumers; for example, a consumer might easily believe that Large, Grade A eggs are the top grade.

There has been an expansion of laws and regulations affecting food safety and quality control in the poultry industry. Federal inspection for wholesomeness and safety is mandatory for interstate poultry sales, and state inspection programs must be equal to the federal standards. The Egg Product Inspection Act of 1970 brought egg processing plants under federal inspection. Uniform federal-state grades for shell eggs were adopted in 1972. New food safety standards for eggs and poultry were put in place in the late 1990s.

Price Cycles and Profits

Like the cattle and hog enterprises, the poultry industries experience cyclical expansions and contractions in output. The resulting price cycles are much shorter than for cattle and hogs because of the shorter biological time period needed to alter the industry's productive capacity. Poultry and egg price cycles are caused by a relatively elastic supply curve and the tendency for producers to base future production plans on current prices. The poultry price cycles are illustrated in Figure 25-4. Contrary to many people's expectations, production concentration, geographic specialization of the industry, and integration have not eliminated the boom-and-bust cycles of the poultry industry. The cycles persist, resulting in considerable risk and income instability. This instability and risk have encouraged growers to enter into production contracts and processor vertical integration.

Nor have specialization and integration raised poultry and egg prices significantly above costs of production and marketing. On the contrary, the price-cost record of the poultry industry resembles that expected of a perfectly competitive industry. Occasional periods of profits stimulate production and are quickly followed by periods of below-cost prices.

From time to time, the poultry industry has considered various ways to control supplies, eliminate the price cycles, and stabilize profits. Marketing orders and agreements, market quotas, mandatory and voluntary supply controls, bargaining associations, two-price systems, and price supports have all been considered and rejected by the industry.

Pricing Problems

Integration and decentralization have complicated the price discovery process for poultry and eggs. Historically, the poultry industry used a base price quotation procedure. Prices were discovered in the large centralized terminal markets such as New York, Los Angeles, Boston, and Chicago, where it was assumed that supply and demand were fully represented. These cash price quotations were reported to the trade by both private and public agencies. Prices at other points in the marketing channel were then determined by applying premiums and discounts to these base

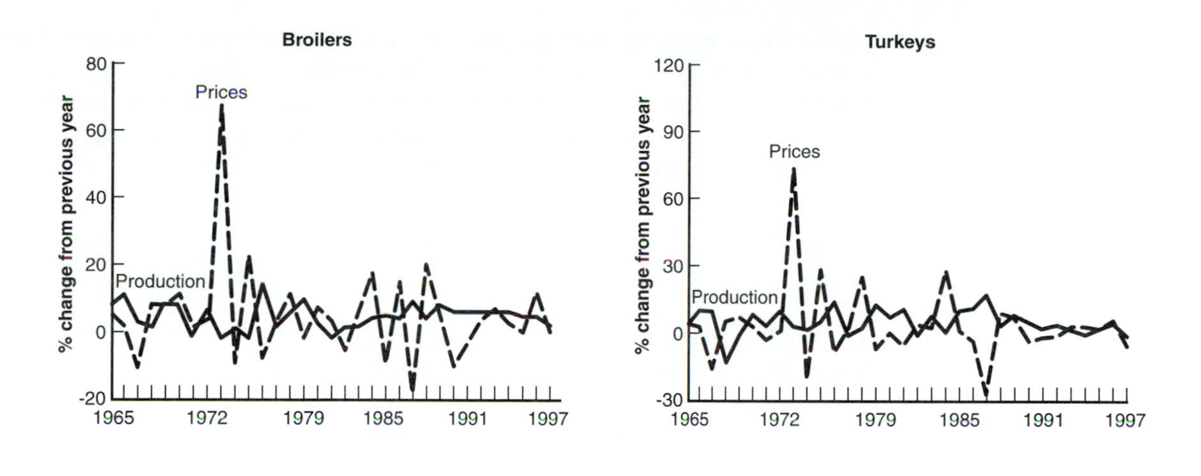

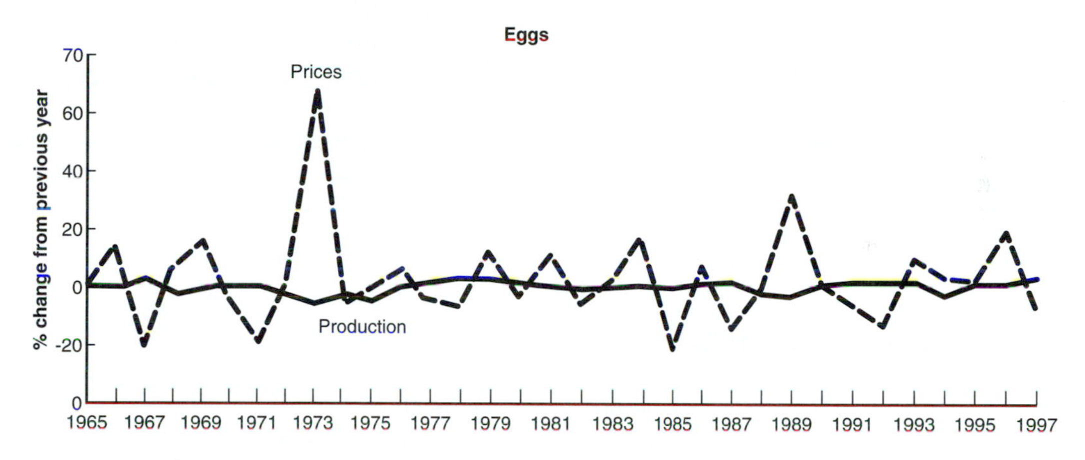

FIGURE 25–4 Poultry and egg cycles: Changes in production and farm prices, 1965–1997. (U.S. Department of Agriculture.)

prices. These base price quotation systems were rapid, low-cost methods of price discovery.

Over time, with integration and decentralization, much of the poultry and egg supplies have bypassed these terminal markets. Daily cash trading of eggs at New York and Chicago was terminated in March 1970, removing two major base price quotation points. As a result, there is concern that base prices discovered at central markets no longer fully reflect demand and supply and therefore are not adequate guides to pricing elsewhere in the marketing channel. However, trade practices change slowly and some prices in the poultry and egg industry are still tied to central terminal market quotations.

Alternative price discovery mechanisms have been experimented with in the poultry industry. One alternative is to shift the base price quotation to another level of the market. For example, because most broilers are produced under contract and not traded in open markets, in 1965 the U.S. Department of Agriculture Market News Service discontinued quoting the price of live birds at farm level and began publishing a nine-city average price for RTC (ready-to-cook) broilers delivered to consuming markets in truckload lots. There is also an electronic egg exchange, the Egg Clearinghouse. This computerized system considers all offers and bids of buyers and sellers and attempts to arrive at a market-clearing price. Pricing eggs continues to be a problem since eggs are so seldom traded in open, competitive markets.

SUMMARY

The poultry and egg marketing channels and processes continue to be shaped by the forces of production concentration, integration, and decentralization. The industry is in the advanced stage of industrialization, with close coordination of production and marketing. Integration extends from the input supply industries to wholesale distribution. Open market negotiating and pricing are rapidly disappearing. Nevertheless, this industry exhibits a relatively high degree of operational and pricing efficiency. The products are available in a wide variety of forms. The growth of broiler production illustrates the consumer-orientation of the industry. Operational efficiencies in production have been passed on to consumers in the form of increased supplies and falling real retail prices. Competitive conditions keep farm and wholesale prices near costs of production and marketing. Major industry problems are production and price instability, safety quality control, and price discovery in integrated, decentralized markets.

KEY TERMS AND CONCEPTS

base price quotations
broilers
egg breakers
ready-to-cook (RTC)

DISCUSSION QUESTIONS

1. What are the reasons for increased concentration of poultry production and marketing? Would you expect this concentration to continue?

2. Discuss the meaning of the word *industrialization* in agriculture and in the poultry industry. Why is this term used to describe the evolution of the poultry industry? What are the pros and cons of industrialized agriculture?

3. If chicken and turkey demand have been shifting rightward as a result of income and other factors, what must happen to the supply curve to result in falling real retail prices? If egg demand has also been shifting rightward with income and population growth, how do you explain falling prices and per capita consumption of eggs?

4. After reading some current trade reports on the broiler industry, identify the major integrated producer-distributors. Which are conglomerate corporations, cooperatives, feed manufacturers, or meat packers? Do the same for eggs.

5. Why does the poultry industry need a base price quotation for price discovery? How would you recommend that prices be discovered in this industry?

6. Examine pricing efficiency in the poultry industry, including responsiveness to consumer demand, relationship of farm and retail prices and profits. How do you explain this performance record?

7. What recommendations would you make to the egg industry in order to stabilize consumption and improve prices?

8. What is the farmer's share for poultry and eggs? Does the share indicate profit levels in poultry production?

9. Why do you think integration in the poultry industry is more often the contract variety rather than the ownership variety?

SELECTED REFERENCES

Frenzen, P. D. et al. *Consumer Acceptance of Irradiated Meat and Poultry Products* (August 2000). Agriculture Information Bulletin No. 757. Economic Research Service, U.S. Department of Agriculture.

Lasley, F. A., et al. (November 1988). *The U.S. Broiler Industry,* Agricultural Economics Report No. 591. Washington, DC: U.S. Department of Agriculture.

Livestock, Dairy and Poultry Outlook. (January 2001). Monthly. Economic Research Service, U.S. Department of Agriculture.
http://www.ers.usda.gov/publications/so/view.asp?f=livestock/ldp-mbb/.

Madison, M. & D. Harvey. (May 1997). U.S. Egg Production on the Sunny Side in the 1990's. *Agricultural Outlook.* U.S. Department of Agriculture.

Ollinger, M., J. McDonald, & M. Madison. (May–Aug. 2000). "Poultry Plants Lowering Production Costs and Increasing Variety." *Food Review,* Vol. 23, Issue 2. U.S. Department of Agriculture.

Ollinger, M., J. McDonald, & M. Madison. (September 2000). *Structural Change in U.S. Chicken and Turkey Slaughter.* Agricultural Economic Report No. 787. Economic Research Service, U.S. Department of Agriculture.

Perry, J., D. Banker, & R. Green. (March 1999) *Broiler Farm' Organization, Management, and Performance.* Agriculture Information Bulletin No. 748. Economic Research Service, U.S. Department of Agriculture.

Poultry and Eggs Briefing Room and Key Topics. (January 2001). Economic Research Service. U.S. Department of Agriculture.
http://www.ers.usda.gov/briefing/poultry/;
http://www.ers.usda.gov/Topics/view.asp?T=103210.

CHAPTER 26

GRAIN MARKETING

No man is qualified to become a statesman who is entirely ignorant of the problem of wheat.
Socrates

OBJECTIVES

After reading this chapter, you will be able to:

1. Describe the major production characteristics, products, and distribution channels of the grain and oilseed sectors.
2. Understand how and why the grain marketing industries are changing and the drivers of these changes.
3. Explain how biotechnology and specialty value-added grains are changing the production and marketing of the nation's grains and oilseeds.
4. Appreciate the effects of government price and income policies and trade policies on the grain industry.

The principal food grains of the United States are wheat, rice, and rye. The feed-grains—corn, oats, barley, and grain sorghum—are primarily grown for livestock feed. The soybean is actually an oilseed, not a grain, but is included here because soybeans are marketed much like the grains.

The most important developments influencing grain marketing over the past fifty years have been (1) the sheer growth in volume of grain and oilseeds to be marketed; (2) changed harvesting and farm storage technology, which have influenced the flow-to-market of supplies; (3) consolidation and integration of firms; (4) the increase in off-farm sales of grain and growth of the commercial mixed-feed industry; (5) the development of specialty and identity-preserved grains; (6) the introduction of biotech, genetically modified organisms; (7) changing government price and income policies influencing grain production, storage, and returns; and (8) the growth of grain export markets.

GRAIN PRODUCTION AND USES

Grains and oilseeds are produced in every American state and in every country of the world. This geographic dispersion of grain production produces a complex grain marketing system. However, in America a few states in the center of the country dominate grain production. For example, Iowa, Illinois, and Nebraska produce 36 percent of American corn. Kansas, North Dakota, and Montana produce 38 percent of American wheat. And Iowa and Illinois produce 35 percent of American soybeans.

The sources of supply and uses of wheat, corn, and soybeans are summarized in balance-sheet format in Table 26-1. These grain balance sheets are constructed so that the total supply exactly matches (or balances) total use during the harvest-to-harvest marketing year. Commodity balance sheets can assist in forecasting grain prices.

As shown in Table 26-1, imports are not an important source of U.S. supplies of corn, wheat, or soybeans. The United States has a comparative advantage in producing these commodities and is largely self-sufficient. Table 26-1 also shows that these commodity supplies were divided differently among their various uses: livestock feed, food, exports, industrial, seed, and ending stocks.

TABLE 26-1 U.S. Commodity Balance Sheets for Wheat, Corn, and Soybeans, 1993-1997

	Percent of Bushels: 1993–1997 Average		
	Wheat	*Corn*	*Soybeans*
Supply sources:			
Carryover of previous years	17%	12	9
Production	80	88	91
Imports	3	*	*
Total supply	100	100	100
(billion bu.)	(2.9)	(9.6)	(2.6)
Uses:			
Livestock, feed, seed	9%	53	5
Food, industrial	34	17	55†
Exports	39	19	32
Ending stocks	18	11	8
Total usage	100	100	100

SOURCE: U.S. Department of Agriculture.

*Less than 0.5 percent.

†This domestic soybean crush produces both oil and meal, some of which is exported. Most of the meal is used in animal feeds.

Product Uses

Corn accounts for over 90 percent of U.S. feed grain production. It is grown in every state, but 8 out of 10 bushels are produced in the Cornbelt, stretching from Iowa to Ohio. A small portion of U.S. corn acreage is used for sweet corn and popcorn. Most corn is grown for livestock feed. In recent years an increasing proportion of the corn crop has been processed into food and industrial products, as illustrated in Table 26-2.

TABLE 26-2 Wet-Milled Corn Products

	Foods	*Industrial*	*Livestock Feed*	*Brewing*
Corn starch	baking powder confections desserts gravies sauces	dextrins adhesives ceramics inks metal castings paints textiles	corn-gluten meal	ale, beer
Corn syrup	bakery products canned fruits confections food sauces frozen fruits frozen desserts ice cream jellies preserves table syrups	leather tanning paper phamaceuticals textiles tobacco curing		
Corn sugar	baby foods bakery goods canned fruits condensed milk frozen fruits	chemicals phamaceuticals	mixed rations	
Crude corn oil	bakery products condiments confections margarine salad dressings	leather tanning paints soaps synthetic rubber textile sizings varnishes	corn-oil meal	
Steepwater concentrates		pharmaceuticals		
Alcohol		ethyl alcohol chemicals		

SOURCE: *Grains: Production, Processing, Marketing.* © 1973, 1977, 1982. Board of Trade of the City of Chicago. Reprinted by permission.

In recent years there has been a notable increase in the processing of corn into high-fructose sweetener, a major sweetener in food and beverages, and in the conversion of corn into ethanol, a gasoline supplement.

Corn, grain sorghum, oats, and barley are the major U.S. feedgrains. Frequently, these are fed on the farm where they are produced or are sold directly to another farmer. In other cases, these feedgrains are purchased by mixed-feed manufacturers and then resold to farmers and livestock feeders. The development of large-scale livestock feeding operations with a need for huge volumes of animal feeds has created an entirely new market for these grains.

The U.S. wheat crop principally is consumed as a domestic food grain or is exported. Wheat may, however, be fed to livestock when supplies are abundant and it is competitively priced relative to other grains. About 50 percent of the U.S. wheat crop is exported.

Each special class of wheat has particular food uses. The hard, red spring and winter wheats that are grown in the Great Plains area are used largely in the bread industry. The soft, red winter and white wheats that are grown in the eastern part of the North Central region and in the Pacific Northwest are used in pastries, crackers, biscuits, and cakes. Durum wheat, grown chiefly in North Dakota, is the wheat used by pasta manufacturers.

Soybeans are grown primarily in the Midwest and the Mississippi River delta states. About 30 percent of the U.S. soybean crop is exported. Soybeans are a versatile agricultural commodity. They are crushed for their oil and meal, and the numerous resulting food, feed, and industrial products are shown in Figure 26-1. The major food products made from soybeans include margarine, salad oils, and cooking oils. Soybean meal is a high-protein animal feed.

Product and Production Characteristics

The grains and oilseeds serve as raw materials for conversion into higher-valued products. This conversion adds form utility and can take place on the farm, as in livestock feeding; in the food processing sector, as in the case of flour or wet-corn milling; or in the household, where flour, shortening, and sugars are combined to bake a cake. This dependence on further-processing results in a relatively low farmer's share for these commodities and a loss of their farm identity in the marketing process.

Some farm-level grain markets probably come closer to perfect competition than any other food or farm market. There are a large number of relatively small producers, the product is reasonably homogeneous (once graded), movement into and out of production is fluid, and market information is quite good. As a consequence, grain farmers encounter all the problems of perfect competitors: price taking status, limited supply control, prices above or below costs of production in the short run, and no effective market power. Grain farmers are highly specialized today.

The effects of weather, disease, and other uncontrollable factors in grain production also complicate farmer efforts to control supply and influence prices. The acreage planted to the various grains does not vary markedly from year to year, but production does. This is largely because of the weather and its effect on yields. Pro-

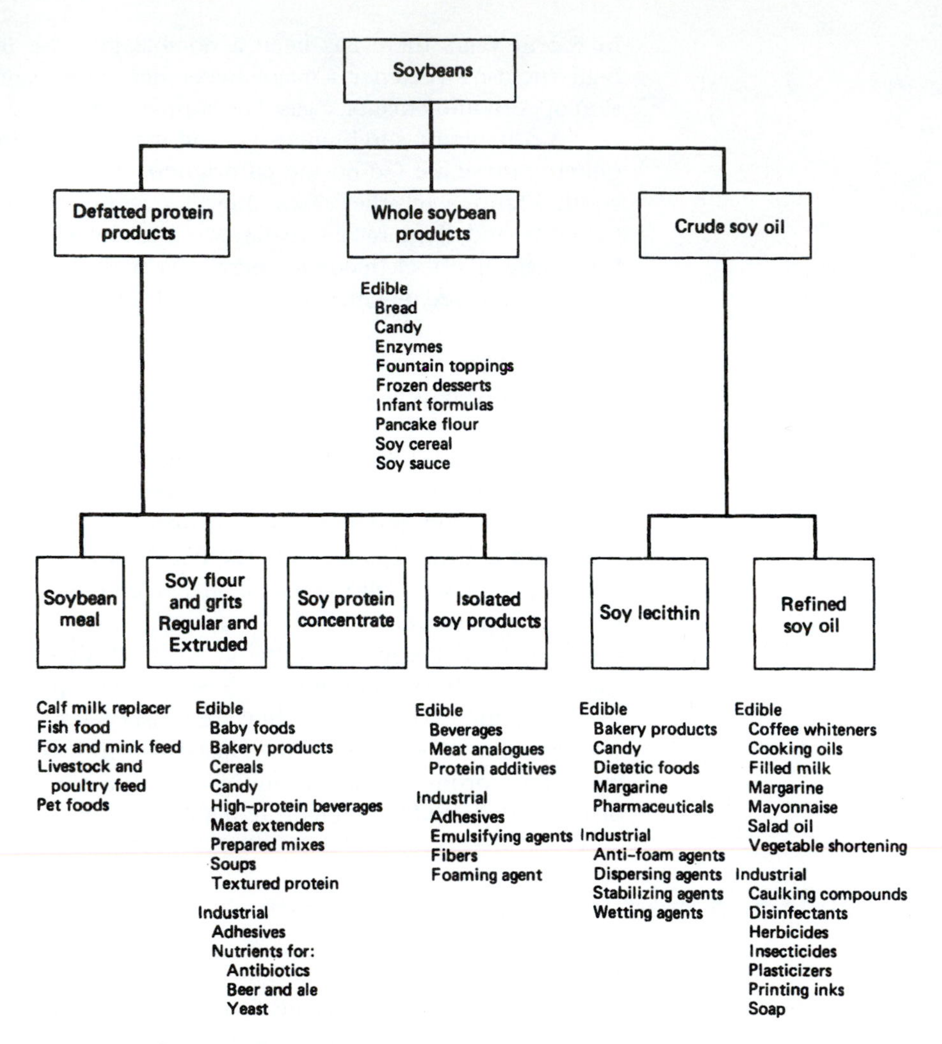

FIGURE 26–1 Soybean products. (U.S. Department of Agriculture.)

duction is also highly seasonal. Most of the U.S. corn crop is harvested during October through November. Wheat, oats, and rye are harvested during the relatively short period of June through September. The soybean crop is harvested during the early fall months.

In contrast to the seasonal nature of production, there is a fairly constant demand throughout the year for grain and oilseed products. Somewhere in the marketing channel, then, there must be facilities for huge amounts of seasonal storage. There must also be a price mechanism to encourage storage and to ration the annual supply over the year.

The unpredictable nature of production means that some stocks must be carried over from one year to another. This is necessary if there is to be any stability of supplies of food and feed from year to year. If the harvest is a large one, the season-end carryover will be large. If the harvest is small, the carryover will be much reduced. Such carryovers are the safety factors that protect dependent industries from violent, feast-and-famine fluctuations.

The various grains and oilseeds have markedly different demand trends. The demand for feedgrains and soybeans has expanded rapidly with the growth in meat consumption, both domestically and worldwide. Soybean demand also has benefited from the consumer trend of substituting vegetable oils for animal fats. In general, the grains and oilseeds have inelastic farm demands and low income elasticities. Per capita grain consumption has risen from its all-time 1972 low point as a result of changes in the American diet.

In the 1990's American grain production was substantially changed by new farming technologies, such as precision farming, minimum tillage, and biotechnologies. These developments affected the costs of grain farming, yields, and the market characteristics and marketing systems of grains. These technologies are also affecting the size and numbers of grain farmers, and in some cases the location of American grain farmers. This is another example of how food production and marketing go hand-in-hand and often influence each other.

THE MARKETING CHANNELS

The geographic dispersion of production, the large number of producers, the interrelationship with the livestock economy, and the high degree of processing involved all contribute to a long and complex grain marketing channel. The corn marketing channel is shown in Figure 26-2. The chief agencies in this channel are farmers—as buyers and sellers—elevators, commission merchants, brokers, processors and millers, exporters, and the grain exchanges.

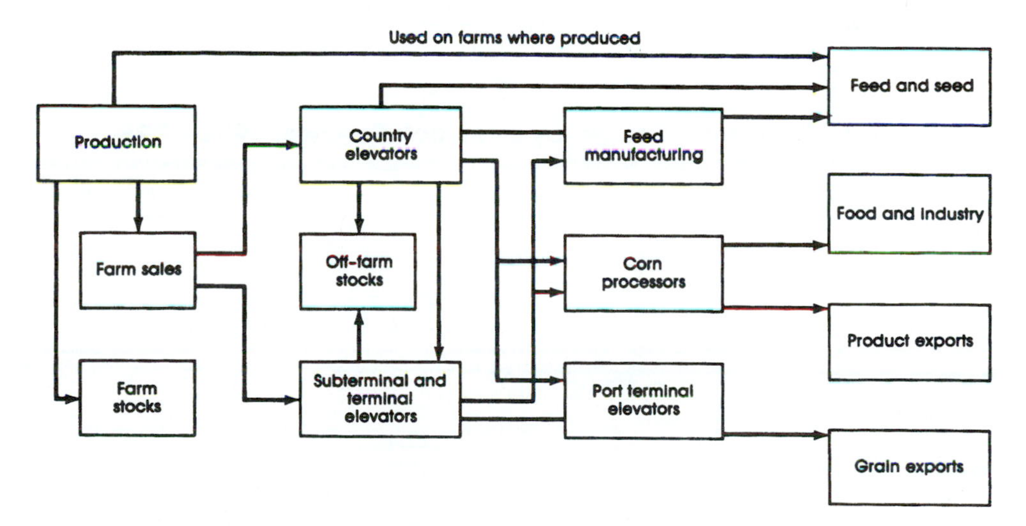

FIGURE 26–2 The corn marketing channel. (U.S. Department of Agriculture.)

Farmers and Country Elevators

Farmers have three general marketing alternatives for their grain: (1) feed it to on-farm livestock or sell to other farmers for feed; (2) sell the grain to the commercial market system at harvest; or (3) store the grain—on-farm or off-farm—for later sale. There are many variations of these alternatives.

Farmers use about one-third of the major feedgrains on the farm where they are produced. However, there has been a decline in the share of corn used on the farm of origin. This reflects continuing trends toward farmer purchase of complete feeds from feed manufacturers and specialization of the livestock and cash grain operations. Table 26-3 shows how the nation's feedgrain supply is allocated to the various types of livestock. As a food grain, a larger share of wheat production enters the commercial marketing channels than is the case for feedgrains.

Farmers have several alternatives for off-farm grain sales. Besides selling to other farmers, they may sell to elevators, dealers, processors, and even itinerant truckers. However, farmers sell most of their grain to the local country elevator. These elevators (so named because the grain must be conveyed, or elevated, to the top of bins for storage and gravity flow) are scattered throughout grain producing areas, usually along railroads. There are about 7000 country grain elevators.

Country elevators fall into three general classes according to their ownership and organization. The independent elevators are under the operational control of their individual owners. Cooperative or farmer-owned elevators are owned and operated cooperatively by the farmers of the area. They may either be organized singly or in statewide groups. The third type is the line elevator. This is a group of elevators owned and operated from a central headquarters as a chain. Such chains may be owned either by grain companies or millers and processors who use them to secure supplies directly for their manufacturing or exporting operations. The ownership of elevators will vary from area to area.

Many studies have been made of the costs of elevator operations. Nearly all have found that operating costs per bushel handled tend to decline with increasing

TABLE 26-3 Feedgrain Usage, by Livestock Species, 1982-1998

	Grain Consuming Animal Units			
	1982		*1998*	
Species	*Million Units*	*Percent of Total*	*Million Units*	*Percent of Total*
Hogs	20.5	27%	23.8	27%
Beef cattle	22.3	30	24.3	27
Dairy	12.4	17	10.2	12
Poultry	18.3	25	29.1	33
All others	0.7	1	0.6	1
Total	75.2	100%	88.1	100%

SOURCE: U.S. Department of Agriculture.

volume. Increased volume for an elevator can be obtained in two principal ways. One is to add other business lines to that of grain handling. The other is to consolidate elevators in order to obtain a larger trade area. Many elevators have taken on various sideline enterprises to boost their volume and offset the highly seasonal nature of their grain operations. The most common sideline is the feed business. In addition, seeds, implements, building materials, petroleum, and farm supplies are common sales departments. In some communities the elevator is not only the outlet for grain but also the principal source of general farm supplies.

Increased volume also can be obtained by the consolidation of existing elevators. The present locational pattern was established back in the era of horse-drawn transportation, poor roads, and poor communication, and the supply area of many elevators is small. As is often the case, the number of firms that will provide an adequate volume for low-cost operation must be balanced against the industry structure that will assure effective competition. Modern farmers, with telephones, daily newspapers, radios, TVs, the Internet, and other information sources have a communication network that covers a larger area than was available to their grandparents. With trucks farmers can sell their grain over wider areas. Some reasonable elevator consolidation seems possible without sacrificing an effectively competitive situation.

Most country elevators sell and ship their grain to larger firms located at a terminal market. Grain may be consigned to commission merchants on the market for sale, or it may be sold on *to-arrive* or *on-track* bids. If the grain is consigned to commission merchants, the elevator ships the grain to the terminal and takes the price secured by the commission merchant, much the same as with the use of livestock commission firms in the livestock yards. In selling on a "to-arrive" basis, the price and the shipment details are agreed upon in advance of delivery. When sold "on-track, country point," the price is agreed upon for grain in cars at the local elevator. The buyer of the grain takes title to the grain at the country elevator and arranges and pays for the cost of transportation.

Terminal Elevators

Terminal elevators are located in the major grain marketing centers such as Chicago, Minneapolis, Fort Worth, Kansas City, and St. Louis and at the major ports, such as New Orleans, the Texas Gulf ports, the Pacific ports, and Baltimore. Like country elevators, terminal elevators may be operated independently, by farmer cooperatives, or by integrated grain marketing companies. The principal distinctions between country elevators and terminal elevators are their location, their size, and their sources of grain. Terminal elevators are located at major grain trading centers, seldom in the country; they are usually much larger than country elevators; and while terminal elevators receive some grain directly from farmers, most of their grain is purchased from the country elevators.

Terminal elevators also do much more grain storage than country elevators. Terminal operators often hold grain for considerable periods of time. These operators are small-margin, large-volume handlers, and price fluctuations might easily turn profits into losses. Therefore, futures market hedging insurance often is very desirable to them.

Cash Grain Commission Merchants

The cash grain merchant is the representative of the country elevator operator in the terminal market. For a fixed charge the cash grain merchant will accept the responsibility of selling the grain. The principal job of the commission merchant is to seek out interested buyers. Because these merchants on most markets must all charge the same commission, they compete for grain consignments on the basis of service they can render to the elevator operator. They specialize in the analysis of market information for their clients. They also take charge of any arrangements that must be made with the buyer of the grain.

The commission merchant also handles the grain sold "on-track" or "to-arrive." For these sales, the merchants make a practice of sending out bids to the elevator operators at the close of each market day.

In soybean marketing the interior carlot dealer is also an important intermediary between the country elevator and the soybean processor. Unlike the commission merchant, the carlot dealer may take title to the soybeans. However, the firm does not take physical possession of the soybeans, but directs and helps finance their movement to the processor.

Processors and Millers

There are two stages to processing food grains. Flour, rice, corn, and oilseed millers first refine or mill basic farm commodities. Their output is then further processed into consumer products by breakfast cereal manufacturers, bakers, distillers, cooking-oil manufacturers, and others. For example, the various flour-using industries are illustrated in Figure 26-3.

The buyers of grain differ widely in their needs and desires. Flour millers desire wheat of a specific gluten or protein content, depending on whether they are producing bread or cake flours. The grains desired for breakfast foods will differ. Corn processors and maltsters look for special grain characteristics.

Some processors maintain their own buyers on large terminal markets and purchase their supplies from the cash grain commission merchants. Others retain the services of brokers who buy grain wherever it can be obtained. Still others have built up their own large country buying operations and operate line elevators in the grain belt. This not only protects their supply position but also control the handling and moving of grain throughout the marketing channel.

Changing Grain Marketing Patterns

In the 1990s, grain and oilseed producers experienced an increase in demand for value-added, specific-attribute grains. Food processors began to encourage and pay growers premiums to produce grains which had desirable attributes in producing certain food products. This included high-oil corns, grains and oilseeds with improved protein and high-lysine, high-starch corns, white corns for use in snack foods, and the like. Because of their special, value-added properties and often higher price, these grains must be segregated from other grains. They are therefore called *identity-*

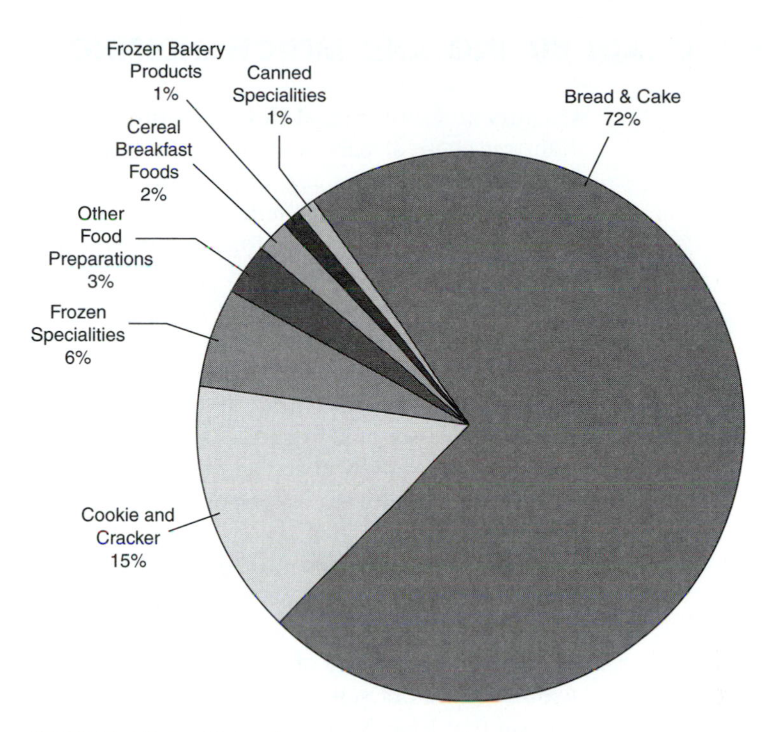

FIGURE 26–3 Major flour-using food industries.

preserved (IP) grains. This means they require separate and special handling and marketing. Many IP grains are produced under contract to grain processors, and the processors often control the genetics of these products. Through grower contracts, the production and marketing of IP grains and oilseeds are becoming highly integrated.

A form of decentralization has altered grain marketing patterns. Many grain processors and soybean oil mills have located their plants in rural areas, and larger country elevators—called *subterminals*—have assumed many of the terminal marketing functions. As a result, farmers are increasingly selling directly to processors or to subterminal elevators, who in turn export or process the products. These sales bypass the traditional terminal elevators and the cash grain commission merchants. The prepared feed industry must also be considered decentralized because most feed manufacturing is done at country points, often at the country elevator.

Grain market decentralization has often been encouraged by farmers. It has provided increased local marketing alternatives to the country elevator. Moreover, although pricing efficiency remains imperfect in grain markets, it is not likely that decentralization has adversely affected pricing efficiency in country markets. There is also evidence of increased operational efficiencies from consolidation of country elevators, subterminal elevator operations, and direct sales by farmers to processors.

Grain marketing has also been influenced by the growth of on-farm storage, the substitution of truck for rail transport, the growth of port facilities to meet export demand, and increasing use of grain and soybean futures contracts for managing market risks.

COUNTRY GRAIN BUYING AND MERCHANDISING

Country and subterminal elevators purchase grain from farmers in several ways. Outright purchase at harvest or from the farmer's storage is the most common type of transaction. Elevators also contract for price in advance of delivery (*forward pricing*), accept grain from the farmer with the price to be set at a later date (*deferred pricing*), and arrange sales for the farmer without actually taking possession of the grain.

Many elevator operators use *basis pricing* in quoting grain prices to farmers. A basis price is an offer price stated in terms of a "nearby" futures price. For example, an elevator may quote a November 10 cash price to the farmer of "20 cents under," meaning that the cash price is 20 cents less than the Chicago December futures price. This basis of 20 cents would be set in such a way as to cover elevator handling, transportation costs, and desired profit. Basis pricing is sometimes preferred by farmers and elevator operators because the basis will change less frequently than cash prices.

Elevator operators must buy their grain from producers in such a manner that when they sell it they have an adequate margin to cover their costs and also some profit. Margins taken by elevators vary widely. The price realized by farmers for their grain is a combination of the quoted price of the elevator plus the grade of grain delivered. Elevators sell grain on the basis of official federal grades but often do not buy on that basis from farmers. Instead, each elevator devises its own system of discounts—usually based on weight and moisture. Some elevator owners have been known to follow a system of high-quoted prices plus a heavy discount ("dockage") program. Others have used underdocking and overgrading instead of price as a competitive weapon. This is similar to the difference in sorting methods used by livestock buyers.

Farmers must evaluate both quoted prices and buying practices before selling their grain. Once an outlet is chosen, the farmer cannot be sure it will remain the most advantageous. Prices quoted to farmers in a particular county may vary substantially, and no single elevator will always pay the high or low price.

Grain merchandising is the term used to describe the marketing activities of elevator operators and other grain handlers. Grain merchandising decisions include when to buy or sell, where to buy or sell, how to sell, quality improvements, and pricing the grain. In addition to their buying and selling margin, elevator operators and other grain merchandisers frequently can profit from storage, conditioning, and blending of grain. Storage profits are earned by correctly anticipating seasonal price rises or unusual market developments. *Grain conditioning* refers to the improvement of the quality and value of grain through cleaning, separating, or drying. *Blending* refers to mixing different lots of grain in a way that raises the grade and value of the total lot.

GRAIN GRADING

Nature does not produce uniform farm products of equal quality. You will recall from Chapter 2 that standardization is a necessary marketing function which can contribute to efficient marketing. Grading of products, such as grains, into uniform classes or grades is important so that products can be identified and properly valued by sellers and buyers, including consumers.

The 1916 U.S. Grain Standards Act established grain quality grades and requires that these grades be used for grain moving in interstate commerce. Most grain sold off-farm is federally graded. Prior to 1977, grain graders were independent operators licensed by the U.S. Department of Agriculture and paid by the users of grading services. A 1976 Amendment to the Grain Standards Act, however, established a new Federal Grain Inspection Service, and graders are now U.S. government employees. The Grain Quality Improvement Act of 1986 further improved the quality of grain marketed to domestic and foreign buyers. This law prohibits adding back *dockage,* or foreign material to grain once it has been removed and reduces the tolerance levels for live and dead insects in grain grades. There were more than a hundred changes in grain grades between 1916 and 2000.

Whenever grain arrives at a terminal market, a sample is drawn and graded. Grain grades range from the highest grade, Number 1, down through Number 5 and sample grade. There is also a division into classes and subclasses for each kind of grain. For example, Class I of wheat consists of hard, red spring wheat. This is further divided into the subclasses of Dark Northern Spring, Northern Spring, and Red Spring. Within each class the requirements for the various grades are specified.

Though the factors considered in grades differ somewhat from one grain to another, in general the following factors are used to determine the grade of a given class of grain: (1) *test weight*—a volume measure of grain density; (2) moisture content; (3) foreign material or *dockage,* such as dust, dirt, broken kernels, or weed seeds; and (4) damaged kernels resulting from fermentating, frost, fungus, mold, insects, or other conditions.

Each grade has minimum or maximum limits for these quality standards. Failure to meet any one of the requirements will reduce the grade. The variation possible within a given grade explains why different carloads of the same grade of grain sell at slightly different prices at the same location and time. Different processors have different requirements. One processor who is particular not about damage but about moisture content might pay more for a particular carload than for another. Moreover, some processors may require certain factors, such as protein content in wheat, that have not yet been incorporated into official grading standards.

The grade of grain is determined by a combination of different factors, each of which can vary within limits, and is determined by the lowest factor. Handlers can mix or blend grain in order to raise the grade. For example, suppose two loads of grain were received. One was graded Number 2 because of its moisture content, but it met all other requirements of Number 1 grain. The other load was graded Number 3 because of excessive cracked kernels and foreign material but in all other aspects it met the requirements for Number 2 grain. Through the proper mixing or blendings of these two loads of Number 2 and Number 3 grain, it might be possible to secure two loads (or nearly so) of grain that would grade Number 2. This would be possible because one load was low in the grade factors in which the other load was high. Country elevators sometimes blend grain, and terminal elevators usually do.

Accurate and reliable grades are particularly important for international trade where grains are sold around the world, through the use of electronic communications. International buyers have often complained about U.S. grain grades. In the early 1990s a number of U.S. grain shipments were refused for grading problems. A

U.S. Department of Agriculture study found that the U.S. grain grades rely too heavily on physical characteristics (weight, moisture, presence of weeds, etc.), and not enough on intrinsic characteristics, such as protein content and gluten quality, which affect the value of the product for the end user. The study recommended incorporating these intrinsic qualities into U.S. grain grades so that producers would have an incentive to grow those varieties of highest value to buyers.

Despite their limitations, grain grades make substantial contributions to the marketing process. They increase the homogeneity of grain lots, facilitate efficient buyer and seller transactions, and differentiate between higher- and lower-valued products.

GRAIN STORAGE

One of the principal marketing decisions of the grain producer is whether to sell at harvest time or to store for sale at a later date. If it is decided to store the grain, the next questions are where and for how long it should be stored.

In answering the question of *whether* to store, the farmer must balance the costs of storing the grain against the possible gains from a rise in price later in the season. Two factors besides the normal seasonal price rise must be considered. One of these is the possibility of taking advantage of the government price-support programs. The nonrecourse loan removed much of the uncertainty about the relationship of the current price to a possible future price. If the market price of grain at harvest time is enough below the available loan rate to offset storage costs, there is very little to lose and much to gain by storing under government loan. The other consideration is the outlook for the general movement of prices during the storage period. If the general price level is expected to move downward, sales at harvest may be much more attractive to the producer than if the general level is expected to move upward.

Another important question for the farmer is *where* to store the grain. On-farm storage capacity has expanded rapidly in recent years. There is now about as much on-farm grain storage capacity as there is off-farm storage capacity of elevators and processors.

The growth of on-farm grain storage is a form of vertical integration for farmers. It provides them with marketing flexibility, gives them additional opportunites for earning conditioning and storage profits, and, overall, increases the farmer's role in the grain marketing process. The 1977 and 1981 farm bills stimulated the construction of on-farm grain storage facilities through lower interest rate loans and provisions for a farmer-held grain reserve.

It is not possible to make a blanket prescription about whether farmers should store their grain, or whether it should be stored on-farm or off-farm. These decisions depend on numerous factors, including the availability of on- and off-farm storage facilities, price expectations, livestock feeding opportunities, and attitudes toward price risk. These will vary for each farmer and each year. Farmers can use storage hedges (Chapter 20) to shift the price risks of storage to others if they choose to store. Moreover, storage hedges can help farmers earn carrying charges and thus permit the use of on-farm facilities year after year regardless of price movements.

For on-farm storage, producers face a number of costs. There are certain fixed costs that must be met whether grain is stored or not. These include the depreciation, maintenance, insurance, taxes, and interest on the capital invested in the available storage facilities and equipment. Then there are several variable costs that will occur only if grain is stored. These will include the costs of shrinkage and loss from damage, interest, insurance and taxes on the grain, any expense of treating or conditioning the grain, and the cost of the labor and transportation expenses resulting from the storage operation. Against these costs of farm storage must be weighed the cost of hiring an elevator to store the grain. Elevator storage availability and charges vary widely.

Farmers without storage facilities can rent or hire storage space from elevators. One form of this is the *grain bank.* Farmers deposit grain to be dried and stored until they need it for animal feed. As the grain is needed the farmers "withdraw" grain as they would money from a checking account. The withdrawal is of the same quality that the farmers deposited, although it need not be specifically the same grain. The farmers then pay for the storage and handling services. The grain bank is often tied to the elevator's feed mixing and milling services.

STRUCTURE AND COMPETITION IN THE GRAIN AND OILSEED PRODUCT INDUSTRIES

The major grain and oilseed products are bread, bakery products, breakfast cereals, vegetable oils, and mixed animal feeds. Market structure and competitive behavior differ substantially in these industries. Firms producing and selling these products tend toward oligopoly and monopolistic competition. With the exception of the livestock feed industries—but including the pet food industry—marketing costs are relatively high, and the farmer's share is lower than in most other food industries. The farmers' share for white bread averages 7 percent, the share for margarine 22 percent, and the share for shortening 30 percent.

Timeliness is the key to marketing costs in the bread and soft bakery goods industries. These products are extremely perishable, and consumers place a high value on freshness. Although older products can be picked up and sold at low-priced outlets, there is a high "stale loss" in these businesses. In order to ensure freshness, bakers have traditionally operated relatively expensive store-door routes, with delivery direct to grocery stores. More recently, however, corporate and affiliated chainstores have achieved some operational efficiencies by substituting direct warehouse purchases and shipments for the driver-salesman method of procurement. Some grocers also operate their own central or in-store bakeries. Frozen bakery products also offer some potential for reducing baked goods' marketing costs. In contrast, the significant increase in the number of bread varieties—some bakers sell more than forty kinds of breads—has probably sacrificed some economics of standardization and size in bread baking. Another example of consumer sovereignty at work.

The bakery, cracker, and breakfast cereal industries are good examples of industries that practice a high degree of product differentiation along with extensive consumer advertising programs. A Saturday morning spent watching television will demonstrate the large advertising efforts that go into the merchandising of the newest cereal. Advertising expenses tend to be higher for cereal manufacturers than for other grain processors. Nonprice competition is widely used in the marketing of these products. The major breakfast cereal manufacturers found themselves in a price war in 1996.

Although grain processing firms have grown in size, there has also been an increase in the number of these firms in many sectors, as shown in Table 26-4. More efficient plant sizes, mergers, business failures, and growth to attain economies of size have all been factors in these trends. Both relatively high and low levels of market concentration can be found for these products. The most concentrated food sectors are malt beverages, cookies and crackers, and breakfast cereals.

Vertical integration has been prevalent in the grain and oilseed industries. Chainstores have integrated into baking and bakery product wholesaling. Many oilseed refining facilities are owned by manufacturers of shortening, salad and cooking oils, and margarine. Processors increasingly are using farm contracts to ensure quantity and the desired quality of grains. There are frequent combinations of soybean processing and livestock feed manufacturing. Feed manufacturers and dealers have also integrated into livestock production as a way to increase markets for their products.

Biotechnology began to revolutionize the crop industries in the 1990's and has presented new marketing problems. Genetic modifications make it possible to develop crops with specific traits, such as herbicide resistant soybeans or corn and cotton that are resistant to pests which reduce yields. There are also efforts to produce grains which increase yields with less nitrogen and have more of the properties desired by food marketing firms. These genetic developments have the promise of lowering farm costs, increasing output, reducing the use of agricultural chemicals and

TABLE 26-4 Number of Companies and Market Concentration of the Grain and Oilseed Processing Industries, 1967-1992

	Number of Companies			*Percent of Sales by the Four Largest Firms*		
Industry	*1967*	*1982*	*1992*	*1982*	*1987*	*1992*
Breakfast cereals	30	32	42	88	87	85
Blended, prepared flour	126	120*	156	68	43	39
Wet-corn milling	32	25	28	68	74	73
Cookies, crackers	256	296	374	59	58	86
Soybean oil mills	60	52	42	55	71	71
Malt beverages	125	67	160	40	87	90
Pet foods	NA	130	102	NA	61	58
Distilled liquors	70	71	43	54	53	62
Edible fats and oils	63	79	72	43	45	35
Rice milling	54	48*	44	46	56	50
Cottonseed oil mills	91	47	22	42	43	62
Macaroni, spaghetti	109	196*	182	31	73	78
Grain mill products	438	251	230	30	44	66
Bread, cakes	3,445	1,869	2,180	26	34	34
Prepared animal feeds	NA	1,182	1,160	NA	20	23
Potato chips, snacks	NA	277*	334	NA	NA	70

SOURCE: U.S. Census of Manufacturers, U.S. Bureau of Census.
*1987 data.

reducing the land needed for crop production to feed a growing world population. However, some consumers' and nations' objections to these crops in the late 1999's have slowed this technological development. It is possible that these crops will need to be segregated from other crops and labeled at both the farm and consumer levels. Some food marketing firms have refused to purchase biotechnology crops and some have also paid a premium price for nonbiotech crops.

THE INTERNATIONAL GRAIN TRADE

Grains have always been important trade products. Wheat has been on the move since the first primitive strains were cultivated in the Middle East in 7000 B.C. The Roman Empire imported grain from conquered territories. Ancient Greece was dependent on the grain trade. The Industrial Revolution in England was fueled by American and Russian wheat. Today, virtual rivers of grain continue to flow around the world influencing international political, social, and economic events.

World production and trade of grains has expanded rapidly as a result of rising world incomes and populations. The United States, Canada, Australia, and Argentina are the major sellers in the world grain trade today, but many other nations are involved. In the 1970s, the U.S. grain sector became internationalized, and exports of these products rose dramatically (see Figure 7-3). Over the 1977-1982 period, the United States accounted for 46 percent of the world wheat trade, over 60 percent of the feedgrain trade, and over 98 percent of soybean exports.

In the 1982-1986 period, U.S. exports of grains and oilseeds declined significantly as the value of the dollar rose and worldwide production of these crops increased. During this period the United States also lost some of its export market share. The portion of the U.S. wheat crop exported declined from 64 percent to 47 percent between 1979 and 1986. The share of U.S. feedgrains exported fell from 31 percent to 22 percent of production over this period. U.S. agricultural exports grew again in the 1986-1996 years and reached a record high of 60 billion dollars in 1996. America's share of world food exports reached 23 in 1995, the highest level in ten years. World population growth, favorable exchange rates, and export promotion programs all played a part in this export growth. Much of this growth was in higher value-added commodities (fruits, vegetables, meats, and processed foods). U.S. grain and oilseed exports declined in the 1996-2001 years because of large worldwide crops, the strong dollar, and weak Asian markets.

The largest markets for U.S. grain and oilseed exports are the Far East (Japan, South Korea, Taiwan), Western Europe, Mexico, and Latin America. The former Soviet Union and China were large but inconsistent importers of U.S. grains and oilseeds in the 1990s.

A few large grain trading firms account for most of U.S. grain and oilseed exports. These include Cargill-Continental, Archer Daniel Midlands, Bunge Corp., Louis Dreyfus, Inc., and André and Co./Garnac Grain Co., Inc. These privately owned, multinational firms operate a worldwide complex of grain purchasing, storage, transportation, and merchandising facilities. Their extensive information and communication systems assist them in buying and selling huge volumes of grain all over the world.

The United States relies on these private grain traders, along with some grain cooperatives, to export its grain surpluses. However, in many other countries the grain trade is monopolized by *state trading agencies.* For example, Exportkhleb, the Russian government trading agency, negotiates all Russian grain imports and exports. The Japanese Food Agency buys all wheat and barley imports for that country. The Australian and Canadian Wheat Boards are monopolistic sellers with potential marketing powers.

The worldwide grain trade is intertwined with international politics. The United States has signed *bilateral agreements* guaranteeing grain supplies with such importers as Russia, the People's Republic of China, Poland, and Mexico. America has also used its grain as a foreign policy tool, notably when it placed an *embargo* on grain and oilseed exports in 1973 and suspended grain sales to the Soviet Union in 1974 and again in 1980. The North American Free Trade Agreement (NAFTA) and the Uruguay Round General Agreement on Tariffs and Trade (GATT) expanded world trade in grains and oilseeds. These bilateral trade agreements and embargoes influence who ships where in the world but probably have little effect on total trade and prices because of the worldwide market for grains.

GRAIN POLICY

The production and marketing of grains and oilseeds has always been strongly influenced by government policies. This continues to be true despite the 1996 farm bill, which reduced, but did not eliminate, government influence over these prices. As indicated in Chapter 21, the 1996–2001 federal farm bill decoupled grain subsidy payments from prices, eliminated acreage restrictions, and replaced nonrecourse loans with marketing loans. This provided grain and oilseed producers with increased crop flexibility, income protection, but increased price risk.

SUMMARY

The grain marketing channel is a complex one, involving long distances, interrelated livestock marketing decisions, and numerous farmer choices. Feedgrains, food grains, and oilseeds undergo more form transformation in the marketing process than most other agricultural commodities. The major markets are livestock feeding, consumer food products, and exports. The somewhat perfectly competitive nature of farm grain markets contrasts with the imperfectly competitive milling and processing industries. Several developments have changed grain marketing patterns, including decentralization and the growth of subterminal elevators, expansion of the commercial mixed feed industries, increased on-farm storage facilities, and expanded world exports. By and large, these developments have increased markets for these products and increased farmers' marketing alternatives.

KEY TERMS AND CONCEPTS

basis pricing
bilateral agreement
biotech
generic certificate
genetically-modified organism (GMO)
grain balance sheet

blending
carryover
cash grain commission merchant
country elevator
deferred pricing
dockage
embargo
forward pricing
grain bank
grain conditioning
grain merchandising
identity-preserved (IP)
subterminal elevator
terminal elevator
test weight
to-arrive, on-track sales

DISCUSSION QUESTIONS

1. Identify farmers' alternative grain storage facilities and sales outlets. What factors influence these farmers' decisions?
2. Why do grain farmers use the futures market? Why don't more use it?
3. How might grain and oilseed farmers escape from their nearly perfectly competitive markets?
4. What are the supply control problems in the grain industry?
5. The breakfast cereal sector is one of the most concentrated and criticized segments of the food industry. Examine the market performance of this sector.
6. What are the major demand trends for grains and oilseeds?
7. How is the decision to build on-farm storage facilities different from the decision to use that facility?

SELECTED REFERENCES

Ali, M. B., N. L. Brooks, & R. G. McElroy. (June 2000). *Characteristics of U.S. Wheat Farming: A Snapshot.* Statistical Bulletin No. 968. Economic Research Service, U.S. Department of Agriculture, Washington, D.C.

Atkin, M. (1996). *The International Grain Trade*. Cambridge, England: Woodhead Publishing, Ltd.

Ballenger, N., M. Bohman, & M. Gehler. (April 2000). Biotechnology: Implications for U.S. Corn and Soybean Trade. *Agricultural Outlook.* U.S. Department of Agriculture, Washington, D.C.

Biotechnology Key Topics. (January 2001). Economic Research Service, U.S. Department of Agriculture. http://www.ers.usda.gov/Topics/View.asp?T=101000.

Commodity Outlooks: Wheat, Feed grains, Rice, Oil Crops.
(wheat) http://www.ers.usda.gov/publications/so/view.asp?f=field/whs-bb/
(feed grains) http://www.ers.usda.gov/publications/so/view.asp?f=field/fds-bb/
(rice) http://www.ers.usda.gov/publications/so/view.asp?f=field/rcs-bb/
(oil crops) http://www.ers.usda.gov/publications/so/view.asp?f=field/ocs-bb/

Commodity Yearbooks: Feedgrains, Oilcrops.
(feed grains) http://www.ers.usda.gov/data/sdp/view.asp?f=crops/88007/
(oil crops) http://www.ers.usda.gov/data/sdp/view.asp?f=crops/89002/

Cramer, G. L., & E. J. Wailes. (1993). *Grain Marketing.* Boulder, CO: Westview Press.

Crop Briefing Rooms, Corn, Soybeans, Wheat, and Rice. (January 2001). Economic Research Service, U.S. Department of Agriculture.
http://www.ers.usda.gov/Briefing/Corn/http://www.ers.usda.gov/briefing/soybeansoilcrops/
http://www.ers.usda.gov/briefing/wheat/
http://www.ers.usda.gov/briefing/rice/

Grain Inspection, Packers and Stockyards Administration. (January 2001). http://www.usda.gov/gipsa/

Krause, K. R. (April 1989). *Farmer Buying/Selling Strategies and Growth of Crop Farms,* Technical Bulletin No. 1756. Washington, D.C.: U.S. Department of Agriculture.

Leath, M. N., L. H. Meyer, & L. D. Hill. (February 1982). *U.S. Corn Industry,* Agricultural Economic Report No. 479. Washington, D.C.: U.S. Department of Agriculture.

Lin, W. W., W. Chambers, & J. Harwood. (April 2000). Biotechnology: U.S. Grain Handlers Look Ahead. *Agricultural Outlook.* U.S. Department of Agriculture, Washington, D.C.

Price, G. K. (May–Aug. 2000) Cereal Sales Soggy Despite Price Cuts and Reduced Couponing. *Food Review,* Volume 23, Issue 2. Washington, D.C.: U.S. Department of Agriculture.

Vocke, G. (August 2000). Forces Shaping the U.S. Wheat Economy. *Agricultural Outlook.* Washington, D.C.: U.S. Department of Agriculture.

Westcott, P. C., & L. A. Hoffman. (July 1999). *Price Determination for Corn and Wheat: The Role of Market Factors and Government Programs.* Technical Bulletin No. 1878. Economic Research Service, Washington, D.C.: U.S. Department of Agriculture.

CHAPTER 27

COTTON AND TEXTILE MARKETING

Cotton, Fabric of Our Lives.
Slogan for cotton industry advertising

OBJECTIVES

After reading this chapter, you will be able to:

1. Explain the changing regional production and marketing patterns for U.S. and world cotton.
2. Appreciate how promotional strategies have led to cotton's comeback in the clothing markets.
3. Understand how government price supports can influence the demand and prices of cotton.

From colonial days to the nineteenth-century era of "King Cotton," cotton has been a primary U.S. agricultural commodity. While its relative importance has declined over time as America's agricultural commodity base has become more diversified, cotton accounts for 4 percent of U.S. cash farm receipts. It is the world's most important textile fiber, accounting for more than 46 percent of all fibers produced. While grown in over 80 countries, the United States, China, India, Pakistan, and Uzbekistan account for nearly 74 percent of the world supplies of cotton. America ranks second in world cotton production producing over 20 percent of the world's cotton, consumes 12 percent of world supply, and held a 25 percent share of world exports during the 1990s.

The cotton industry has experienced several changes in market organization and marketing trends in recent years, including erosion and recapture of markets by manmade fibers and worldwide competition, changes in the geographic location of cotton production, decentralization and integration, and a more market-oriented government policy.

COTTON PRODUCTION

From the end of the Civil War until the mid-1920s, U.S. cotton acreage increased from less than 8 million acres to more than 46 million acres. Cotton production reached a peak of 18 million bales in 1926. U.S. cotton acreage declined from 40 million acres in 1930 to 10 million acres in 1966 and remained at about that level through 1988 (Figure 27–1). Acreage expansion then occured in the 1990s. Cotton yields have increased dramatically as a result of improved production and harvesting technologies and irrigation.

Cotton is produced in seventeen states with major concentrations of production in the Mississippi delta states (Mississippi, Arkansas, Louisiana), Georgia, the Texas high plains, Central Arizona, and the San Joaquin Valley of California. Cotton production shifted from the South to the Southwest following its introduction in Arizona and California in 1912. Between 1965 and 1985, the western states' share of U.S. cotton production rose from 17 percent to 31 percent. Cotton is irrigated in the West and yields there are more than double yields elsewhere. In the 1990s, cotton acreage shifted back to the Southeast and Mississippi delta states as the boll weevil came under control and water supplies became limited in the West.

Technological and economic pressures have resulted in larger and fewer cotton farms. The number of U.S. farms producing cotton declined from 1.1 million in 1949 to 31,000 in 1997 while the average size of these farms grew from 24 to 315 acres. Western cotton farms average twice the acreage of cotton farms elsewhere.

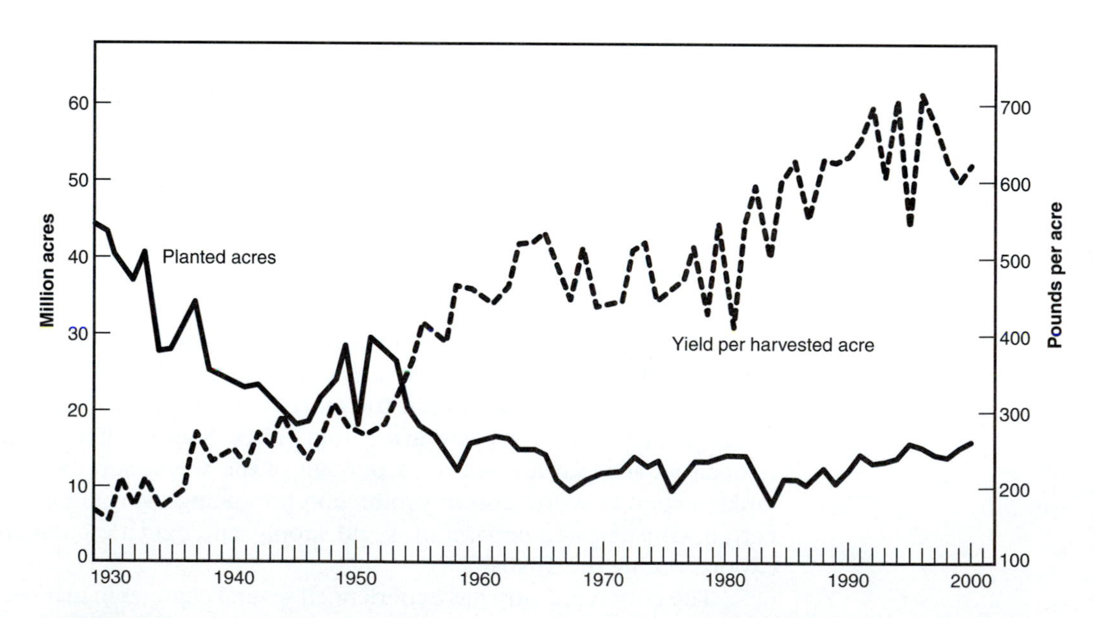

FIGURE 27–1 U.S. acreage and yield of cotton, 1930–2000. (U.S. Department of Agriculture.)

For most of the twentieth century, the U.S. cotton industry was marked by excess capacity, high stock levels, and frequently low prices. From 1933 to 1965, government cotton programs included acreage allotments, nonrecourse loans, marketing quotas, and price supports. Since 1965, rigid acreage and marketing controls have been relaxed, price supports have been lowered to world levels, and direct payments to producers have replaced government purchase and price support programs. Recent farm bills have reversed the upward trend in government cotton price supports and provided more industry flexibility to promote international competitiveness. In exchange for government loans and direct payments, producers are required to comply with an Acreage Reduction Program (ARP). Marketing loans (Chapter 21) have replaced nonrecouse loans and allowed cotton to move into markets rather than into government inventories.

CONSUMPTION TRENDS

The demand for cotton fiber is derived from the demand for textile and nonwoven cotton products. Textiles are used in apparel and household and industrial products such as clothing, carpets, towels, upholstery, conveyor belts, and air filters. The demand for cotton in these uses has been affected by the prices of cotton and competing fibers as well as changes in consumer preferences for fiber and textile products.

Total U.S. fiber consumption increased from 39 pounds percapita in 1962 to 80 pounds in 2000. However, U.S. cotton consumption generally fell from the 1940s into the 1980s as consumer preferences shifted to such manmade fabrics as nylon, rayon, and acetate. There were several reasons for this. First, the manmade fibers were better suited for many of the industrial and consumer products developed during this time. The manmade fibers were often easier to care for, more durable, and easier to handle in mills. In addition, government price support programs kept cotton prices higher than manmade fibers during the 1970–1984 period.

Nevertheless, cotton has some important advantages over manmade fibers in such areas as breathability and absorbency. These characteristics have helped cotton maintain its dominance in denim, underwear, and towels. The development of permanent press cotton garments, cotton/polyester blends, a cotton promotion program sponsored by producers, and the lowering of government price supports have improved the competitive position of cotton in recent years. As a result, U.S. cotton demand increased during the 1990s when its share of U.S. mill usage rose from 35 percent in 1990 to 40 percent in 2000. In 1997 U.S. mills used almost as many bales of cotton as the record set during WWII. U.S. per capita consumption of cotton more than doubled from the 1975–1980 period to 1999.

Cotton's marketing comeback in recent years is not an accident. It is the result of a promotional campaign coordinated by Cotton Incorporated, the research and promotional agency of U.S. cotton growers and importers. Begun in 1973 when the U.S. cotton industry was at a low point, the program includes the trademark *Seal of Cotton* and the copyrighted slogan, *Fabric of Our Lives.* This branding effort has successfully promoted the benefits and qualities of cotton. Cotton, Inc. licences its seal to over 1,000 clothing and textile firms, and also promotes industrial uses of cotton.

These changes in fiber usage are dramatic examples of how different factors can affect the market position of a product. Government pricing policy and technology can bring sharp changes in supply. New products, promotion, and shifting substitutability can bring major changes in demand. The resulting price and quantitity developments in the market are reflected back from retailers, through manufacturers to farmers.

The Marketing of Cotton

Similar to other agricultural commodities, the seed cotton that is harvested is a raw product with little direct use. It must be further processed into products that industry and consumers will buy.

There are two major raw products obtained from the cotton plant: cottonseed and cotton fiber. Cottonseed is the raw product from which cottonseed oil and cottonseed meal is obtained. The fiber is the raw product for the textile industry. The products of cottonseed compete with other edible oils and protein feeds in the oil and feed markets. The value of cottonseed as a contributor to the value of the farmers' cotton crop fluctuates widely.

Figure 27–2 shows the flow of cotton as it moves from the farms through the various marketing institutions to the consumer. The cotton marketing channel consists of assembly markets, processors, wholesalers, and retailers.

Cotton ginning is the critical first step in producing raw cotton fibers and starting them through the marketing process. At the gin, seed is separated from the fiber, and the task of cleaning out the field trash is done. The cleaned fiber is then packaged into bales of universal density by presses, wrapped, and banded. On the average, about 1800 pounds of raw cotton are required to produce a 480-pound bale of lint.

Because of the bulk and waste involved with seed, cotton gins are located near production areas. The trend has been toward fewer, more efficient, larger capacity gins. Most of the modern high-capacity gins are located in the newer production areas of the West and the Mississippi delta states.

Textile mills need fairly constant fiber supplies throughout the year. However, the cotton harvest and ginning is highly seasonal. Therefore, large amount of storage capacity is needed for this very bulky product. To perform this storage function over 300 cotton warehouses operate throughout the production areas, many fewer than in the days of government storage and purchase programs. About two-thirds of the cotton crop moves from gins to these warehouses while the other third of the crop moves directly from gin to mill or for sale abroad. Cotton grading also occurs at this stage, and government grading is mandatory for cotton moving in interstate and foreign commerce.

After ginning, farmers either sell their cotton directly to private cotton merchant shippers, 25 percent is handled by cooperatives, and the rest moves direct to mills or through other intermediaries. These merchants and cooperative associations assemble the cotton into larger uniform lots and arrange for sale to domestic mills and foreign buyers.

A few cotton trading centers, strategically situated throughout the Cotton Belt, play a key role in cotton marketing. Cities such as Memphis, Dallas, Little Rock, Fresno, Lubbock, and Montgomery have large-scale facilities for storing and merchandising cotton and have concentrations of cotton merchants. Trading is done according to the rules of the local cotton exchange serving these market areas. These "spot markets" serve as focal points for the discovery of cotton prices.

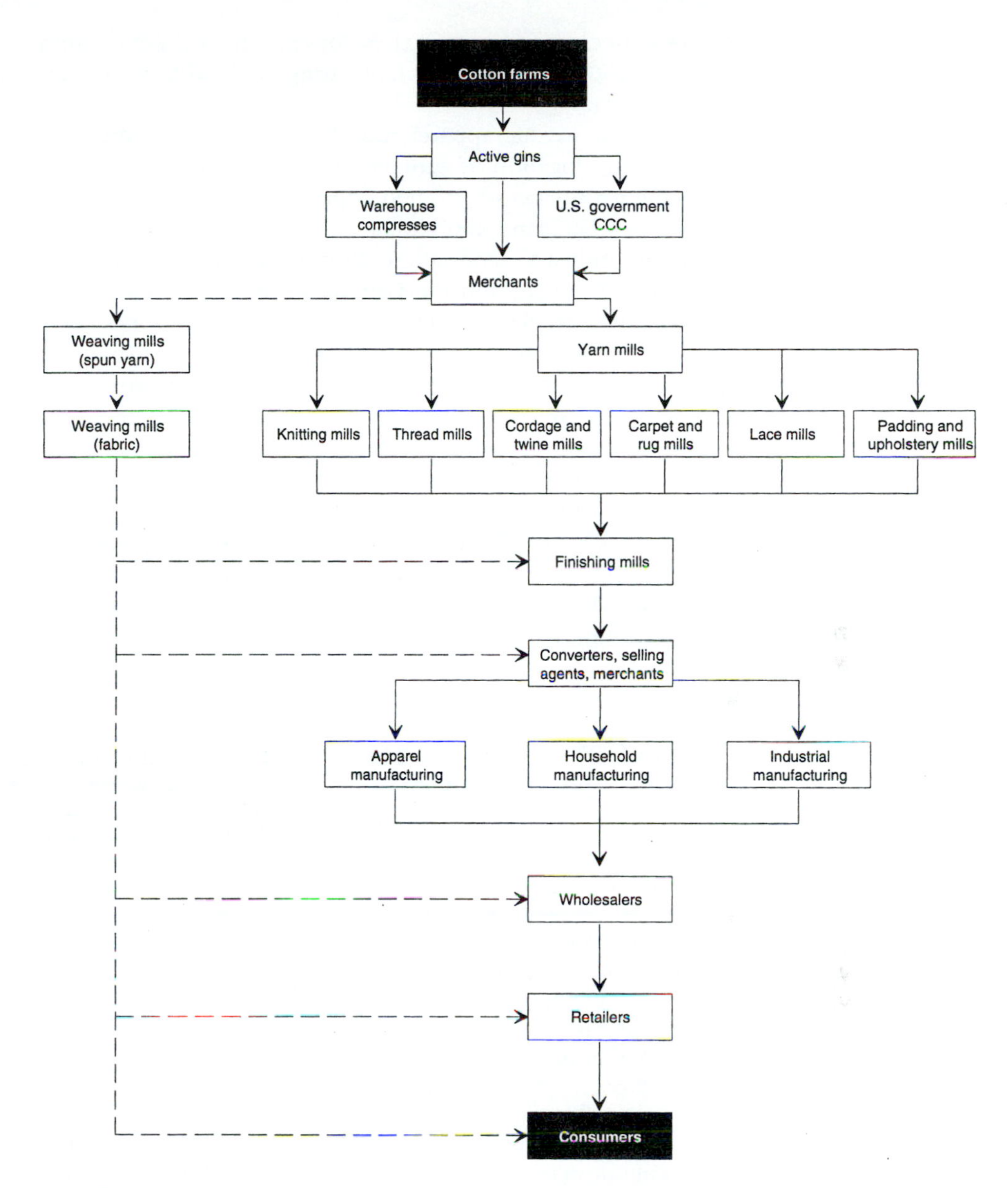

FIGURE 27–2 Cotton industry flowchart. (Economic Research Service, U.S. Department of Agriculture.)

Textile and Clothing Manufacturing

The textile and apparel industries transform the cotton into finished consumer and industrial products. The textile/apparel complex is made up of three types of firms: (1) textile mills that spin yarns and weave cloth; (2) finishing mills that bleach, dye, print, and prepare fabric for apparel makers; and (3) garment cutters and apparel manufacturers who make the final consumer products. Many companies specialize in one of these operations; others combine various phases of spinning, weaving, finishing, fabricating, and even wholesaling and retailing of textile products.

The textile/apparel industry presents an interesting example of how the changing location of agricultural production can influence the marketing channel. Many producers of finished apparel and household textile products are located in the New England and Middle Atlantic states, the historical center of the American textile industry. Domestic cotton mills, on the other hand, are concentrated in Alabama, Georgia, and the Carolinas—the early centers of cotton production. However, over 50 percent of the cotton crop is now produced in the West and the Southwest. Hence, the geographic locations of the textile and apparel industries have not caught up with changing cotton production patterns, and there has been a lengthening of the cotton-textile distribution channel. In recent years a growing share of the Western cotton has been exported to the Far East rather than sold to the eastern U.S. textile mills.

In 1992, there were 4,115 textile milling companies and 3,565 firms producing apparel (Table 27-1). The number of these companies and plants has generally declined over the years with mergers, consolidations, and plant closings. Despite labor-saving technological changes, the textile and apparel industries are major employers in their market areas.

TABLE 27-1 The U.S. Textile and Apparel Industries, 1982-1992

	Number of Companies		*Number of Plants*		*Employment (000)*	
Principal Product	***1982***	***1992***	***1982***	***1992***	***1982***	***1992***
Textile mills:						
Yarn and thread mills	290	261	525	461	86	75
Broad woven fabric mills:						
Cotton	209	281	269	323	79	56
Manmade	342	321	523	422	141	87
Wool	115	87	131	99	13	14
Dyeing, finishing plants	521	317	571	348	46	41
Narrow fabric mills	241	224	281	258	17	17
Knitting mills	2,103	1,801	2,399	2,096	207	193
Textile finishing mills:						
Cotton	266	154	275	168	12	16
Manmade	265	163	296	180	34	25
Others	177	123	182	133	12	10
Floor covering mills	NA	383	505	447	42	49
Apparel manufacturers:						
Mens, boys*	1,982	2,065	2,561	2,590	335	307
Women, children*	1,680	1,500	1,859	1,608	1,564	1,261

SOURCE: *Census of Manufacturing,* U.S. Bureau of the Census.
*1987 data. Comparable 1982 data not available. 1997 census data not available.

Cottonseed Processing

Cottonseed is a bulky product with a low value, so cottonseed oil mills are generally located in cotton-producing areas. Ginners usually purchase cottonseed from farmers and many ginners operate their own processing facilities. Cottonseed is processed into oil, cakes or meal, and hulls, all of which have commercial value.

Cottonseed oil and meal compete with soybeans as sources of edible vegetable oils and animal feeds. The soybean crushing industry has grown more rapidly than the cottonseed-processing industry, but both have benefited from the substitution of vegetable fats for animal fats in the consumers' diet.

COTTON TRADE

Cotton is produced in about 75 countries and is a key U.S. export commodity. The U.S. accounts for 25 percent of world cotton supplies, and more than 50 percent of America's cotton crop is exported. U.S. cotton is sold to Western Europe, Canada, Japan, South Korea, Taiwan, Hong Kong, China, and India. Ports such as Houston, Galveston, Los Angeles, and San Francisco are the major gateways for American cotton exports. China, the United States, India, and Pakistan produce 60 percent of world cotton supplies.

World production of cotton has been expanding more rapidly than the rate of growth in the U.S. production. In 1950, the United States accounted for 50 percent of world cotton production, but only 20 percent in 2000. In the 1950s and early 1960s, high cotton price supports reduced U.S. cotton exports. However, subsequent changes in the cotton price support programs and export subsidies resulted in a rise in the U.S. share of cotton exports.

U.S. imports of foreign textile products have also increased rapidly. More than 80 percent of U.S. apparel imports originate in Asia, particularly Korea, Hong Kong, Taiwan, People's Republic of China, Pakistan, and India. These countries also purchase most of the U.S. cotton exports, so this cotton travels across the Pacific Ocean twice! U.S. imports of textiles and clothing from India, Pakistan, and Egypt have also increased. Thus, U.S. cotton has faced three competitive challenges: competition from domestic as well as foreign manmade fibers and competition from foreign cotton producers and textile manufacturers.

World cotton consumption stagnated in the 1990s because of slower economic growth, the collapse of the Soviet Union's textile industry, growing polyester consumption, and the economic recession which devastated the Asian textile industry. During this time, textile production and apparel exports grew rapidly in the lower-income countries causing losses of American textile factories and jobs. However, American cotton exports to these countries expanded as a result of the North American Free Trade Agreement and the Caribbean Basin Initiative. Thus, American textile and clothing imports contained a larger share of American cotton. Mexico is now the world's largest cotton importer and the U.S.'s largest cotton customer. Turkey has also become an important U.S. cotton customer. U.S. trade in textiles and apparel is regulated by the Multi-Fiber Arrangement (MFA), a set of bilateral treaties among the major textile importing and exporting countries designed to reduce trade barriers.

COTTON MARKETING MARGINS AND PRICES

Because considerable value must be added to raw farm cotton to make it useful to consumers, the farmer receives an average of less than 10 percent of the consumer's cotton dollar. This share varies by products; farmers receive only 6 percent of the retail price of a cotton shirt. Manufacturing and retailing activities take about two-thirds of the cotton dollar. As with food products, labor is the largest marketing cost, representing more than 50 percent of the spread between retail cotton product prices and the farm value of cotton. Overall, marketing adds about 8 to 10 cents per pound to farm prices for domestically sold cotton and 13 to 15 cents per pound for exported cotton.

The U.S. farm price of cotton is influenced by domestic and international trends in demand; cotton production, supply, and stocks; and the federal cotton price programs. The Commodity Credit Corporation supports the farm price of cotton through marketing loan and target price-deficiency payment programs.

The cotton industry's experience with the federal farm price support program illustrates the complexity—and sometimes unintended effects—of government intervention in the farm pricing process. The market price of cotton closely followed the Commodity Credit Corporation's loan rate or support level over the 1948–1966 period. Cotton prices were supported at the 27 to 35 cents per pound level at a time of declining domestic demand for U.S. cotton and falling world cotton prices. The result was that U.S. farmers received higher prices than they otherwise would have, but U.S. cotton and cotton products lost their competitive position in world markets. In order to encourage cotton exports, the U.S. government granted a subsidy to cotton exporters, amounting to the difference between the United States "supported" price and the competitive world price. This, however, put American textile mills, which were paying the higher support price for U.S. cotton, at a competitive disadvantage in world markets. To correct this situation, the American government granted another subsidy, called the *equalization payment,* to domestic cotton users—again, the difference between the United States and the world price of cotton.

In recent years, government intervention in cotton pricing has been reduced. The cotton loan rate was dropped to world price levels, and the export subsidy and equalization payments were no longer necessary. Cotton farmers were given deficiency payments to support their incomes. Cotton prices moved above support levels in 1967 and have generally remained there. At the same time, the U.S. share of world cotton exports has risen, cotton stock levels have fallen, and the CCC is no longer the major market outlet for U.S. cotton.

PROBLEMS IN COTTON MARKETING

The cotton industry has undergone considerable change in recent years at all market levels. New production patterns, improved ginning and textile manufacturing technologies, and altered marketing practices have all influenced the marketing of cotton. Associated with these changes have been adjustments in the market organization and in the competitive environment.

Cotton demand and market development have been major concerns of this industry. Changes in government price support programs affecting the competitive position of cotton in domestic and international markets, research into new cotton products, and cotton promotion have been directed toward improving the cotton demand picture.

Another area of concern is that it is increasingly difficult to discover cotton prices in the spot markets. Even though less and less of the cotton is passing through them, these markets sometimes serve as the basis for pricing cotton throughout the marketing channel. There is also some concern for the effects of cotton contracts between producers and cotton buyers on the open market competitive processes. In response to these concerns an electronic trading system—Telcot—was established in 1975 to assist in cotton price discovery. Computer terminals at hundreds of cotton gins and in buyers' offices permit buyers and sellers to make bid and offer prices on cotton and to negotiate with each other electronically. The New York Cotton Futures Exchange also serves as a price discovery point. This futures market is used both for hedging and as a reference point for cash market sales.

Cotton grades and standards are another concern. U.S. Department of Agriculture cotton grading is provided as a free service to producers through Cotton Classing Offices in the Cotton Belt. More than 97 percent of the cotton crop is USDA-graded. Grades are determined by color, trash content, smoothness, fiber length, and fiber fineness. Cotton grades must be revised periodically to accommodate changing buyer needs and production practices.

SUMMARY

Cotton marketing patterns are changing as a result of shifts in regional production, competition from synthetic fibers and from other nations that produce cotton and cotton products, the growth of contract integration and of cotton cooperative selling activities, and direct mill purchases of cotton. Cotton has regained some of its markets lost to manmade fibers. Cotton exports are growing and are important to the U.S. economy. Demand for U.S. and world cotton has increased significantly with reduced trade barriers, increased world incomes, and a more market-oriented government policy.

KEY TERMS AND CONCEPTS

Multi-Fiber Arrangement
cotton gin
Cotton Incorporated
cotton lint or fiber

DISCUSSION QUESTIONS

1. What were the reasons for the shift of United States cotton production from the Southeast to the West and back? What do you think has happened to the land formerly used to grow cotton in the South?

2. What market development strategies could the cotton industry have used in the past fifty years to prevent declining consumption of cotton?

3. Under what circumstances might it be efficient to ship U.S. cotton to the Far East to be made into clothing for sale to U.S. consumers?

SELECTED REFERENCES

Briefing room: Cotton. (February 2001). Economic Research Service, U.S. Department of Agriculture. http://www.ers.usda.gov/briefing/cotton/.

Cotton and Wool Yearbook (annual). U.S. Department of Agriculture. http://www.ers.usda.gov/publications/so/view.asp?f=field/cws-bby/.

Glade, E. H. Jr., L. A. Meyer, & H. Stults. (July 1996). *The Cotton Industry In the United States.* Agricultural Economic Report No. 739. Economic Research Service, U.S. Department of Agriculture.

Key Topics: Cotton. (February 2001). Economic Research Service, U.S. Department of Agriculture. http://www.ers.usda.gov/topics/view.asp?T=101206.

MacDonald, S. and L. Meyer. (December 2000). World Cotton Market: A Decade of Change, *Agricultural Outlook.* Economic Research Service, U.S. Department of Agriculture.

CHAPTER 28

TOBACCO AND TOBACCO PRODUCT MARKETING

OBJECTIVES

After reading this chapter, you will be able to:

1. Explain the production and marketing patterns for tobacco and tobacco products.
2. Understand the complex controversies surrounding the production and marketing of tobacco products.
3. Analyze the public farm policies influencing the tobacco sector.

Tobacco is a controversial farm commodity. Since its introduction to the civilized world over four hundred years ago, it has been under constant attack. Though tobacco consumption has fallen in recent years, the commodity has shown a remarkable resiliency, despite legislation, taxation and zealots. Here, we will not pass judgment on the health issue connected with tobacco. We will merely review the effect of this controversy on the marketing of the commodity.

There can be no doubt that tobacco is an important farm and consumer product. In 1998 Americans spent $59 billion for tobacco, about 1 percent of their disposable income. Tobacco is the sixth-largest cash crop, just behind corn, soybeans, wheat, hay, and cotton, and it accounts for about 1.5 percent of total U.S. cash farm receipts. There are 90,000 U.S. tobacco farms, and 67,000 of these produce tobacco as their primary commodity. Tobacco is also important to the American economy. Tobacco accounted for 0.3 percent of federal tax receipts in 1998, and state tobacco tax receipts have exceeded federal collections since 1986. Tobacco contributes about $600 million per year to the U.S. balance of trade. Finally, tobacco generates directly or indirectly an estimated 2 million jobs.

TOBACCO PRODUCTION

There are six major classes of tobacco grown in the United States. However, two of these, burley and flue-cured, account for about 90 to 95 percent of total production. These tobaccos are chiefly used in making cigarettes, the largest tobacco product. Cigarettes normally contain 30 to 35 percent U.S. flue-cured tobacco, 25 to 30 percent U.S. burley tobacco, and the rest is imported flue-cured burley and oriental tobacco. Other classes are Maryland (primarily a cigarette tobacco); cigar types (for cigars and chewing tobacco); fire-cured (used for making snuff, plug, twist, cigar, and smoking tobacco); and air-cured (for cigarettes and cigars). In curing, the green tobacco leaves are hung in barns and dried by metal flues, open fire, or by air circulation. In addition to curing the tobacco, the heat produces sugar, which gives the leaves a sweet taste.

Tobacco is grown in 16 states, but the 6 southeastern states of North Carolina, Kentucky, South Carolina, Virginia, Georgia, and Tennessee account for more than 90 percent of all U.S. tobacco production. As shown in Table 28-1, the importance of tobacco has declined in the major producing states, but is still an important farm product in some areas.

Tobacco is a labor-intensive commodity that requires up to 260 man-hours of labor per acre, compared to 3 man-hours per acre for wheat. As a result of this and government limitations on each farmer's production, the average tobacco farm unit is small, averaging 9.3 acres. This means that the marketing system must accommodate small quantities of product from numerous producers. There are specialized tobacco farms, but usually tobacco is grown along with other crops.

As shown in Figure 28-1, annual tobacco production fluctuated with acreage over the 1963-1998 period, then grew more rapidly in the 1990s. The growth in

TABLE 28-1 Importance of Tobacco Farming, by States, 1966-1999

	Tobacco's Percentage of Farm Cash Receipts			
State	*1966–1970*	*1981–1985*	*1990–1992*	*1997–1999*
North Carolina	38	28	21	13
Kentucky	35	29	26	25
South Carolina	23	19	16	12
Virginia	16	13	9	8
Tennessee	13	13	10	10
Georgia	8	5	4	3
U.S.	2.8	2.1	1.7	1.4

SOURCE: U.S. Department of Agriculture.

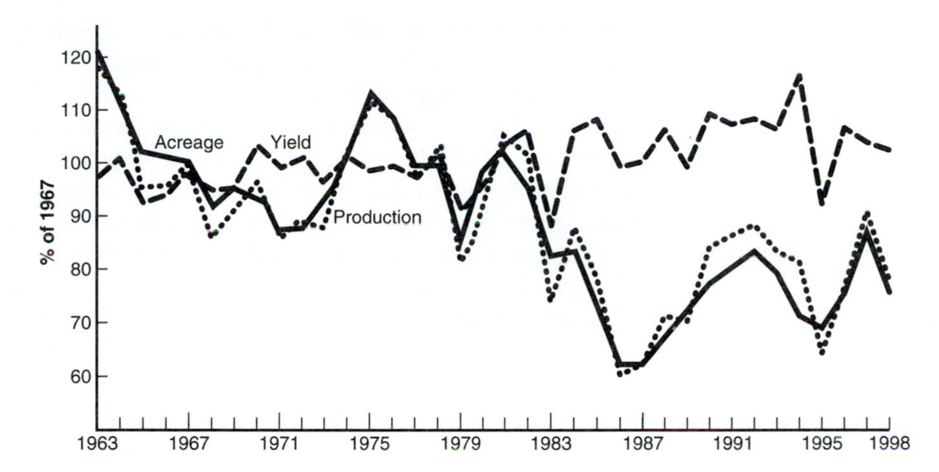

FIGURE 28–1 Tobacco acreage, yield, and production, 1963–1998. (U.S. Department of Agriculture.)

output was due primarily to increased exports of tobacco. Yields did not change appreciably over this period.

TOBACCO CONSUMPTION

The United States is a major producer and consumer of tobacco in the world, whether measured in total quantity, value, or in per capita terms. With 6 percent of the world's population, the United States consumes nearly one-sixth of the world's tobacco. The relative importance of various tobacco products has changed over the years. In the 1920s, cigarettes accounted for only 23 percent of tobacco use; pipe tobacco, cigars, and snuff accounted for the remainder. Cigarettes now account for 94 percent of domestic tobacco use.

Tobacco has a unique demand situation. Because its use is apparently governed by habit, and there are few close substitutes, the price elasticity of demand for smokers is quite low ("essential" products normally have inelastic demands). But for the nonsmoking population there is *no demand* for tobacco, presumably at any price. Indeed, the foes of tobacco believe the demand is unwholesome and excessive. There is also a latent demand for tobacco products that are perceived to be less harmful to health. This is suggested by the growth of low-tar tobacco products in the 1970s and the popularity of smokeless tobacco products (snuff and chewing tobacco) in the 1980s.

Because many believe tobacco is an unwholesome and unhealthy product, tobacco has received more adverse publicity than any other agricultural product. This has included the 1964 U.S. Surgeon General's Report on Smoking and Health, which discussed the carcinogenic properties of tobacco, a 1966 law requiring a warning label on cigarette packages and advertisements, a change in the wording of this warning in 1970 from "may be hazardous to your health" to "is dangerous to your health," a 1971 ban on radio and TV cigarette advertising, and extensive antismoking campaigns conducted by health organizations and the federal government. A 1972 U.S. Surgeon General's Report broadened the attack on cigarette smoking by suggesting that high concentrations of carbon monoxide in smoke-filled rooms and autos pose a health hazard to smokers and nonsmokers alike. Beginning with Arizona in 1973, most states have enacted laws banning smoking in certain public places. In 1988, the U.S. Surgeon General labeled tobacco an addictive drug. Many businesses have also restricted smoking by workers and patrons.

The tobacco industry continues to be under investigation and attack. In 1997, the industry agreed to an unprecedented $368 billion settlement with all 50 states providing funds for health-related damages, removing outdoor smoking advertisements, and retiring Joe Camel. In 2000, a President's Commission On Improving Economic Opportunities in Communities Dependent on Tobacco Production While Protecting Public Health was appointed. There are additional bills in the Congress and court cases which tobacco industry executives claim would bankrupt the industry.

How has this affected the demand for tobacco products? Many people have quit smoking or smoke less frequently. As shown in Figure 28–2, per capita consumption of cigarettes peaked in 1963 and was at one-half that level in 2000, years of intense adverse publicity. Not all of the decline in tobacco usage can be attributed to the health issue. Tobacco taxes and prices rose significantly, moving smokers along the demand curve. Consumer expenditures for tobacco products increased 30 percent in the 1990s as higher prices offset lower usage. The proportion of consumers' disposable income spent on tobacco products held constant at .04 percent during the 1990s. Tobacco manufacturers have also increased the efficiency of tobacco use, resulting in less tobacco waste and less tobacco used per cigarette.

The tobacco industry has attempted to develop "safer" tobacco products over the years. Filter-tip cigarettes are an example, as is the low-tar cigarette, which accounted for 55 percent of all cigarettes sold in 1987. In the mid-1980s smokeless

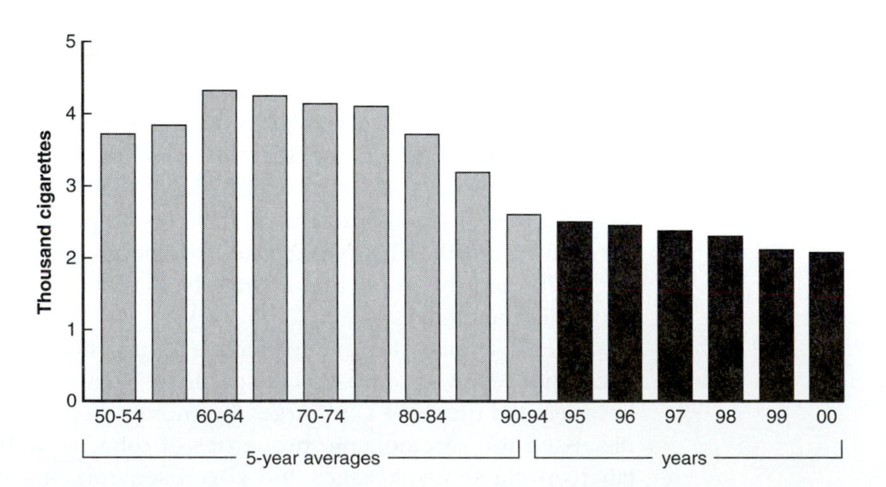

FIGURE 28–2 Per capita cigarette consumption, 1950–2000, 18 years or older. (U.S. Department of Agriculture.)

tobacco (chewing tobacco and snuff) sales increased. There are periodic reports that the industry is developing a nontobacco cigarette.

MARKETING CHANNELS AND METHODS

The tobacco distribution channels are shown in Figure 28–3. In 2000, this channel consisted of about 50 auction houses, 9 tobacco manufacting firms, 2000 tobacco wholesalers, hundreds of thousands of retail tobacco outlets, and millions of vending machines.

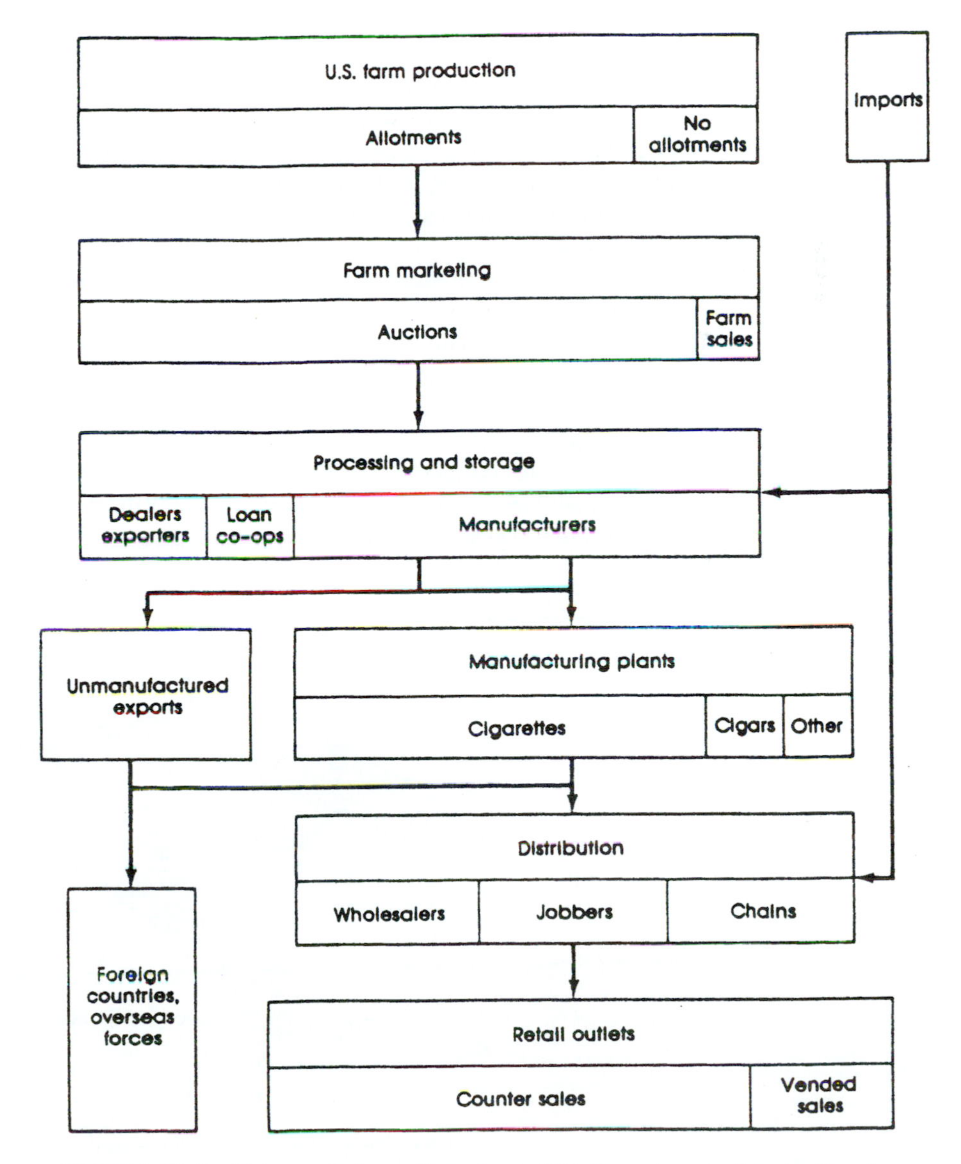

FIGURE 28–3 Tobacco industry flow chart. (U.S. Department of Agriculture.)

Farm Tobacco Markets

Most U.S. tobacco is auctioned at sales warehouses scattered throughout the tobacco belt. These major production and marketing areas are shown in Figure 28–4. These auction markets have changed little since the first one was organized in Danville, Virginia, in 1863. The sales warehouses are one-story buildings, usually well lighted with skylights, and sometimes have as much as 100,000 square feet or more of unobstructed floor space. Sales warehouses are located in the major production areas so that most of the tobacco to be sold is usually brought in by the farmers themselves. The tobacco is weighed and tagged and displayed on the floor in round piles with the stem end out. Before the sale the tobacco is graded by a federal grader.

Sales are conducted according to a prearranged schedule that permits buyers to move among the several warehouses in a given market area. Small markets with only two or three sales warehouses may have only one group of buyers, whereas a larger market may have several sets of buyers. Buyers include representatives of tobacco manufacturing and exporting companies and speculators. The sale is conducted as the auctioneer moves about the various piles of tobacco on display in the warehouse.

To the uninitiated observer sales are a scene of utter confusion. One can scarcely tell auctioneer from buyer from warehouseperson, and it is almost impossible to determine who bought which pile of tobacco for what price. The owner may "no sale" any basket and place it elsewhere in the warehouse, remove it to another warehouse, hold it for later sale, or take out a nonrecourse loan from the Commodity Credit Corporation.

The sales dates vary with production areas. The Georgia–Florida markets open first, usually in mid-July, and continue through most of August. Markets in South Carolina open a week or so later and continue through October. North Carolina markets

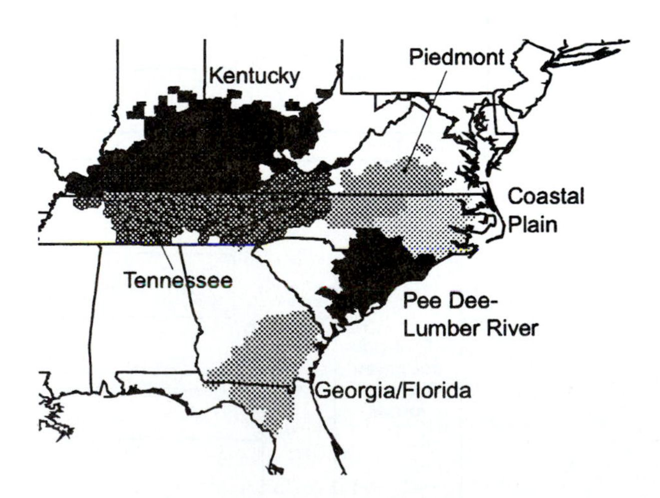

FIGURE 28–4 Tobacco markets and producing areas. (U.S. Department of Agriculture.)

follow and conduct sales through November. Burley and fire-cured markets operate from December to February or March. Southern Maryland markets begin sales in March and continue through early April. Thus, some type of tobacco is being sold for ten months of every year.

Cooperative associations do not make up a separate type of market organization, but function largely as farmers' representatives in conjunction with the auction system. Tobacco cooperatives have a long and turbulent history. By 1923 almost half of the nation's tobacco was marketed through various cooperative associations. But the movement declined, and by 1930 only about 2 percent of the production was marketed by cooperative organizations. At the present time *cooperative stabilization corporations* perform the storage functions in connection with the government loan programs, which are the mechanism of price supports. If the bid price falls below the government support price, the tobacco goes to the cooperative and the farmer receives the government loan price. This loan must be repaid by the farmer when the cooperative finally sells the tobacco.

Tobacco, as delivered to the market by farmers, is in a semiperishable condition because of high moisture content. Except for certain cigar types, aging and fermentation occur after leaving the farm. Sometimes a period of two or three years is necessary for the complete fermentation process to take place. This situation means that much of the current available supply of tobacco is carried over from one season to the next in storage. Prior to storage, tobacco is cleaned, stemmed, reclassified, and redried to obtain the moisture necessary for fermentation. The weight loss from cleaning, redrying, aging, and stemming is such that the manufacturer's net yield is about two thirds of the farm sales weight.

Tobacco Product Manufacturing

Tobacco product manufacturing is subject to substantial economies of scale and is a highly concentrated industry. The four largest tobacco companies account for about 90 percent of total cigarette sales. The number of plants making cigarettes declined from 19 in 1958 to 9 in 1997, but these plants accounted for 90 percent of the total value added by all tobacco manufacturing plants. Cigarette plants are concentrated in North Carolina, Virginia, and Kentucky. Most of the cigar plants are located in Pennsylvania and Florida.

Cigarette manufacturing is an excellent example of oligopoly—a few firms produce and sell nearly all the U.S. cigarette output. Competition tends to focus on product differentiation, such as length, product image, filters, and the like. Mass-media advertising is used to build brand perferences. Cigarette manufacturers are among the largest advertisers of consumer products. A few cigarette brands dominate. The major U.S. brand, for example, has about a 25 percent market share. In the early 1990s, the major tobacco companies engaged in a price war and introduced discount cigarette brands.

The price differential of cigarettes sold in the United States and those exported has presented an interesting arbitrage opportunity for independent traders. Because American cigarettes sold for export are priced so low, it is possible to reimport them into the country and still make a profit even after paying import duties and excise taxes. In 1999, perhaps 1 percent of all cigarettes sold in the United States passed through this gray market.

Reconstituted tobacco sheets represent an example of efficient technological change in the tobacco industry. These sheets use tobacco stems, scrap, and dust that were formerly discarded. Since the mid-1950s, reconstituted tobacco sheets have replaced most of the natural binder in cigars, and have also permitted total cigarette production to rise although the amount of domestic leaf used in cigarettes has fallen. A 1993 federal law requires U.S. tobacco manufactureres to use at least 75 percent American-grown tobacco in their products. In 1999, American cigarettes contained a record 48 percent foreign leaf.

TOBACCO TRADE

The United States is a leading producer of tobacco, one of the largest exporters of tobacco, and one of the largest tobacco importers. America accounts for about 10 percent of world tobacco production and 33 percent of world tobacco trade. Other large producers are the People's Republic of China (35 percent of world production) and India (8 percent of world production). Tobacco production is increasing rapidly in Brazil, Malawi, and Zimbabwe. In recent years, about 33 percent of the U.S. tobacco crop has been exported with cigarettes accounting for most of these exports. About 35 percent of the tobacco used in the United States is imported.

U.S. tobacco exports have generally been larger than imports, making a positive contribution to the U.S. balance of trade. However, U.S. tobacco imports are growing and more foreign tobacco is being used in making domestic cigarettes. Japan and the countries of the European Union are leading buyers of American tobacco. U.S. tobacco imports come primarily from Turkey, Greece, Brazil, Malawi, and Zimbabwe.

SUPPLY CONTROL AND PRICE SUPPORT PROGRAMS

Federal tobacco price support programs have existed since the early 1930s. These programs have been designed to maintain tobacco supplies in line with demand, to promote the orderly marketing of tobacco, and to support and stabilize farm prices. The principal components of the program are producer supply controls and a nonrecourse loan program. The two are related: Tobacco growers agree to restrict marketings in return for guaranteed minimum prices of tobacco.

In the original 1933 supply control program, each tobacco grower was allotted a share of the national acreage, which would result in a reasonable balance of supply and demand. Predictably, these acreage allotments encouraged farmers to boost yields, and tobacco yields increased rapidly. As a result, a poundage allotment and marketing quota were added to the acreage allotment for burley and flue-cured tobacco in 1938. This slowed yield increases, forced farmers' attentions on quality rather than quantity, and provided more effective tobacco supply control.

The tobacco poundage marketing quotas are a much stronger form of supply control than the acreage allotments or set-asides used on other crops. Peanuts are the only other crop with marketing quotas. Each year the Secretary of the Department of Agriculture determines the national tobacco needs based on intended purchases by cigarette manufacturers, recent imports, and desired tobacco stock levels. Each eligible producer is then given a poundage quota. There is a heavy penalty for growers who exceed their market allotment.

Under the terms of the CCC nonrecourse tobacco loan program, the Secretary of Agriculture also establishes a support price for each grade of tobacco. As shown in Figure 28-5, this support price has placed a floor under tobacco market prices, which are usually higher than the support price. If the auction bid is below the support price, the tobacco can be delivered to the local cooperative association, which pays the farmer the loan support price. The cooperative is authorized by CCC to purchase, handle, and eventually market the tobacco. In most years, the cooperative has been able to market the tobacco at a price sufficient to repay the CCC loan rate; therefore, the program has been self-financing. The No-Net-Cost Tobacco Program of 1982 requires tobacco growers, as well as buyers, to bear the costs of operating these stabilization cooperatives.

These government and industry supply-control and price support programs have contributed to the profitability of tobacco production. Between 1971 and 1998 tobacco price supports more than doubled, but these supply restrictions kept grower prices above support levels in most years. As a result, the government tobacco program has been largely financed by consumers rather than taxpayers. The effects of tobacco policies, then, are to enhance the incomes of growers with allotments by restricting production through government controls.

The tobacco allotment and price support programs have frequently been criticized. Allotments of any kind on a historical basis tend to freeze production patterns, limit entry and freedoms, and provide profits to allotment holders. Tobacco land prices are more affected by the land's allotment than its fertility and productivity. There can be no guarantee that the allotments will be held by the most efficient producers, although they are transferable and are sold and rented. There is also concern that the price support program has kept U.S. tobacco prices above world levels and thus eroded the competitive position of American tobacco in world markets.

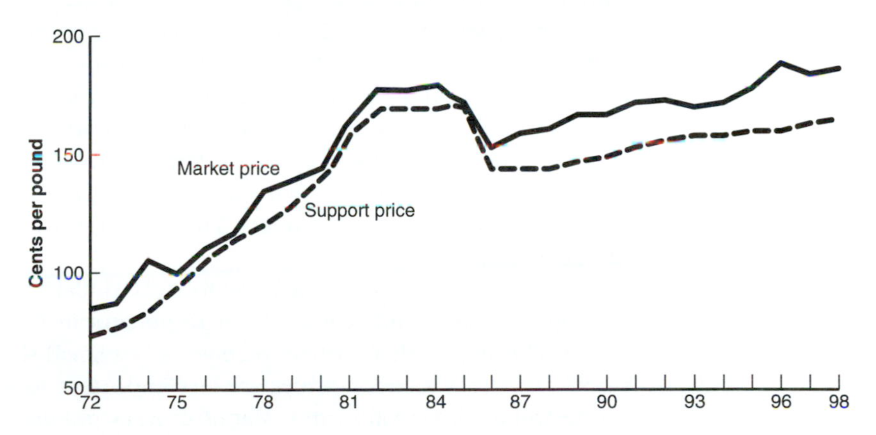

FIGURE 28–5 Flue-cured price and support levels, 1972–1998. (U.S. Department of Agriculture.)

The tobacco program, however, has proved to be popular with growers. Except in 1939, they have overwhelmingly voted for the allotments in triennial referendums, often approving them by 90 percent of all growers. Apparently tobacco growers consider the sacrifice of production freedoms to be a small price to pay for the price supports and supply restrictions.

TOBACCO MARKETING PROBLEMS

The public-health issue overshadows all other tobacco marketing problems. The possibility of legislation that would further restrict tobacco use, and changes in consumer attitudes toward tobacco, create considerable uncertainty for the industry. It appears that, rightly or wrongly, demand for tobacco would continue reasonably strong if consumers are given a free choice, but there is no guarantee of this freedom in the case of tobacco.

The very substantial differences in the sizes and numbers of tobacco growers and tobacco product manufacturers is another source of industry concern. The large numbers of relatively small tobacco producers selling in open auction markets represents a situation quite close to perfect competition. Tobacco manufacturing, however, is among the most concentrated, oligopolistic industries. The resulting imbalance of market power has consequences for tobacco prices and other terms of trade. The federal government frequently has applied the antitrust laws to the tobacco industry. The most famous case was the 1911 dissolution of the American Tobacco Company (which had accounted for 90 percent of industry sales) into four firms: American Tobacco Co., R. J. Reynolds Tobacco Co., Liggett and Myers, and P. Lorillard Co. Tobacco manufacturers have diversified their operation in recent years into food processing and other areas.

The large number of tobacco sellers in relation to buyers presents problems concerning the fair allocation of sales opportunities among growers and tobacco-producing areas. There have been charges of preferential treatment of some growers. In the 1960s, many growers engaged in extensive tobacco shipping when they were unable to sell their tobacco at local markets as rapidly as they would have liked. As a result, in 1974 the industry adopted a program whereby each grower designates a sales warehouse, within a 100-mile radius of his or her county, where his or her tobacco will be sold. The Flue-Cured-Tobacco Advisory Committee then sets opening dates, sales schedules, and sales opportunities for each market area. This program has reduced cross-state tobacco shipment costs, provided more equitable treatment for all producers, and generally resulted in more orderly tobacco marketing.

There is also concern that tobacco manufacturers will vertically integrate into production using grower contracts. This would substantially alter the tobacco auction market system.

The farmer's share of the retail tobacco dollar is quite low compared with other agricultural commodities. It fell from 16 percent in 1950 to 2 percent in 1998 (Figure 28-6). This means that farmers receive only a small share of increased retail tobacco prices, and farm tobacco prices are less important to retail prices than the manufacturing and marketing activities. Manufacturers and distributors account for two-thirds of the retail tobacco dollar. A large share of the tobacco dollar goes for taxes. Federal, state, and local excise taxes account for about 25 percent of the tobacco dollar.

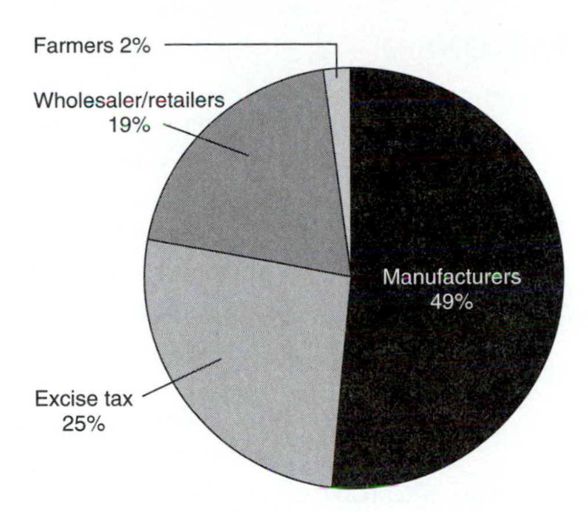

FIGURE 28–6 Components of the Cigarette User's Dollar, 1998. (U.S. Department of Agriculture.)

Tobacco quality continues to be an industry problem. Currently, there are 152 grades of flue-cured tobacco and 112 grades of burley. Weather conditions cause wide variations in the proportion of each year's crop falling into the various grade categories. In recent years, loan rates have been adjusted to give above-average prices for the tobacco grades that are in strongest demand. Mechanical harvesting of tobacco—and less leaf handling by farm workers—has increased the amount of foreign matter on tobacco, thereby reducing its value to manufacturers. As a result, in 1977 tobacco grades were revised, imposing more stringent regulations regarding the presence of foreign matter.

SUMMARY

The tobacco industry is facing a serious demand problem because of the public health issues regarding smoking. The tobacco issue is complicated by the large number of producers involved; the strategic role tobacco plays in the economy of some states; the contribution tobacco taxes make to federal, state, and local treasuries; and the critical position of tobacco in the U.S. trade picture. Tobacco is marketed at open auctions in the tobacco belt. Almost perfectly competitive producers face strongly oligopolistic buyer-manufacturers in these markets. Tobacco prices are supported by a producer acreage and marketing allotment program and a nonrecourse loan price support program. Tobacco stabilization cooperatives and industry advisory committees play strategic roles in tobacco marketing.

KEY TERMS AND CONCEPTS

cooperative stabilization corporations
tobacco allotment
tobacco class
tobacco curing
tobacco quota

DISCUSSION QUESTIONS

1. How does tobacco consumption differ from the consumption of other agricultural commodities? What is a *habit?* And how do *habits* differ from *preferences?*

2. How should the public weigh the economic benefits of the tobacco industry (taxes, trade, farm income, etc.) against the "costs" of smoking? How do the notions of "risk" and "consumer sovereignty" enter into this policy question?

3. Why is tobacco produced on such small farms? How does this affect farm marketing?

4. What unique role do cooperatives play in tobacco marketing?

5. Why do you think there is little integration of tobacco product manufacturers into tobacco production? Why have these firms diversified in recent years?

6. How could tobacco farmers increase their share of the retail tobacco dollar?

SELECTED REFERENCES

Capehart, T. (December 1999). The Changing Tobacco User's Dollar. *Tobacco Situation and Outlook.* TBS-245. Economic Research Service, U.S. Department of Agriculture.

Capehart, T., Jr. (January–February 2001). Cigarette Consumption Continues to Slip. *Agricultural Outlook.* U.S. Department of Agriculture.

Dimitri, C., & E. Jaenicke. (December 1999). Contracting in Tobacco? Tobacco Situation and Outlook. TBS-245. Economic Research Service, U.S. Department of Agriculture.

Gale, H. F., Jr., L. Foreman, & T. Capehart, Jr. (September 2000). *Tobacco and the Economy: Farms, Jobs, and Communities.* Agricultural Economic Report No. 789. Economic Research Service. U.S. Department of Agriculture.

Grise, V. N., & K. F. Griffin. (September 1988). *The U.S. Tobacco Industry,* Agricultural Economic Report No. 589. Washington, DC: U.S. Department of Agriculture.

Johnson, P. R. (1984). *The Economics of the Tobacco Industry.* New York: Praeger Publishers.

Tobacco: Briefing Room. (February 2001). Economic Research Service, U.S. Department of Agriculture. http://www.ers.usda.gov/briefing/tobacco/.

Tobacco: Situation and Outlook, Quarterly and Yearbook. (December 2000). Economic Research Service. U.S. Department of Agriculture.

Tobacco: Topics. (February 2001). Economic Research Service, U.S. Department of Agriculture. http://www.ers.usda.gov/Topics/view.asp?T=101226.

U.S. Department of Agriculture. (Quarterly). *Tobacco Situation and Outlook Report.* Washington, DC: U.S. Department of Agriculture, Economic Research Service.

CHAPTER 29

FRUIT AND VEGETABLE MARKETING

Sell it or smell it.

Motto in the fruit and vegetable industry indicating the importance of rapid marketing for such a perishable product.

OBJECTIVES

After reading this chapter, you will be able to:

1. Better understand how the geographic concentration of production and the long marketing channels have influenced the development of the fruit and vegetable marketing system.
2. Appreciate how changing food consumer preferences and industry promotional programs have altered the demand for various fruits and vegetables and how they are marketed.
3. Contrast the marketing of fresh fruits and vegetables with their processed products.

The fruit and vegetable industry consists of a wide array of crops and products, each with very different supply conditions, marketing needs, and demand trends. While these products share common marketing channels and experience somewhat similar trends and problems, the market uniquenesses of individual fruits, nuts, and vegetables—apples, artichokes, grapes, almonds, potatoes, and others—should be kept in mind. In particular, fresh fruits and vegetables are marketed quite differently from the processed products. This marketing system has changed substantially in recent years. The major trends involve decentralization and direct marketing, geographic concentration and specialization of production, market development, interregional competition, industry consolidation, increased imports and exports, and vertical integration of production and marketing in shipping point markets.

FRUIT AND VEGETABLE PRODUCTION

Fruits and vegetables are important farm products. The farm value of these commodities amounted to $28 billion in 1999. This represented 15 percent of total farm sales and 30 percent of total crop sales. There were 160,000 fruit and vegetable farms in 1997. The 20 million acres devoted to fruit and vegetable production was about equally divided between fruits and vegetables.

While there are hundreds of commercially produced fruits and vegetables, a few major crops dominate in this industry. Four vegetable crops (potatoes, lettuce, tomatoes, and sweet corn) and three fruit crops (oranges, grapes, and apples) accounted for more than 45 percent of the farm value of fruits and vegetables in 1997 (Table 29-1).

Although there are many small fruit and vegetable growers, there are also several large farm operations in this industry, particularly in the specialized producing areas. Some of these large operations are owner-operated farms, but many are also owned by nonfarm corporations, including food processors, marketing firms, and conglomerate corporations.

New fruit and vegetable technologies are influencing the marketing of these crops. The trend toward mechanical harvesting has altered the location of fruit and

TABLE 29-1 Major U.S. Fruit and Vegetable Crops, 1997

	Farm Value	
Product	*($ million)*	*Percent of Total*
Vegetables		
Potatoes	2,622	17
Lettuce	1,744	12
Tomatoes	1,645	11
Onions	929	5
Sweet corn	669	4
Broccoli	481	3
Carrots	497	3
All others	7,195	45
All vegetables	15,086	100
Fruits and nuts		
Grapes	3,122	25
Oranges	1,834	14
Apples	1,575	12
Almonds	1,161	9
Strawberries	903	7
Peaches	444	3
Grapefruit	283	2
All others	3,468	28
All fruits, nuts	12,790	100

SOURCE: Agricultural Statistics, 1997, U.S. Department of Agriculture.

vegetable production and in some cases the quality of these products. Vegetables can now be produced in nutrient-water mediums in greenhouses (hydroponics). Plant growth stimulants and retardants can be used to influence yields and timing of crops. And controlled-atmospheric storage has extended the season and keeping quality of produce.

FRUIT AND VEGETABLE CONSUMPTION

Americans spent $71 billion at retail for fruits and vegetables in 1999. Fruits and vegetables account for about 18 percent of consumers' at-home food budget and represent 10 percent of foodstore sales. Supermarket produce departments carry over 300 different items today. In 1997 fresh fruit and vegetable sales to the foodservice (away from home) market exceeded sales to grocery stores for the first time.

After declining in the first half of this century, per capita consumption of fruits, vegetables, and potatoes climbed in the 1960s and 1970s, and increased at an accelerating rate in the 1980s and 1990s. Most of the recent increase in consumption of these products has been due to rising demand for fresh, frozen, and juice forms (Figures 29-1 and 29-2). Per capita consumption of fresh fruits and vegetables increased 12 percent from 1987 to 1997. Per capita consumption of canned fruits and vegetables has continued its long-term decline. These changes in consumer preferences reflect higher incomes; a renewed interest in diet, health, fitness, and natural foods; an aging population that eats more fruits and vegetables; a growing immigrant and ethnic population whose diet contains more fruits and vegetables; and perhaps the success of the industry's "5-a-Day" promotional program. The growing ethnic population has also

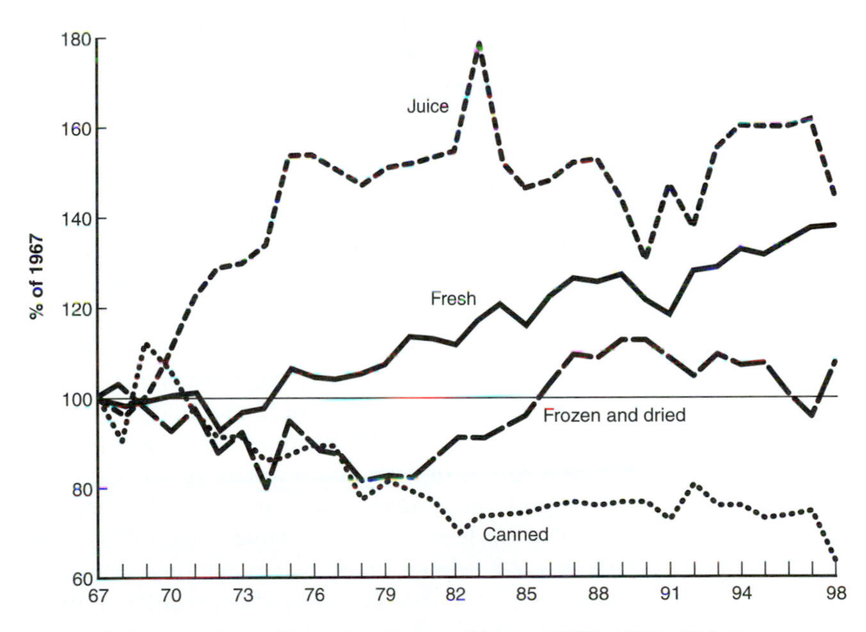

FIGURE 29–1 Per capita consumption of fruits, 1967–1998. (U.S. Department of Agriculture.)

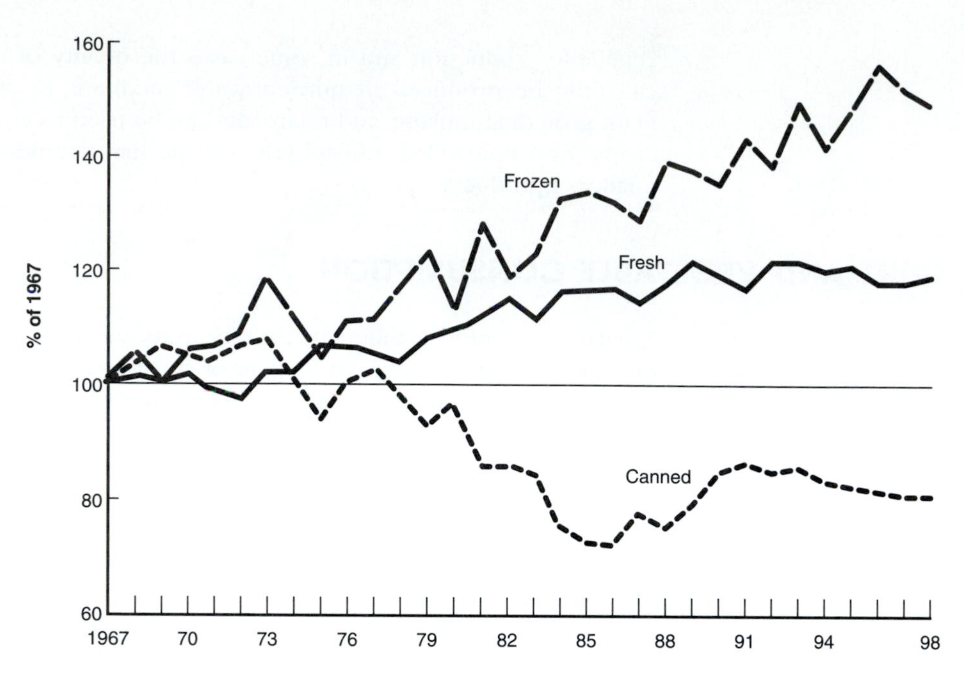

FIGURE 29–2 Per capita consumption of vegetables, 1967–1998. (U.S. Department of Agriculture.)

increased the demand for specialty vegetables, such as casaba, Santa Claus melons, Chinese kale, French shallots, and shiitake mushrooms.

Market development and promotional efforts have significantly affected the demand for some produce items. For example, the introduction of fresh-cut baby carrots increased carrot consumption from 8 pounds per capita in 1987 to 14.5 pounds in 1997. Per capita consumption of melons doubled from 1980 to 1998 as a result of quality improvements, including seedless hybrids, industry promotional programs, increased sales in restaurant and grocery store salad bars, and the further development of pre-cut, prepackaged products. Many firms have also developed strong branding programs with premium prices for fresh fruits and vegetables.

There have been significant recent changes in consumer preferences for individual fruits and vegetables. As shown in Table 29-2, consumption of certain vegetables—tomatoes, broccoli, and cauliflower—have been the recent growth stars of the fresh produce department. Conversely, consumption of such traditional items as fresh oranges, lettuce, celery, and sweet corn has faltered. In other cases, consumption of the processed forms of a commodity are growing while fresh consumption declines. Figure 29-3, for example, illustrates the growth in processed potato products as a source of demand for potato producers. The same thing has happened with orange juice and processed tomato products such as tomato sauce, paste, and puree.

The fresh fruit and vegetable "desired product" bundle of attributes is quite complex. Consumers emphasize nutritional values, appearance, firmness, color, convenience, and freshness as desirable qualities of these products. The assortment and quality of the produce department is also very important in consumers' food store selections. Some shoppers judge the quality of fresh produce by its price. Higher

TABLE 29-2 Changing per Capita Consumption of the Major Fresh Fruits and Vegetables, 1970-1997

	Per Capita Consumption (pounds)				*Percentage Change*
	1970	*1980*	*1990*	*1997*	*1980–1997*
Fresh Fruits					
Grapes	2.5	3.3	7.9	8.9	+170
Strawberries	1.6	1.9	3.2	4.4	+131
Pineapples	.7	1.4	2.1	1.9	+36
Pears	1.9	2.3	3.2	3.1	+35
Bananas	17.6	20.8	24.4	25.0	+35
Apples	16.3	18.3	19.6	19.3	+5
Plums and prunes	1.4	1.5	1.5	1.4	−7
Oranges	15.7	15.3	12.4	12.8	−16
Grapefruit	8.0	7.8	4.4	5.8	−26
Nectarines, peaches	6.2	7.1	5.5	4.3	−40
Fresh Vegetables					
Broccoli	.5	1.5	3.4	5.2	+347
Green peppers	2.0	2.7	4.5	7.2	+167
Carrots	5.7	6.0	8.3	12.5	+108
Onions	11.5	9.3	15.1	17.9	+92
Cucumbers	2.6	3.6	4.7	6.3	+75
Tomatoes	10.5	11.4	15.5	18.9	+66
Cauliflower	.6	1.0	2.2	1.6	+60
Cabbage	10.6	7.5	8.8	10.2	+36
Corn	7.3	6.6	6.7	8.1	+23
Fresh Potatoes	59.3	49.0	47.0	48.9	0
Green beans	1.6	1.4	1.1	1.4	0
Lettuce	20.8	24.9	27.8	24.3	−2
Celery	6.7	7.3	7.2	6.0	−18

SOURCE: U.S. Department of Agriculture.

prices are associated with higher quality and vice versa. In fact, the opposite is often true. When supplies are short and prices high, more of the marginal quality crop will be harvested and sent to market than when prices are lower.

The organic produce industry has grown rapidly in recent years, though it is still a small share of total output. In 2000, the U.S. Department of Agriculture issued labelling standards for organic foods. Many foodstores now handle these products, and they are still sold through alternative outlets, such as farmers' markets and natural food stores. Led by California's Certified Organic Farmers Program, some sixteen states have developed certification programs for organically grown crops. Although many consumers report a willingness to pay a higher price for organic produce, the typical price premiums of 30 to 100 percent sometimes encounter consumer price resistance.

As with other foods, inefficient fruit and vegetable marketing techniques lead to the marketing of lower quality and higher priced products. Inadequate tempera-

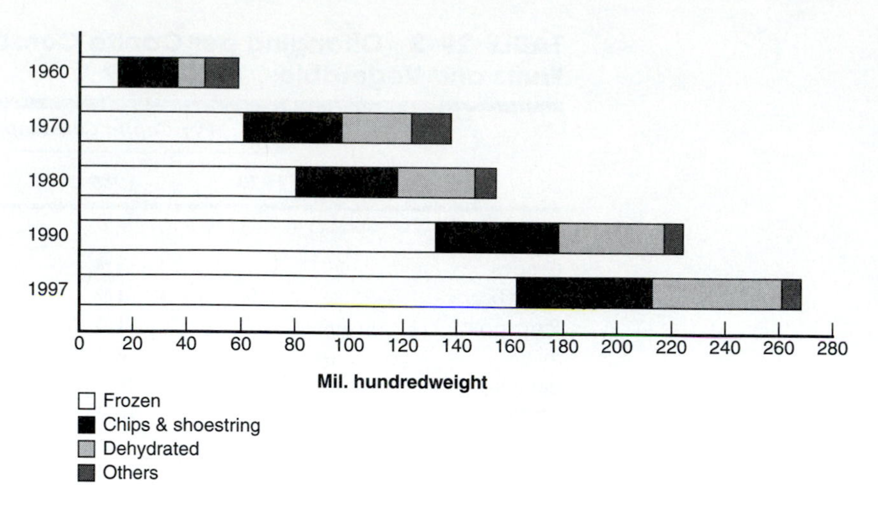

FIGURE 29–3 Raw potatoes used for processed potato products, 1960–1997.

ture and humidity conditions can result in loss of vitamins in fruits and vegetables. The more complex distribution channels may encourage the substitution of produce that is more resistant to the harsh treatment of long-distance shipments for less hardy, but perhaps more nutritious and flavorful, varieties. For example, romaine, red, and butter lettuce are less tolerant of modern marketing techniques than iceberg lettuce, the dominant crop today, but they contain five times the vitamins.

FACTORS INFLUENCING FRUIT AND VEGETABLE MARKETING

The fruit and vegetable marketing system has been influenced by a number of production, product, and market characteristics: (1) perishability, (2) large price and quantity variations, (3) seasonality, (4) alternative product forms, (5) bulkiness of product, and (6) geographic specialization of production. An examination of these will illustrate how a food marketing system evolves in response to unique product and industry characteristics.

Perishability

Fruits and vegetables are highly perishable commodities. Their quality begins to deteriorate from the moment of harvest and continues throughout the marketing process. There is an urgency to processing and marketing these products as quickly and efficiently as possible to maintain their farm-fresh value. The entire distribution process is geared toward rapid marketing, and this affects every phase of fruit and vegetable marketing.

It is estimated that up to 10 percent of the value of fresh fruits and vegetables is lost in the marketing process. This is due to improper storage and handling, spoilage, trimming to improve appearance, careless handling by shoppers, and theft. Doubtless, further losses occur in home storage and preparation of these products. It has been said that the garbage can and disposal are among the largest "markets" for fresh produce.

Elaborate and expensive marketing channels, facilities, and equipment are necessary for maintaining and enhancing the quality of fresh fruits and vegetables during marketing. This includes a wide array of marketing firms, facilities, and equipment. Often, expensive but more rapid transportation will be selected instead of slower but cheaper forms. California strawberries, for example, are flown to East Coast markets to preserve their quality. Trucks have largely replaced rail transport because of their greater speed and flexibility in marketing highly perishable products.

Produce perishability also affects price negotiations. These products cannot be held for long periods while sellers wait for or attempt to discover a better price. A common saying in the trade is, "Sell it or smell it." Fresh produce from California, Texas, and Florida is often sent to distant markets without a firm buyer or price. Prices for these "rollers" are negotiated while the commodities are en route, and they are frequently diverted from their original destination if a better price can be found. Sellers may have little market power in determining a price for a carlot of lettuce that is heading east without a buyer and rapidly deteriorating.

As a result, a great deal of trust and informal agreements are involved in marketing fresh fruits and vegetables. There is not always time to write everything down and negotiate the fine details of a trade. A grower with a surplus of carrots might send them to a New York agent with directions to "get what you can before these spoil" and not know the price until the carrots have been sold.

This urgent, informal marketing process often leads to disputes between buyers and sellers of fresh fruits and vegetables. The Perishable Agricultural Commodities Act (1930) prohibits a number of unfair practices of produce commission merchants, dealers, and brokers, licenses these firms, and provides for arbitration of disputes.

Price/Quantity Risks

Related to perishability and the biological nature of the production process is the difficulty of scheduling the supply of fruits and vegetables to market demands. These crops are subject to high price and quantity risks with changing consumer demands and production conditions. Unusual production or harvesting weather or a major crop disease can seriously disrupt fruit and vegetable marketing patterns. The long production periods and high fixed costs of orchard crops also present price and marketing problems. Producers of these are not very sensitive to short-run price changes and may produce at prices below total costs for rather long time periods.

The modern food marketing system requires price and supply stability for market planning and merchandising programs. A number of marketing arrangements have evolved to provide stability in the fruit and vegetable industry. Processors use grower supply and price contracts to ensure capacity utilization of their plants and to assist in sales programs. Many large chainstores operate buying offices in the major producing areas in order to ensure a steady source of supplies for their merchandising programs. Grower cooperatives and marketing orders and agreements (Chapter 15) assist in orderly marketing of these biologically sensitive, perishable products.

The biology and product perishability also prevent effective supply control in this industry. There are large numbers of unorganized growers of many fruits and vegetables in several states. There are no government programs limiting acreage or production of these crops, and only a few marketing orders permit growers to limit production. However, for some of these crops there are relatively few, large growers concentrated in one or a few states and marketing collectively through central sales or bargaining agencies. These growers may well have some supply control, and price-enhancing powers, though these would be somewhat limited by uncontrollable weather and yield variations.

Seasonality

Most fruits and vegetables have seasonal production and demand patterns that also influence their marketing. The canning and freezing crops have single-period harvests and year-round demands. The entire crop is picked and processed within a short period. Processing facilities are strained to capacity at this time and may sit idle for the remaining months of the year. Attempts are being made to use these facilities more efficiently, not only through postharvest technologies that allow deferred processing of the crops but also by adding complementary products to the processing line, such as fruit juices or carbonated beverages.

Fruits and vegetables are often planted, harvested, and marketed in sequence from a number of fields and production areas during a season. This is done to provide an even flow of produce over a longer season. For example, the spring peach harvest begins in Georgia and then moves northward through South Carolina, New Jersey, and Pennsylvania. Early sweet corn originates in Florida while local northern supplies become available in all states later in the summer. The seasonal sequencing of these crops is very specific and a supply disruption in one area affects the others.

The major fruit and vegetable producing areas in California, Texas, and Florida harvest and market early in the season and are generally out of the market when local summer supplies become plentiful. Thus, emphasis shifts seasonally from early long-distance fruit and vegetable shipments to local marketing—roadside markets, U-pick operations, and farmers' markets—later in the season. These, then, are complementary, but often competitive, supply sources.

Some chainstores have been criticized for inflexibility in shifting from distant to local produce sources. Many large retail-wholesalers continue to purchase and ship long distance, even when local supplies are plentiful. They defend this practice on the grounds that central procurement and warehousing facilities are necessary to provide a large assortment of fruits and vegetables year-round, and the efficiencies of those systems are reduced when a few items are procured locally. Many supermarkets handle both local and shipped-in produce, and of course the consumer has freedom of choice in purchasing produce from either a local roadside market, a farmers' market, or a supermarket.

There are efforts to reduce the seasonality of fruit and vegetable supplies. This often involves either a shift in production to an area of favorable climate, perhaps even a foreign country, or to greenhouse production. This shift can increase production and marketing costs, but consumers may be willing to pay the price. The year-round availability of salad vegetables—lettuce, tomatoes, cucumbers, etc.—is an example of this.

Alternative Product Forms and Markets

For most fruits and vegetables there are alternative markets. These include form markets (fresh, canned, frozen, dried), time markets (winter, spring, summer, fall), and place markets (different U.S. cities, foreign markets). The lemon market is an illustration. Lemons are produced in California, Arizona, and Texas and are sold in every U.S. city as well as exported to many other countries. About one-half of the crop is sold fresh while the other half is sold as processed lemon products. Fresh lemons may be produced and sold in the winter or summer months or stored for other months. These alternative market opportunities greatly complicate the fruit and vegetable marketing task and contribute to the complexity of fruit and vegetable marketing channels.

How are fruits and vegetables such as apples, peaches, and carrots allocated to the various markets? The decisions are influenced by the costs of servicing each market (processing costs, transport costs, etc.) and the relative prices of products in each market. The general rule is that a fixed supply should be allocated to alternative markets in such a way that the net returns (prices minus costs) will be equal in all markets. In practice it is not a simple matter to discover this desired marketing pattern. Grower cooperatives and fruit and vegetable marketing orders and agreements can assist the industry in achieving this pattern.

These alternatives provide opportunities for market development in the fruit and vegetable industry. Producer organizations, private firms, and cooperatives have invested heavily in new product development and promotional programs to expand sales and improve prices. An example has been the rapid growth in frozen fruit and vegetable products, including potato products and frozen orange juice concentrate.

These processing markets have greatly expanded the potentials for differentiating fruit and vegetable products through brands, packaging, coupons, and other means. There are many attempts to develop consumer brand loyalty for fresh fruits and vegetables—long considered nonbrandable products. Such market development programs in turn have altered traditional marketing and competitive patterns. Multiple product forms, extensive branding, and advertising have changed the competitive environment of this industry. Federal fruit and vegetable grades and standards are less used today. And a declining share of these commodities pass through the central, public markets, making price discovery more difficult.

Product Bulkiness

Opportunities to process and concentrate fruits and vegetables have influenced the location of production and processing facilities in this industry. Since water is a major component of these farm products, they are bulky, low value-per-unit commodities that are expensive to ship in fresh form. Substantial economies of transportation can be achieved by concentrating these products near production areas and shipping higher-valued products to market. This is the reason that the orange juice processing industry is located in Florida and tomato sauce/puree processing takes place in California. The California dried fruit industry is another example.

Transportation economies have led to more complex products. Frozen concentrated orange juice and dried soups illustrate situations where water is taken out in processing and put back by the consumer. The breakfast cereal industry adds dried fruit flakes to cereals; the flakes assume the fresh product form when bathed in milk.

Geographic Specialization of Production

Fruits and vegetables are produced in all states, but a few states dominate in the production of these crops (Table 29-3). California produces 30 percent of all U.S. fruits and vegetables, over 50 percent of U.S. commerical vegetables, and more than

Table 29-3 Leading States in Fruit, Nut and Vegetable Production, 1997

	Vegetables (excluding potatoes)		
State	*Farm Receipts, Million $*	*Percent of U.S. Vegetable Sales*	*Major Commodities*
California	4,019	48%	Tomatoes, lettuce, broccoli, carrots, celery
Florida	1,083	13	Tomatoes, peppers, potatoes, cucumbers, sweet corn
Washington	270	3	Potatoes, asparagus, onions
Oregon	213	2	Onions, potatoes, sweet corn, snap beans
Michigan	183	2	Dry beans, potatoes, cucumbers
Wisconsin	149	1	Potatoes, sweet corn, peas, snap beans
Total U.S.	8,401	100%	
	Fruits and Nuts		
State	*Farm Receipts, Million $*	*Percent of U.S. Fruit, Nut Sales*	*Major Commodities*
California	7,823	62%	Grapes, almonds, oranges, strawberries, walnuts
Florida	1,493	12	Oranges, grapefruit, strawberries, peanuts
Washington	1,240	10	Apples, pears, cherries, grapes
Oregon	308	2	Pears, cherries, apples
Michigan	231	2	Apples, cherries, blueberries
New York	185	2	Apples, grapes, strawberries
Total U.S.	12,660	100%	

SOURCE: U.S. Department of Agriculture.

60 percent of U.S. vegetables for processing. Florida, Michigan, Texas, New York, New Jersey, Washington, Oregon, Minnesota, and Wisconsin are also important producers. This geographic specialization results from natural comparative advantages due to climate, soil, and other factors. However, regional specialization has also been influenced by such manmade factors as irrigation, farm labor, and transportation policies.

Regional specialization has fundamentally altered marketing patterns in the fruit and vegetable industry. In addition to lengthening and complicating the marketing channels and increasing transportation costs, it has transferred the production of these commodities from small, diversified farms to larger, specialized operations. It has also shifted production to lower cost areas; speeded up the mechanization of fruit and vegetable production and harvesting; facilitated grower organizations such as cooperatives, advertising groups, and bargaining associations; lengthened the marketing season of most crops; and concentrated capital for new product development, advertising, and other market expansion activities. Critics charge this geographic specialization of production also eliminated an important source of income for many small farmers, added unnecessary costs to marketing, and resulted in less flavorful fruits and vegetables.

In some cases this trend has resulted in single state production of products. Hawaii is the only state producing bananas, and California produces almost all of the canning peaches. The market power of growers in these states is limited, however, by the existence of foreign sources of supply and substitute products. In most cases there is intense interregional competition in this industry. The orange industry illustrates regional division of labor and market complementarity. California-Arizona producers market two-thirds of their oranges in fresh form and account for about 75 percent of U.S. fresh oranges. Florida producers market 20 percent of U.S. fresh oranges but process 95 percent of their oranges, mostly into juice products. The California fresh orange is prized for its appealing skin appearance. Florida oranges are not as attractive but have a juice content 30 percent higher than the western orange. The potato and apple industries also illustrate interregional competition between highly specialized producing areas.

Other industries have been influenced by the geographic concentration of fruit and vegetable production. These products are an important source of truck and railroad revenues. Numerous brokers, commission agents, shipping point firms, and freight forwarders make up the complex distribution network serving this industry. Processing facilities and labor markets have also been influenced by the location of fruit and vegetable production.

FRESH FRUIT AND VEGETABLE MARKETING CHANNELS

Fresh fruit and vegetable marketing channels are illustrated in Figure 29-4. There are three principal markets: (1) shipping point markets, (2) wholesale markets, and (3) retail markets. This marketing system is undergoing change as a result of vertical integration, decentralization, new handling and transportation methods, and the growth of the away-from-home and direct farmer-consumer markets.

Shipping point markets are located in fruit and vegetable producing areas. Their purposes are to aggregate large volumes of produce from numerous growers, prepare the products for market (performing such functions as sorting, grading,

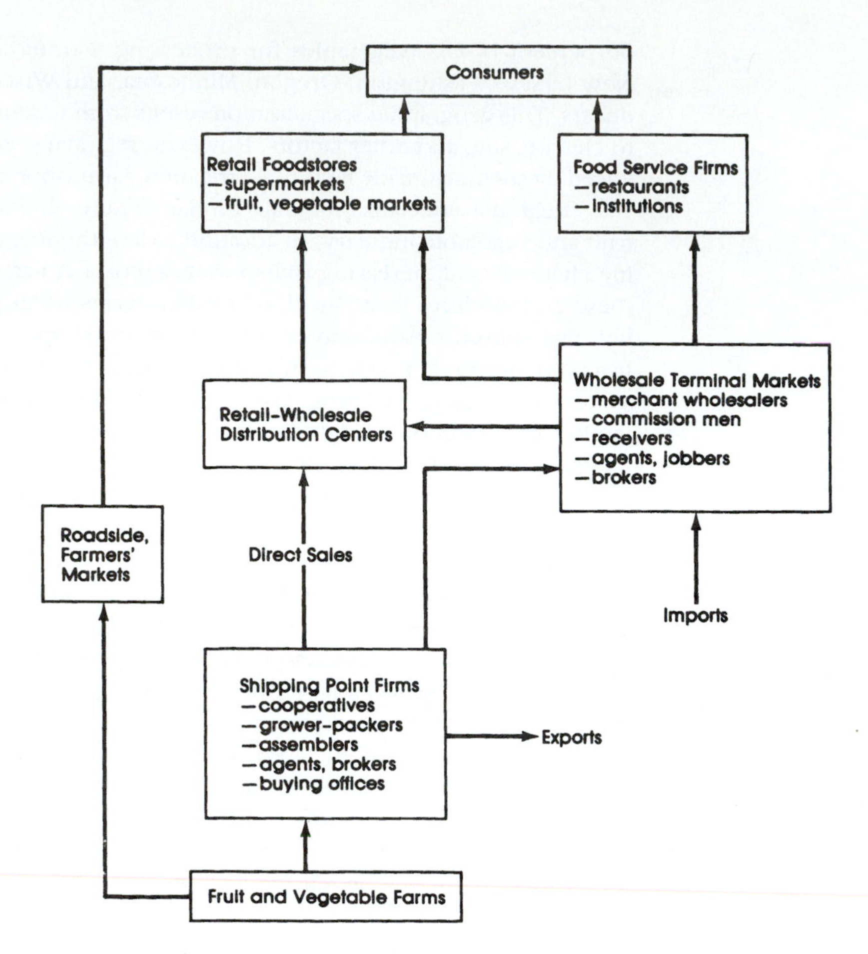

FIGURE 29–4 Fresh fruit and vegetable marketing channels.

cleaning, packaging, and storing), and allocate the products to the various fruit and vegetable markets.

There are a variety of shipping point firms. Fruit and vegetable packing plants often represent integrated combinations of production and marketing functions: grower-packers, packer-shippers, and grower-packer-shippers. About 50 percent of fresh fruits and vegetables are marketed under some form of grower-shipping point vertical integration, including contracts. Producer cooperatives also operate in shipping point markets, and co-ops handle an estimated 20 percent of fruit and vegetable farm sales. There are also a number of agents and brokers in shipping point markets who may not physically handle the product or take title to it but serve to bring buyers and sellers together, arrange for shipping, and provide valuable market information to traders. Finally, many large retail chain-store and wholesale firms operate buying offices in shipping point markets.

The bulk of fruit and vegetable crops is first delivered to these shipping point markets. However, a small but growing share of these crops is being sold direct to consumers at roadside markets, U-pick operations, and farmers' markets. The number of farmers' markets grew from about 100 in 1974 to more than 2,000 in 1998. These markets require growers to take a more active part in the marketing process, including the performance of food retailing functions.

Shipping point firms ship fresh fruits and vegetables to three alternative markets: (1) the export market, (2) wholesale terminal markets, and (3) integrated wholesale-retail markets.

Wholesale fruit and vegetable terminal markets are located in population centers. They take long-distance shipments from shipping point firms, break them into smaller lots, and sell to food retail firms and the institutional away-from-home market. Traditionally, these central markets have served as the focal point for price discovery in the fresh produce industry. This role has declined somewhat with the growth of decentralized, direct sales from shipping point to large wholesale and retail buyers. Newer wholesale produce firms markets precut vegetables that are shredded, diced, and packaged for consumers.

There have been several produce terminal market adjustments to the competitive development of direct sales and self-distributing retailers. First, service wholesalers combining more marketing functions replaced the traditionally highly specialized terminal market firms. Secondly, these newer firms provided additional buyer services such as prepackaging, repacking, credit, and merchandising assistance that are valued by the smaller retailers they specialize in servicing. Third, the terminal market operators have aggressively cultivated the growing away-from-home food markets and firms that generally cannot purchase direct from shipping point. Finally, there have been major improvements in the location and physical layout of many produce terminal markets. These have been moved out of congested downtown areas to more efficient locations. An example was the 1966 relocation of the largest produce terminal market in the country, in New York City, from downtown Manhattan to Hunt's Point in Queens.

Perhaps a quarter of all fresh fruit and vegetable wholesaling firms are located in California and Florida. The number of these firms is declining at all market levels while their average size is increasing. The Census of Business indicated that the number of fresh fruit and vegetable merchant wholesale establishments fell from 6,520 in 1954 to 5,474 in 1997; the number of agents and brokers fell from a high of 1,085 establishments in 1963 to 647 in 1997. Although national market sales concentration of fruit and vegetable wholesalers is low, for some crops and local areas the proportion of sales controlled by a few wholesalers may give them some market power.

The buying offices, central distribution centers, and affiliated retail chain-stores of large food wholesale and retail firms represent a vertically integrated marketing channel. This procurement system was established to provide these organizations with dependable supplies and large volumes of fresh fruits and vegetables on a year-round basis at favorable prices. The firms purchase from either shipping point markets, terminal markets, or importing firms. Most wholsale produce sales are F.O.B. shipping point; however, some produce is sold on contract, through auctions, or on consignment. In 2000 retail grocery firms purchased about 50 percent of their fruits and vegetables directly from shipper-growers, up from 38 percent in 1987.

Fruit and vegetable marketing orders and agreements (Chapter 15) play an important role in the fresh produce channel. These cooperative agreements between growers and shipping point firms are authorized and enforced by the Secretary of Agriculture. Growers and shippers use these devices to influence the flow-to-market of their products, to standardize containers, to facilitate market development and promotion programs, to enforce fair trading practices, and to gather market information. The purpose of these orders is to facilitate "orderly marketing" of fresh produce through the use of storage pools, quantity allocations to various markets, and promotional and research programs.

Authorized since 1937, there are about 40 federal fresh fruit and vegetable marketing orders currently in effect, primarily in the major producing areas. These orders cover more than one-half of all fruits and nuts marketed and about 15 percent of the vegetables produced. Various states also authorize marketing orders that reinforce and extend the powers of the federal orders.

By influencing the flow-to-market of fresh fruits and vegetables, marketing orders and agreements probably raise producer and consumer prices somewhat. However, a 1981 Department of Agriculture task force report concluded that fresh produce marketing orders have both positive and negative effects on grower prices and revenues, consumer prices, price stability, product quality, and the overall allocation of resources in this industry. The market power afforded growers by these orders is limited by the existence of substitute products and the long-run supply response to higher prices.

The fresh produce marketing channel is experiencing rapid firm consolidation through mergers, acquisitions, and the decline of smaller firms. As a result, market concentration is increasing at all levels in the industry. Individual farmers and grower-shippers now face many fewer, larger and more powerful buyers for their products. These larger food wholesaling and retailing firms may be able to exert their influence over contract terms, prices, and services in negotiations with growers and shippers.

There are two potentially significant areas for improving marketing efficiency in the fresh fruit and vegetable industry: transportation and loss prevention.

Transportation efficiency is vital to fruit and vegetable marketing. Intercity transportation can represent one-half of the packer-to-retailer marketing time, and transport costs may exceed the farm value of fruits and vegetables hauled across the country. Trucks and railroads are the dominant fresh produce carriers. Only a small fraction of these crops are transported by air and water. Truck transportation has become more important than rail in recent years. Trucks haul about 90 percent of fresh fruits and vegetables while railroads move the rest. Shippers accept the sometimes higher costs of truck transportation in return for its speed, flexibility, and dependability.

There have been major changes in the regulation of transportation in recent years, and the railroads have revised their rates and improved services in an attempt to regain lost produce freight. One approach—the trailer-on-flat-car (TOFC)—combines the energy efficiency of railroads with the flexibility of trucks. Railroad mergers allowing uninterrupted coast-to-coast hauls have also reduced delivery times. Innovations in truck technology such as windscreens, radial tires, and diesel engines have also improved fresh fruit and vegetable marketing efficiency.

Fruit and vegetable loss prevention is another potential area for improvement in marketing efficiency. Quantity, quality, and economic losses occur at all stages of produce marketing: in harvesting, handling, and preparation for eating. Fruit and vegetable losses result from poor temperature and humidity control, improper packaging and handling, slow delivery and waiting periods, trimming of product, and poor coordination of market supplies and demand. Produce shelf life and keeping quality can be enhanced through modified atmosphere packaging and radiation.

Merchandising considerations may conflict in some cases with the objective of minimizing produce losses. For example, moisture loss and product deterioration are accelerated when the outer leaves of some vegetables are removed to improve product appearance. Prepackaging of produce at shipping point can reduce product losses and extend shelf life but is resisted by some consumers and grocers. Fresh fruits and vegetables can also be packed in bulk bins that move directly from field to retail stores without individual product handling.

Electronic commerce is also growing in this industry. Fresh fruits and vegetables are now marketed electronically through business-to-business websites. This can increase marketing efficiency and lower costs.

PROCESSED FRUIT AND VEGETABLE MARKETING CHANNELS

The processed fruit and vegetable industry is composed of canners, bottlers, dehydrators, and freezers. These firms process and market a wide variety of products ranging from preserved whole farm products (e.g., canned corn, frozen peas), to transformed farm products (e.g., catsup, pickles), to multi-ingredient foods (e.g., soups, pre-prepared dinners). The demand for some of these products is stable while others are experiencing rapid growth or decline. The freezing industry is growing more rapidly than the more mature canning and dehydrating industries.

In 1997, some 2000 fruit and vegetable processors operating many more plants, marketed products valued at $50 billion. Although the number of canners has declined in recent years, the number of dehydrating and freezing plants has grown slightly. Overall sales market concentration in this industry is relatively low and has been stable in recent years. The four-firm market concentration ratios for fruit and vegetable processors generally averages less than 30 percent but is higher than this in selected product categories, such as pickles, catsup, and canned vegetable juice.

These processing plants are located in the major fruit and vegetable producing areas, concentrated in California, New York, Wisconsin, Michigan, and Florida. There are well-known firms and brands: Campbell Soup, Heinz, Del Monte, Green Giant, Birdseye, Stokely-Van Camp, and Tropicana. And there are numerous smaller firms with less prominent brands. There are also some large grower cooperatives—California Canners and Growers and Tri-Valley Growers—and cooperative processors account for perhaps 20 percent of canned fruit and vegetable sales. There have been some large mergers of these processors. R. J. Reynolds acquired Del Monte, Pillsbury purchased Green Giant, Coca-Cola purchased Minute Maid, and Quaker Oats bought Stokely-Van Camp.

Packers secure their raw product supplies in three ways: (1) open market purchases; (2) production and marketing contracts; and (3) own production. A small portion of fruits and vegetables for processing are sold on the open market at harvest time and perhaps 20 percent of these crops is grown on the packers' land. For the remaining 80 percent, growers sign preharvest marketing contracts with packinghouses. These contracts may specify delivery time and price, financing arrangements, production and harvesting techniques, and other terms of trade. Grower cooperatives, bargaining associations, and a few processing marketing orders influence contract negotiations in canner procurement markets.

Competition between fruit and vegetable processors focuses on new product development, product ingredients and formulas, brand advertising, such consumer promotions as coupons and cents-off deals, and trade promotions (merchandising assistance, trade discounts, and allowances). These have elements of both price and nonprice competition. Fruit and vegetable processors are generally classified as oligopolists or monopolistic competitors.

Staple products such as canned corn and peaches and frozen peas are relatively homogeneous products marked by a high degree of price competition. New and smaller processors will typically pack and sell these items under private labels. Larger packers will also market these staple products as nationally advertised labels or privately labeled brands. In addition, large packers with consumer brand loyalty also market more complex, highly differentiated products that are marked by less rigorous price competition. Both the large and small processors pack the generic, no-frills products that provide consumers with an economical alternative to nationally advertised brands.

Economies of size in the production and marketing of processed fruits and vegetables also influence competition. Studies have indicated that moderately sized firms can enter this industry and compete on a production cost basis with larger firms. This has occurred for the staple items and particularly in the freezing industry. However, marketing economies of size may disadvantage the smaller packer. The larger processors have extensive distribution networks and sales forces as well as a degree of brand loyalty fostered by years of brand advertising that provides them with some protection against new rivals. However, these have not proven to be an invulnerable defense for the large processors.

Because processed fruits and vegetables are harvested and packed in a short time period and sold year-round, storage is an important marketing function. Inventories are held by processors, distributors, and consumers. These products may be stored for more than a marketing year when the annual pack exceeds annual consumption or disappearance. Stocks from the year-earlier pack may constitute anywhere from 10 to 35 percent of a year's total supply of these products. Some pipeline stocks are necessary to ensure an orderly transition from one pack to the next, but excessive carryover stocks can depress fruit and vegetable prices for growers, processors, and consumers.

INTERNATIONAL TRADE IN FRUITS AND VEGETABLES

U.S. horticultural exports are growing faster than other farm products, and they are high-value products. Horticulture's share of U.S. agricultural product exports increased from 10 percent in 1989 to 18 percent in 1999–2000. The major growth products have been onions, carrots, celery, lettuce, apples, oranges, potato chips, and vegetable combinations.

Markets for fresh fruits and vegetables are now global with improved transportation and refrigeration. U.S. imports and exports of fruit and vegetable products are playing an increasingly important role in this industry. Fruit, vegetable, and nut exports have been growing rapidly in recent years (Figure 29-5). In the first half of the 1990s U.S. exports of processed fruits and vegetables grew at a rapid 10 percent annual rate while exports of fresh produce increased at a 7 percent annual rate. This was a result of market-opening trade agreements, rising world incomes, favorable exchange rates, and expansion in overseas fast-food businesses.

The U.S. exports a wide variety of horticultural products, including fresh oranges, grapefruit, lemons, and apples as well as almonds, raisins, prunes, lettuce, onions, and fresh and processed tomatoes. Canada, Mexico, South America, Europe, Japan, and Hong Kong are the largest buyers of U.S. products.

The United States also imports large quantities of fruits, nuts, and vegetables. Latin America provides over 85 percent of U.S. fruit and vegetable imports. Chilean exports of fruits to America have increased significantly in recent years. One study found that Chilean grape imports actually increased total U.S. grape consumption by making grapes more widely available thoughout the year. In most years the United States is a net importer of fresh vegetables and often of canned and frozen vegetables. Bananas, a noncompetitive import, account for about 17 percent of total U.S. fruit and vegetable imports. Imports of Asian canned pineapples, Chinese canned mushrooms, Caribbean oranges, Canadian potatoes, and Portugese and Spanish tomato products have all made significant inroads into the U.S. market in recent years.

Imports from Mexico of fresh strawberries, tomatoes, peppers, cucumbers, squash, and cantaloupes have also expanded, and these compete directly with U.S. products grown in Florida, Texas, and California. Although these Mexican imports must bear a tariff and meet U.S. market order quality standards, production costs are somewhat lower in Mexico. Many U.S. growers feel these Mexican imports depress U.S. winter vegetable prices.

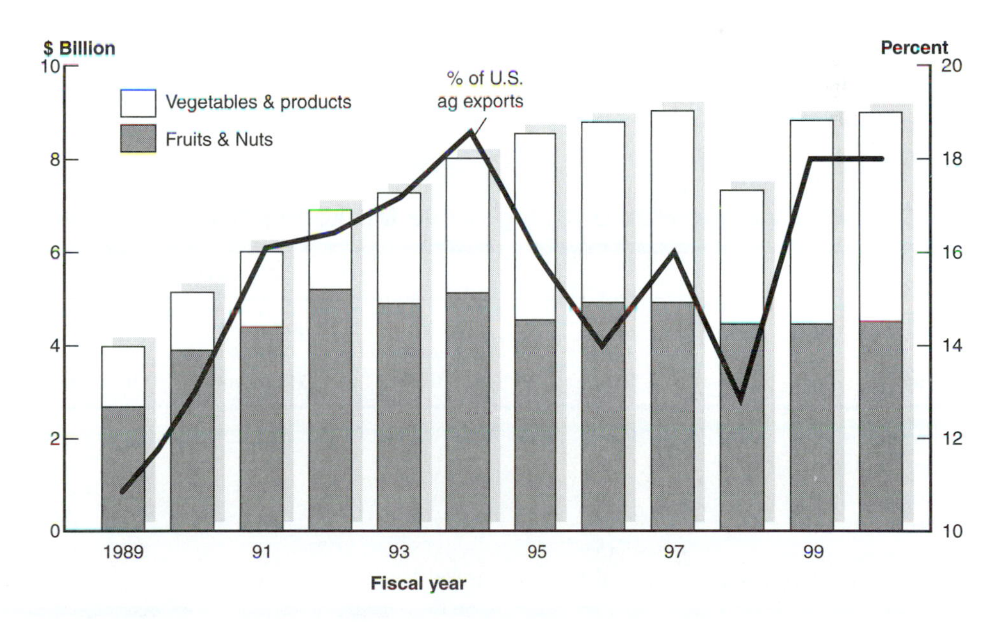

FIGURE 29–5 U.S. fruit, vegetable, and nut exports, 1989–1995.

Asia and Mexico are likely to continue as the most significant growth markets for U.S. produce. Recent NAFTA, GATT, and WTO trade agreements lowered trade barriers, reduced quotas and resolved phytosanitary disputes. Latin America will continue to be the major seller of fruit and vegetables to the United States.

MARKETING COSTS AND FARM SHARE

The perishability and special handling needs contribute to high produce marketing costs. Fruit, nut, and vegetable marketing costs account for about 25 percent of the total costs of marketing all domestically produced food products. The farm–retail spread for fresh fruits and vegetables increased by 200 percent between 1982 and 1999 while the spread for processed fruits and vegetables increased by 67 percent. During this period, the farm–retail spread for a market basket of all foods increased by 104 percent.

The importance of various production and marketing functions for fruits and vegetables is illustrated in Table 29–4. The retail marketing functions account for a large share of the consumers' food dollar because of the high labor costs in handling these products at retail and because of their below-average sales per square foot of retail selling space. Packing and transportation are also important cost elements. Processing costs dominate the marketing bill for canned and frozen fruit and vegetable products. Packaging is one of the largest processing costs, and the costs of containers equal the farm value for some processed fruit and vegetable products.

The farmer's share of the consumer's food dollar averages about 18 percent for both fresh and processed fruits and vegetables. This is influenced by the amount of processing services added to the product, the distance to market, and handling costs. Fresh market products generally provide a higher farmer's share than products for the processing market.

Future fruit and vegetable prices, marketing costs, and spreads will be influenced by the fresh/processed product mix, the geographic concentration of production, the growth of gardening, direct marketing, trends in freight rates and costs of

TABLE 29–4 Average Fruit and Vegetable Marketing Costs and Farm Value, 1986–1990

	Marketing Functions					
Product	*Farm Value*	*Packing, Processing*	*Intercity Truck Transport*	*Wholesaling*	*Retailing*	*Retail Price*
Potatoes, northeast, round white	25	11	6	5	53	100%
California oranges	20	16	11	13	40	100
California lettuce	16	12	10	8	54	100
Frozen orange juice	37	25	3	13	22	100
California canned tomatoes	9	56	10	5	20	100

SOURCE: U.S. Department of Agriculture. This series is no longer available.

other marketing services, and new techniques for handling and marketing these products.

SUMMARY

The fruit and vegetable industry consists of a long, complex distribution channel and a variety of firms distributing these products to multiple time, form, and spatial markets. This channel and marketing process has been shaped by the geographic specialization of production, product perishability and bulkiness, seasonality, and rather large price and quantity variations. There is significant packer and processor integration into production. Marketing orders and agreements, cooperatives, and grower bargaining associations all influence fruit and vegetable marketing. Both price and nonprice forms of competition are evident at the product, regional, brand, and international levels. Per capita consumption of fresh fruits and vegetables is rising, dramatically for some items. Because of the long marketing channels, the highly processed products, and the large amount of manual labor involved, the farmer's share is lower for fruits and vegetables than for many other fresh farm products.

KEY TERMS AND CONCEPTS

buying offices
direct marketing
orderly marketing
Perishable Agricultural Commodities Act
prepackaging retail-wholesale distribution centers
roadside marketing
rollers
shipping point markets
trailer-on-flat-car (TOFC)
U-pick operations

DISCUSSION QUESTIONS

1. What effects would a new chemical process that delays deterioration and spoilage of fresh fruits and vegetables have on their production and marketing patterns?
2. How would a promotional strategy for a declining consumption product like fresh potatoes differ from that of an increasing demand product like romaine lettuce?
3. How does the fruit and vegetable industry illustrate the principle of comparative advantage?
4. It is often observed that products that can be sold in different time, form, and spatial markets provide opportunities for price discrimination. How would this apply in fruit and vegetable markets?
5. How could the development of a new processed fruit and vegetable product (such as frozen orange juice concentrate) expand orange producers' total market?
6. Why are market orders and agreements so prevalent in the fruit and vegetable industry?

7. Why is contract production so common for processed fruits and vegetables and comparatively rare for fresh crops?

8. List the alternative consumer sources of fruits and vegetables. How do these affect market competition in this industry?

9. Why do you think California oranges have a higher farmer's share than California lettuce?

SELECTED REFERENCES

Calvin, L., and R. Cook, coordinators. (2000). *U.S. Fresh Fruit and Vegetable Marketing: Emerging Trade Practices, Trends and Issues.* Agricultural Economic Report No. 795. Economic Research Service, U.S. Department of Agriculture. www.ers.usda.gov/aer795.

Dimitri, C. (March 1999). Integration, Coordination, and Concentration in the Fresh Fruit and Vegetable Industry. *Fruit and Tree Nuts,* FTS-285. Economic Research Service, U.S. Department of Agriculture, pp. 23-31.

Fruits and Tree Nuts, Key topics. (February 2001). Economic Research Service, U.S. Department of Agriculture. http://www.ers.usda.gov/Topics/view.asp? T= 101210.

Fruit and Tree Nuts, Situation and Yearbook. (February 2001). Economic Research Service, U.S. Department of Agriculture. http://www.ers.usda.gov/publications/so/view.asp?f=specialty/fts-bb/.

Handy, C. R. et al. (May-Aug. 2000). Evolving Marketing Channels Reveal Dynamic U.S. Produce Industry. *Food Review.* Economic Research Service, U.S. Department of Agriculture, pp. 14-20.

Kaufman, P. R. et al. (2000). *Understanding the Dynamics of Produce Markets; Consumption and Consolidation Grow.* Agricultural Information Bulletin No. 758. Economic Research Service, U.S. Department of Agriculture. www.ers.usda.gov/publications/aib758/.

Powers, N. J. (August 1994). *Marketing Practices for Vegetables,* Agriculture Information Bulletin No. 702. Washington, D.C.: U.S. Department of Agriculture.

———. (March 1990). *Federal Marketing Orders for Fruits, Vegetables, Nuts, and Specialty Crops,* Agricultural Economic Report No. 629. Washington, D.C.: U.S. Department of Agriculture.

Vegetables and Melons, Briefing room. (February 2001). Economic Research Service, U.S. Department of Agriculture. http://www.ers.usda.gov/briefing/vegetables/.

Vegetables and Specialties, Outlook Reports. (February 2001). Economic Research Service, U.S. Department of Agriculture. http://www.ers.usda.gov/publications/so/view.asp?f=specialty/vgs-bb/.

GLOSSARY

Acreage Allotment An individual farm's share, based on previous production, of the national acreage needed to produce sufficient supplies of a crop and a government-determined price.

Acreage Controls A provision of many farm programs that attempts to reduce farm output by limiting the acreage that farmers plant.

Acreage Reduction Program A goverernment program in which farmers agree to reduce crop acreage to be eligible for CCC loans and government payments.

Advertising Nonpersonal sales presentations addressed to large numbers of consumers for the purpose of altering the demand curve.

Advertising-to-Sales Ratio A firm's or industry's advertising expenditure expressed as a percent of its sales; a measure of the importance of advertising to the firm's or industry's marketing effort.

Affiliated Chainstore A contractually integrated retail and wholesale operation; the affiliation can either be between a wholesaler firm that is cooperatively owned by independent retailers or a wholesaler-sponsored retail chain.

Agent Middleman A food marketing firm that represents buyers and sellers in the market place; agents do not take title to goods for their own account and may not physically handle the food products.

Agribusiness The sum total of all institutions, firms, and activities involved in the commercial production and marketing of food.

Agricultural Adjustment Act (of 1933, 1937, and 1938) Early farm legistation that contained the first provisions designed to aid farmers through supply control, government purchase programs, and orderly marketing tools.

Agricultural Exemption A clause of the Interstate Commerce Act that exempts from Interstate Commerce Commission regulations motor carriers transporting raw agricultural commodities.

Agricultural (Farm) Structure Refers to the number, size, ownership, specialization, and other characteristics of farming.

Agricultural Productivity Refers to operational efficiency in farming; usually measured by the ratio of farm output to farm inputs.

Agricultural Stabilization and Conservation Service (ASCS) The agency of the U.S. Department of Agriculture that administers the farm price support programs.

Agricultural Substitute A product that is manufactured from farm commodities but that is a substitute for a traditional farm food product (e.g., corn-oil margarine or soy-protein steaks).

Agricultural Trade Development and Assistance Act of 1954 Also known as the "food for peace" program, or PL-480; the federal program designed to increase United States farm product exports by selling commodities on low-interest loans, exchanging commodities for local currencies, or donating commodities to needy countries.

Agricultural Treadmill Refers to the situation in which farmers find themselves producing food at a lower price with no increase in long-run profits after having been encouraged to adopt new output-increasing and cost-reducing production technologies.

Analog, Food A food product that resembles, and is a substitute for, a traditional farm food product (e.g., a steak made out of soybean meal, with a plastic bone).

Assembly Market; Assemblers Markets and firms that consolidate the produce of individual farms and prepare it for the marketing process.

Auction Market A production-area wholesale market that sells farm products to buyers on an auction basis.

Away-from-Home Food Market The market where consumers buy food for consumption away from home; includes restaurants, cafeterias, hotels, motels, and other food service operations; sometimes abbreviated "FAFH."

Backhaul Freight that a carrier hauls in returning to its point of origin; without a backhaul, a truck or train must run empty when returning to its departure point.

Balance of Trade The value of a nation's exports less its imports.

Balance Sheet, Commodity A summary table showing the sources of supply and the uses of a commodity, with total supply equal to total use.

Bargained Prices Prices arrived at by joint negotiation of a group of buyers and sellers.

Bargaining Association A farm cooperative having as its principal function the influencing of farm prices and other terms of trade.

Bargaining Power A form of market power denoting the relative strength of buyers and sellers in influencing the terms of exchange in a transaction.

Base Price Quotation A price that is discovered in a central market and that serves as a reference point for pricing elsewhere in the marketing channel.

Basic Formula Price (BFP) The farm price of milk that is used in producing manufactured dairy products, such as cheese and butter. Fluid milk prices are based on this price plus a differential in each milk market.

Basis The difference between a cash price and a futures price.

Basis Pricing A price quotation technique whereby the current cash price of a commodity is described by indicating the basis; for example, "30 cents under" would indicate a $2.70/bu. cash price if the futures price were $3.00.

Battle of the Brands The competitive rivalry between food processors' nationally advertised brands and the unadvertised (private) brand products of food wholesalers and retailers.

Bear A trader who anticipates falling prices.

Behavioral System A set of related marketing institutions and functions, their goal being the solution of marketing problems.

Bid Price The price buyers offer to sellers.

Bilateral Trade Agreement An agreement between two countries on trading patterns; as, for example, the U.S.-U.S.S.R. grain trading agreements.

Biological Lag The time period necessary for a changed production decision to influence market supplies, owing to the biological nature of agricultural products.

Biotechnology Also referred to as *genetic engineering.* Generally, the use of biological processes of microbes and of plant or animal cells for changing the nature of farm and food products in order to make them more useful for humans. Specifically, the manipulation of genetic materials to improve their production or food values.

Blend Milk Price The farmer's price of milk as determined by a weighted average of Class I and Class II milk prices.

Blending A grain marketing strategy whereby two different qualities of grain are blended in such a way as to raise the total value of both lots.

Boxed Beef Beef that is sold by the meatpacker in primal, subprimal, or final retail cuts rather than as a carcass; this beef is usually frozen and boxed for shipping convenience.

Boycott, Consumer A group tactic whereby a number of consumers refuse to purchase a product (as in the 1967 and 1972 meat boycotts) in an attempt to influence the price or to influence other market conditions.

Brand Any name, sign, symbol, or design used to identify the products of one firm and set them apart from competitors' offerings.

Breaker An egg processor who makes dried, frozen, and liquid egg products.

Broiler A young meat-type chicken; also called a *fryer.*

Broker An agent middleman who facilitates trades but does not usually physically handle food products; usually paid a set service fee by buyers and sellers.

Bulk Handling Any procedure that permits handling large concentrations of products rather than small lots.

Bull A trader who anticipates rising prices.

Business Cycle Periodic expansions and contractions in the real value of the gross national product, often accompanied by periods of rising and falling price levels and employment.

Buyer's Market A market situation whereby a large amount of products provides buyers some advantage in price bargaining with sellers.

Capper-Volstead Act (1922) The federal law that sanctions and encourages agricultural cooperatives to serve as instruments for maintaining reasonable competition in the marketing and purchasing of agricultural products and supplies.

Carcass Basis Purchase An alternative to pricing live farm animals; in carcass weight and grading, the final value of the animal is determined after slaughter.

Carryover Stocks Supplies of commodities left over from one marketing season to the next.

Cartel An association of firms that attempts to regulate industry output and prices through mutual agreement.

Central Meat Processing The movement of the meat cutting and processing operations to a central point (usually a retail chain warehouse) in the meat distribution channel.

Chainstore A set of eleven or more related grocery stores, operating under a similar name.

Chainstore Movement The process of combining retail stores into a single firm that may also operate food wholesaling facilities.

Channel Captain A dominant member of an industry who assumes responsibility for the coordination of all members' decisions in the marketing process.

Class Freight Rate A common freight rate that applies to a class of products.

Classified Milk Pricing The system whereby farmers receive a higher price for Grade A milk used in producing fluid milk products than for milk used in making processed dairy products.

Clayton Act (1914) A supplement to the 1890 Sherman Antitrust Act; this act further identified market activities considered to be monopolistic and in restraint of trade.

Commission Firm, Merchant An agent middleman who physically handles products for buyers and sellers and is paid a percentage of the selling price for his services.

Commodity Balance Sheet See Balance sheet, commodity.

Commodity Credit Corporation (CCC) The government agency that buys, sells, and stores farm commodities in connection with federal farm price and income support programs.

Commodity Exchange A marketplace where cash commodities and futures contracts are traded.

Commodity Futures Trading Commission The federal agency responsible for regulating futures trading in the United States.

Common Carrier Any regulated transportation firm that offers its services to all shippers at a fee that follows scheduled routes.

Common Market A set of countries with a common trading policy; the European Economic Community, for example.

Comparative Advantage The relative efficiency of producing one commodity as compared with producing another using the same set of resources.

Conservation Reserve Program A government program in which farmers agree to take highly erodable cropland out of production for ten years and devote it to conservation uses.

Competitive Fringe The small firms in an industry that account for a relatively small share of total industry sales but that nevertheless furnish competition for the dominant core firms.

Complementary Import An imported product that does not compete directly with domestically produced products of the importing nation; also known as a *noncompetitive import.*

Complementary Products Products that are usually consumed together (e.g., hamburgers and french fries).

Component Pricing Pricing farm milk according to its components of butterfat, protein, and other solids so that the farmer's price more closely reflects the value of the milk.

Conglomeration The combining of unlike kinds of economic activities into a single firm; also known as *diversification.*

Consumer The ultimate buyer and user of food products; firms that buy products for resale rather than for direct consumption are referred to as *middlemen* or *market intermediaries.*

Consumer Boycott See Boycott, consumer.

Consumer Franchise Refers to consumer loyalty to a brand or to a store as a result of past experiences or promotional efforts.

Consumer Sovereignty The proposition that ultimately the consumer should, or does, direct all production and market activities in the economy.

Consumer Want Goods or services purchased by consumers and therefore presumed to be satisfying in some sense; wants may be physiological or sociopsychological; most products (e.g., food) are purchased to satisfy both wants and needs.

Consumerism A wide range of government, business, and consumer activities designed to assist and protect consumers and to foster consumer rights and responsibilities.

Consumption Pattern The set of products that consumers purchase, as well as the processes by which these products are purchased and prepared for use.

Contract Carrier A transport carrier that deals with a few customers and does not offer its services to the general public.

Contract Integration An agreement between two independent firms that involves coordination and transfer of managerial decisions and responsibilities.

Contract Maturity The time at which the futures contract promise is due; usually seven to eight trading days before the end of the month for which the contract is named.

Convenience Food A product that reduces the time, effort, or ingredients required of the consumer in home preparation; for example, frozen orange juice is a convenience form of beverage orange juice; convenience foods are said to have "built-in maid service."

Convenience Store A neighborhood, limited-line food retail outlet.

Cooperative A business voluntarily owned and controlled by its member–patrons and operated for them on a nonprofit or cost basis.

Cooperative Patron A member of a cooperative who uses the co-op's services.

Corner An illegal situation where a single buyer or seller holds a large share of the outstanding long or short positions in the cash and futures markets, and therefore can influence to his or her advantage prices in these markets.

Corporate Chainstore Multiple retail units operating under a single corporate identity; e.g., A&P, Kroger, or Safeway.

Corporate Farm Any farm organized under corporate law as compared to a proprietorship or a partnership. Family farms may or may not be corporations; some corporate farms are owned by food marketing firms or nonfood firms.

Cost–Price Squeeze A situation wherein price levels are persistently equal to, or even occasionally below, costs of production.

Countervailing Power A market power position that nullifies another position in the market; for example, farmers might gain countervailing power against imperfectly competitive buyers by selling, as a single agent, through cooperatives.

Crop Insurance, Federal A voluntary risk management tool, available to farmers since the thirties, that protects them from the economic effects of unavoidable adverse natural events; some percent of these insurance costs are federally subsidized.

Cross Elasticity of Demand The response of quantity demanded for one commodity to a price change of another commodity.

Cross-Subsidization The practice of using the profits from one industry or market to subsidize aggressive price competition in another.

Dairy Compact, Northeast A government program by which a number of states agree to fix retail diary product prices in order to increase farm milk prices.

Dashboard Food Food made to be eaten in an automobile, often with one hand.

Decentralization A market trend that has replaced central market trading by direct sales of farmers to buyers in production areas; also refers to the movement of food processing plants from cities to farm production areas.

Deferred Pricing A price that is determined sometime after the product has been transferred from the seller to the buyer.

Deficiency Payment A government payment made to farmers who participate in crop programs that pays the farmer the difference between a target price and the national average price for the first five months of the marketing year.

Demand A schedule of the quantities of products that consumers will buy at alternative prices.

Demand Expansion A marketing effort that seeks to shift the demand curve for a product or industry to the right so that more can be sold at the same price, or so that a higher price can be obtained for a given quantity of sales.

Demography The study of human populations—numbers of people, where they live, how they live, and their characteristics.

Depression A period of time when the real value of the gross national product declines sharply, as in the 1930s depression.

Derived Demand The relationship of a demand schedule at one market level to a schedule at another market level; for example, the farm demand for hogs is derived (or results) from the consumer demand for pork chops.

Differentiated Marketing The strategy whereby firms attempt to make their products different from competitors' offerings.

Differentiated Products Products that, in the eyes of consumers, have significant differences, because of either price, quality, advertising, service, or other characteristics.

Direct Buying The practice whereby food marketing firms purchase directly from farmers or shipping-point markets rather than from terminal markets.

Direct Sale A farm sale to buyers that does not go through, or use, the facilities of an intermediate market or marketing agency.

Disaster Payment Federal aid provided to farmers when either planting is prevented or crop yields are abnormally low because of adverse weather or related conditions.

Diversification Refers to performing more than one unrelated market activity or producing more than one product; the opposite of *specialization.*

Diversion Privilege Permission for a shipper to divert a product to an alternative destination while en route.

Dockage Extraneous material found in grain, such as dirt, weed seeds, other grains, or broken kernels.

Dominant Core The largest firms in the industry, those that account for a significant share of industry sales.

Dumping The practice of pricing products for sale in foreign markets at prices considerably below the selling price in the domestic market.

Economies of Scale or Size A situation wherein efficiency gains accrue from increasing the size of operations; for example, large food processors may have lower average unit costs than smaller processors.

Effective Demand Consumer needs and wants backed up by purchasing power or a willingness to sacrifice money for products.

Efficiency A ratio of market output (satisfaction) to marketing input (cost of resources); an increase in this ratio represents improved efficiency, a decrease denotes reduced efficiency.

Electronic Marketing The use of electronic communications such as computers, TV, or the Internet (also known as e-commerce) to facilitate the matching of buyers' and sellers' offers and to discover prices.

Elevator A marketing facility for purchasing and storing grain and for mixing animal feeds; many elevators also sell farm supplies.

Embargo A prohibition against exports on the part of an exporting country.

Engel's Law A tendency for the share of a family's (or nation's) income spent for food to fall as income rises. This suggests that the income elasticity of food is lower than that of other products.

Entrepreneur An individual or firm that commits resources to productive activities in pursuit of a profit; a risk-taker and profit-seeker.

Equilibrium Price That market price at which the quantity supplied is equal to the quantity demanded; also called the *market clearing price.*

Exchange An economic process by which two traders voluntarily arrive at a mutually advantageous bargain.

Exchange Rate The rate at which one nation's currency can be exchanged for another nation's currency; for example, dollars for yen.

Exempt Carrier A motor vehicle that hauls raw agricultural commodities and is therefore exempt from Interstate Commerce Commission regulations.

Export Enhancement Program (EEP) A government program that helps U.S. exporters meet competitors' prices in subsidized markets.

Export Subsidy A government grant, made to a private enterprise, for the purpose of facilitating exports. In Europe, it is often termed *restitution.*

European Community (EC) A group of 15 European nations with some common economic policies.

Export Credit Guarantee Program GSM-102. The largest U.S. agricultural export credit program which guarantees repayment of private, short-term credit for up to three years.

Extender, Food A product that can be added to a traditional food to expand its volume (e.g., "Hamburger Helper").

Family Farm A farm enterprise in which the owner-operator's family supplies most of the land, capital, and labor.

Farm Problem A complex set of problems associated with agriculture's perfectly competitive structure, the inelastic demand for food, and the high ratio of fixed-to-variable costs. The farm problem is usually identified as instability of farm prices and incomes, periodically declining farm prices, and farm prices sometimes below costs of production.

Farm–Retail Price Spread The difference between the retail price of a food product and the farm value of an equivalent quantity of food sold by farmers.

Farm Service Agency The U.S. Department of Agriculture agency that administers the farm price and income programs through 2,500 local service centers and 51 state offices.

Farm Specialization Any specialization of production activity on the farm, either by commodity, processes, personnel, or geography.

Farmer Cooperative Service See Rural Business–Cooperative Service.

Farmer-Direct Sales Farm sales of produce made directly to consumers, without the use of traditional middlemen; the farm roadside market is an example.

Farmer Cooperative Service The division of the U.S. Department of Agriculture with responsibility for fostering and assisting agricultural cooperatives.

Farmer Market Power Problem Problems associated with the relative bargaining power of farmers and their suppliers and buyers.

Farmer-Owned Reserve A federal program designed to provide farmers with protection against grain shortages and to provide a buffer against sharp price movements.

Farmer's Share The farm value of food expressed as a percentage of its retail price.

Fast Food Industry The restaurant segment of the away-from-home food market that emphasizes standardized menus, rapid service, and minimum table service.

Fed Beef Beef produced from cattle that have been finished on grain rations, as compared with grass-fed or range cattle.

Federal Trade Commission An independent federal regulatory agency with broad powers over commerce, including the regulation of competition, trade practices, and mergers.

Feeder Animals Livestock bought for the purpose of finishing to market weights.

Feedlot A specialized form of farming that involves the feeding of livestock, under confined conditions, to market weights.

Financing A marketing function involving the advancing of money to carry on the various aspects of marketing.

Finish A term referring to final quality of livestock.

Fixed Costs Those costs that do not vary as a firm's output changes, such as overhead, insurance, and management salaries.

F.O.B. Abbreviation for "free on board"; an F.O.B. factory price is a price quotation that does not include shipping charges.

Food and Drug Act A 1906 federal law regulating the healthfulness and wholesomeness of the food supply.

Food Marketing System The collection of product channels, middlemen, and business activities involved in the physical and economic transfer of food from producers to consumers.

Food Reserves Food stocks that are held in order to balance supplies with demand over the long run.

Food Share of Income The percentage of consumers' income (usually after-tax income) used for the purchase of food.

Food Stamp Program A publicly supported food program that attempts to increase the food purchasing power and dietary adequacy of certain groups of people; food stamps are a form of currency which can be used only to purchase food.

Foodway A society's behavior pattern of producing, selling, buying, sharing, and consuming food.

Form Utility The value added to basic commodities when they are converted into finished products.

Formula Pricing A pricing technique whereby an individual transaction is priced according to an agreed-upon method; for example, much wholesale meat is priced "at the market," as indicated by a privately collected market price quotation.

Forward Price Contract A contract farmers may enter into, prior to the harvesting of a crop, that fixes the price in advance.

Forward Pricing Any technique that permits a seller or buyer to fix the price of a commodity prior to the actual physical exchange.

Freedom to Farm Act The Federal farm bill for the 1996–2001 period.

Free Market A market place with minimum direct involvement of government in market decisions.

Free-Rider Problem A problem common to any group effort where each person must make some sacrifice so that the group will benefit. Unless the benefits can be restricted to only those making the sacrifice, individuals will have no incentive to cooperate, but may nevertheless enjoy a "free ride" on the group.

Free Stocks Commodity stocks held by farmers or the trade, rather than by government.

Freight Forwarder A wholesaler who specializes in aggregating small lots of products into larger shipments in order to benefit from more favorable rates.

Function, Marketing A specialized business activity necessary to the food marketing process.

Functional Food Foods with health and medicinal benefits beyond the nutritional and caloric values of all foods; these foods are thought to enhance overall health and/or help prevent or treat a disease or a health condition.

Further Processing Adding value to a commodity by changing its form utility, for example, cutting up chicken into parts for added convenience.

Futures Contract A legal contract certifying agreement to either make or take delivery of a commodity in the future at a specified price.

Futures Market The arena for exchanging futures market contracts; consists of all traders, the trading exchanges, and all attendant trading operations.

Futures Price The current price of a futures contract.

Gatekeeper Any firm that controls a strategic resource in the food marketing system, such as shelf-space in retail stores, a sales force, or any other key marketing variable.

General Agreement on Tariffs and Trade (GATT) A continuing agreement negotiated in 1947 among 23 countries, including the United States, to increase international trade by reducing tariffs and other trade barriers. Now called the World Trade Organization (WTO).

General-Line Wholesaler A wholesaler offering a reasonably complete assortment of products to retailers, as opposed to a limited-line wholesaler.

Generic A word used to describe a general class of products, such as meats, vegetables, or grains; also refers to an unadvertised brand.

Generic Advertising Advertising that attempts to alter the demand for an entire class of products; for example, all meat or all beef.

Genetically Modified Organism (GMO) A plant or animal whose genes have been altered through deliberate genetic engineering for the benefit of humans; may lower costs of food production or enhance certain desirable food properties.

Government Purchase Program A government program intended to raise farm prices by purchasing and storing agricultural commodities; these commodities may be sold later, diverted to other markets, or destroyed.

Grading The sorting of unlike lots of same product into uniform categories, according to quality standards.

Grain Bank An elevator service to farmers who lack on-farm storage space for their grain, which is to be fed to their livestock; farmers deposit grain with the elevator and withdraw it as needed over the marketing year.

Grain Merchandising General term referring to the marketing activities of grain handlers, including timing of buying and selling, grain conditioning, storage, and grain blending.

Grazing Also called *snacking.* A popular American foodway whereby consumers have a number of small food encounters throughout the day rather than three larger meals.

Gross Farm Income The sum of the value received by farmers from sales of products, government payments, the value of farm products consumed on-farm, the rental value of farm dwellings, and other farm-related income.

Gross Margin The difference between the price a firm pays for products and the price it charges its customers; it is also termed *gross profit,* and may be expressed as a percentage of the firm's selling price.

Hazard Analysis Critical Control Point (HACCP) A food safety system of identifying important areas of food production, marketing, and cooking where foodborn pathogens can be controlled or reduced.

Hedge A risk-management, profit-oriented marketing tool involving the temporary substitution of a futures market transaction for a cash transaction; involves equal and opposite positions on the cash and futures markets.

Homogeneity of Products A characteristic of a set of products denoting that, in the eyes of traders, they are perfect substitutes for each other.

Horizontal Integration Combining similar marketing functions and decisions at the same market level into a single firm; for example, one food processor buying another food processing company.

HRI Market The hotel, restaurant, and institutional market for food.

Home-Meal Replacement A convenience food product sold in grocery stores, designed to be eaten at home, and intended to compete with restaurant meals.

Import Quota A restriction on the amount of a specific category of goods that may be imported into a country.

Income Elasticity The responsiveness of food consumption (measured in quantity or expenditures) to changes in consumers' income.

Identity-Preserved Products (IP) Farm products which are identified by source or production practice so that the commodities can be traced back to their production source and values can be assigned to products of differing end-product qualities.

Independent Retailer Any retailer who is not formally affiliated with other retailers or a food wholesaling operation.

Inferior Good A product for which there is an inverse relationship between consumers' income and consumption; i.e., consumption declines as income rises; sometimes referred to as "poor people's products."

Inflation A persistent and general rise in the price level that reduces consumers' purchasing power.

Innovation A new product, a new business technique, or a new way of doing things.

Input Industry Also called the farm supply industry; the industry includes markets that supply resources for farm production, such as chemicals, seeds, feed, energy, and other farm supplies. Farmers procure production supplies from the input markets.

Input-Output System Any activity that transforms resources into more valuable products; the production and marketing processes are input-output systems.

Integration The consolidation under a single management of previously independent production or marketing functions and decisions.

Intermediate Export Credit Guarantee Program GSM-103. A complement to the Export Credit Guarantee Program (GSM-102), which guarantees repayment of private credit for three to ten years.

Institution, Marketing Any public or private entity involved in the marketing process: an individual business, a corporation or group of firms, a government agency, or the legal system.

Integrator A firm that takes the initiative in consolidating various levels of the marketing system under a single management.

Intentions Report A report on the expected future behavior of farmers, for example, the planting intentions report.

Interregional Competition Rivalry between two areas or markets, e.g., oranges from Florida and oranges from California.

Interstate Commerce Commission (ICC) The federal regulatory agency that has authority over freight rates, transportation routes, and other facets of the nation's transportation system.

Jobber A wholesaler of foods; also called *distributor* and *wholesaler*.

Law of Demand An economic principle stating that, everything else being equal, consumers can be expected to buy more of a product as its price falls and less as its price rises.

Law of Market Areas (LOMA) An economic principle holding that the boundary between two markets or sellers is a locus of points such that the final selling prices, including plant price plus transportation costs, are equal for all sellers.

Law of One Price (LOOP) A marketing principle holding that under perfectly competitive market conditions, all prices within a market will be uniform after the costs of adding place, time, and form utility are taken into consideration.

Law of Supply An economic principle suggesting that, everything else being equal, producers will offer to sell more of a product at a higher price than at a lower price.

Length of Run A period of time during which production, marketing, and consumption decisions are influenced; the proportion of fixed and variable costs, as well as the decision makers' ability—and willingness—to alter decisions, will vary during the length of run.

Leverage The ability to control a large amount of money with a small amount of funds.

Loan Deficiency Payment (LDP) Also called a *marketing loan.* A federal government subsidy for certain farmers, including grain producers; farmers receive a government payment equal to the difference between the desired national price and the market price. Unlike government purchase and storage prices, this does not generally result in the government buying and storing commodities.

Loan Rate The price the government is willing to loan the farmer for commodities under the terms of a nonrecourse storage loan; a form of price support; intended to assist the farmer in avoiding sale at harvestime low prices.

Long A trader who has purchased a futures contract without an offsetting sale.

Long Hedge A hedge that initially is established by buying in the futures market.

Loss Leader A product priced at less than the seller's cost, in the hope of attracting into the store customers who will then purchase more profitably priced merchandise.

Mailbox Price The price of milk which dairy farmers receive after some marketing costs are deducted from their checks.

Manufacturer's Sales Branch An establishment operated by a processor or manufacturer that serves as a wholesale outlet for a market territory.

Margin The amount of earnest money that must be posted with a broker when a futures commodity position is first taken (buy or sell).

Market An arena for organizing and facilitating business activities and for answering the economic questions: What to produce? How much to produce? How to produce? How to distribute production? It may be defined by a location, a product, a time, a group of consumers, or a level of the marketing system.

Market Concentration A measure of the dominance of large firms in an industry or market; usually measured by the share of total sales made by the largest companies.

Market Conduct The manner in which firms within an industry adjust prices, output, product quality, and promotional efforts in response to competitive pressures.

Market Development Marketing activities and efforts designed to enhance the value of food products to consumers and in the process expand sales and profits.

Market Information Any form of information relevant to a market decision, a marketing function.

Market Intelligence Any market information relevant to decisions made by firms.

Market News Descriptive information on current market conditions, including prices, stocks, demand, and so on.

Marketing Loan A federal program that permits producers to repay their nonrecourse loans at rates lower than the market price.

Marketing Orders and Agreements A means for farmers to collectively influence the supply, demand, price or quality of certain commodities.

Market Performance The economic results that market participants (farmers, consumers, middlemen) and society expect from the food marketing system.

Market Power The ability to influence markets, market behavior, or market results.

Market Risk The possibility of loss—through product deterioration in quantity or quality or value change—while a product is being produced, stored, or marketed.

Market Segmentation The marketing technique of separating consumers into classes of people who respond similarly to market programs, including product, promoting, and pricing.

Market Structure The environmental characteristics of an industry that influence the behavior of firms in the marketplace; includes size and number of firms, product differentiation, and barriers to firm entry.

Market Tone A market news term referring to the general attitude of buyers and sellers toward prevailing prices.

Marketable Surplus The production of an individual or a society exceeding that needed or desired for personal consumption; this surplus is then available for sale to other individuals or countries.

Marketing The performance of all business activities involved in the flow of food products and services from the point of initial agricultural production until they are in the hands of consumers; a value-adding process that adds time, form, place, and possession utility to farm commodities; the study of exchange processes and relationships.

Marketing Agreement A self-help farmer marketing tool that permits farmers and handlers of farm commodities to enter into agreements about how products will be marketed.

Marketing Bill The total dollar expenditures going to food marketing firms to pay for all marketing activities.

Marketing Board A marketing agency granted wide powers over the production and marketing of a commodity.

Marketing Channels Alternative routes of product flows from producers to consumers.

Marketing Concept A management philosophy holding that all company planning begins with an analysis of consumer wants, and that all company decisions should be based on the profitable satisfaction of consumer wants.

Marketing Function See Function, Marketing.

Marketing Institution See Institution, Marketing.

Marketing Loan A provision of farm bills that allows eligible farmers to repay their nonrecourse government loans at the market price. This encourages producers to redeem their loans and sell their crops on the open market rather than forfeit them to the government.

Marketing Machinery A synonym for the food marketing system.

Marketing Margin The portion of the consumers' food dollar paid to food marketing firms for their services and value-adding activities; the "price" of all food marketing activities.

Marketing Mix The unique way in which a firm or industry combines its price, promotion, product, and distribution channel strategies to appeal to consumers.

Marketing Myopia A term referring to the tendency of some firms to define their business too narrowly in terms of a specific product; for example, a dairy firm may define its business as *milk* or, more broadly, as *fluid beverages.*

Marketing Order A legally binding instrument that specifies how a particular farm product shall be marketed; the purpose of an order is to foster orderly marketing of the commodity over time, form, and space.

Marketing Plan, Farm A set of objectives, strategies, and tactics that guide a farmer's production and marketing decisions.

Marketing Process The sequence of events and actions that coordinate the flow of food and the value-adding activities in the food marketing system.

Marketing Quota A government-determined quantity of a crop that can be legally sold. Growers are penalized when their supplies exceed the national marketing quota for tobacco and peanuts, which are set by the secretary of agriculture.

Marketing Strategy A plan to achieve a market goal; for example, one grocery store may use the strategy of low prices to attract consumers; another might employ the strategy of high quality.

Merchant Middleman A food marketing firm that provides a variety of marketing functions, including taking title to products.

Merchant Wholesaler A wholesaling middleman who physically handles and takes title to products.

Merger The combining of the assets and operations of two independent firms into a single company.

Middleman A person or business firm operating between producers and consumers and performing marketing functions; a term frequently used to describe the nonfarm firms in the food industry; food processors, wholesalers, and retailers are middlemen.

Middleman Bias A commonly held attitude that middlemen are not productive, and that they frequently exploit farmers and consumers.

Milkshed A designated geographic area of milk production and consumption.

Minnesota–Wisconsin Milk Price The price of Grade B milk used for manufacturing purposes in Minnesota and Wisconsin.

Mix Pricing A pricing tactic of multiproduct retailers whereby each product is assigned a different gross profit, thus attempting to differentiate one retail chain store from others.

Monopolistic Competition A competitive situation wherein a relatively large number of firms sell products that are slightly differentiated.

Monopoly A market situation where there is only one seller of a product, and where there are no close substitutes for the product.

Monopsony A firm that is the only buyer of a product.

Net Farm Income Net profit from farming operations; gross farm income less the costs of farm production.

New Generation Cooperative (NGC) A farm coooperative which has moved further downstream in the marketing and processing of farm commodities and added more value than more traditional farm cooperatives; these co-ops may also have different managerial and patronage rules than traditional coops.

Niche Products Food products that have a narrower target market than most food products; these foods are said to fill a market niche which other products might overlook.

Nonrecourse Storage Loan A government price support program that allows the farmer to store commodities at harvest and use them as collateral for a loan; the farmer may redeem the commodities later by paying back the loan, or keep the loan by forfeiting the commodities to the government.

Normal Good A product with a positive income elasticity; the quantity consumed increases with income.

Normal Profit A profit level or return to resources just equal to what they could earn in other uses.

Nutritional Labeling Labels that provide consumers with information about products' nutritional levels.

Offer Price A seller's asking price.

Oligopoly A market situation with relatively few sellers who are mutually interdependent in their marketing activities; many food marketing industries are oligopolistic.

Oligopsony A market situation where there are a few large buyers of a product.

One-Stop Shopping The tendency for some consumers to purchase all desired products from one store in one trip.

Open-Code Dating Food labels providing consumers with information on when food was processed and packaged, when it should be sold or withdrawn from the market, or when the product is no longer acceptable for sale.

Operational Efficiency The ratio of output to inputs or costs. A measure of the degree to which any output is produced in the least costly way. Also referred to as productive efficiency.

Opponent–Gain Bargaining Power Bargaining power gained by offering compensation to another market party.

Opponent–Pain Bargaining Power Bargaining power gained through threat or coercion of another market party.

Option The right to buy or sell a futures contract at a predetermined price.

Orderly Marketing Coordination of the total supply of a commodity over time, form, and spatial markets, in such a way as to achieve some market objectives.

Organic Farming Products Methods of producing and marketing foods which exclude biotechnology, synthetic fertilizers and pesticides, and antibiotics. Sometimes referred to as *natural farming*.

Outlook A forecast of future market events, such as supplies, prices, and so on.

Ownership Integration A complete combination of the assets of two firms; the same as a merger.

Paid Land Diversion A voluntary land retirement system in which farmers are paid for ideling land.

Parasitic Advertising Advertising by one group that takes sales away from another group. For example, beef advertisements may reduce pork sales; orange advertisements may reduce apple sales.

Parity Price A concept that attempts to define a "fair price" for farmers such that farm products have a purchasing power equivalent to what they had in an earlier period.

Parity Ratio The ratio of farm prices received to farm prices paid; a 100 percent parity price would be a farm price that gives a commodity the same purchasing power it had in a base year (usually 1910–1914).

Patronage Refund The earnings of a cooperative that are paid to member-patrons, either in the form of stock or cash.

Payment-in-Kind A provision of the 1983 farm program whereby producers of certain commodities were paid in government-owned supplies of the commodity to idle acreage.

Per Capita Food Consumption The average quantity of food eaten per person within a time period, usually a year; calculated by dividing the total food available for consumption by the population.

Perfect Competition A market situation wherein there are so many firms that no single one of them has a significant influence on price; other prevailing conditions are homogeneous products, ease of new firms' entry into the market, and perfect market information; also termed *pure competition* and *atomistic competition.*

Perfect Hedge A hedge in which the profit and losses from the cash and futures markets are exactly equal.

Physical Distribution Refers to all physical market activities (such as transportation, warehousing, materials handling, inventory control, and so on) involved in moving products from producer to consumer.

Phytosanitary Regulations Prohibitions against shipping diseased or unhealthy plants or animals across state or national boundaries.

Place Utility Also referred to as *spatial utility;* the value added to products by transporting them in space to make them more valuable to users.

Point-of-Purchase Advertising An advertising message to which consumers are exposed in the store, such as shelf labels or store banners.

Possession Utility The value added to products by transferring title from sellers to buyers; also referred to as *ownership utility.*

Preharvest Hedge A hedge that is in effect prior to the harvesting of a commodity; a hedge designed to lock-in a favorable harvest price prior to harvest.

Price Ceiling A legally imposed price above which market prices cannot rise.

Price Cycle Regular, periodic fluctuations in prices resulting from changes in production.

Price Discovery The process by which buyers and sellers attempt to arrive at the equilibrium price consistent with supply and demand conditions.

Price Discrimination A marketing technique that allocates the total supply of a commodity to alternative elasticity markets in a way that will maximize total seller revenues.

Price Elasticity The relationship of changes in quantity supplied, or demanded, to price changes.

Price Floor A legally imposed price below which market prices cannot fall.

Price-Taker Any firm that cannot influence the general level of commodity prices by altering the quantity produced; price-takers are often called "quantity adjusters" because their chief decision is to adjust output to a given price; perfectly competitive firms are price-takers.

Price War A price-cutting situation wherein firms attempt to rival each other in lowering prices.

Pricing Efficiency The capability of prices to allocate resources efficiently and in accordance with consumer preferences. Also termed allocative efficiency.

Principle of Comparative Advantage An economic principle that holds that economic gains are to be had, in the form of reduced costs of production and/or increased standards of living, if, under free trade conditions, each nation will specialize in and export those products it can produce relatively most efficiently, by virtue of its resource endowment, and import commodities for which it has a comparative disadvantage.

Private Label A brand used exclusively by a wholesaler or retailer and usually not widely advertised.

Private Treaty A market agreement arrived at by a buyer and seller in private negotiations.

Product A bundle of physical, service, and symbolic attributes that satisfies consumer wants and needs.

Product Bundle of Attributes All of the physical characteristics and sociopsychological values that consumers associate with a product.

Product Differentiation See Differentiated Products.

Product Life Cycle A set of stages that most new products pass through; these include the development stage, the introductory stage, the growth stage, the mature stage, and the declining sales stage.

Production The process of adding value to resources; an input-output process. Production is not restricted to farming, because value also is added to food in the marketing system.

Protectionism Government trade policies that limit the volume of trade in order to protect domestic industries from international competition.

Psychological Lag The time period necessary to convince producers or consumers that changed economic conditions warrant a change in market behavior.

Public Food Program Publicly supported programs that attempt to increase the demand for food, improve the level of the diet, or otherwise influence food consumption; examples are the food stamp program, the school milk program, and the temporary food assistance program.

Public Law-480 The 1954 Agricultural Trade Development and Assistance Act that authorizes shipment of United States food to foreign countries under special conditions; also known as the "food for peace" program.

Quality A property of a product relating to its usefulness, desirability, and value to customers.

Quality Standards Commonly agreed-upon yardsticks for measuring differences in product quality.

Quota Any restriction on the quantity of product that can be produced or marketed; a protectionist trade device.

Real Price The current price of a commodity, adjusted for its change in purchasing power over time; for example, today's real price of corn in 1970 dollars is the current price of corn divided by the current index of all prices (1970 = 100).

Relative Price The price of one product stated in terms of another; a ratio of two prices.

Resale Price Maintenance (Fair Trade) Laws that permit manufacturers to set and enforce the retail prices at which their products are sold; there are few such laws today.

Reservation Demand The "demand" of a seller for his or her own product, either for current use or for storage and later sale; farmers, for example, may reserve part of their production for later sale by storing at harvest time.

Residual-Income Claimant A firm that has little control over prices received, or income, in the marketplace, and accepts whatever payment is offered for services.

Resource Allocation The shifting of resources between alternative uses to improve output or reduce costs.

Resource Endowment A nation's, firm's, or individual's unique set of productive resources, including land, labor, capital, and management skills.

Risk Bearing A marketing function performed by any firm that is subject to price or quality change risks.

Risk Management The process by which farms and firms attempt to reduce their exposure to economic risks; hedging, diversification, and crop insurance are examples of risk management tools.

Roadside Market A food marketing system in which farmers sell directly to consumers, thereby transferring the food marketing functions from middlemen to consumers and farmers.

Robinson-Patman Act (1936) A federal law that prohibits certain kinds of price discrimination among buyers and sellers.

RTC Ready-to-cook; refers to a broiler or turkey marketed as eviscerated, with blood and feathers removed.

Rural Business–Cooperative Service The U.S. Department of Agriculture agency which provides research and technical assistance to agricultural and food cooperatives.

Sales Branch A manufacturer's establishment that serves as a warehouse and sales office for a particular territory.

Scratch Food A form of food that requires considerable preparation on the part of consumers before it can be eaten; a "scratch" cake is one made by formulating the basic ingredients in the home, as opposed to a ready-to-bake cake.

Seasonal Price Variation A regular price variation occurring within a marketing season.

Seasonal Stocks Stocks of commodities held to balance out supply and demand over a marketing season when production is not continuous over the season.

Seller's Market A market situation wherein a small quantity of products provides sellers some advantage in bargaining for prices with buyers.

Set-Aside A supply control provision of some farm programs that entails farmers withdrawing some acreage from production, either voluntarily or on a required basis.

Sherman Anti-Trust Act (1890) A federal law prohibiting restraint of trade and monopolistic practices.

Short A trader who sells a futures market contract without an offsetting purchase.

Short Hedge A hedge that is initially established by selling in the futures market.

Shrinkage The loss in physical volume or weight of a commodity while it is being held or marketed.

Social Capital The societal resources that are available to facilitate food marketing, such as the transportation, legal, and communication systems.

Sorting The process of classifying a mixed lot of produce into homogeneous categories, as in the grading of farm products.

Speciality Food Store A food retail store with a limited and specialized product line; for example, a dairy store, a bakery, or a fruit and vegetable store.

Specialization A fundamental economic process whereby division of labor and concentration of effort on a single task can increase efficiency and output.

Speculative Middleman A marketing firm specializing in the risk-bearing marketing function; an individual who profits from uncertain changes in food prices.

Speculative Stocks Inventories of commodities held in anticipation of a future price rise.

Speculator A trader who holds a market position in anticipation of a favorable price movement.

Spot Price The cash price offered for immediate delivery.

Standardization The grouping of unlike items into uniform lots on the basis of qualitative criteria, such as a food grade.

State Trading Monopoly A trading agency granted a government monopoly in foreign sales or purchases of agricultural commodities.

Sticky Marketing Margin Refers to the fact that the dollar marketing margin does not readily adjust to changing farm and retail food prices.

Stockholders' Equity The total financial interest that stockholders have in a firm; also called *net worth.*

Storage Hedge A hedge that is in effect during the time period when a commodity is stored; this hedge can assist firms in earning storage returns.

Subsector Analysis A commodity approach to studying the food marketing process; in this method the entire set of institutions and relationships involved in transforming basic resources into satisfying food products is examined.

Substitute Products Products that, in the eyes of buyers, have some characteristics and utilities in common (e.g., apples and oranges).

Subterminal Elevator A large grain elevator located in producing areas; the owner buys directly from farmers or country elevators and sells directly to exporters or grain processors.

Supermarket A large-scale, departmentalized retail store with a variety of merchandise that usually is operated on a self-service basis.

Supermarket Movement A trend toward replacing small, neighborhood high-service food retail stores with large, departmentalized, self-service grocery stores.

Supplementary Import An imported product that competes directly with a domestically produced product; also referred to as a *competitive import.*

Supply A schedule indicating the quantity of product that producers are willing to produce and sell at alternative price levels. See Law of Supply.

Supply Control Any attempt to reduce marketable supplies by restricting either production or flow-to-market.

Support Price A price level above the equilibrium supply and demand point; support prices are used by government to increase or stabilize farm prices or farm incomes.

Synthetic Food/Fiber A food or fiber produced from a nonagricultural raw material, for example, a nondairy coffee creamer, a synthetic orange juice, an imitation shoe leather, or a manmade fiber.

Targeted Export Assistance (TEA) Program A federal program that assists U.S. producer groups or regional organizations whose exports have been adversely affected by foreign government policies.

Target Marketing A marketing strategy whereby the firm develops specific products for different segments of the population; for example baby foods, teenage foods, or geriatric foods.

Target Price A support price level set by government that determines the deficiency payment to farmers.

Tariff A government tax on imported products that has the effect of reducing imports and increasing the consumer prices of imported products; a protectionist trade device.

Technology The machinery, state-of-the-art, and level of ingenuity that determine the efficiency and output of any productive process.

Terminal Market A central wholesale market, usually located at the terminus of a transportation line in an area of high population.

Texas Hedge Not a true hedge; involves either short or long positions in both the cash and futures markets at the same time.

Thin Market A market in which so little volume is traded that it is questionable whether the supply and demand forces are adequately represented.

Time Utility The value added to products by changing their time of availability to users through storage operations.

Trader A general term for a buyer or seller; one who engages in buying and selling for a profit.

Transactions Costs Those costs incurred by buyer-seller search, negotiation, and contract-enforcement activities.

Transit Privilege Permission for a shipper to halt the movement of a product for processing or handling and then continue shipment to its destination, at a through rate.

Transportation Mode A means of transportation, such as rail, truck, or water.

Truth-in-Packaging A 1967 federal law (the Fair Packaging and Labeling Act) that attempts to facilitate consumer comparisons of grocery products by standardizing package sizes and label nomenclature (e.g., reducing the number of "jumbo" sizes).

Unit Pricing Comparative pricing of items in standard units of measurement, for example, cents per ounce or dollars per pound.

Unit Train Entire trains carrying a single commodity with a single point of origin and point of destination.

United States Warehouse Act (1916) A federal law that regulates public warehouses.

Utility A characteristic of goods and services that describes their want-satisfying power for consumers; economists differentiate between form, time, place, and possession utility.

Value-Added The difference between the cost of goods purchased by a firm and the price for which it sells those goods; this difference represents the value-added by the productive activities of the firm.

Value Adding, Value-Added Foods The management and marketing processes by which firms attempt to improve their competitive positions and satisfy their target markets; may involve processing of products, adding services, or any markeing activity.

Variable Costs Those costs of the firm that change when output changes, for example, raw material costs, wages, packaging, and so on.

Variable-Price Merchandising A retail pricing strategy whereby selective price reductions are used to attract consumers to the store.

Vertical Integration Centralizing vertically related marketing functions and decisions either by contract or by ownership.

Vertical Market Coordination The process of directing and harmonizing the several interrelated and sequential decisions involved in producing and marketing farm and food products.

Warehouse Food Outlet A discount supermarket that offers lower prices and less services than traditional supermarkets.

Weak Seller A seller of a product who has little bargaining power or influence on price, either because of product perishability, lack of control over supply, or other factors.

Webb-Pomerene Act (1918) A federal law authorizing monopolies in the export trade.

Wheel of Retailing A theory that retailing passes through alternating cycles of low service, low prices and high service, high prices.

Wholesale-Sponsored Chainstore A set of retail food stores that are supplied by a single wholesaler.

World Trade Organization (WTO) A multinational organization with the mission of lowering global trade barriers and promoting freer trade; the successor organization to GATT.

Workable Competition A set of market performance criteria that represent a compromise between perfect competition and the real world of competitive imperfections.

Working Stocks The stock levels necessary for the marketing system to function efficiently at full capacity and without supply disruptions; also referred to as *pipeline stocks.*

Yardage Fees charged by livestock merchants and marketing organizations for services to buyers and sellers.

Yield Grade A livestock grading term that refers to the meat yield of the carcass.

INDEX

G

H

I

J

K

L

M

N

O

P

Q

R

S